# CHEMISTRY OF ORGANIC
# FLUORINE COMPOUNDS

# CHEMISTRY OF ORGANIC FLUORINE COMPOUNDS

*by*

MILOŠ HUDLICKÝ

THE MACMILLAN COMPANY

NEW YORK

1962

THE MACMILLAN COMPANY
*60 Fifth Avenue, New York 11, N.Y.*

BRETT-MACMILLAN LTD.
*132 Water Street South, Galt
Ontario, Canada*

PERGAMON PRESS LTD.
*4 & 5, Fitzroy Square, London W. 1
Headington Hill Hall, Oxford*

CHEMISTRY

Library of Congress Card. No. 61-9782

Printed in Poland to the order of Państwowe Wydawnictwo Naukowe
by Zakłady Graficzne im. M. Kasprzaka, Poznań

# CONTENTS

185

# PREFACE TO THE ENGLISH EDITION

A RAPIDLY increasing pace in the field of organic fluorine compounds has forced a thorough revision of the text before translation. The three-year gap between the Czech and English editions has left enough time to note defects and inadequacies in the original book. Although the frame was left almost intact, some three hundred references up to the end of 1959 were added to keep pace with the ever-growing results in research and technology. Attendance at the International Symposium on Fluorine Chemistry in Birmingham, 1959, enabled me to obtain a more personal touch with fluorine chemistry than had been possible from the literature. After discussions with experienced fluorine chemists I have been able to correct some of my erroneous opinions and to incorporate in the book results which had not yet been published. In the course of these three years I have acquired some more laboratory experience and have been able to add a few more procedures. I must state that I have preferred to quote my own results wherever I have tested recorded procedures. The yields given are therefore not to be considered as maxima.

In connexion with the English translation I would like to express my most sincere thanks to my teacher, Professor A. L. Henne, who kindly revised enough of the manuscript to translate my personal English into a language understandable to English speaking people.

My thanks also belong to my wife, Mrs. A. Hudlická, for her efficient help in preparing the English manuscript and the indexes. I am much obliged to Captain I. R. Maxwell of Pergamon Press, for the ready acceptance of my book, and to the printers, for the care in printing the text and especially the formulas.

M. HUDLICKÝ

*Prague*

# PREFACE TO THE CZECH EDITION

THE most competent person to write a chemical compilation is no doubt a specialist with the largest possible experimental experience in the field, who is therefore predestined to the task of passing his knowledge on to his less experienced colleagues. This book was built by the reverse process: It was written by an apprentice intending to get acquainted with the problems and technique of a field in which he started to work about ten years ago. It is an old adage that one learns best by teaching. One is thus forced to think more thoroughly and to formulate many problems instead of merely recording the new knowledge passively. This somewhat egoistic point of view was my original incentive for writing, and it necessarily colours the whole book: the book presents a subjective selection of facts believed to have a direct relation to laboratory practice. It was my intention to include only the data most essential for experimental work, so that the reader is only roughly informed about possibilities of development in the chemistry of organic fluorine compounds, or the laboratory technique which in many respects differs from common laboratory practice. The book does not always respect the chronology of methods and reactions but tries to give priority to papers which describe experimental details most completely. I have refrained from theoretical considerations except where they were needed for understanding. On the other hand, quite a number of procedures included as concrete examples are more instructive than general directions.

The material for this book was collected mostly from the original literature up to the end of 1956. Physical data were taken from reference books on fluorine chemistry to limit the number of references. As already stated, the literature included is a selection and the book is therefore incomplete in some aspects and overloaded in some others. The author will therefore welcome any suggestions from the readers which would aim at improving the book, since its first edition is viewed as a dress rehearsal rather than a first night performance.

I would like to extend my sincere thanks to all who participated in some way in the creation of this book. I am obliged to my Professor R. Lukeš for his recommendation on the basis of which I was granted a UNESCO fellowship. My knowledge and experience in the field of organic fluorides I owe to Professor A. L. Henne, Ohio State University, in whose laboratory I was trained for almost one year, and also to Professor E. T. McBee, of Purdue University, and Dr. G. C. Finger, of the State Geological Survey of Illinois. For thorough review of the manuscript I am indebted to Professor O. Wichterle and Ing. V. Reinöhl. Thanks are also due to my wife, Mrs. A. Hudlická, for efficient help in the completion of the manuscript and indexes.

*Prague*

# I

# INTRODUCTION

## DEVELOPMENT OF THE CHEMISTRY OF ORGANIC COMPOUNDS OF FLUORINE

THE history of fluorine and its compounds did not begin with ancient Egypt nor Phoenicia nor Arabia of the Middle Ages but with the preparation of hydrogen fluoride by Scheele in 1771 and elementary fluorine by Moissan in 1886. Systematic study of organic compounds of fluorine started however only about 1900 with Swarts' fundamental research in this field. A further important milestone in the chemistry of organic compounds of fluorine was the application of fluorinated derivatives of methane and ethane in refrigeration (Midgley, Henne, 1930). This discovery initiated a new wave of research into the preparation and properties of these compounds. About 1940 the importance of perfluoro derivatives was recognized, and methods for their preparation and manufacture were worked out, for instance noncatalytic and catalytic fluorination with elementary fluorine and fluorination by means of high-valency fluorides of silver and cobalt. At about the same time, polytetrafluoroethylene was discovered (Plunkett, 1940), and a few years later, industrial production of polychlorotrifluoroethylene was started. One of the most important discoveries of recent years is electrolytic fluorination, i.e. preparation of fluorinated derivatives by electrolysis of organic compounds in anhydrous hydrogen fluoride (Simons, 1948). This method is an easy way to perfluorocarboxylic acids from which perfluoro alkyl halides are prepared. Addition of these compounds to olefins and acetylenes, preparation of perfluoroalkylmagnesium halides and their synthetic applications, and finally preparation of perfluoroolefins by the pyrolysis of perfluorocarboxylic acids, are examples of applications of these electrolytically prepared compounds.

Numerous interesting reactions of fluorinated derivatives discovered recently bear witness to the increasing importance and development of the chemistry of organic compounds of fluorine, and show that compounds and reactions of fundamental importance may still be discovered in this field.

The most important of the organic fluorine compounds are those which contain a large number of fluorine atoms in their molecules (polyfluoro and perfluoro derivatives). These compounds owe much of their interest to their inertness and to their thermal as well as chemical stability. They are theoretically surprising since they very often deviate from the behaviour of chemical compounds of other halogens.

To date, the chemistry of organic compounds of fluorine includes several thousand derivatives and so many publications that it is impossible for one person to master this branch to its full extent. There exists no exhaustive monograph registering all the described organic compounds of fluorine, and the reviews of an exhaustive nature are narrowly specialized. In addition to the reviews published in various books and journals (*106, 352, 432, 433, 758, 918*) the following publications should be quoted:

| | |
|---|---|
| Advances in Fluorine Chemistry | Stacey M., Tatlow J. C., Sharpe A. G. (editors) Vol. I(1960) *(966)* |
| Aliphatic Fluorine Compounds | Lovelace A. M., Rausch D. A., Postelnek W. (1958) *(624)* |
| Fluorine Chemistry | A reprint from *Ind. Eng. Chem.* *39* (1947), and *Anal. Chem. 19* (1947) |
| Fluorine Chemistry I. | Simons J. H. (editor) (1950) *(947)* |
| Fluorine Chemistry II. | Simons J. H. (editor) (1954) *(948)* |
| Fluorine and its Compounds | Haszeldine R. N., Sharpe A. G. (1951) |
| Fluorocarbons | *(387)* |
| Organische Fluorverbindungen in | Rudner M. A. (1958) *(892)* |
| ihrer Bedeutung für die Technik | Schiemann G. (1951) *(919)* |
| Phosphorus and Fluorine. The | Saunders B. C. (1957) *(910)* |
| Chemistry and Toxic Action of | |
| Their Organic Compounds | |
| Preparation, Properties and | Slesser C., Schram S. R. (1951) *(959)* |
| Technology of Fluorine and Organic | |
| Fluoro Compounds | |
| Toxic Aliphatic Fluorine Compounds | Pattison F. L. M.; Elsevier, 1959 |

## NOMENCLATURE OF ORGANIC COMPOUNDS OF FLUORINE

The nomenclature of organic compounds of fluorine which contain one or only a few fluorine atoms in the molecule does not create any difficulty. The position of the fluorine atom in the name of the compound is designated by numerals or Greek letters according to current rules:

1          $CH_2F—CF_2—CH_3$          1,2,2-Trifluoropropane

$CH_3—CO—CH_2—CH_2F$          Methylβ-fluoroethyl ketone

A serious nomenclature problem arose with compounds in which the number of fluorine atoms predominated over the number of other elements bound to the carbon skeleton, and with compounds in which all the hydrogen atoms are replaced by fluorine. In these cases the designation of the position of the individual atoms of fluorine would lead to names which are too long and cumbersome.

2      $CHF_2\!-\!CF_2\!-\!CF_2\!-\!CHF\!-\!CF_3$      1,1,2,2,3,3,4,5,5,5-Decafluoropentane

         $CHFCl\!-\!CF_2\!-\!CF_3$      1-Chloro-1,2,2,3,3,3-hexafluoropropane

In polyfluoroderivatives in which the number of hydrogen atoms in the molecule does not exceed 4, and the ratio of the hydrogen: halogen atoms is not more than 1 : 3, the name involves the total number of fluorine atoms, and the number and positions of the hydrogen or halogen atoms are expressed according to the following model (*18, 645*):

3          1H,4H-Decafluoropentane

         1H,1-Chlorohexafluoropropane

For compounds whose molecules do not contain hydrogen atoms or other halogen atoms, a nomenclature was adopted in which the name of the basic non-substituted compound is preceded by a prefix "perfluoro" (*16, 18*):

4          $C_7F_{16}$          Perfluoroheptane

$$\begin{array}{c} \diagup CF_2\!-\!CF_2 \diagdown \\ CF_2 \qquad\qquad CF\!-\!CF_3 \\ \diagdown CF_2\!-\!CF_2 \diagup \end{array}$$      Perfluoromethylcyclohexane

The Greek letter may be substituted for the prefix "perfluoro" so that the above-mentioned compounds would be called (*327*):

5          Φ-Heptane          Φ-Methylcyclohexane

This nomenclature can also be used for these polyfluoroderivatives which contain but one or a few atoms of hydrogen (*327*). The first named compound in scheme 2 would then read:

6          1,4-Dihydro-Φ-pentane

Since the "perfluoro" nomenclature system was not unambigous in some cases, e.g. did not differentiate exactly, whether the prefix "perfluoro" applied to the whole molecule or only to a part as in the following example:

7          $C_6H_5C_2F_5$

               Perfluoroethylbenzene

         $C_6F_5C_2F_5$

an entirely new system of names has been worked out using the name of the basic non-substituted compounds and inserting "for" (abbreviation of fluor) before the suffix designating the chemical nature of the compound (*946*). According to this nomenclature, the last mentioned derivatives of ethylbenzene would read:

**8**                       $C_6H_5C_2F_5$    Ethforylbenzene

                            $C_6F_5C_2F_5$    Ethforylbenzforene

This nomenclature seems to be unambiguous and practical for perfluoro derivatives. On the other hand, it is not common and is absolutely incomprehensible to a chemist who is not acquainted with fluorine chemistry. Examples of various types of nomenclature are shown below.

**9**

| Formula | Nomenclature According to | | "Perfluoro" Nomenclature | "for" Nomenclature |
| | Chem. Abstracts | International Union of Chemistry | | |
| --- | --- | --- | --- | --- |
| $CF_2{=}CF{-}CF{=}CF_2$ | Hexafluoro--1,3-buta-diene | | Perfluoro-1,3-butadiene | Butadi-forene |
| (cyclohexane ring structure) | Decafluoro--1,3-bis (tri-fluorome-thyl) cyclo-hexane | 1,3-Bis (tri-fluorome-thyl) deca-fluorocyclo-hexane | Perfluoro (1,3-dimethyl-cyclohexa-ne) | 1,3-Dimeth-forylcyclo-hexforane |
| $CF_3CF_2CF_2OCF_2CF_2CF_3$ | Bis (hepta-fluoropro-pyl) ether | Heptafluo-ropropoxy-heptafluoro-propane | Perfluoro-n-propylether | Dipropforyl ether |
| $CF_3COOC_2F_5$ | Pentafluoro-ethyl trifluo-roacetate | Pentafluoro-ethyltrifluo-roethanoate | Perfluoroet-hyl perfluo-roacetate | Ethforyl acetforate |

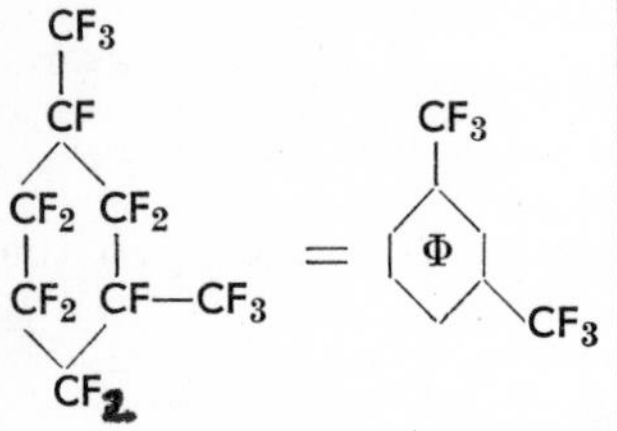

In the field of fluorinated organic compounds of phosphorus, a new, more consistent yet quite unusual nomenclature is being adopted illustrated by the following examples (*910*):

**10**

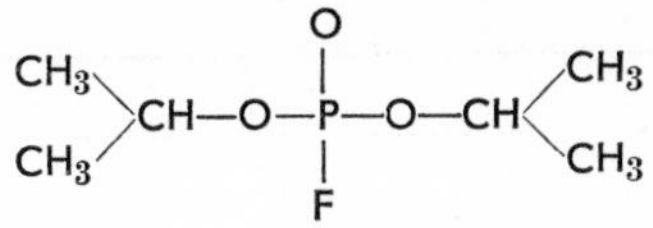

$$\begin{array}{c} CH_3 \\ \diagdown \\ CH_3 \diagup \end{array} CH - O - \overset{\displaystyle O}{\underset{\displaystyle F}{\overset{\|}{P}}} - O - CH \begin{array}{c} CH_3 \\ \diagdown \\ \diagup \\ CH_3 \end{array}$$

Diisopropylfluorophosphate

Diisopropylphosphorofluoridate

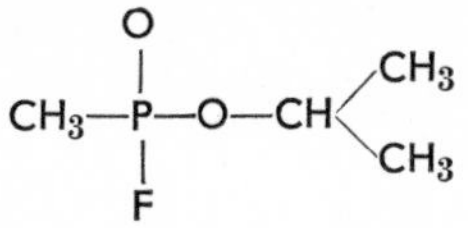

$$CH_3 - \overset{\displaystyle O}{\underset{\displaystyle F}{\overset{\|}{P}}} - O - CH \begin{array}{c} CH_3 \\ \diagdown \\ \diagup \\ CH_3 \end{array}$$

Isopropylmethanefluorophosphonate

Isopropyl P-methylphosphonofluoridate

Discussion of the nomenclature of fluorinated compounds should also include the commercial designation of refrigerating agents whose composition is characterized by special names and numerical symbols. Fluorinated derivatives of methane and ethane are generally called Freons and marked with numerals according to the following key:

The first numeral means the number of carbon atoms minus one,
the second numeral means the number of hydrogen atoms plus one,
the third numeral relates to the number of fluorine atoms;
the number of chlorine atoms is not considered.

The derivatives of methane will therefore carry two numerals (zero is omitted), the derivatives of ethane numbers beginning with one. As can be seen, this system is not unambiguous and universal. The position isomers of ethane cannot be properly distinguished, since the basis for the symbols is formed by empirical formulae, and no allowance is taken for differentiating between symmetric and asymmetric derivatives. For brominated derivatives, B followed by the number of bromine atoms has to be added to the numerical symbol of the Freons.

This nomenclature is now generally accepted and frequently used in patent literature and commercial terminology. The most common trade names are listed in Table 1.

TABLE 1

*Nomenclature of Freons*

| Chemical Name | Formula | Trade Name |
|---|---|---|
| Trichlorofluoromethane | $CCl_3F$ | Freon 11 |
| Dichlorodifluoromethane | $CCl_2F_2$ | Freon 12 |
| Chlorotrifluoromethane | $CClF_3$ | Freon 13 |
| Dichlorofluoromethane | $CHCl_2F$ | Freon 21 |
| Chlorodifluoromethane | $CHClF_2$ | Freon 22 |
| Trifluoromethane (Fluoroform) | $CHF_3$ | Freon 23 |
| Tetrachlorodifluoroethane | $C_2Cl_4F_2$ | Freon 112 |

*continued*

(TABLE *1.—continued*)

| Chemical Name | Formula | Trade Name |
|---|---|---|
| Trichlorotrifluoroethane | $C_2Cl_3F_3$ | Freon 113 |
| Dichlorotetrafluoroethane | $C_2Cl_2F_4$ | Freon 114 |
| Chloropentafluoroethane | $C_2ClF_5$ | Freon 115 |
| Dichlorodifluoroethane | $C_2H_2Cl_2F_2$ | Freon 132 |
| Bromotrifluoromethane | $CBrF_3$ | Freon 13B1 |
| Dibromodifluoromethane | $CBr_2F_2$ | Freon 12B2 |
| Perfluorocyclobutane | $\begin{array}{c} CF=CF \\ \vert \quad\ \vert \\ CF_2-CF_2 \end{array}$ | Freon C318 |

## HANDLING OF FLUORINE, HYDROGEN FLUORIDE AND FLUORIDES

Fluorine, hydrogen fluoride, and some fluorides are very reactive compounds. Their handling therefore requires suitable care. Their action upon various materials will be discussed in the chapter on production and in the chapter on reactions with organic compounds. In this paragraph only safety measures for their handling will be briefly outlined (*717*).

Undiluted gaseous *fluorine* reacts even at room temperature with many organic and inorganic materials, and even with such substances as asbestos and water. Utmost care is therefore necessary when working with it. Inhalation of fluorine resembles in its effects that of chlorine or ozone. An atmosphere containing fluorine is therefore to be avoided. When operating equipment containing fluorine, rubber gloves should be worn since fluorine burns the skin like hydrogen fluoride. The treatment of such burns is the same as that of hydrogen fluoride burns (*602*) (see below).

Working with *hydrogen fluoride* is much more frequent than with fluorine. Even dilute aqueous hydrofluoric acid causes painful burns especially under the fingernails. Anhydrous hydrogen fluoride is in this respect extraordinarily dangerous, for it reacts immediately with the skin before it can be neutralized, and causes painful and slowly healing wounds. It is therefore necessary to protect the hands with undamaged rubber gloves (made of synthetic rubber), to shield the face with a protective shield of polystyrene or methyl methacrylate the body with a rubber or plastic apron, and the feet with rubber boots. It is advisable to test the gloves and boots for leakages from time to time. Even an insignificant leak in the glove may be a cause of serious trouble.

The author of this book came in touch with anhydrous hydrogen fluoride which penetrated throught a pinhole in the thumb of a rubber glove. Since the skin of the thumb was little sensitive the presence of the hydrogen fluoride on the skin was not noted in time. After a quarter of an hour when the hydrogen fluoride diffused sufficiently deep to reach the nerve the finger started to ache. A throbbing pain lasted almost 24 hr after which a painful subcutaneous blister was formed on the thumb. The recovery took more than a week.

Work with anhydrous hydrogen fluoride must be carried out in an efficient hood or in well ventilated rooms since its vapours attack the lungs and mucous membranes (especially those of the eyes).

Anhydrous hydrogen fluoride reacts violently, almost explosively, with water. Disposal of spent hydrogen fluoride must therefore be carried out with caution and with protective devices against splattering. Smaller amounts are disposed off by pouring into the sink and flushing rapidly with a strong stream of water, larger amounts by treatment with a lime slurry.

When the skin is stained with anhydrous hydrogen fluoride, the spot is washed as soon as possible with large amounts of water (602). Immersion of the burned spot into ice-cold saturated magnesium sulphate solution or 70% ethyl alcohol for about half an hour is recommended (717). Finally a paste of magnesium oxide and glycerol is applied prepared by mixing 20% magnesium sulphate, 6% magnesium oxide, 18% glycerol, 55% water, and 1·2% procaine hydrochloride (603). More extensive burns are treated by applying calcium gluconate subcutaneously (717).

The eyes must be washed for a quarter of an hour with water, and then exposed for another quarter of an hour to the action of 2–3 drops of 0·5% pantocaine. Ointments are not recommended (717).

Inorganic compounds of fluorine, especially those existing normally in a liquid state, must be handled with special caution. These are *fluoroboric acid, fluorosulphonic acid, antimony pentafluoride* and *chlorine* and *bromine trifluoride*. Solid fluorides, though usually toxic, are no more dangerous than other chemicals.

Of the organic compounds, some polyfluorinated acids attack the skin and cause burns. Of special danger is inhalation of *fluoroacetates, fluorophosphonates* and some pyrolysis products of chlorodifluoromethane and tetrafluoroethylene. Their biological effects are described in a special paragraph (p.310).

Many of the substances mentioned are quite rare in current practice. The most rigid precautions have to be taken when handling hydrogen fluoride.

# II

# APPARATUS AND MATERIAL

The current laboratory equipment is usually unsatisfactory for the preparation of organic fluorine compounds because many reagents attack glass. Although many reactions which were originally carried out in metal are nowadays performed in glass which is, in the worst case, superficially etched, there are still instances in which metals are indispensable. When the reactions do not require very high temperatures some plastics may also be used. The following chapter will describe the fundamental devices including some special apparatus needed for the preparation of fluorinated organic compounds. In addition, the chemical resistance of the main materials from which the apparatus and their parts are constructed, will be discussed.

## LABORATORY DEVICES AND APPARATUS

*Weighing* and *measuring* of fluorinated compounds corrosive to glass, i.e. aqueous and anhydrous hydrogen fluoride, are carried out in glass vessels whose inner walls are coated with a thin film of paraffin. Since this kind of glass protection has some disadvantages such as cracking of the paraffin coating and relatively low melting temperature of the paraffin, plastic vessels are preferred nowadays which are made of transparent or translucent materials such as polystyrene or polyethylene.

*Evaporation* of aqueous solutions of hydrofluoric acid is carried out in copper, silver, or platinum dishes. Common stainless steel pots are unsuitable since they are seriously corroded by aqueous hydrofluoric acid. If perfect purity is not required copper devices are satisfactory. A copper dish is thus suitable for preparing all of the inorganic fluorides that are prepared from aqueous hydrofluoric acid, e.g. sodium fluoride, potassium fluoride, fluoroboric acid, alkali fluoroborates, antimony trifluoride etc. Recently, evaporation dishes made of polytetrafluoroethylene proved to be of great advantage (*933*).

Cold solutions of hydrofluoric acid or some fluorides can be *filtered* through glass funnels whose inner surface is coated with a paraffin layer. Hot solutions require the use of a polyethylene funnel. Filter paper and thick cotton tissue are suitable materials for filtering dilute solutions of hydrofluoric acid and fluorides. Filtration of more concentrated hydrofluoric acid or anhydrous hydrogen fluoride is carried out by means of metal crucibles with perforated bottoms covered with metal wool (*893*), or polyethylene funnels fitted with perforated disks covered with fine polyethylene shavings (*417*).

Special filtering plates and cones to be inserted into polyethylene funnels can be made by mixing one part of powdered polyethylene with four parts of sodium chloride, pressing the mixture between two Petri dishes or two funnels (the inner one with its stem sealed off), sintering the material at 140-145°, and washing out the sodium chloride with water after cooling (*561*).

*Distillation* of compounds dissolving or corroding glass is carried out in equipment made of steel, copper, silver or platinum (Fig. 1). A copper apparatus will do for most preparation purposes. It is suitable for instance for the distillation of aqueous as well as anhydrous hydrogen fluoride and for the distillation of some corrosive fluorides such as antimony trifluoride,

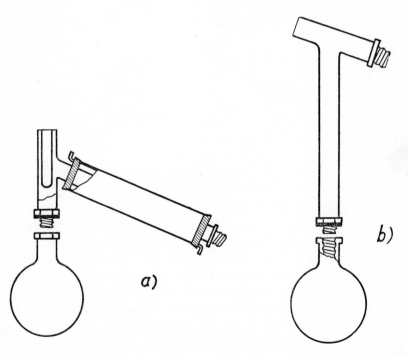

FIG. 1. Copper distillation apparatus: a) distillation, b) heating under reflux.

antimony pentafluoride, and others. It consists of a copper flask with an inner thread and of a condenser. The condenser is a copper tube fitted with a thermometer-well and threaded at both ends. It may be used as a descending condenser (Fig. 1a), or as a reflux condenser (Fig. 1b). For water-cooling the condenser is fitted with a cooling jacket.

Analytical distillations of hydrogen fluoride have to be carried out in silver or platinum equipment or in a polyethylene outfit at temperatures below 85° (*188*).

For *stirring* solutions of hydrogen fluoride or other fluorides corrosive to glass, iron, copper or silver wires, a stainless steel or plastic spatula and similar devices are used. The same materials may be used for constructing mechanical stirrers.

Very often gaseous fluorine or hydrogen fluoride must be *passed through reaction mixtures* of solid or liquid substances. Although pure and dry fluorine does not attack glass and consequently may be handled in glass apparatus (*96, 104, 289, 293, 1047*), safer work is guaranteed by using an apparatus and equipment made of steel or copper (*152, 1050*). In the case of anhydrous hydrogen fluoride the use of steel or copper tubing and reactors is imperative.

Organic fluorine derivatives are often prepared by *refluxing organic compounds with inorganic fluorides*. Alkaline fluorides, thallium fluoride, silver monofluoride, mercurous and mercuric fluoride, antimony trifluoride and pentafluoride and iodine pentafluoride can be worked with in glass if they are free of hydrogen fluoride and water. Reactions with silver difluoride, cobaltic fluoride, chlorine and bromine trifluoride are usually carried out in steel or copper. The simplest apparatus for this purpose consists of a metal flask and condenser which are connected by means of threaded joints fitted with lead, copper or polytetrafluorethylene gaskets (Fig. 1).

A great deal of fluorination is carried out by *heating organic compounds with metal fluorides* or *anhydrous hydrogen fluoride* at high temperatures under pressure. So long as the pressure does not exceed approximately 15 atm universal steel apparatus can be used which are improvised "bombs" made from seamless steel pipes by welding at one end a convex bottom and at the other end a lid which has a boss fitted with a threaded hole. The pipe which has to be inserted into the hole is also fitted with threads which are cut by a standard plumber thread-cutter. These threads have a conical shape so that no packing is necessary and a gas-tight connection is produced by screwing the two joints tightly together. However winding one turn of a polytetrafluoroethylene tape round the threaded pipe, or smearing the threads with a viscous cement is recommended to prevent leakage through the threads' imperfections. For reactions involving hydrogen fluoride, a cement

of phenol-formaldehyde or urea-formaldehyde resins is satisfactory since it
solidifies in contact with acidic fumes, For fluorine reactors a luting has been
recommended composed of graphite, clay, castor oil, and denatured alcohol
with a gum base (*1070*).

By a suitable combination of pipes and fittings almost any complicated
apparatus can be assembled. Devices for refluxing a mixture under pressure

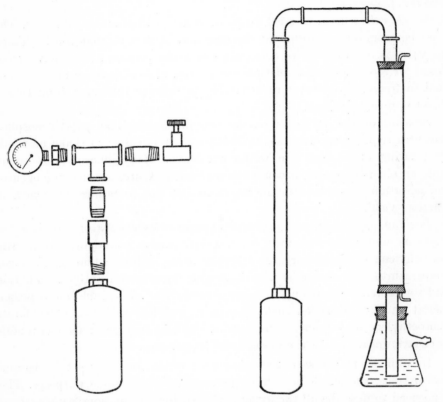

FIG. 2. Iron apparatus for pressure
reactions.

FIG. 3. Iron apparatus for reactions
with antimony fluorides.

which allow introduction of a gas, or simultaneous removal of reaction components by distillation, and similar operations are illustrated on Fig. 2 and 3
as combinations of the individual parts.

Reactions which require pressures of some tens of atmospheres or higher
must be carried out in suitable autoclaves made of steel or stainless steel.

The pressure is measured by current brass *manometers*. Before inserting
the manometer into the apparatus the manometer tube should be filled with
oil which protects, at least to some extent, the inside of the manometer

against corrosion. Protective coils filled with oil are not satisfactory. Where corrosion of the inside of the manometer must be prevented, the tube of the manometer is filled with oil so that no residual gas bubble is included, and the manometer is separated from the apparatus by a resilient metal or plastic membrane. The pressure in the apparatus deforms the membrane, and the oil being incompressible transfers the pressure to the manometric tube (*877*).

*Needle valves* made of steel, brass or stainless steel are satisfactory. They must be mounted according to the direction of flow so that the bushing of the valve is not exposed to corroding gas under pressure. After having been closed for too long periods of time the valves, especially those made of steel, tend to freeze. It is therefore advisable to operate the valve from time to time to prevent sticking.

*Stirring* of the mixtures during the fluorinations at superatmospheric pressures requires special resistant bushings. It is easier to shake or tumble by rotating or rocking the whole autoclave. Many fluorinations require quite an efficient mixing because the reaction mixtures are not homogeneous and agitation not only cuts the reaction time but sometimes influences the reaction yield.

*Packing* (*gasketing*) of the metal apparatus especially one designed for reactions under super-atmospheric pressure causes considerable difficulties. Polyethylene is satisfactory for reactions with hydrogen fluoride at normal temperatures. For reactions with hydrogen fluoride and antimony fluorides, lead stands temperatures up to approximately 200°. When fluorine is present, special packing is needed such as copper, modified asbestos, polychlorotrifluoroethylene or polytetrafluorethylene. The last two materials also function as electric insulators in the electrolytic preparation of fluorine.

In addition to the standard equipment which may be used for common fluorinations, special devices have been designed for some purposes. There is no need to describe all the apparatus used for fluorinations with elemental fluorine since many of them are now obsolete. One remarkable apparatus was developed for *non-catalytic fluorination with elemental fluorine*. Its main feature is a concentric orifice in which a stream of organic vapours is separated by a stream of an inert gas from a stream of fluorine (Fig. 4).

*Catalytic fluorination with elemental fluorine* is carried out over copper turnings coated with silver or gold in a special apparatus shown in Fig. 13. (p.75).

For fluorinations with inorganic fluorides, two types of reactors have been developed: one consisting of several flat trays (Fig. 14, p.79), the other of cylindrical shape and fitted with revolving paddles (Fig. 15, p.81).

Cells for *electrolytic production of fluorine* in laboratory and industrial scale are illustrated in Figs. 10 and 11 (p.45,47), respectively. Special mention should be made of the apparatus in which *electrolytic fluorination* by electrolysis of organic compounds in anhydrous hydrogen fluoride is carried out. These are substantially cells with additional modifications which allow drain-

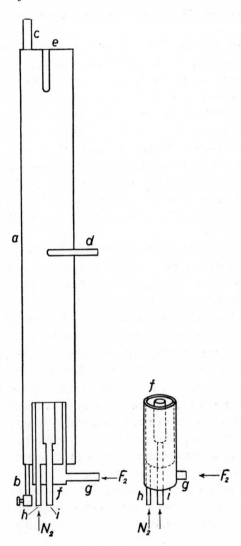

FIG. 4. Apparatus for non-catalytic fluorinations: a) reactor, b) liquid product outlet, c) gas product outlet, d,e) thermometer wells, f) concentric burner, g) fluorine inlet, h) nitrogen inlet, i) organic vapour inlet.

ing of the liquid products from the bottom, taking off of the gaseous products from the top, condensation of hydrogen fluoride in a reflux condenser, and control and maintenance of the hydrogen fluoride level. One of the simplest apparatus of this kind is shown in Fig. 16 (p.85).

## CHEMICAL RESISTANCE OF MATERIALS

The following paragraph will deal with the chemical resistance of the most important materials of which equipment for the preparation of organic fluorine compounds is constructed.

### Glass

Glass stands the action of *elemental fluorine* as long as it is dry and free of hydrogen fluoride. Quite a series of fluorinations using elemental fluorine has been carried out in glass apparatus. The reason for preferring metal apparatus is safety, since the glass apparatus may easily break and fluorine then escapes into the atmosphere; another reason is that in the majority of direct fluorinations of organic compounds, hydrogen fluoride is liberated which attacks glass.

The corrosion of glass by *hydrogen fluoride* is based on its dissolving of silicon dioxide and silicates and transforming them to silicon tetrafluoride. The intensity of the action of hydrogen fluoride depends on its concentration and on the specific resistance of the glass used. Softer glass is attacked more easily than the hard species. Hydrogen fluoride of low concentration does not damage ordinary glass to a considerable extent. Quartz stands its action much better (*1084*). Even if dilute (4%) hydrofluoric acid is detrimental to volumetric vessels and devices (*924*), a sufficiently thick glass apparatus will stand quite a few experiments before its walls become thin and liable to break (*259*). However the contact of glass with hydrofluoric acid of current concentrations (40, 48 and 70%) must be avoided since the dissolution takes place very rapidly. The same holds true for the work with anhydrous hydrogen fluoride which dissolves glass instantly with evolution of gaseous silicon tetrafluoride.

Free *fluoroboric acid* can be handled in glass vessels when its contact with glass does not last too long. For storing, glass is unsuitable since fluoroboric acid of 35% strength dissolves approximately 75 g/m² of a borosilicate glass per day.

Glass apparatus stand the action of most of the *metal fluorides* commonly used in the fluorination reactions. Usually surface etching occurs whose intensity depends on how much hydrogen fluoride was contained in the starting fluoride or was formed in the reaction from moisture. For the work with chlorine and bromine trifluorides glass is not recommended even if a glass apparatus was used for a reaction with the latter compound (*30*). Unless pressure is required, the common fluorinations with antimony tri- fluoride (*107, 688*), pentafluoride (*675*) and mercuric fluoride (*461*) can be run in glass apparatus.

The reactions with *potassium fluoride* are commonly carried out in glass apparatus. Steel autoclaves have been used (*336, 912*) in this method of fluo- rination only in those instances when the working temperature exceeded the boiling points of the organic components. Nowadays the use of super-atmos- pheric pressure is obviated by using proper high-boiling solvents.

## Iron and Steel

Cast iron and soft steel do not stand *aqueous hydrofluoric acid* of concentra- tion lower than approximately 58%, and silicon containing steels do not stand even the more concentrated acid. Unfortunately more accurate data about the corrosion of these materials by hydrofluoric acid are still missing. Hydrofluoric acid of higher than 75% concentration does not attack iron at normal temperatures provided air is excluded.

Anhydrous or almost *anhydrous hydrogen fluoride* is innocuous to cast iron and steel even at higher temperatures. Commercial anhydrous hydrogen fluoride is commonly stored in steel tanks. The majority of reactions involv- ing anhydrous hydrogen fluoride can be carried out in apparatus made of carbon steel (*1038*). Strong corrosion takes place in fluorinations by means of hydrogen fluoride and antimony pentachloride if even negligible amounts of water are present in the system. The most intensive corrosion was observed at the interphase surface of the liquid and vapour. It is probable that the air oxygen which enters the apparatus during its opening plays a part in this corrosion.

In the transformation of hexachloroethane to trichlorotrifluoroethane by means of anhydrous hydrogen fluoride in the presence of antimony pentachloride, a relatively thick autoclave wall was finally perforated. In each routine experiment, the level of the liquid components had reached approximately the same height in the reaction ves- sel, and the most intensive corrosion was observed at this spot along the whole circum- ference (*877*).

Steel and iron can further be used in contact with *elemental fluorine* at temperatures which do not exceed approximately 400°C. Big industrial electrolytic cells for the production of fluorine are constructed of steel. The same material can be used for preparing perfluoro-derivatives by the process which uses *cobalt trifluoride* continuously regenerated by elemental fluorine

*(646)*. Steel tanks are used for storing compressed *(291, 599)* or even liquid fluorine *(5, 614)*.

For exact work in which even negligible contamination of the products with iron compounds must be avoided and in which catalytic influence of the metal upon the reaction must be excluded, the use of iron is not suitable since some corrosion always takes place.

## Copper

Copper is suitable for the reactions involving *elemental fluorine*, both anhydrous and aqueous *hydrogen fluoride* and perhaps all the *inorganic fluorides*, at least those which are useful in the preparation of organic fluorine compounds. It is more resistant than iron or steel, but is nevertheless corroded by antimony pentachloride with hydrogen fluoride containing even small amounts of water *(877)*. Most common applications of copper are in the form of tubing for elemental fluorine and liquid and gaseous hydrogen fluoride and of vessels for reactions with aqueous hydrofluoric acid. It is however unsuitable for analytical experiments with hydrogen fluoride.

## Nickel

Nickel is one of the most resistant and most suitable materials for the construction of apparatus especially for work with *elemental fluorine* at higher temperatures. It resists the action of fluorine in the electrolysis of acid fluorides *(218)* and in the fluorination of organic compounds by electrolysis in anhydrous hydrogen fluoride. Consequently it serves as a material for fabrication of electrodes or at least anodes for electrolytic processes *(175, 218, 321, 546)*. Beside steel it is used for the construction of tanks for storing elemental fluorine *(291)*. Nickel resists the action of *anhydrous hydrogen fluoride*, but is attacked by aqueous hydrofluoric acid, particularly at higher temperatures *(833)*. A reaction mixture containing antimony pentachloride, hydrogen fluoride, and traces of water also strongly corrodes nickel equipment *(877)*.

## Aluminium, Magnesium, Lead

Aluminium and magnesium stand *elemental fluorine* relatively well. Their resistance is inferior to that of nickel, but superior to iron *(602)*. Aluminium vessels are recommended for the work with *antimony pentafluoride (832)*. Lead is resistant to aqueous *hydrofluoric acid* below about 70% concentration *(21)*. It is badly corroded by *anhydrous hydrogen fluoride (21)*.

## Precious Metals

Precious metals, silver, gold and platinum, resist excellently the action of *hydrogen fluoride* and *corrosive fluorides* and are therefore used in all instances where any impurities which may arise by corrosion have to be excluded. Very often coating of the apparatus by any of the above mentioned metals is satisfactory (*938*).

## Alloys

*Anhydrous hydrogen fluoride* and *fluorine* may be handled in brass or bronze, for both resist corrosion in mild conditions. Their resistance to aqueous hydrofluoric acid depends largely on the way of their use.

The best material, both for contact with *elemental fluorine* at low and high temperatures and for contact with anhydrous and aqueous *hydrogen fluoride*, is *Monel metal* which is the most suitable material for the preparation of inorganic as well as organic fluorine derivatives in almost any conditions (*599, 1105*). It is noteworthy that this alloy (67% Ni, 30% Cu, 1·4% Fe, 0·1% Si, 0·15% C) is more resistant than both of its main components (*21*).

## Plastics

From the broad field of plastics only *polytetrafluoroethylene* and *polychlorotrifluoroethylene* are completely resistant to *fluorine*. *Polyethylene* is superficially fluorinated, but more accurate data are missing. Anhydrous or highly concentrated *hydrogen fluoride* leaves unattacked only those plastics which contain saturated hydrocarbon or halohydrocarbon chains; esters (methacrylates, terephthalates), amides (polyamides, polyurethanes) and formaldehyde resins are usually out of the question. For work with aqueous *hydrofluoric acid* at lower temperatures, the following materials are satisfactory: *polyethylene* (the high-pressure material stands reliably temperatures up to 80°C, the low-pressure material up to 110°C), *polystyrene, polyvinylchloride* (up to 50°C), and a copolymer *vinyl chloride-vinylidene chloride* (up to 80–90°C). *Polychlorotrifluoroethylene* and *polytetrafluoroethylene* stand hydrofluoric acid even at temperatures above 100°C. The above mentioned plastics resist also the action of anhydrous hydrogen fluoride though not all equally well. So far, only polychlorotrifluoroethylene and polytetrafluoroethylene are available for contact with anhydrous hydrogen fluoride at temperatures higher than 100°C. It is possible that new isotactic polymers of aliphatic olefins will also be satisfactory.

Rubber loses resilience and becomes hard and brittle when exposed to hydrogen fluoride. However it stands low concentrations of both hydrogen fluoride and fluorine so that it can be used for a short contact with both of the compounds. Chloroprene rubber is much more resistant.

Generally speaking, all the data on the resistance of individual materials are to be judged with much reserve. Systematic study which would follow exactly the behaviour of so many materials under so many various conditions is lacking. Estimation whether a material is or is not suitable for the purpose is considerably subjective and specific. Sometimes even negligible corrosion must be avoided, whereas in another instance any material which stands the necessary time, irrespective of what happens to it or the compounds contained in the apparatus, is satisfactory. The behaviour described above of certain materials toward hydrogen fluoride, fluorine and some of their compounds should give only a rough orientation. It is quite possible that in an individual case even such material will satisfy, whose resistance is generally considered low, and on the other hand, even a very resistant material may fail if exposed to conditions at which its resistance was not yet evaluated. Tables 2 (602), 3 (21) and 4 (19, 833) should serve as examples of numerical expression of chemical resistance of several materials.

TABLE 2

*Corrosion by Gaseous Fluorine at Atmospheric Pressure (21,602)*

| Material | Corrosion Rate (Inches Penetration per Month) at Temperature °C | | | |
|---|---|---|---|---|
|  | 200 | 300 | 400 | 500 — |
| Armco iron | Nil | 0·009 | 0·024 | 11·6 |
| Sheet steel (0,007%Si) | Nil | 0·004 | 0·012 | 7·4 |
| SAE 1030 (trace of Si) | 0·002 | 0·009 | 0·015 | 19·8 |
| Copper |  |  | 0·16 | 0·12 |
| Nickel |  |  | 0·0007 | 0·005 |
| Monel metal |  |  | 0·0005 | 0·002 |
| Aluminium |  |  | Nil | 0·013 |
| Magnesium | Nil | Nil |  |  |

TABLE 3

*Corrosion of Metals and Alloys by Aqueous and Anhydrous Hydrofluoric Acid (21)*

| Conditions | Corrosion Rate (Inches per Year) | | | | | | | |
|---|---|---|---|---|---|---|---|---|
|  | Carbon Steel | Lead | Silver | Copper | Nickel | Monel | Inconel | Bronze (10%Al) |
| 38% HF, 110°C Presence of Air | 2·90 | 0·146 | 0·004 | 0·047 | 0·110 | 0·047 | 0·113 | 0·046 |
| 98% HF, 38°C Presence of Air | 0·005 | 1·48 | 0·002 | 0·006 | 0·002 | 0·002 | 0·002 | 0·006 |

## TABLE 4

*Corrosion of Materials (19,833)*

| Material | F₂ | HF 100% | HF 40—65% | NH₄F NaF KF aq | NH₄HF₂ KHF₂ aq | SbF₃ |
|---|---|---|---|---|---|---|
| Iron | 0 | 0 | 3 | 0 | | + |
| Carbon steel | 0 | + | | | | + |
| Chromium-nickel steel | | | 3 | | | |
| Stainless steel | | | 3 | | 0—3 | |
| Nickel | 0—1 | 0 | 0 | | | |
| Copper | 0 | 0 | 1—3 | | | + |
| Monel metal | 0 | 0 | 0—2 | | | |
| Bronze | | | 0 | 3 | | |
| Lead | 1 | + | 0—3 | | | |
| Aluminium | 3 | | | 0 | | |
| Magnesium | 0 | | 0—1 | | | |
| Electron | 0 | | 0—1 | | | |
| Silver, gold, platinum | 0 | 0 | 0 | | | |
| Glass | 0—3 | 3 | 3 | | | + |
| Rubber | | | 0 | 0 | | |
| Polyethylene, paraffin | | 0 | 0 | | | |
| Polytetrafluoro-, poly-chlorotrifluoroethylene | + | 0 | 0 | | | |
| Polyvinyl chloride, poly-vinylidene chloride | | + | 0 | | | |
| Cellulose | 0—1 | | 0 | | | |
| Polyamide, polyester | | + | | | | |

+ Applicable, but no data available

*Explanations*

| Degree of Resistance | 0 Practically Stable | 1 Sufficiently Stable | 2 Applicable | 3 Unsuitable for Chemical Thermal or Mechanical Reasons |
|---|---|---|---|---|
| Loss in g per m² and day for aluminium and alloys | 0—0·8 | 0·8—8 | 8—24 | More than 24 |
| Loss in g per m² and day for all other materials | 0—2·4 | 2·4—24 | 24—72 | More than 72 |

# III

# FLUORINATING REAGENTS

IN the chapter on fluorinating reagents, compounds which are used for the preparation of organic derivatives of fluorine will be considered. In the first place, hydrogen fluoride and fluorine will be mentioned. Of the inorganic fluorides only those which are used in preparative practice will be described. A brief mention will be made also of organic fluorine compounds which act as fluorinating agents.

In dealing with individual fluorinating reagents the attention will be directed toward their preparation or production, their physical properties, and their applications in the preparation of organic fluorine derivatives.

## HYDROGEN FLUORIDE

Hydrogen fluoride is, from the point of view of preparation of organic fluorine compounds, the most important substance since it is used not only for the direct preparation of fluorinated organic derivatives, but also for the production of elemental fluorine and inorganic fluorides, which are themselves fluorinating reagents. Therefore, its preparation, properties and applications will be described in sufficient detail.

### Preparation of Hydrogen Fluoride

Industrial production of hydrogen fluoride is based on the reaction between fluorspar with concentrated sulphuric acid. The fluorspar suitable for the production of hydrogen fluoride must be of the so called "acid grade" which means a high grade material containing at least 97% of calcium fluoride, less than 1·5% of silicon dioxide, and less than 1% of calcium carbonate. The decomposition is carried out in rotary kilns made of mild steel. The crude gaseous hydrogen fluoride is either directly liquefied, or else absorbed in weak hydrofluoric acid to raise the concentration to 70–80%. Distillation separates anhydrous hydrogen fluoride and leaves a residue of approximately 38% hydrofluoric acid which is re-used for absorption of crude hydrogen fluoride (*614*).

A redistillation of the anhydrous hydrogen fluoride removes silicon tetra-fluoride, sulphur dioxide and carbon dioxide so that pure hydrogen fluoride of 99·95% purity is obtained (*975*).

Anhydrous hydrogen fluoride is shipped in steel tanks or cylinders. For laboratory use cylinders containing 10–20 kg of hydrogen fluoride are most convenient.

Hydrogen fluoride is drawn from the container either in liquid or in gaseous state. In the first case, the cylinder valve must be fitted with a metal tube reaching below the surface of the liquid. Otherwise the cylinder has to be turned upside down or placed in an inclined position. When gaseous hydrogen fluoride is taken from the container, the valve is operated with the cylinder in its normal upright position. The cylinder has to be warmed with hot water or steam especially when large amounts of hydrogen fluoride are consumed and the laboratory temperature is low (b.p. of hydrogen fluoride is 19·5°C).

If high purity hydrogen fluoride is needed, the gas which is drawn off the tank is led through a condenser attached directly to the tank valve and is received as a liquid. The material obtained in this way is usually purer than

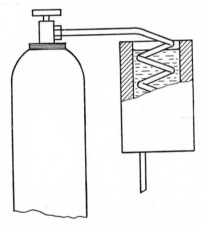

FIG. 5. Equipment for distilling anhydrous hydrogen fluoride out of a steel cylinder.

that obtained by pouring from the tank. First of all it does not contain any mechanical impurities, no possible residue of sulphuric acid, fluorosulphonic acid, or fluorosilicic acid, and no water, because this way of drawing-off represents actually an additional distillation. The manner just described for obtaining pure anhydrous hydrogen fluoride is illustrated in Fig. 5.

When tanks of anhydrous hydrogen fluoride are not available, the compound can be prepared in the laboratory by distillation of aqueous hydrofluoric acid of 70% content, or by repeated distillation of hydrofluoric acid of 48% content of hydrogen fluoride. The distillation is carried out in a copper (in the case of 70% acid even steel) apparatus provided with ice-water or brine cooling.

If a high-grade hydrogen fluoride for analytical purposes is required, the distillations must be carried out in a polyethylene or better a platinum apparatus. The purest anhydrous hydrogen fluoride is prepared in the laboratory by thermal decomposition of potassium bifluoride at 500—600°C after previous drying by electrolysis at 220°C in a copper apparatus (949).

Aqueous hydrofluoric acid is commercially available in concentrations of 40%, 48%, and 65–70%. The acid containing 40% or 48% of hydrogen fluoride is shipped in containers made of plastics such as guttapercha, ebonite, polyvinyl chloride, polystyrene or polyethylene, the acid of 70% content in steel tanks.

The content of hydrogen fluoride in aqueous hydrofluoric acid may be roughly determined by measuring the density whose relation to concentration is not, however, unambiguous, as shown in Fig. 6.

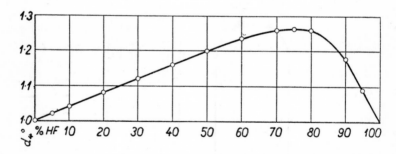

FIG. 6. Dependence of the density of aqueous hydrofluoric acid on the content of hydrogen fluoride.

Hydrogen fluoride and water form an azeotrope with maximum boiling point of 112·0°C at 750·2 mm, which contains 38·26% of hydrogen fluoride and 61·74% of water (755). From the chart in Fig. 7 (755) it may be seen that practically pure hydrogen fluoride (98%) can be obtained from a 48% acid by two distillations and from a 70% acid by one distillation only. The liquid-vapour equilibrium of the mixture hydrogen fluoride–water is thus for this purpose extraordinarily favourable.

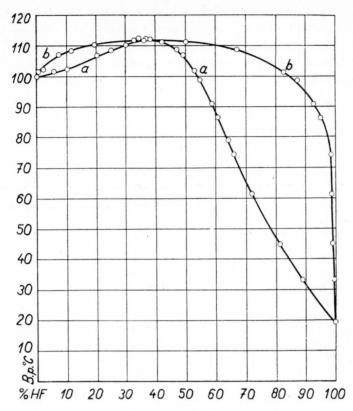

F<small>IG</small>. 7. Liquid-vapour diagram in water-hydrogen fluoride system: a) liquid, b) vapour (both in weight per cent).

## Properties of Hydrogen Fluoride

Anhydrous hydrogen fluoride is a colorless mobile liquid forming thick fumes in contact with air moisture. It has a pungent smell and strongly attacks the mucous membranes of eyes, nose and throat. It causes painful burns in contact with the skin. It is therefore imperative to protect the eyes with a shield, the hands with heavy rubber gloves and the body with a plastic apron during any operations in which handling of anhydrous hydrogen fluoride is encountered. Although aqueous hydrofluoric acid is not nearly as dangerous, its contact with the skin must also be avoided.

Anhydrous hydrogen fluoride is "no more corrosive than distilled water" (*614*) (except for glass, of course). It is handled in mild steel material. However glass is attacked instantly and destroyed completely in a short time.

Aqueous hydrofluoric acid attacks glass a little more slowly. It may be handled in copper, lead and, as long as the content of hydrogen fluoride is not lower than approximately 58%, in steel equipment. Good service is given by plastics: polyethylene (*188*), polychlorotrifluoroethylene (*904*) and polytetrafluoroethylene (*933*).

Physical properties of hydrogen fluoride are listed in Table 5 (*947*) and some of them illustrated in Charts 8 and 9.

TABLE 5

*Physical and Physico-Chemical Constants of Hydrogen Fluoride (15,947)*

| | |
|---|---|
| Molecular weight .................... | 20·01 |
| Melting point ...................... | −83° |
| Boiling point ....................... | 19·54°C |
| Critical temperature ............... | 230·2°C |
| Heat of fusion (at −83°C) .......... | 1094 cal/mole |
| Heat of evaporation (at 19·4°C) ..... | 6025 cal/mole |
| Specific heat of liquid HF between 200− −270°K....................... | $15·51 − 0·0538T + 0·000218T^2$ cal/mole |
| Density of liquid HF .............. | $1·0020 − 0·0022625t + 0·000003125t^2$ g/ml. |
| at | −60°          −30°          0°C |
| | 1·1660        1·0735        1·0015 |
| Viscosity of liquid HF ............. | $\log \eta = − 4·282 + 0·006372T + \dfrac{806,1}{T}$ mP |
| at | −25°    −12·5°   −6·25°   0°    +6·25°C |
| | 0·350    0·296    0·274   0·256   0·240 cP |
| Surface tension .................... | $40·7 \left(1 − \dfrac{T}{503·2}\right)^{1·78}$ dyne/cm |
| at | −60°      −40°      −20°     0°C |
| | 15·4      13·6      11·7    10·1   dyne/cm |
| Dielectric constant at ............... | −73°      −70°      −42°     −27°      0°C |
| | 174·8     173·2     134·2    110·6    83·6 |
| Specific conductivity of the purest HF .. | $2·6 − 5·7 . 10^{−6}$ ohm$^{−1}$ cm$^{−1}$* |
| Vapour tension .................... | $\log P_{mm} = 7·3739 − \dfrac{1316·79}{T}$ |
| Thermodynamic functions for ........ | $H_2(g) + F_2(g) = HF(g)$ |
| $H_{298·1}$ ............................ | −64·100 cal/mole |
| $F_{298·1}$ ............................ | −64·910 cal/mole |
| $S_{298·1}$ ............................ | 1·68   cal/mole |

* Reference *899*.

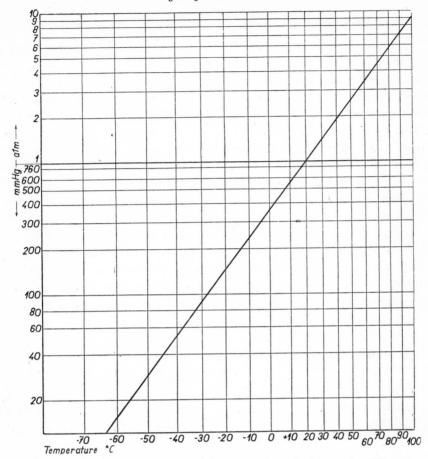

FIG. 8. Dependence of vapour pressure of anhydrous hydrogen fluoride on temperature.

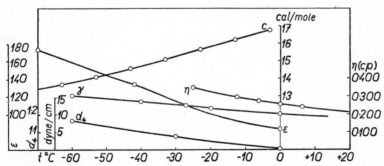

FIG. 9. Dependence of density, viscosity, surface tension, dielectric constant and specific heat of anhydrous hydrogen fluoride on temperature: $d_4$ density, $\eta$ viscosity, $\gamma$ surface tension, $\varepsilon$ dielectric constant, $c$ specific heat.

## Applications of Hydrogen Fluoride

Anhydrous hydrogen fluoride is one of the compounds which developed during the last thirty years from laboratory curiosities to first-class large-scale products. One important application is its use as a catalyst in alkylation of lower olefins by paraffins in the production of high-octane gasoline. It is used for the preparation of aluminium fluoride, cryolite, uranium fluoride and many inorganic fluorides. It is the only source for the production of fluorine, and at the same time an almost universal fluorinating reagent by means of which the majority of technically important organic fluorine compounds are prepared. The distribution of hydrogen fluoride into individual applications is shown in Table 6 (8). A more detailed description of modern applications of hydrogen fluoride, can be found in reviews (15, 947, 1086).

On the laboratory scale of preparation of organic fluorine compounds, anhydrous hydrogen fluoride adds to olefins and acetylenes, replaces halogens by non-catalytic and catalytic processes in liquid-phase and vapor-phase reactions, exchanges diazotized aromatic aminogroups for fluorine, and replaces hydrogen by fluorine in the electrolytic fluorination process.

TABLE 6

*Consumption of Anhydrous Hydrogen Fluoride in Industry (8)*

| | | | |
|---|---|---|---|
| Fluorocarbons | 31% | Inorganic Fluorides | 6% |
| Aluminium Fluoride | 30% | Metal Cleaning | 5% |
| Uranium Fluorides | 12% | Alkylation Process | 4·5% |
| Synthetic Cryolite | 7% | Glass Etching | 1·5% |

## FLUORINE

The use of elemental fluorine cannot be avoided in the preparation of certain metal fluorides, especially fluorides with the highest valency. Some of them are used in the preparation of organic fluorine compounds. Organic perfluoroderivatives, for instance, could not be prepared without elemental fluorine. It has been in recent years only that the use of fluorine has decreased because of competition by the electrochemical fluorination of organic compounds.

## Preparation of Fluorine

The first to succeed in preparing elemental fluorine was Moissan (1886) who electrolysed anhydrous hydrogen fluoride in a platinum U-tube with platinum electrodes (745). The main interest in fluorine started, however, much later, in the time when its need in the production of uranium was recognized. Anhydrous hydrogen fluoride was found an unsuitable electrolyte for the generation of fluorine since its conductivity is very low (Table 5, p. 42). Therefore acid potassium fluoride or better still its mixture with hydrogen fluoride of the composition of KF.2KF is preferred and is very satisfactory because it melts around 100°C and the electrolytic cell may be heated with steam. Cathodes and diaphragms are made of iron or steel, anodes of carbon or nickel. Nickel electrodes have higher mechanical resistance, greater possibility of extremes in temperature and hydrogen fluoride concentration, and lesser sensitivity towards water in the electrolyte (*218*). The cell having carbon anodes allows more economical exploitation of the electrolyte and gives higher current yields (*218*). It also does not produce so much sludge as the nickel-anode cell.

For laboratory-scale preparation of fluorine, cells have been designed which are simple in construction and can be improvised by simple means (*206, 429*). The only requirement is the use of "spectrographic" graphite for electrodes.

The apparatus shown in Fig. 10 (*429*) is fabricated by welding two copper pipes of 5 cm in diameter and 30 cm in length in the shape of a 70° V. No diaphragm is needed in this construction since the cathodic and anodic spaces are separated by the higher level of the electrolyte. The cathodic branch is open and the electrode is fixed by means of a cork stopper which is notched several times on its periphery to allow the hydrogen evolved in the electrolysis to escape. In addition the stopper may be fitted with a cap-

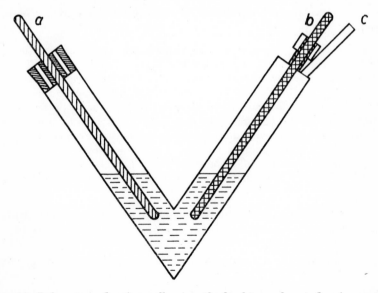

FIG. 10. Laboratory fluorine cell: a) cathode, b) anode, c) fluorine outlet.

per thermometer well and a thermometer to read the temperature of the electrolyte. The anodic arm is closed by a copper disk fitted with the female half of a force joint and an eccentrically welded copper delivery tube for fluorine. The anode is inserted in the male half of the force joint and cemented with Portland cement or polymerized varnish. It is then screwed into the female force joint using a copper gasket. The cell is insulated with asbestos and mica, and heating wire is wound around both branches, which again are insulated with asbestos or any other heat insulating material.

Graphite rods made of spectrographic graphite are used as electrodes, and a mixture of potassium fluoride and hydrogen fluoride corresponding to an empirical formula of KF.3HF is used as an electrolyte which allows carrying out of the electrolysis at 70–100°C.

At the start of the electrolysis, oxygen is evolved in the anodic space since the electrolyte contains some water, and the electrodes become coated with silicates which form a film preventing passage of the current. When the electrolysis stops, the electrodes are replaced by new ones. The used electrodes are recovered by removing the silicate layer by means of sand paper. As soon as all the moisture of the electrolyte is removed by electrolytic decomposition to hydrogen and oxygen, the electrolysis goes smoothly and produces fluorine at a potential of 12 V or 18 V and the corresponding current intensity of 5 or 10 A, respectively (depending on the construction of the cell). The heat evolved in the electrolysis suffices to maintain the electrolyte in liquid state. The current efficiency is around 75%.

A suitable means for replenishing the electrolyte is to connect the anodic space of the still hot cell with a hydrogen fluoride tank; after opening the tank valve the hydrogen fluoride vapours are sucked into the cell up to an equilibrium concentration of KF.3HF (*429*).

For the electrolytic production of fluorine on a large scale, a series of apparatus of various constructions, sizes and operating characteristics was tested. Those operating around 100°C seem to prevail nowadays; one of them is shown in Fig. 11 (*929*).

A steel cylinder surrounded by a steam-heated jacket is surmounted by a steel lid with holes for hydrogen and fluorine outlets and holes for electric cables. The body and the lid are screwed together, using a copper gasket. The inner side of the lid is fitted with two concentric steel cylinders. The narrower one reaches below the surface of the electrolyte and acts as a diaphragm. The wider one reaches almost to the bottom of the cell and forms the cathode. The anode is a carbon rod screwed on to a copper rod which is fixed in the center of the cell and serves simultaneously as a current lead. A mixture of potassium fluoride with hydrogen fluoride of a composition corresponding approximately to KF.1.8 HF with a small amount of lithium fluoride acting as a melting point depressant is used as the electrolyte. The cell is operated at 100°C under a potential drop of 7,5–8 V and current density of 6,5–11 A/dm².

Still larger fluorine cells have a whole series of anodes suspended from one lid. Usually several such cells are combined to form batteries with common fluorine delivery (*218*).

Crude fluorine always contains some hydrogen fluoride. This is entirely removed by passing the crude gas over granulated sodium fluoride so that fluorine of 99·9% purity results (*291, 614*).

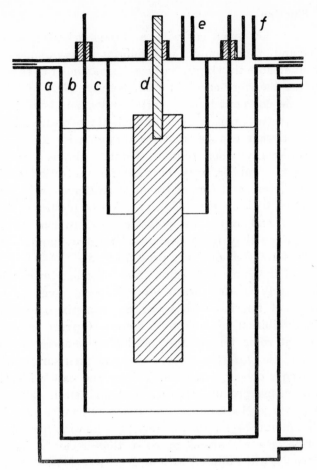

FIG. 11. Industrial fluorine cell: a) cell body surrounded by heating jacket, b) iron cathode, c) iron diaphragm, d) carbon anode, e) fluorine outlet, f) hydrogen outlet.

It is more practical to consume fluorine where it is made. Storing and shipping are more difficult than for other gases. On account of its extraordinary reactivity fluorine cannot be compressed mechanically. To be put into steel or nickel cylinders, it is liquefied by means of liquid nitrogen and allowed to evaporate up to the pressure of 28 atm (*291, 614*). Heating fluorine up to 300°C in order to remove ozone and other impurities has been recommended (*599*). Recently huge quantities of fluorine have been shipped in liquid state. The fluorine is stored in a tank with two jackets filled with liquid nitrogen (*5*).

## Properties of Fluorine

Fluorine is a yellow-green gas with a strong irritating smell resembling that of chlorine and ozone. Contact with fluorine is very dangerous. At higher concentrations it attacks the skin and causes burns similar to those of hydrogen fluoride. They are therefore cured in a similar way (602) (p.24).

Among organic compounds only carbon tetrafluoride resists entirely the action of concentrated fluorine even at high temperatures. However quite a few organic materials can be used for contact with fluorine at ordinary temperatures, especially when diluted. Rubber gloves, particularly those made of chloroprene rubber, may be used for protection against fluorine provided they are clean and free of traces of any grease which would easily catch fire in contact with fluorine (602). Fluorinated plastics such as polychlorotrifluoroethylene and especially polytetrafluoroethylene are successfully used as gaskets since they resist even concentrated fluorine, the latter up to temperatures of 150°C at ordinary pressure (878).

In the field of inorganic compounds fluorine reacts even with materials which are completely stable toward other halogens, such as asbestos and water. Reactions of fluorine with water are sometimes explosive because of their initial inhibition by unknown factors. Current metals react with fluorine already at room temperature: at higher temperatures the reaction may be very energetic (291). Some metals form at their surface an impervious film which prevents deeper corrosion. This is the case with Monel metal, nickel, aluminium, magnesium, copper, iron and steel. Surface treatment with fluorine is purposely used for imparting resistance and insulation to copper and aluminium wires (10). Entirely pure fluorine free of moisture and hydrogen fluoride does not attack glass, so that it can be handled in glass apparatus especially when diluted with an inert gas (104).

Fluorine is the element with the highest relative electrode potential (−2·85V) (the corresponding value for chlorine is −1·36V). The main physical properties of fluorine are listed in Table 7 (947).

TABLE 7

*Physical Constants of Fluorine (947)*

| | |
|---|---|
| Atomic Weight | 19·00 |
| Melting Point | −218·0°C |
| Boiling Point | −187·99°C |
| Critical Temperature | −129·2°C |
| Critical Pressure | 55 atm |
| Heat of Fusion | 372 cal/mole |
| Heat of Evaporation | 1581 cal/mole |
| Viscosity at 0° | 0·0002093 P |

## Applications of Fluorine

Fluorine played and still plays a very important role in the production of nuclear energy. It is indispensable for the preparation of the higher fluorides of uranium and its isotopes, which are then separated from each other by thermodiffusion. In addition, the fluorinated compounds used as media or materials for carrying out the thermodiffusional separation were fabricated by means of fluorine or higher metal fluorides which in turn were prepared from fluorine.

Recently fluorine has been produced in quantities as a very efficient rocket fuel. In the combination with kerosene liquid fluorine is 22–40% more effective than liquid oxygen (7).

Fluorinated organic compounds are prepared by various methods, such as by addition of fluorine to some halogenated olefins and to aromatic compounds, by replacement of hydrogen in paraffins, of halogen in halogen derivatives, and of other elements or groups by fluorine. Usually several of these reactions occur simultaneously. Final products are polyfluoro-and perfluoroderivatives. All these reactions are effected by non-catalytic or catalytic fluorination with elemental fluorine, or by the action of silver difluoride or cobalt trifluoride which are prepared and regenerated by means of fluorine. Recently some of these methods of preparation of polyfluoro- and perfluoroderivatives have been replaced by the electrochemical fluorination process which is carried out by electrolysis of organic compounds in anhydrous hydrogen fluoride and which obviates handling of free fluorine and also makes use of the hydrogen fluoride produced by the fluorination reaction (p. 84).

## INORGANIC FLUORIDES

In addition to hydrogen fluoride and fluorine, a whole series of inorganic fluorides find a use in the preparation of organic fluorine compounds. Their action upon organic substances is not so universal as that of hydrogen fluoride or fluorine, but is quite specific. The majority of the metal fluorides can replace only some other halogen by fluorine in an organic molecule. Only a few of them can replace hydrogen or cause the addition of fluorine across double bonds. Though new fluorides increase from time to time the series of practical inorganic fluorinating reagents, commonly only twenty to thirty inorganic fluorides are used for this purpose, out of which approximately ten are really important.

Fluorides used most frequently in the preparation of organic fluorine compounds will be described more thoroughly and eventually procedures will be given for their laboratory preparation. With others only reference to their formation will be made. Physical properties and ways of application in the production of organic fluorine derivatives are listed in Table 8 (p. 62). The individual fluorides will follow according to their sequence in the periodical table of elements.

## Ammonium Fluoride

Ammonium fluoride is obtained by neutralizing hydrofluoric acid with ammonia and evaporating the solution to dryness. It is very soluble in water.

Ammonium fluoride is suitable only for replacement of a very reactive halogen such as chlorine linked to silicon or phosphorus, and its use is very limited.

## Lithium Fluoride

Lithium fluoride is prepared by dissolving lithium carbonate in aqueous hydrofluoric acid and evaporating the solution.

Its use in the preparation of organic compounds of fluorine is rare.

## Sodium Fluoride

The preparation of sodium fluoride is described in Chapter IX (p. 322) since the pure material is used as an analytical standard. It is obtained by neutralization of aqueous hydrofluoric acid with sodium carbonate or sodium hydroxide and fusion which removes any excessive hydrogen fluoride.

Sodium fluoride can replace a sufficiently reactive halogen such as chlorine in ester-chlorides of phosphoric acid It is however less active and less soluble in organic solvents than potassium fluoride and therefore much less in use.

## Potassium Fluoride

Potassium fluoride is the most popular alkali fluoride. It is prepared by neutralization of aqueous hydrofluoric acid with potassium carbonate or potassium hydroxide and evaporation. It forms hydrates containing two and four molecules of water.

Potassium fluoride for the preparation of organic fluoro-derivatives must be perfectly dry. After drying for 24 hours at 125°C, the fluoride is finely powdered and dried again for 24 hours at 150°C (*501*).

Potassium fluoride is an almost universal chemical for the replacement of halogens by fluorine in organic compounds. In addition to reactive halogens activated by nitro groups, potassium fluoride can substitute fluorine for even as unreactive a halogen as the chlorine in alkyl chlorides.

## Acid Potassium Fluoride

Acid potassium fluoride is obtained in the following way: aqueous hydrofluoric acid is divided into two equal volumes, one is neutralized with potassium carbonate and blended with the other volume of the acid. After sedimentation of potassium fluorosilicate which is usually present as an impurity, the solution is decanted into a platinum dish, evaporated to crystalization, and after cooling the crystals are separated from the mother liquor by filtration through a cotton tissue. The acid potassium fluoride is dried at 120–150°C in a copper vessel while a stream of air dried with sulphuric acid and phosphorus pentoxide is passed through the melt (*893*).

The application of acid potassium fluoride is limited to the exchange fo reactive halogens such as those in acid chlorides.

## Silver Fluoride

In the preparation of silver fluoride, it is possible to start with silver oxide or silver carbonate. The latter is obtained by pouring a solution of silver nitrate into a dilute solution of sodium bicarbonate, decanting, and washing the precipitate with distilled water until it does not contain nitrate ions. The silver carbonate thus obtained is dissolved in an excess of pure aqueous hydrofluoric acid, the solution is evaporated to dryness, and the residue finally dried by heating on a sand bath. The silver fluoride is described as a fine-grain hygroscopic powder (*893*).

Silver oxide necessary for the preparation of the fluoride is conveniently obtained by oxidation of freshly precipitated silver with hydrogen peroxide; dissolving the product in 20% hydrofluoric acid gives a tetrahydrate of silver fluoride (*1014*). More. frequently, a solution of silver nitrate is treated with alkaline hydroxide, the precipitate is washed free of alkali, dissolved in an excess of 40% hydrofluoric acid, the solution is evaporated to dryness and dried by heating to 200°C *in vacuo* (*243*).

Silver oxide 80 g (0·35 mole) is added to 80 ml (1 mole) of 25% hydrofluoric acid, placed in a polyethylene beaker; after 30 minutes, the mixture is filtered, the filtrate evaporated in a porcelain dish, and the residue dried at 150° (*88*).

In the reactions with organic halogen derivatives, a considerable excess of silver fluoride must be used, since only one half is available for the reaction, while the other half combines with the appearing silver chloride to form a complex compound AgF.AgCl (*694*, *998*). Silver fluoride is suited for the exchange of halogens in aliphatic compounds. Its importance nowadays is much decreased since the same reactions can be carried out with potassium fluoride.

## Silver Difluoride

Silver difluoride is formed by passing fluorine over molecular silver spread in a platinum boat in a quartz tube. The temperature is 60°C at the start and 150-200°C toward the end of the reaction (896). More frequently the preparation is based on the reaction between fluorine and halides of monovalent silver at temperatures of 150–240°C (896, 974).

The reaction is carried out in a flat reactor (Fig. 14, p. 79) in which fluorine is passed over a layer of silver chloride dried at 110°C. The temperature is maintained at 200°C until no chlorine is present in the effluent gas, and three additional hours. The yield of silver difluoride ranges from 90 to 95% (852).

The same reaction is used for regeneration of spent catalyst in the fluorinations of organic compounds (974). Local overheating of the reactor, as when using a free flame, is to be avoided, otherwise an exothermic reaction may occur between the silver difluoride and the metal, resulting in melting the reactor (974). The regeneration is best carried out by passing fluorine at successive temperatures of 25°C, 125°C and 250°C (646).

Applications of silver difluoride to the preparation of fluorinated organic derivatives are very wide. It transforms saturated, unsaturated and aromatic hydrocarbons and their halogen derivatives to polyfluoro- and perfluorocompounds. Its range of application is approximately the same as that of cobalt trifluoride and elemental fluorine. Compared with cobalt trifluoride, silver difluoride is more suited for the fluorination of halogen derivatives and less suited for the fluorination of hydrocarbons.

## Zinc Fluoride

Zinc fluoride is obtained by dissolving zinc carbonate in hot and sufficiently concentrated hydrofluoric acid, evaporating the solution to dryness and drying the residue at 300°C with exclusion of moisture (893).

The use of zinc fluoride in fluorinations is limited to the exchange of reactive halogens such as chlorine in chlorosilanes.

## Mercurous Fluoride

Two procedures have been developed for the preparation of mercurous fluoride, one starting with mercurous nitrate (520, 893, 1055), the other with mercuric oxide (475).

a) One hundred and fifty grams (0·5 mole) of mercurous nitrate dihydrate is dissolved in 450 ml. of hot water acidified with 10 ml. of nitric acid, and the solution is poured to an agitated solution of 75 g (0·75 mole) of potassium bicarbonate in 1·4 l. of water. The precipitate of mercurous carbonate is washed several times with water saturated with carbon dioxide, allowed to stand overnight to decompose any basic nitrates formed, filtered with suction, and dissolved, still wet, in 75 ml. (1·5 moles) of 40% hydrofluoric acid diluted with 25 ml. of water. Mercurous fluoride settles as a yellow powder. After decantation a few ml. of 40% hydrofluoric acid is added and the mixture is evaporated to dryness on the steam- or sand-bath. The residue is dried by heating for 2–3 hours at 120–150°C in a drying box yielding 103 g (0·47) (93·5%) of mercurous fluoride. The dissolution in hydrofluoric acid, as well as evaporation and drying, are best carried out in a platinum dish.

b) Forty grams (0·185 mole) of red mercuric oxide is dissolved in a mixture of 28 ml. (0·42 mole) of concentrated nitric acid and 60 ml. of water. The solution is vigorously shaken with 40 g (0·20 mole) of mercury as long as it dissolves readily. When mercurous nitrate starts to deposit, a solution of 4 ml. of concentrated nitric acid in 45 ml. of water is added to redissolve the crystals. The solution is decanted from the unreacted mercury into a solution of 48 g (0·48 mole) of potassium bicarbonate in 200 ml. of water. The precipitated mercurous carbonate is filtered by suction and washed with 1400 ml. of water saturated with carbon dioxide. The wet material is added portionwise to a stirred solution of 100 ml. (2·9 moles) of 48% hydrofluoric acid in 260 ml. of water in a platinum dish. The mixture is evaporated to dryness on the steam bath, and the residue is dried by heating on the steam bath for an additional hour. The yield of mercurous fluoride is approximately 80 g.

All the operations are to be carried out in a subdued light. The product is stored in tightly stoppered metal or plastic containers. Mercurous fluoride is a relatively weak fluorinating reagent, which is able to replace only a sufficiently reactive halogen (bromine or iodine). However, its efficiency is multiplied by partial or total transformation to mercuric fluoride halide by the reaction with iodine or chlorine *in situ*.

## Mercuric Fluoride

One of the possible ways of preparing mercuric fluoride is the reaction of mercurous fluoride with chlorine and separation of the formed mixture of mercuric fluoride and mercuric chloride by sublimation (*520, 893*).

The reaction is conveniently carried out by passing chlorine over a thin layer of mercurous fluoride spread in a flask, or a horizontal tube heated in an electric furnace to approximately 350° (*520*).

On the bottom of a one-litre Walter flask, 100 g (0·45 mole) of dry mercurous fluoride is spread, the flask is stoppered with a cork stopper fitted with a water-cooled finger, a calcium-chloride tube, a thermometer and an inlet tube for chlorine, and heated in an oil bath to 150°C. At this temperature, chlorine is passed into the flask just above the surface of the mercurous fluoride. During 1·5–2 hours 7 l. of chlorine (20% excess) is consumed. After the stream of chlorine has been shut off the temperature is raised to 280°C and during 1–1·5 hours mercuric chloride is sublimed off and deposited on the condenser. Mercuric fluoride (35–40 g, 0·14–0·17 mole, 65–75%) remains on the bottom of the flask. It is still contaminated with sublimate.

A more modern way of preparing mercuric fluoride is the reaction of mercuric chloride with elemental fluorine. It is carried out by passing a stream of fluorine through a horizontal copper tube into a steel or copper vessel whose bottom is covered with mercuric chloride (Fig. 12). The transformation is quantitative (*461*).

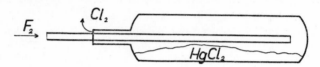

FIG. 12. Equipment for the preparation of mercuric fluoride from mercuric chloride and fluorine.

Another method of preparing mercuric fluoride is the reaction of one mole of yellow or red mercuric oxide with ten to twelve moles of anhydrous hydrogen fluoide. The reaction is carried out at $-80°C$, gives 85–90% yield of mercuric fluoride, and is especially advantageous for the preparation of organic fluorine derivatives since the mercuric fluoride does not have to be isolated: The mixture of mercuric oxide and anhydrous hydrogen fluoride is cooled with a dry-ice bath, an organic halogen derivative is added, the reaction vessel is closed and heated to the required temperature (p. 103).

Similar formation of mercuric fluoride *in situ* is possible by mixing mercuric chloride with anhydrous hydrogen fluoride (*866*) (p. 103).

Mercuric fluoride is a very strong and useful fluorinating reagent suitable for replacing halogens by fluorine in aliphatic mono- and polyhalogen derivatives. According to conditions it can replace one or two and sometimes even three halogen atoms linked to one carbon atom. It may be used in such cases where antimony trifluride or hydrogen fluoride fail or must be avoided because of their detrimental effects on the rest of the molecule, as in the case of some compounds containing oxygen or nitrogen functions.

## Boron Trifluoride

Boron trifluoride is prepared by heating a mixture of an alkaline fluoroborate with boric oxide and sulphuric acid (*113*) or fuming sulphuric acid (*901*) in a ground-glass apparatus consisting of a flask and a water-cooled condenser. Boron trifluoride formed by the decomposition according to the

11         $6 NaBF_4 + B_2O_3 + 6 H_2SO_4 \rightarrow 8 BF_3 + 6 NaHSO_4 + 3 H_2O$

equation is led through glass tubing with a mercury safety valve into a scrubber containing concentrated sulphuric acid, saturated in the cold with boric oxide, and further through a safety trap directly into a liquid for absorption (*113*).

Boron trifluoride is a by-product in the Schiemann reaction (p. 115). If the decomposition of the diazonium fluoroborates is sufficiently slow, the escaping boron trifluoride can be caught by absorption. Boron trifluoride thus prepared is usually contaminated with small amounts of aromatic fluorinated derivatives, especially when obained in preparing low-boiling aromatic fluorides. It may however be used in reactions which do not require too pure a material such as the synthesis of $\beta$-diketones from ketones and acid anhydrides.

A very convenient form of boron trifluoride is its addition compound with diethyl ether, the so called boron-trifluoride etherate, $BF_3.(C_2H_5)_2O$ (*493*). This compound, a colorless fuming liquid having b.p. of 125–126°C, may easily be obtained in the Schiemann reaction, if the escaping gas mixture of nitrogen and boron trifluoride is passed through ether cooled with dry ice. The amount of ether is 1–1·5 mole per mole of the boron trifluoride to be evolved. The pure product is obtained by fractination in a distilling column.

Applications of boron trifluoride or its etherate in the preparation of organic fluorine derivatives are very scarce. However, both compounds are useful catalysts in many reactions.

## Fluoroboric Acid

Fluoroboric acid is formed by dissolving boric acid in hydrofluoric acid. Boric acid is added in small portions to an equivalent amount of hydrofluoric acid containing 40% or more hydrogen fluoride and placed in a platinum or copper dish or a dish made of a resistant plastic. The reaction mixture is stirred with a metal rod and cooled with ice or an ice-salt bath. The resulting solution is allowed to stand for several hours at room temperature and used as such for the reaction, e.g. the preparation of diazonium fluoroborates.

An approximately 50% solution of fluoroboric acid suitable for preparing diazonium fluoroborates is prepared by adding over a period of 30 minutes 140 g (2·25 moles) of boric acid into 300 g (9 moles) of 60% hydrofluoric acid in a dish cooled with ice and salt (*897*).

## Fluoroborates

Salts of fluoroboric acid, ammonium fluoroborate, sodium fluoroborate and potassium fluoroborate, are prepared by neutralizing ice-cooled fluoroboric acid with 25--30% alkali hydroxides using methyl orange as the indicator.

*Potassium fluoroborate* is slightly soluble and crystallizes easily. The crystals are filtered by suction, washed free of potassium fluoride with distilled water and finally washed with alcohol. The yield is around 90% (*705*).

For the preparation of diazonium fluoroborates, ammonium or sodium fluoroborates are more convenient. Because of their higher solubilities in water, the yields are accordingly lower. Evaporation of the mother liquors contaminates the products with alkali fluorides since the fluoroborates decompose hydrolytically at higher temperatures. Solutions of *sodium fluoroborate* can be obtained by the reaction of boric acid with hydrochloric acid and sodium fluoride (*884, 980*). This procedure has the advantage of avoiding the use of hydrofluoric acid. Before using the solution in the reactions with diazonium salts, it is advisable to determine the content of the fluoroborates analytically, since the formation is not quantitative.

Boric acid (340 g) (5·5 moles) and sodium fluoride (950 g) (22·6 moles) are added to 1400 ml. (16·5 moles) of concentrated hydrochloric acid with vigorous shaking and occasional cooling. After standing for two hours, the mixture is filtered and gives 1350 ml. of a solution containing 5·5 moles of sodium fluoroborate (*980*).

Sodium fluoroborate can be obtained as a by-product in the Schiemann reaction, when the escaping gas mixture containing nitrogen and boron trifluoride is passed into a ice-cooled suspension of sodium fluoride in water. The amount of sodium fluoride needed is calculated for a quantitative decomposition of the aromatic diazonium fluoroborate, and the amount of water is three times as large. The product is contaminated with sodium fluoride and the content of the fluoroborate has to be determined analytically.

Fluoroboric acid and its salts are used only for the preparation of aromatic diazonium fluoroborates, whose decomposition leads to the compounds having fluorine linked to the aromatic nucleus.

## Thallium Fluoride

The starting material for the preparation of thallium fluoride is thallium formate (*762*).

Thallium formate (100 g) (0·4 mole) is dissolved in 150 ml. of water, the solution is alkalized with ammonia, then treated with hydrogen sulfide, the precipitated thallium sulphide is filtered, washed with water, and dissolved in an excess of 50% hydrofluoric acid. The solution is filtered through a filter paper placed in a polyethylene funnel, evaporated to dryness, the residue is redissolved in 33% hydrofluoric acid, filtered again, evaporated to dryness, and dried by heating at 110°C for 8 hours. The yield is 87 g (0·39 mole) (97·5%).

A similar procedure is used for the regeneration of spent thallium fluoride after the fluorination reactions (after boiling with ethanol and washing with acetone) (*762*).

## Silicium Tetrafluoride

Silicium tetrafluoride is a by-product in the production of hydrogen fluoride by the decomposition of fluorspar with sulfuric acid (*893*), and also in the manufacture of superphosphate. The pure compound is best prepared by thermal decomposition of barium fluorosilicate at 500°C and by condensation of the gas in a trap cooled with liquid air (*499*).

Application of silicium tetrafluoride to the preparation of organic fluorine derivatives is only exceptional.

## Fluorosilicic Acid and Fluorosilicates

Fluorosilicic acid is formed by dissolving silicon dioxide in aqueous hydrofluoric acid. Its salts, the fluorosilicates, are obtained as by-products in the production of superphosphate, therefore they are usually not prepared in the laboratory. Both sodium and especially potassium fluorosilicates are very slightly soluble in water.

Their use in the preparation of organic compounds of fluorine is very limited.

## Cerium Tetrafluoride

A monohydrate of cerium tetrafluoride was prepared by treatment of hydrated cerium dioxide with aqueous hydrofluoric acid (*129*). Nowadays, however, it is produced by passing fluorine over cerium trichloride (*1069*) or trifluoride (*667*, *1069*) at higher temperatures. Cerium trifluoride, for instance, gave cerium tetrafluoride after reacting with fluorine for 3 hours at 230°C and several hours at 470°C (*667*).

Cerium tetrafluoride acts upon organic compounds like cobalt trifluoride or silver difluoride: it can substitute fluorine for halogen or hydrogen, and can cause the addition of fluorine across double bonds, In all instances it proved to be inferior to the other two fluorides.

## Lead Tetrafluoride

Lead tetrafluoride is obtained in the same fashion as cerium tetrafluoride, viz. by passing fluorine diluted with carbon dioxide over lead difluoride at 250–300°C (*1069*). The starting lead difluoride is prepared by dissolving lead oxide or carbonate in hydrofluoric acid (*1069*).

Applications of lead tetrafluoride to organic compounds are scarce for the same reasons which hold for cerium tetrafluoride.

### Antimony Trifluoride

For preparing antimony trifluoride, antimony sesquioxide is dissolved in an excess of aqueous hydrofluoric acid and the solution is evaporated to dryness by heating with a free flame in a platinum or copper dish (893). With heating at a sufficiently high temperature, the residue melts and compact antimony trifluoride is obtained, which may be crushed to pieces and stored in tightly closed containers. It is suitable for the majority of fluorination reactions.

The crude product may be purified by distillation from a copper flask fitted with air-cooled copper condenser. First portions have to be discarded, since they are usually contaminated with copper compounds. Still purer products be obtained by sublimation.

Antimony trifluoride is a classical fluorinating reagent, whose merit lies in the preparation of the first organic fluorinated compounds. It is most suitable for treatment of polyhalogen derivatives having at least two halogen atoms bound to the same carbon atom; according to the conditions used, one or both of these halogens are replaceable by fluorine. In a trichloromethyl group neighbouring to a double bond or an aromatic ring, all three halogen atoms can be exchanged for fluorine without difficulties. Still more efficient than the plain antimony trifluoride is a reagent prepared from antimony trifluoride and chlorine or bromine.

### Chlorofluorides of Antimony

Mixed chlorofluorides of antimony having a varying ratio of chlorine to fluorine are usually prepared in situ before or during the fluorination of organic compounds. The ability to replace organic halogens by fluorine is a function of the content of pentavalent antimony in the mixture. The more intensive the fluorination that is needed, the larger is the proportion of antimony which has to be converted to the pentavalent state, by adding chlorine, bromine or antimony pentachloride. Since the mixed chlorofluorides of antimony thus obtained are usually not isolated but used immediately for fluorination of organic compounds, their detailed description will be given in the preparation of organic fluorine derivatives (p. 96).

A similar mixture of chlorofluorides of antimony results also when heating antimony trichloride and pentachloride with anhydrous hydrogen fluoride. In the reaction, fluorine is substituted for chlorine in the antimony salts, so that mixed chlorofluorides having a varying composition are formed. An exact study of the reaction conditions is still lacking, though some progress has already been made by isolating a compound of the formula $SbCl_4F$ (580, 877).

12
$$SbF_3 + Cl_2 \rightarrow SbF_3Cl_2$$
$$SbF_3 + Br_2 \rightarrow SbF_3Br_2$$
$$SbF_3 + SbCl_5 \rightarrow SbF_3Cl_2 + SbCl_3$$

The fluorides containing pentavalent antimony cause the same transformations as antimony trifluoride but under milder conditions.

A very efficient fluorinating reagent is formed by heating one mole of antimony pentachloride with 5 moles of anhydrous hydrogen fluoride at 150°C in an autoclave. Its composition is expressed by a formula $SbF_3Cl_2.2HF$ (*1081*).

13
$$SbCl_5 + 5HF \rightarrow SbF_3Cl_2 \cdot 2HF + 3HCl$$

This compound is able to replace even three halogen atoms linked to the same carbon atom.

### Antimony Pentafluoride

Antimony pentafluoride is produced by an exchange reaction between antimony pentachloride and anhydrous hydrogen fluoride according to the equation:

14
$$SbCl_5 + 5HF \rightarrow SbF_5 + 5HCl$$

The reaction is started with cooling and the temperature is raised progressively to 150°C until all of the hydrochloric and hydrofluoric acid is distilled off (*893*).

Anhydrous hydrogen fluoride is passed with stirring at a temperature of 10–35°C into 1500 g (5 moles) of antimony pentachloride placed in an aluminium vessel fitted with an aluminium reflux condenser kept at −45°C. The temperature of the reaction mixture is slowly raised to 50°C and kept there for 3–4 hours after introducing 900 g (45 moles) of hydrogen fluoride. It is then raised to 60–70°C and maintained at this value until hydrogen chloride stops escaping. After cooling the reaction vessel to 40°C the temperature of the condenser is raised to 12–25°C and the vessel is heated at 140–150°C in order to strip the excessive hydrogen fluoride. Crude antimony pentafluoride is thus obtained in a yield of 1084 g (5 moles). Its distillation from an aluminium flask gives a compound of 80–90% purity (*832*).

The action of antimony pentafluoride upon organic compounds is not limited to the replacement of halogens. Antimony pentachloride is able even to add fluorine across double bonds and to aromatic systems.

### Bismuth Pentafluoride

Bismuth pentafluoride is obtained by passing fluorine at 460–500°C over bismuth trifluoride (*1069*), prepared by dissolving bismuth hydroxide in hydrogen fluoride.

Bismuth pentafluoride may be used for the transformation of higher paraffins to perfluoroderivatives.

## Sulphur Tetrafluoride

Sulphur tetrafluoride is obtained by heating a mixture of sulphur dichloride and sodium fluoride in acetonitrile at 70–80°C and subsequenty distillation. The crude product (90%) is contaminated with thionyl fluoride which boils at −43.8°C (964).

## Chlorine Trifluoride

The reaction of chlorine with fluorine in a copper or nickel reactor at 200°C (947) or 250–300°C (614) gives chlorine trifluoride.

The action of chlorine trifluoride upon organic compounds is too energetic to be of practical value. Usually not only fluorinated but also chlorinated derivatives are formed. However, chlorine trifluoride is very useful for preparing some other inorganic fluorides of higher valency, such as cobalt trifluoride or silver difluoride (883).

## Perchloryl Fluoride

An interesting compound is formed by the action of fluorine upon chlorates or by double decomposition of potassium perchlorate with fluorosulphonic acid (614). It is also prepared by electrolysis of 10% solution of sodium perchlorate in anhydrous hydrogen fluoride under a potential drop of 4–7 V (255). The compound is formulated as a fluoride of perchloric acid, $FClO_3$.

15                    $NaClO_3 + F_2 \rightarrow NaF + FClO_3$

16                    $KClO_4 + FSO_3H \rightarrow KHSO_4 + FClO_3$

In the reactions with organic compounds, perchloryl fluoride acts in two ways. Reacting as a pseudohalogen, it replaces certain individual atoms of hydrogen by fluorine (6, 529) (p. 83). Its nature as a fluoride of perchloric acid is manifested by a Friedel–Crafts reaction with aromatic compounds in which $ClO_3$ group is entered (528) (p. 210).

## Bromine Trifluoride

Bromine trifluoride is obtained by passing fluorine into liquid bromine in a copper (950) or nickel reactor (614) until fluorine starts to escape.

The use of bromine trifluoride in the fluorination of organic compounds is very limited.

## Iodine Pentafluoride

Iodine pentafluoride is prepared by passing fluorine diluted with three volumes of nitrogen over iodine in a quartz tube. At 20°C a spontaneous reaction takes place (*894*).

Out of all halogen fluorides iodine pentafluoride has the greatest importance for the preparation of fluorinated derivatives. It is used for the transformation of $-CF_2I$ and $-CI_3$ groups to $-CF_3$ group.

## Manganese Trifluoride

The starting material for the preparation of manganese trifluoride is manganese dichloride, which is transformed to white manganese difluoride by treatment with hydrogen fluoride. Its reaction with elemental fluorine gives red manganese trifluoride (*286*).

Manganese trifluoride seems to be useful only in the preparation of perfluoroderivatives from higher boiling paraffins.

## Cobalt Trifluoride

Cobalt trifluoride may be obtained in yields of 90–95% by passing fluorine over anhydrous cobalt dichloride at 250°C (*851*), or by passing fluorine at the same temperature over cobalt difluoride prepared from anhydrous cobalt dichloride and anhydrous hydrogen fluoride at 400–450°C (*287*). Another alternative is the reaction with fluorine of cobalt sesquioxide at a temperature slowly raised to 300°C until fluorine starts to escape (*646*). All these transformations are carried out in the apparatus shown in Figs. 14 and 15 (p. 79,81). In the same manner and in the same apparatus the spent cobalt trifluoride is regenerated (*646*).

Cobalt trifluoride is one of the most important and universal fluorinating reagents and is used for transforming hydrocarbons or their halogen derivatives to polyfluoro- and perfluoro derivatives. In addition to the replacement of halogens and hydrogens by fluorine it can produce addition of fluorine to double bonds and aromatic systems.

### ORGANIC FLUORINE DERIVATIVES AS FLUORINATING AGENTS

Of organic fluorine derivatives only a few act as fluorinating reagents. Among these are the aromatic *iodide fluorides* obtained by the action of iodoso- derivatives with hydrogen fluoride. Their importance is negligible since their action is limited to very reactive systems and the yields are poor. Their preparation is dealt with in the following chapter (p. 69,84). Another scarcely used fluorinating compound is *trifluoromethylhypofluorite* (p. 84).

## TABLE 8

*Physical Constants and Applications of Inorganic Fluorides*

| Formula | Mol. Wt. | M.P.°C | B.P.°C | Density per°C | Solubility g. per 100 g. of water | Addition to $C=C$ | Replacement of H | Replacement of Hal. | Replacement of OH | Replacement of $NH_2$ | Page |
|---|---|---|---|---|---|---|---|---|---|---|---|
| $NH_4F$ | 27·04 | subl. | | 1·315 | very sol. | | | + | | | 110 |
| $LiF$ | 25·94 | 870 | 1676 | 2·295/21·5 | 0·27/18 0·135/35 | | | + | | | 110 |
| $NaF$ | 42·00 | 992 to 1040 | 1705 | 2·79 | 4/0 4·22/18 5/100 | | | + | | | 110 113 |
| $KF$ | 58·10 | 885 | 1500—5 | 2·48 | 92·3/18 | | | + | + | | 104 |
| $KHF_2$ | 78·10 | decomp. | | | | | | + | | | 109 |
| $CsF$ | 151·91 | 684 | 1251 | 3·586 | 366/18 | | | + | | | 104 |
| $AgF$ | 126·88 | 435 | ~1150 | 5·852/15·5 | 182/15·5 205/108 | | | + | | | 98 |
| $AgF_2$ | 145·88 | ~690 | | 4·57-4·78 | decomp. | + | + | + | | | 77 |
| $ZnF_2$ | 103·38 | 872 | ~1500 | 2·567/10 | | | | + | | | 111 |
| $Hg_2F_2$ | 439·22 | 570 | decomp. | 8·73/15 | decomp. | | | + | | | 100 |
| $HgF_2$ | 238·61 | 645 | <650 | 8·95/15 | decomp. | | | + | | | 101 |
| $BF_3$ | 67·82 | −127 | −101 | 2·999/1 | 106 ml | | | | | | 66 |
| $BF_3 \cdot (C_2H_5)_2O$ | 125·94 | | 125—126 | | | | | | | | 118 |
| $HBF_4$ | 87·83 | | 130 decomp. | | ∞ | | | | | + | 115 |
| $NH_4BF_4$ | 104·86 | subl. | | 1·851/17 | 25/16 95/100 | | | | | + | 115 |
| $NaBF_4$ | 109·82 | 384 decomp. | decomp. | 2·47/20 | 108/26 210/100 | | | | | + | 115 |
| $KBF_4$ | 125·92 | 530 | | 2·498 | 0·44/20 6·27/100 | | | | | | |
| $TlF$ | 223·39 | 298 | | | 78·6/15 | | | | + | | 110 |
| $SiF_4$ | 104·06 | −95·7 | −65/1810mm | 3·57 g/l 4·67 g/l | decomp. | | | | | | |
| $CeF_4$ | 216·13 | 1460 | | | | + | + | + | | | 82 |
| $PbF_4$ | 283·21 | ~600 | | | | + | + | + | | | 82 |
| $SbF_3$ | 178·76 | 292 | 319 | 4·379/20.9 | 384·7/0 563·6/30 | | | + | | | 93 |

TABLE 8—*continued*

| Formula | Mol. Wt. | M.P.°C | B.P.°C | Density per°C | Solubility g. per 100 g. of water | Addition to C=C | Replacement of H | Replacement of Hal. | Replacement of OH | Replacement of NH₂ | Page |
|---|---|---|---|---|---|---|---|---|---|---|---|
| $SbF_5$ | 216·76 | 7 | 142,7 | 3·145/15.5 2·993/22.7 | sol. | + | | + | | | 98 |
| $BiF_5$ | 304·00 | | 550 subl. | | | + | + | + | | | 82 |
| $SF_4$ | 108·01 | | −38 to −37 | | | | | | + | | 112 |
| $FSO_3H$ | 100·02 | −87·3 | 165·5 | 1·743/15 | decomp. | | | | | | |
| $FClO_3$ | 102·46 | −147·74 | −46·67 | | | | + | | | | 83 |
| $ClF_3$ | 92·46 | −82·6 | 12·1 | 1·77/13 | decomp. | + | | + | | | 103 |
| $BrF_3$ | 136·92 | 8·8 | 135 | 2·49/135 | decomp. | | | + | | | 103 |
| $IF_5$ | 221·92 | 8·5 | 98 | 3·5 | | | | + | | | 104 |
| $MnF_3$ | 111·93 | | | 3·54 | decomp. | + | + | + | | | 82 |
| $CoF_3$ | 115·94 | ∼1200 | | 3·88 | decomp. | + | + | + | | | 79 |

# IV

## METHODS FOR INTRODUCING FLUORINE INTO ORGANIC COMPOUNDS

In the following chapter, methods will be discussed which are available for introducing fluorine into organic compounds, by the addition of hydrogen fluoride or fluorine across double bonds, by the substitution of fluorine for other halogens*, oxygen or the amino group, or else by some special ways. Attention will be directed toward the scope and limitation of the individual methods. General rules, which apply to these methods, will be outlined and experimental details will be shown on selected typical examples written as laboratory procedures. Main reaction conditions will be illustrated in an abridged form in the reaction equations. A table is included showing the reagents recommended for individual types of fluorination.

TABLE 9

*Action of Fluorinating Reagents on Organic Compounds*

| Method of Fluorination / Fluorinating Reagent | Addition of | | Replacement by F of | | | |
|---|---|---|---|---|---|---|
| | HF | $F_2$ | H | Hal | O, OH OSO$_2$R | NH$_2$ |
| HF | + | + | | + | + | + |
| Electrolysis in HF | | + | + | + | | |
| $F_2$ | | + | + | + | + | |
| ClF$_3$, BrF$_3$, IF$_5$ | | + | + | + | | |
| AgF$_2$, CoF$_3$ | | + | + | + | | |
| (MnF$_3$, CeF$_4$, BiF$_5$, PbF$_4$) | | | | | | |
| SbF$_3$ | | | | + | | |
| SbF$_5$ | | + | | + | | |
| AgF, HgF | | | | + | | |
| HgF$_2$ | | | | + | | |
| KF | | | | + | + | |
| (NaF, TlF, LiF, CsF, ZnF$_2$) | | | | | | |
| SF$_4$ | | | | | + | |
| FClO$_3$ | | | + | | | |
| CF$_3$OF | | + | + | | | |

* Throughout the book the term "halogen" will, for practical reasons, be applied to chlorine, bromine and iodine, not to fluorine.

## ADDITION OF HYDROGEN FLUORIDE

Anhydrous hydrogen fluoride adds across carbon-carbon double and triple bonds and also to some other unsaturated compounds. Double bonds of aromatic nuclei resist hydrogen fluoride. In contrast with other hydrogen halides, the addition of hydrogen fluoride is generally less convenient and often requires special conditions.

### Addition of Hydrogen Fluoride to Olefins

Olefinic hydrocarbons react with anhydrous hydrogen fluoride in two ways: Hydrogen fluoride adds across the double bond, and forms monofluorides, or it causes polymerization of the olefin to high-molecular compounds. The first type takes place preferably at lower temperatures and with the olefin added to the hydrogen fluoride; otherwise the second reaction prevails (*329, 679*):

17 $\quad$ $CH_3-CH=CH_2 \xrightarrow[0°]{HF} CH_3-CHF-CH_3$ $\quad$ 61% $\qquad$ [*329*]

18 $\qquad\qquad\qquad\qquad\qquad\qquad\qquad\qquad\qquad\qquad$ [*679*]

$$\text{(cyclohexene)} \xleftarrow{\substack{HF \\ -78° \to 20°}} \quad \xrightarrow{\substack{HF \\ 100°}} \text{polymers}$$

(fluorocyclohexane) 70%

From the preparative point of view, more importance is attached to the addition of hydrogen fluoride to halogenated olefins, because their polymerization tendency in the presence of hydrogen fluoride decreases with the increasing number of halogen atoms linked to the doubly bonded carbons. The addition of hydrogen fluoride is usually accompanied by replacement of a chlorine atom by fluorine. At higher temperatures, the replacement reaction prevails over the simple addition (*452*).

19 $\qquad\qquad\qquad\qquad\qquad\qquad\qquad\qquad\qquad\qquad\qquad\qquad$ [*452*]

$$CCl_2=CH-CH_2-CH_3 \xrightarrow{HF} CCl_2F-CH_2-CH_2-CH_3 + CClF_2-CH_2-CH_2-CH_3$$

| | 33% | 5.3% |
|---|---|---|
| 65°, 18 hrs. | 33% | 5.3% |
| 100°, 6 hrs. | 10.3% | 49.6% |

*Procedure 1.*

*Addition of Hydrogen Fluoride to Olefins*
*Preparation of 1,2-Dichloro-2-fluoropropane from 1,2-Dichloro-2-propene\**

---

\* Carried out by the author in the laboratory of A. L. Henne.

A two litre steel bomb cooled with dry ice is charged with 500 g (4·5 moles) of 1,2-dichloro-2-propene and an equivalent amount (90 g, 4·5 moles) of anhydrous hydrogen fluoride. The bomb is closed with a screwed assembly bearing a manometer and a needle valve (Fig. 2., p.29), and is heated for 18 hours in a water bath at 50°C. After cooling with dry ice and opening the valve the bomb is dismounted, the content is poured over crushed ice, the mixture is neutralized with a dilute solution of sodium hydroxide, and the organic layer is separated and steam-distilled. The crude product (about 450 g) thus obtained is dried with calcium chloride and fractionally distilled to yield approximately 200 g (1·5 moles) of 1,2-dichloro-2-fluoropropane distilling over a range of 85–90°C, and some 200 g of a fraction boiling between 90 and 94°C consisting of a mixture of product and unreacted starting material. This portion is subjected to a new treatment with anhydrous hydrogen fluoride and yields an additional 75 g (0·6 mole) of 1,2-dichloro-2-fluoropropane. The total yield of the distilled product is 46·7%.

The more halogen atoms bound to the double bond, the more difficult the addition of hydrogen fluoride (*470*). Trichlorethylene does not react with hydrogen fluoride even at 160°C (*470*). Only the use of boron trifluoride as a catalyst makes this reaction feasible (*438*).

**20**
$$\begin{array}{c} \xrightarrow{160°} \text{No reaction} \end{array} \quad CHCl=CCl_2 + HF \xrightarrow[120°]{BF_3} \begin{array}{c} CH_2Cl-CCl_2F + CH_2Cl-CClF_2 \\ 35\% \qquad\qquad 22\% \end{array} \qquad [438, 470]$$

Olefins carrying fluorine atoms at the double bond are sometimes very resistant to the electrophilic addition of hydrogen fluoride, but are prone to nucleophilic addition of the fluoride ion by means of potassium fluoride in formamide (*720*).

**21**
$$CClF=CF_2 \xrightarrow[55°, 30 \text{ hrs.}]{KF, HCONH_2} CHClF-CF_3 \qquad 72\% \qquad [720]$$

The course of the addition of hydrogen fluoride is determined by the electron density at the double bond. The fluorine atom joins the carbon having lower electron density. The addition of hydrogen fluoride to hydrocarbons follows the Markownikoff rule (equation 17, p. 65). In halogenated olefins, the fluorine atom combines with the carbon atom bearing the larger number of halogens (equation 20)*. Exceptions are rare.

The addition of hydrogen fluoride to dienes has to be carried out at very low temperatures (*22*); otherwise polymerization prevails.

**22**
$$CH_2=C=CH_2 + 2HF \xrightarrow[30 \text{ min.}]{-76° \text{ to } -69°} CH_3-CF_2-CH_3 \qquad 50\% \qquad [22]$$

---

\* Compare with a (somewhat modified) Roman proverb: Quo plura *carbo* habet, eo plura optat.

## Addition of Hydrogen Fluoride to Acetylenes

The addition of hydrogen fluoride to acetylenes is carried out either by adding anhydrous hydrogen fluoride to acetylenes at low temperatures (*328*), or by introducing acetylenes into a solution of anhydrous hydrogen fluoride in oxygenated solvents such as ether or acetone (*469*). With these, oxonium compounds are formed in which two moles of hydrogen fluoride are combined with one mole of the solvent. To prepare geminal difluorides, five moles of hydrogen fluoride are required for one mole of alkyne and one mole of solvent (*469*):

23
$$CH\equiv C-R \xrightarrow[\substack{5\,HF,\,CH_3COCH_3}]{\substack{2HF \\ -70° \rightarrow 5 -15°}} CH_3-CF_2-R \qquad \begin{array}{l} 46-76\% \\ 85-90\% \end{array} \qquad [328,\ 469]$$

The first member of the acetylene series differs in its behaviour toward hydrogen fluoride from its homologues. Under ordinary pressure, it does not react with hydrogen fluoride either at $-70°C$, or at $300°C$. At an elevated pressure, a mixture of vinyl fluoride with ethylidene fluoride is obtained (*328*). Better results were reached when the reaction was catalyzed with activated charcoal impregnated with a solution of mercuric chloride and barium chloride (*769*).

24
$$CH\equiv CH \xrightarrow[\substack{HF,\,HgCl_2,\,BaCl_2,\,C \\ 97-104°}]{\substack{HF \\ 20°,\,13\,atm,\,72\,hrs.}} \begin{array}{cc} 35\% & 65\% \\ CH_2{=}CHF + CH_3{-}CHF_2 \\ 82\% & 4\% \end{array} \qquad [328,\ 769]$$

*Procedure 2.*

*Addition of Hydrogen Fluoride to Acetylenes*
*Preparation of 2,2-Difluorooctane from 1-Octyne and Hydrogen Fluoride (469)*
One mole of ether (74 g) or acetone (56 g) is placed in an ice-cooled half-litre copper flask into which anhydrous hydrogen fluoride is fed through a copper tube until the weight increase reaches 5 moles (100 g). Then 1 mole (110 g) of 1-octyne is slowly added and the mixture allowed to come to room temparature. After 1 hour the reaction mixture is poured over crushed ice, the organic layer is separated and distilled. 2,2-Difluoro-octane is obtained in a yield of 0·7–0·9 mole (70–90%). This procedure is suitable for non-volatile acetylenes and difluorides.

## Addition of Hydrogen Fluoride to other Unsaturated Compounds

Additions of hydrogen fluoride to other compounds containing double bonds are pretty scarce. The reaction of hydrogen fluoride with cyanic acid is

5*

claimed to give carbamyl fluoride *(619)*, its identity is, however, doubtful since its properties do not conform with those of other carbamyl halides.

25 $\qquad$ H—N=C=O + HF $\xrightarrow[-80°]{(C_2H_5)_2O}$ H$_2$N—C=O $\qquad$ [619]
$$\phantom{H_2N—C=O}\;|$$
$$\phantom{H_2N—C=O}\;F$$

The addition of hydrogen fluoride to perfluoro(alkylalkyleneimines) gives di(perfluoroalkyl)amines *(44)*.

26 $\qquad$ CF$_3$—N=CF$_2$ $\xrightarrow[150°]{HF}$ CF$_3$—NH—CF$_3$ $\qquad$ 89% $\qquad$ [44]

The reaction of diazoketones with anhydrous hydrogen fluoride yields fluoroketons *(85, 572)*, whereas aqueous hydrofluoric acid gives ketols.

27 $\qquad$ ⟨◯⟩—CO—CH=N≡N $\xrightarrow[(C_2H_5)_2O]{HF}$ ⟨◯⟩—CO—CH$_2$F $\qquad$ 66% $\qquad$ [85, 572]

## ADDITION OF FLUORINE

In contrast with the addition of hydrogen fluoride, addition of fluorine across double bonds is far less practical. Fluorine and fluorinating reagents which are able to add fluorine to double bonds, are usually so reactive that they cause deeper chemical changes in the organic molecules, especially the replacement of hydrogen or halogen atoms by fluorine (p. 72). With aromatic compounds which are perfectly unreactive toward anhydrous hydrogen fluoride, fluorine and strong fluorinating reagents cause both addition and substitution reactions concurrently.

### Addition of Fluorine to Olefins

Fluorine has been added to the double bond of crotonic acid at low temperatures to give a mixture of stereoisomeric $\alpha,\beta$-difluorobutyric acids *(104)*. Elemental fluorine diluted with nitrogen or carbon dioxide reacts at $-80°C$ with tetrachloroethylene dissolved in dichlorodifluoromethane and yields 25% of sym-tetrachlorodifluoroethane *(104)*. The yields of the straight addition products are usually poor because replacement of hydrogen and chlorine by fluorine, addition of chlorine to double bonds, and cleavage of the carbon chain occur simultaneously. The reaction of trichloroethylene with fluorine in vapour phase is illustrative of the intricacy of such reactions *(401)*:

28 $\qquad$ CHCl=CCl$_2$ $\xrightarrow[71°]{F_2,N_2}$ CClF$_2$—CCl$_2$F + CCl$_2$F—CCl$_2$F $\qquad$ [401]
$$\phantom{CHCl=CCl_2 \to}24\% \qquad 23\%$$

$$C_2Cl_2F_4 + C_2ClF_5 + C_2Cl_2F_2$$
$$11\% \qquad 1\% \qquad 1\%$$

Better results are obtained by indirect addition of fluorine from anhydrous hydrogen fluoride with lead tetraacetate (*115, 214*) (equation 539) or, better still, with lead dioxide (*485*). This way of carrying out the reactions with fluorine in statu nascendi leads to clean addition of fluorine across the double bond with almost no side-reactions. So far only polyhalogenated olefins have been used as starting materials for this process.

29 $\qquad CCl_2{=}CCl_2 \xrightarrow{\text{HF,PbO}_2} CCl_2F{-}CCl_2F \quad 58\%$ [*485*]

*Procedure 3.*

*Addition of Fluorine across the Double Bond*
*Preparation of Tetrachloro-1,2-difluoroethane from Tetrachloroethylene, Hydrogen Fluoride and Lead Dioxide (485)**

A 1·5 litre steel bomb is charged with 240 g (1·0 mole) of lead dioxide, 83 g (0·5 mole) of tetrachloroethylene, and after cooling with dry ice, 400 g (20 moles) of anhydrous hydrogen fluoride, also precooled with dry ice. The bomb is closed with an assembly bearing a manometer for 20 atm. and a needle valve (Fig. 2, p.29), and allowed to come to room temperature while rocked mechanically. After defrosting the temperature in the bomb rises rapidly in the course of approximately 5 minutes and the pressure reaches 3–3·5 atm. With continued rocking, the pressure drops to 1·5 atm. After cooling with ice-water, the pressure is released, the bomb is opened and the contents poured on 1·5 kg of crushed ice in a 10 litre flask. The bomb is flushed several times with water, the washings are poured into the flask, 800 g of sodium hydroxide is added to the mixture, the flask is fitted with an efficient water-cooled condenser, and the mixture is steam-distilled. The organic layer in the distillate is separated and dried with calcium chloride to yield 78 g (0·38 mole) (76%) of practically pure tetrachloro-1,2-difluoroethane, solidifying at 20°C and showing $n_D^{20}$ 1·4170.

Another indirect way of adding fluorine across double bonds is the reaction of olefins with iodobenzene difluoride in the presence of anhydrous hydrogen fluoride or silicon tetrafluoride (*105*). This reaction is difficult to carry out and therefore not much used.

Addition of fluorine across double bonds can also be effected by metal fluorides at their maximum valency such as antimony pentafluoride (*768*), silver difluoride (*768*) and cobalt trifluoride (*459*). In all these instances side reactions may result from replacement of other elements by fluorine (*768*).

30 $\qquad\begin{array}{c} CCl{=}CCl \\ CF_2 \quad CF_2 \\ CF_2 \end{array} \xrightarrow[90° \to 200°]{CoF_3} \begin{array}{c} CClF{-}CClF \\ CF_2 \quad CF_2 \\ CF_2 \end{array}$ [*459*]

---

* Carried out by the author in the laboratory of A. L. Henne.

Simultaneous addition of fluorine and bromine takes place when treating an olefin with an equimolecular mixture of anhydrous hydrogen fluoride and N-bromoacetamide preferably in tetrahydrofuran as a solvent (*123, 124, 881*).

31

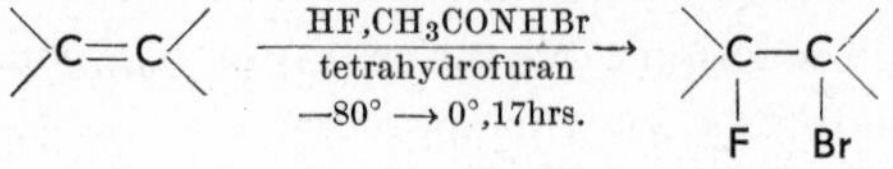

$$\text{>C=C<} \xrightarrow[\substack{\text{tetrahydrofuran} \\ -80° \rightarrow 0°, 17\text{hrs.}}]{\text{HF,CH}_3\text{CONHBr}} \underset{\substack{| \quad | \\ F \quad Br}}{\text{>C—C<}} \sim 50\% \qquad [124]$$

This entirely new method is very useful, since the bromine atom may subsequently be replaced by hydrogen or another element or group.

## Addition of Fluorine to Aromatic Systems

The addition of fluorine across the double bonds of aromatic systems is usually accompanied by partial or total replacement by fluorine of the other elements attached to the aromatic nucleus. Exclusive addition is very rare, but was observed during the reaction of hexachlorobenzene with fluorine (*96*).

32

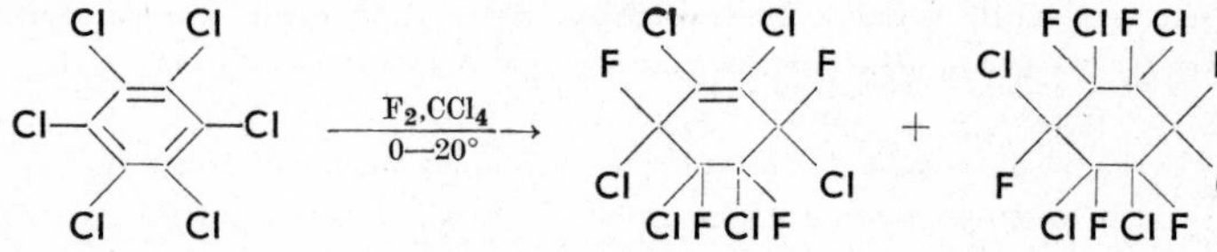

An incomplete addition leaving one unsaturated double bond was noted in the reaction of hexachlorobenzene with fluorine (*96*) or antimony pentafluoride (*678*), and an incomplete replacement of the hydrogens in the reaction of benzene with cobalt trifluoride (*263, 316, 961*):

33

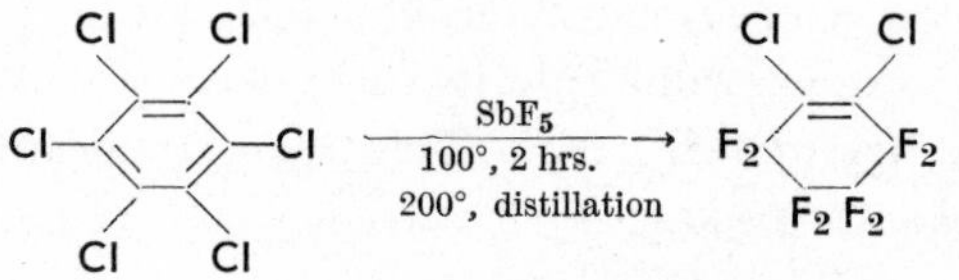

34

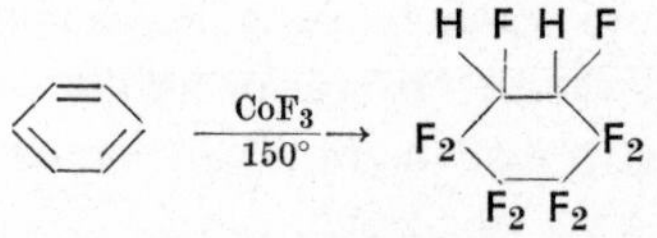

The final products of the reaction between aromatic compounds and fluorine or strong fluorinating metal fluorides are usually saturated cyclic perfluoro-derivatives. The same sort of products is obtained by electrolytic fluorination effected by electrolysis of aromatic compounds in anhydrous hydrogen fluoride. All these reactions will therefore be discussed in the paragraph on the replacement of hydrogen by fluorine (p. 72).

## Addition of Fluorine to Other Double and Triple Bonds

The nitrile group accepts one or two molecules of fluorine depending on the conditions used. Acetonitrile and mercuric fluoride give a mixture of unusual aminoderivatives (515, 765). When the same compound reacts with fluorine, the addition across the carbon-nitrogen triple bond is accompanied by replacement of the methyl hydrogens by fluorine so that perfluoroethyldifluoramine results as a final product (in addition to compounds formed by the cleavage of the organic molecule) (190).

$$\mathrm{CH_3-C{\equiv}N} \xrightarrow[\substack{150-180^\circ \\ 2\,h}]{\mathrm{HgF_2}} \mathrm{CH_3-CF{=}NF} + \mathrm{CH_3-CF_2-NF_2}$$

**35**
$$+ \quad \mathrm{CH_2{=}C{=}NF} + \mathrm{CH_2{=}CF-NF_2} \qquad [190,515,765]$$

$$\xrightarrow[275^\circ]{\mathrm{F_2,N_2,Cu}} \mathrm{CF_3-CF_2-NF_2} + \mathrm{CF_4} + \mathrm{C_2F_6} + \quad \text{polymers}$$

$$20\%$$

The reaction of malonic dinitrile with fluorine gives, in addition to products containing lower number of carbon atoms, a cyclic perfluoro compound, perfluoropyrazolidine (23).

**36**
$$\begin{array}{c} \mathrm{C{\equiv}N} \\ | \\ \mathrm{CH_2} \\ | \\ \mathrm{C{\equiv}N} \end{array} \xrightarrow[250^\circ]{\mathrm{F_2,N_2}} \begin{array}{c} \mathrm{CF_2-NF} \\ | \\ \mathrm{CF_2} \quad | \\ | \\ \mathrm{CF_2-NF} \end{array} + \mathrm{C_2F_5NF_2} + \mathrm{CF_3NF_2} + \mathrm{C_3F_8} + \mathrm{C_2F_6} + \mathrm{CF_4} \qquad [23]$$

Cyanogen chloride and silver difluoride (315) and cyanogen iodide and iodine pentafluoride (854) give perfluoroazomethane:

**37**
$$\mathrm{Cl-C{\equiv}N} \xrightarrow{\mathrm{AgF_2,\,N_2}} \mathrm{F_3C-N{=}N-CF_3} \xleftarrow{\mathrm{JF_5}\atop{130^\circ}} \mathrm{J-C{\equiv}N} \qquad [315,854]$$

The reaction of carbon monoxide with an excess of fluorine yields an interesting compound, perfluoromethyl hypofluorite, a fairly stable gas with strong oxidizing and fluorinating properties. The same compound results from the reaction of fluorine with carbonyl fluoride, ethanol or tert-butylalcohol (154), and from the action of oxygen difluoride on tetrafluoroethylene, perfluoropropene or trifluoroacetyl fluoride (192).

**38**
$$\mathrm{CO} \xrightarrow[170^\circ]{\mathrm{F_2}} \searrow \quad \overset{\mathrm{CH_3OH}}{\underset{\mathrm{CF_3OF}}{\mathrm{F_2,\,N_2 \big\downarrow Ag/Cu,\,160-170^\circ}}} \quad \overset{\mathrm{F_2O}}{\nearrow} \quad \mathrm{C_2F_4} \text{ or } \mathrm{CF_3CF{=}CF_2} \qquad [154,192,552]$$
$$\mathrm{COF_2} \xrightarrow{\mathrm{F_2}} \nearrow \qquad \qquad \searrow^{\mathrm{F_2O}} \quad \mathrm{CF_3COF}$$

Carbon disulphide reacts differently with various fluorinating reagents (376, 753, 944, 1050). Interesting compounds containing sulphur in its highest valency are obtained, in addition to the compounds produced by cracking of the molecule at the carbon-sulphur bond.

39

$$
\begin{array}{c}
\overset{\text{F}_2, \text{N}_2 / \text{Cu}}{\longleftarrow} \quad \text{CS}_2 \quad \overset{\text{IF}_5 / 195°}{\longrightarrow}
\end{array}
$$

SF$_3$—CF$_2$—SF$_5$     CoF$_3$ /250°    250°\ HgF$_2$     CF$_3$—S—S—CF$_3$     76%

SF$_5$—CF$_2$—SF$_5$                                                CF$_3$—S—S—S—CF$_3$     7%

CF$_3$—SF$_3$     CF$_3$—SF$_5$     CF$_3$—S—S—CF$_3$     CF$_4$ + SF$_4$     65%

CF$_3$—SF$_5$

CSF$_2$

CF$_4$   SF$_4$

SF$_6$   S$_2$F$_{10}$                                                                              [376,753,944,1050]

Almost the same diversity of products is obtained by electrolysis of carbon disulphide in anhydrous hydrogen fluoride (175, 945).

40     $\text{CS}_2 \xrightarrow[6.5 \text{ V, } 3.5 \text{ A}]{\text{HF, NaF}}$ SF$_5$—CF$_2$—SF$_5$ + SF$_3$—CF$_2$—SF$_3$ + CF$_3$—SF$_5$ + SF$_6$     [175]

          0,5%               0,5%               >90%

## REPLACEMENT OF HYDROGEN BY FLUORINE

Hydrogen can be replaced by fluorine only by the action of strong fluorinating agents such as elemental fluorine, silver difluoride, cobalt trifluoride, a few fluorides having a large number of fluorine atoms in their molecules, and by electrolytic fluorination, viz. electrolysis in anhydrous hydrogen fluoride. Contrary to similar reactions with other halogens, the replacement of hydrogen by fluorine represents a very difficult problem. Thermodynamic data show that the reaction of organic compounds with fluorine liberates much more energy than similar reactions with the other halogens (104) (Table 10).

TABLE 10

*Reaction Heats of Addition and Substitution Reactions of Halogens (104)*

| Reaction Type | $\Delta H°$ (Kcal/mole) | | | |
| --- | --- | --- | --- | --- |
| X = | F | Cl | Br | I |
| C=C + X$_2$ = CX—CX | −107·2 | −33·1 | −18·8 | +1·2 |
| C—H + X$_2$ = C—X + HX | −102·5 | −22·9 | −6·2 | +13·7 |

Consequently a controlled reaction, which would replace one specific hydrogen atom at a time, is extremely difficult to obtain. Usually several hydrogen atoms are replaced at the same time, and splitting of the carbon chain takes place since the carbon-fluorine bond energy exceeds that of the carbon-carbon bond (58·6 kcal/mole).

Several instances of individual replacement of hydrogen by fluorine have been reported, however. In sufficiently reactive aromatic compounds, fluorine replaces hydrogen selectively in a hydrogen fluoride and lead tetraacetate treatment (*214*).

**41** [*214*]

Careful treatment of butyric acid and butyryl chloride, respectively, with elemental fluorine at low temperatures yielded β-fluoro-, γ-fluoro- and γ,γ,γ-trifluorobutyric acid, and butyryl fluoride, β-fluorobutyryl fluoride and γ-fluorobutyryl fluoride, respectively (*104*).

Direct fluorination at −80°C transformed cyclohexane to fluorocyclohexane (*104*), and methyl ethyl ketone to methyl β-fluoroethyl ketone (*1031*).

**42** [*104*]

**43** $CH_3-CO-CH_2-CH_3 \xrightarrow[-80°]{F_2, N_2(1:4)} CH_3-CO-CH_2-CH_2F$ 22,8% [*1031*]

The yields of all of these reactions are far from being satisfactory. The reactions merely show that the replacement of the individual hydrogen atoms by fluorine is possible. Even though recently a new specific reagent for the preparative replacement of single hydrogen atoms by fluorine has been discovered in perchloryl fluoride (*529*) (Equations 54, 55, 56), practical preparation of a fluorine derivative with its fluorine in a definite spot usually starts from the corresponding halogen derivative (p. 87).

The substitution of fluorine for hydrogen is a very important way to make polyfluoro and perfluoro derivatives, compounds in which many or all hydrogen atoms are replaced by fluorine. Such compounds possess unusual chemical, physical and physiological properties (p.310), for which they have been coveted in the last twenty years. The preparation of these compounds by fluorination of the corresponding hydrocarbons is the simplest way. Therefore much effort has been put into working out an efficient procedure. The methods of preparation of polyfluoro- and perfluoro-derivatives will be discussed according to the reagent used. Except in the most advanced method, electrolytical fluorination, elemental fluorine has always been indispensable, either for direct fluorination, or else for the preparation of the fluorinating reagents, the high-valency metallic fluorides. (p. 77).

## Fluorination with Elemental Fluorine

The first attempts to fluorinate organic compounds directly with undiluted elemental fluorine were completely unsuccessful. They were accompanied by explosions or at least charring, and usually carbon was the only isolated product when the reaction was carried out in gas or liquid phase (746). The failure was due to the exceedingly high reaction heat which caused thermal decomposition of both the starting material and the reaction products. Special arrangements of the experiments were needed to keep the reaction temperature within reasonable limits. One of the modifications of the reaction conditions is the dilution of fluorine with an inert gas, usually nitrogen. The contact of the diluted fluorine with the organic compound can take place either in gaseous, or in liquid state. In the first case, the dilution of the organic vapours with the inert gas is desirable. In the latter case, the organic liquid is diluted with an inert solvent such as dichlorodifluoromethane, and cooled to a low temperature, for instance by a dry-ice bath. Special design of the reaction vessel may be of advantage. The inlets for fluorine and the organic compound have to be separated so that the components mix gradually (294). The inlets for the organic compound, nitrogen and fluorine are sometimes placed concentrically, so that the compound and the fluorine are separated by nitrogen and gradual blending occurs by diffusion (1050) (Fig. 4, p. 31). An efficient modification is the application of a copper gauze surrounding the fluorine inlet tube (289, 340). The gauze acts as a contact surface where the reaction of the organic compound with fluorine takes place. At the same time its heat capacity and conductance causes efficient dissipation of the reaction heat and prevents hot spots.

By all these arrangements, the vigorous reaction between fluorine and organic compounds was controlled to such an extent that quite a number of organic compounds were successfully fluorinated: methane (340), ethane (1051), benzene (289, 295), its homologues (289), and even more complicated compounds such as acetone (294) and nitrogen derivatives, methyl, dimethyl and trimethyl amines (302). However, numerous side reactions considerably lower the yields of the desired products. Since fluorination proceeds by the free radical mechanism (295), products containing the same number of carbon atoms as the starting material are always accompanied by compounds of both lower and higher molecular weight. Fluorination of methane, for instance, resulted in the formation of all possible fluorinated methanes, perfluoroethane and perfluoropropane (*340*); fluorination of benzene led, in addition to perfluorocyclohexane, to the whole spectrum of perfluoroderivatives containing one through five carbon atoms, and moreover to a compound $C_{12}F_{22}$ (*295*). All these side reactions decrease substantially the value of direct fluorination with elemental fluorine, and its use nowadays is restricted.

All of the described methods of fluorination were non-catalytic. Great progress was made by introducing catalysts into the fluorination of organic compounds. When the reaction is carried out at the surface of some metals, the fluorine reacts primarily with the metal and forms a metal fluoride with

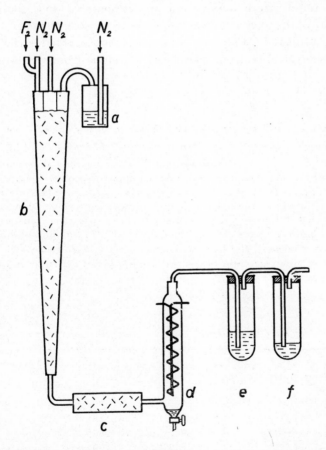

FIG. 13. Apparatus for catalytic fluorinations: a) evaporator of organic compound, b,c) electrically heated reactors, d) water-cooled condenser, e) dry-ice trap, f) liquid-air trap.

the highest metal valency. This fluoride then fluorinates the organic compound and is degraded to a fluoride with a lower metal-valency, which is reconverted to the high valency salt by long contact with fluorine. The suitability of some metals for the catalytic process is expressed by the following sequence (760):

$$\text{Au} > \text{Co} \gg \text{Ag} > \text{Ni} > \text{Cu} \gg \text{Hg, Cr, Rh, Fe.}$$

Practically only silver and gold are used. The reaction tube of the fluorination apparatus (Fig. 13, p.75) is filled with copper turnings which are coated with gold or silver. The rare-metal surface acts as a catalyst, whereas the mass of the copper substrate provides good dissipation of the reaction heat. When silver is used as the catalyst the optimum operating temperature lies around 200°C. At a temperature lower than approximately 140°C, polymers settle on the surface of the catalyst, while temperatures higher than 325°C deteriorate the catalyst properties. Catalytic fluorination is suitable for the preparation of perfluoro-derivatives from paraffins as well as from aromatic hydrocarbons (152).

*Procedure 4.*

*Replacement of Hydrogen by Fluorine by means of Elemental Fluorine*
*Preparation of Perfluoroheptane from Heptane and Fluorine (152)*

The catalytic fluorination is carried out in the apparatus shown in Fig. 13. The fluorine and organic compound inlets are placed approximately 7·5 cm. apart. Nitrogen at a controlled rate is bubbled through an evaporator containing heptane whose vapours are thus carried along to the reactor; fluorine diluted with double volume of nitrogen is fed at a rate which ensures a slight excess of fluorine over the organic compound. The reaction temperature is kept at 135°C. The main product, perfluoroheptane, is condensed in water-cooled condenser, the lower boiling by-products are collected in traps cooled with dry-ice and liquid-air bath, respectively. The yield of perfluoroheptane is approximately 60% (Table 11).

TABLE 11

*Examples of the Catalytic Fluorination Using Silver as the Catalyst (152)*

| Starting Material | Product | Temp. °C | Yield of the Product, % | |
|---|---|---|---|---|
| | | | Crude | Pure |
| $C_7H_{16}$ .................... | $C_7F_{16}$ | 135 | 75 | 62 |
| $C_6H_6$ .................... | $C_6F_{12}$ | 265 | 75 | 58 |
| $C_6H_5CF_3$ ................ | $C_6F_{11}CF_3$ | 200 | 90 | 85 |
| $C_6H_4(CF_3)_2$ .............. | $C_6F_{10}(CF_3)_2$ | 200 | 93 | 87 |
| Anthracene .............. | $C_{14}F_{24}$ | 300 | 73 | 43 |

Catalytic fluorination using the gold catalyst is carried out at 250–280°C and with fluorine diluted with 8–10 volumes of nitrogen (*389*).

44     $\begin{array}{c} \text{F}_2,\ \text{N}_2,\ \text{Ag/Cu} \\ 265° \end{array}$   58%    [*152,389*]

$\begin{array}{c} \text{F}_2,\ \text{N}_2,\ \text{Au/Cu} \\ 250° \end{array}$

$$\begin{array}{c} CF_2{-}CF_2 \\ CF_2 \qquad CF_2 \\ CF_2{-}CF_2 \end{array}$$

In the preparation of cyclic perfluoro-derivatives, aromatic as well as hydroaromatic compounds may be used as starting materials. The latter afford higher yields, as can be seen from Table 12 (*390*).

TABLE 12

*Results of Catalytic Fluorination at 250—280°C Using Gold as Catalyst (390)*

| Starting Material | Perfluoro Derivative | Starting Material | Perfluoro Derivative |
|---|---|---|---|
| $C_6H_5C_2H_5$ ........... | 9·6% | $C_6H_{11}C_2H_5$ ........... | 21% |
| $m\text{-}C_6H_4(CH_3)_2$ ....... | 4·5% | $1,3\text{-}C_6H_{10}(CH_3)_2$ ...... | 15·4% |
| $sym\text{-}C_6H_3(CH_3)_3$ ..... | 5·9% | $1,3,5\text{-}C_6H_9(CH_3)_3$ ..... | 11·2% |
| Tetralin ........... | 11·4% | Decalin ........... | 18·8% |

Both silver and gold catalyzed fluorinations are good methods for preparing perfluoro-derivatives, especially the fluorocarbons. Nevertheless they have been eclipsed by fluorination with silver difluoride or cobalt trifluoride, and in some cases by electrolytic fluorination (p. 84).

## Fluorination with Silver Difluoride

The fluorination of organic compounds with silver difluoride actually takes place during the fluorination with elemental fluorine on a silver catalyst. The silver combines with fluorine to silver fluoride which is converted by the excess of fluorine to silver difluoride. Silver difluoride reacts with the organic compound, and is degraded to silver fluoride which is retransformed to silver difluoride by an excess of fluorine.

45          $-CH_2- + 2\,AgF_2 = -CF_2- + 2\,AgF + 2\,HF$

           $2\,AgF + F_2 = 2\,AgF_2$

In some cases fluorination with ready-made silver difluoride proved superior to catalytic fluorination over a silver catalyst. The reaction with silver difluoride is not carried out in one stage but in several consecutive reactors heated to progressively higher temperatures. The advantage of such an arrangement is that the pure hydrocarbon meets the fluorinating reagent at a relatively low temperature (225°C). The partly fluorinated product, which is less sensitive to the fluorinating reagent, is fluorinated in the next reactor at a higher temperature, and the final fluorination is accomplished on a product so highly fluorinated that it withstands the final temperature (330°C) without appreciable decomposition (628).

After the reaction of the silver difluoride with the organic compound, the fluorinating salt is regenerated in the same apparatus by a stream of fluorine at 220–250°C.

*Procedure 5.*

*Addition of Fluorine and Replacement of Hydrogen by Fluorine by Means of Silver Difluoride*

*Preparation of Perfluoro(methylcyclohexane) from Toluene and Silver Difluoride\**

In an apparatus (Fig. 14) consisting of four flat quadrangle reactors with dimensions of $140 \times 20 \times 5$ cm, 3–3·5 kg of silver fluoride or silver chloride is spread in a thin layer, the apparatus is evacuated, and fluorine is passed through at a temperature not exceeding 100°C. The reaction heat of the transformation of the silver fluoride or chloride to silver difluoride raises the temperature so that the maximum shifts to the following reactor as soon as the transformation in the previous one is completed. The preparation of the silver difluoride is completed by raising the temperature in all the reactors to 250°C. The fluorine is passed through as long as its concentration in the end-gases is sufficient for igniting a wood splinter; this requires 1–2 days.

In the span of two hours, 600 g (6·5 moles) of toluene is dropped into the first reactor whose initial temperature (260°C) is raised to 350°C by the reaction heat. In the second, third and fourth reactor the respective temperatures are 290–320°C, 275–290°C, and 285–290°C. After the addition of the organic material has been completed, the reactors are heated an additional 15 minutes, then allowed to cool in a flushing stream of nitrogen. The products are collected in two nickel traps, of which the first contains 500 ml. of water and is immersed in an ice-bath, while the second is empty and cooled in a dry-ice bath. The contents of both traps are poured over crushed ice, the heavy organic layer is separated, washed with sodium hydroxide, and steam-distilled. The crude, dried product (800 g) is recycled through the apparatus after regeneration of the salt by a stream of fluorine at 100–250°C. The recycling takes two hours and is carried out at 290–305°C, 300–335°C, 275–325°C, and 205–215°C, respectively, in the individual reactors. The contents of the traps are again poured onto ice, the organic layer is se-

---

\* Carried out by the author in the laboratory of E. T. McBee.

parated, washed with aqueous sodium hydroxide, dried and distilled. From 500 g of the crude recycled product approximately 400 g (1·14 moles) (17·5%) of perfluoro(methyl-cyclohexane) and approximately 40 g (0·13 mole) (2%) of perfluorocyclohexane, formed by fluorolysis, are obtained.

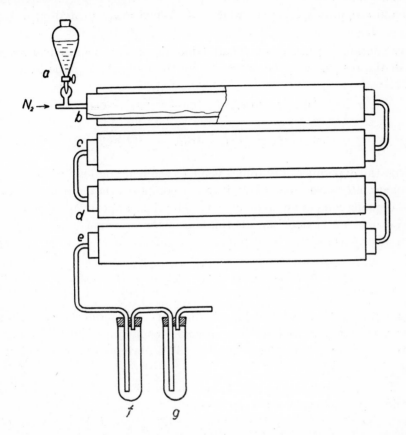

FIG. 14. Apparatus for fluorinations by means of silver difluoride or cobalt trifluoride: a) dropping funnel, b,c,d,e) electrically heated reactors, f) ice trap, g) dry-ice trap.

Though silver difluoride is a very good fluorinating reagent, it has been displaced in practice by the equally efficient but cheaper cobalt trifluoride.

## Fluorination with Cobalt Trifluoride

The fluorinations with cobalt trifluoride are based on a similar principle as the fluorinations with silver difluoride. The reaction yields are equal to, or better than those of the silver difluoride process, and far better than in catalytic fluorination (*888*). On account of its easy and economical preparation, cobalt trifluoride has become a favourite fluorinating reagent for the production of polyfluoro- and perfluoro-derivatives either from saturated, or unsaturated compounds. It is able to accomplish the addition of fluorine across double bonds and aromatic systems as well as the replacement of hydrogen by fluorine.

The fluorinating action of cobalt trifluoride and the subsequent regeneration of the reagent may be expressed by the following equations:

47 $$\text{>CH}{-} + 2\,CoF_3 = \text{>CF}{-} + HF + 2\,CoF_2$$

48 $$2\,CoF_2 + F_2 = 2\,CoF_3; \quad \Delta H_{473°} = -52\,\text{kcal/mole}$$

From the experimentally found value of the reaction heat of the regeneration equation (52 kcal/mole) (*287*), it can be computed that, during the reaction of the organic compound with cobalt trifluoride, approximately one half of the total reaction heat of the fluorination of the organic compound with elemental fluorine (102–104 kcal/mole) is liberated:

49 $$\text{>CH}{-} + F_2 = \text{>CF}{-} + HF; \quad \Delta H_{298} = -104\,\text{kcal/mole}$$

Fluorination with cobalt trifluoride or silver difluoride is thus a roundabout process which exposes the organic compounds to only half the thermal stress of direct fluorination. This brings easier handling and higher yields.

As in the fluorination with silver difluoride, fluorination with cobalt trifluoride can be carried out in a set of reactors (Fig. 14, p. 79) with the first heated only to 150°C. The cobalt trifluoride is spread on the bottom of the reactors in a thin layer over which vapours of organic compounds, sometimes diluted with nitrogen, are passed at suitable temperatures (*287*).

Another type of apparatus, designed for fluorination with cobalt trifluoride, is a tubular reactor with a central rotating shaft with paddles which stir the reagent inside the reactor and improve the contact with the organic vapours (*36, 69*). The reactor is made of nickel and is surrounded by several heating segments which allow heating to different temperatures. The part of the reactor, where the organic material meets the cobalt trifluoride and where the exothermic reaction raises the temperature, can thus be heated less than the more distant parts where the reaction is being completed.

Local overheating is avoided and the temperature may be equally distributed along the whole reactor; this has a favourable effect on the fluorination yields.

Of the total amount of fluorinating reagent present in the reactor, only 25–30% is actually consumed by the fluorination: a great excess of reagent must always be present, or else the yields of perfluorinated

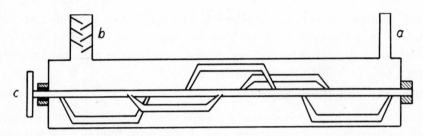

FIG. 15. Rotary apparatus for fluorinations by means of silver difluoride or cobalt trifluoride: a) inlet tube, b) outlet tube with baffle plates, c) rotating shaft with paddles.

products drop (*36*). The spent fluorinating reagent is regenerated by a stream of fluorine passed through the reactor at temperatures around 250–300° (*36, 69*).

The fluorination with cobalt trifluoride is illustrated in several examples and in Table 13.

$$\textbf{50} \qquad CH_3(CH_2)_5CH_3 \xrightarrow[\text{150—165°, 275—300°}]{CoF_3} CF_3(CF_2)_5CF_3 \qquad 91\% \qquad [287]$$

TABLE 13

*Fluorination with Cobalt Trifluoride in Two Reactors in Tandem at 150 and 300°C (287)*

| Starting Material | Product | Yield of Pure Perfluoro Compound |
|---|---|---|
| $C_7H_{16}$ | $C_7F_{16}$ | 68(79)* |
| $C_6H_5CF_3$ | $C_6F_{11}CF_3$ | 77 |
| $C_6H_4(CF_3)_2$ | $C_6F_{10}(CF_3)_2$ | 83(88)* |
| $C_5H_{12}$ | $C_5F_{12}$ | 58 |
| *cyclo*-$C_5H_{10}$ | *cyclo*-$C_5F_{10}$ | 28 |
| $C_6H_4(CH_3)_2$ | $C_6F_{10}(CF_3)_2$ | 42–50 |
| Indane | Perfluorohydrindane | 20 |
| Cetane | Perfluorocetane | 29 |

* After recycling

With aromatic compounds, replacement and addition take place simultaneously and the end product is the same as that obtained from the corresponding cycloparaffin.

51

The perfluorocyclohexane obtained in the reaction of benzene with cobalt trifluoride is accompanied by varying amounts of incompletely fluorinated products such as undecafluoro-, decafluoro-, nonafluoro- and octafluorocyclohexanes (37, 317, 961) (Equation 218, p. 134).

## Fluorination with Other Fluorides

In contrast with fluorinations by means of silver difluoride and especially cobalt trifluoride, the fluorinations with other metal fluorides are not nearly as frequent. The reason is that none of the tested fluorides gave better results in the whole spectrum of applications than cobalt trifluoride.

The fluorination of higher boiling hydrocarbons was succesfully carried out with *manganese trifluoride* or *cerium tetrafluoride*. At comparable conditions and in some cases, these fluorides gave even somewhat higher yields than cobalt trifluoride (286) (Table 14).

TABLE 14

*Relative Yields Obtained in Fluorination of High-Boiling Aromatic–Aliphatic Hydrocarbons (B.P. 325 – 410°C) with Various Metal Fluorides (286)*

| Metal Fluoride | $CoF_3$ | $MnF_3$ | $CeF_4$ |
|---|---|---|---|
| Yield of the Fluorinated Product, % | 50·1 | 59 | 58·3 |

Similar fluorinating properties are shown by *lead tetrafluoride* (666) and *bismuth pentafluoride* (286). However their practical applications are negligible. Experimental technique is the same as that of cobalt trifluoride.

Interesting results were obtained in fluorinations carried out with *chlorine trifluoride*. This compound can replace individual hydrogen atoms by fluorine, its wider use is however prevented by side reactions, mostly chlorinations attributed to the formation of chlorine monofluoride (235, 236).

52

A reagent suitable for direct introduction of individual fluorine atoms into the aromatic ring is claimed to be formed by mixing chlorine trifluoride with a Lewis' acid such as boron trifluoride (*826*). Formation of an ionic complex is assumed, which would contain a $ClF_2^+$ cation able to effect electrophilic reactions. Actually a mixture of chloro- and fluoro-derivatives has been obtained, the occurrence of which is explained in the following way;

53

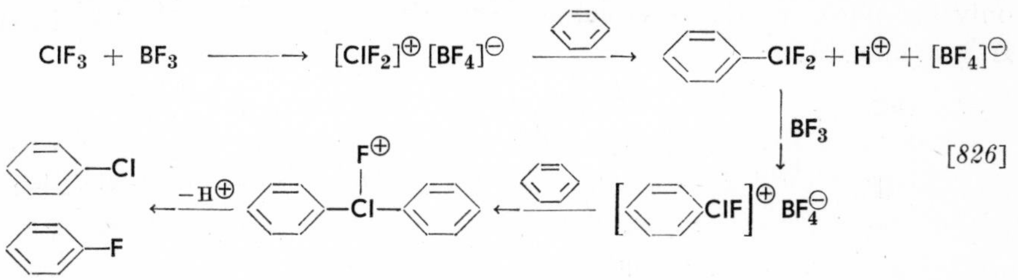


Quite recently, a very interesting and useful fluorinating reagent has been introduced into the laboratory practice. *Perchloryl fluoride* formulated as a fluoride of perchloric acid reacts with $\beta$-dicarbonyl, $\beta$-carbonylcarboxy or $\beta$-dicarboxy compounds so that it replaces the acidic hydrogens of these compounds by fluorine (*529*):

54 [*529*]

$$CH_3{-}CO{-}CH_2{-}COOC_2H_5 \xrightarrow{C_2H_5ONa} CH_3{-}CO{-}CNa_2{-}COOC_2H_5 \xrightarrow{FClO_3} CH_3{-}CO{-}CF_2{-}COOC_2H_5 \quad 59\%$$

Another application is its reaction with enamines or enol ethers to yield $\alpha$-fluorinated ketones (*761*).

55 [*761*]

$$-CH_2{-}CO{-} \xrightarrow{R_2NH} -CH{=}C(NR_2){-} \xrightarrow{FClO_3} -CHF{-}CO{-} \xleftarrow[\text{pyridine, } 0^\circ]{FClO_3} -CH{=}C(OR){-}$$

These surprising reactions are very important for the introduction of individual fluorine atoms into defined places of the organic molecule. They are frequently used in the chemistry of fluorinated steroids (*103*, *761*, *764*).

56

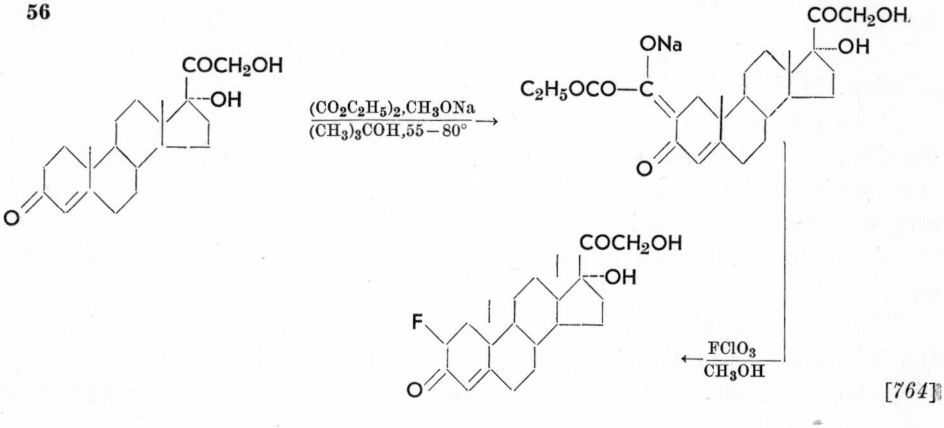

As minor fluorinating agents, mention could be made of *iodobenzene difluoride* (*105*) and *p-iodotoluene difluoride* (*26*), which are prepared from the corresponding iodoso-derivatives and aqueous or anhydrous hydrogen fluoride; they replace aromatic hydrogen by fluorine in sufficiently reactive systems. Since only the most reactive systems are sensitive to this reaction, its application is insignificant.

57 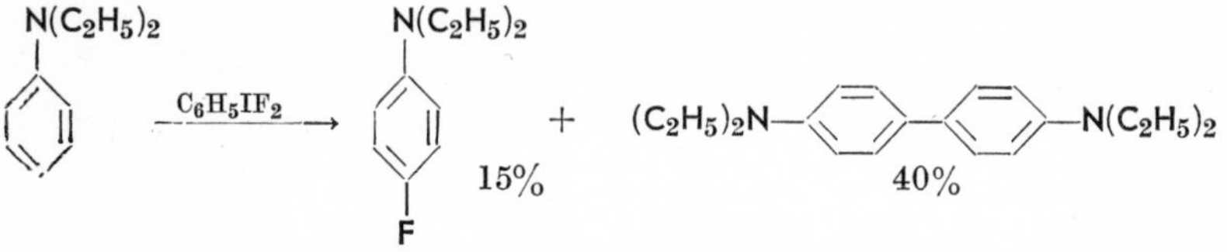    [*105*]

A somewhat more attractive organic fluorinating reagent seems to be *trifluoromethylhypofluorite*, which can be considered as "canned fluorine". It replaces hydrogen atoms by fluorine both in aliphatic and aromatic hydrocarbons (*3*).

58    $$CH_4 + CF_3OF \xrightarrow[\text{12 hrs.}]{h\nu,\ 25^\circ} CH_3F + CH_2F_2 + CHF_3 + CF_4$$    [*3*]

              10        5        5      trace

### Electrolytic Fluorination Process

The most important recent discovery in the field of the preparation of organic fluoro-derivatives is electrolysis of organic compounds in anhydrous hydrogen fluoride. It is carried out in iron or nickel cells equipped with a reflux condenser, a device for replenishing of the hydrogen fluoride and for draining of the fluorinated products. Anodes are made of nickel, cathodes of nickel or steel (*144, 175, 321, 546*). A sketch of such an apparatus appears in Fig. 16.

The organic compound is dissolved or dispersed in anhydrous hydrogen fluoride. In the case of nitrogen, oxygen or sulphur-containing compounds, onium salts are formed which impart sufficient electrical conductance to the solution. In the case of hydrocarbons, which are soluble in hydrogen fluoride only to a very limited extent, inorganic fluorides, such as lithium or potassium fluorides, must be added to render the solution conducting. The solution is then subjected to electrolysis at a current density of approximately 0·02 amp/cm² and at a potential drop lower than 8 V, usually 5–6 V. Under such conditions fluorine is not evolved, but fluorination of the dissolved organic compound occurs.

The electrochemical process has caused mild additions of fluorine across the double bonds in 1,1-diphenylethylene, trichloroethylene and tetrachloroethylene (*923*), and replacement of individual hydrogen atoms in aliphatic acids such as in the preparation of α-fluoro- and β-fluoropropionic acid or α-fluoro-, β-fluoro- and γ-fluorobutyric acids (*922*). The yields and importance of these reactions are negligible.

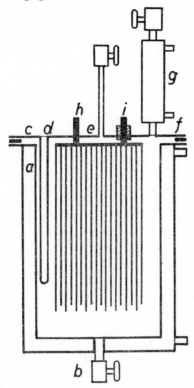

Fɪɢ. 16. Apparatus for electrolytic fluorinations: a) electrolytic cell with water jacket, b) liquid product outlet, c) cell lid, d) thermometer well, e) electrolyte inlet, f) gaseous product outlet, g) reflux condenser, h) cathode, i) anode.

However, the electrochemical fluorination is extraordinarily suitable for the preparation of compounds with a large number of fluorine atoms in their molecules, and especially for the production of the perfluoroderivatives. Starting materials for such compounds are not only hydrocarbons, which their low solubility in anhydrous hydrogen fluoride renders difficult to handle, but mainly nitrogen, oxygen or sulfur compounds which dissolve readily. The electrolysis of amines yields perfluoroamines in addition to some nitrogen-free by-products (*547*).

**59**     $(CH_3)_3N \xrightarrow[\text{4-8 V, 2 amp/dm}^2]{\text{HF}} \underbrace{(CF_3)_3N + CHF_3 + CF_4}_{\text{400 g}} + NF_3$     [*547*]

175 g                    400 g

Great industrial importance is attached to the electrolysis of carboxylic acids and their anhydrides or chlorides with formation of perfluoroacyl fluorides (546, 548).

60           $(CH_3-CO)_2O$ $\xrightarrow[\substack{5.2 \text{ V, 50 amp} \\ 20°, 1 \text{ atm}}]{HF}$ $CF_3-COF$           [546]
                4% solution in HF

*Procedure 6*

*Replacement of Hydrogen by Fluorine by Means of Electrolysis in Anhydrous Hydrogen Fluoride*
*Preparation of Trifluoroacetic Acid by Electrolysis of Acetic Anhydride (546)*
The cell shown in Fig. 16 (p. 85) is charged with 2000 g of a 4% solution of acetic anhydride in anhydrous hydrogen fluoride, the cell is shut by a lid bearing suspended electrodes and a reflux condenser maintained at $-30°C$; the cell's contents are electrolyzed by direct current at 50 amp with a potential drop of 5·2 V and at a temperature of 20°C maintained by a cooling liquid circulating through an outer jacket. Trifluoroacetyl fluoride formed in the electrolysis is separated from the refluxing hydrogen fluoride by distilling past the condenser. It is absorbed in water and dissolves to trifluoroacetic and hydrofluoric acid, separable by ether extraction.

Electrolysis of methanesulphonylchloride or methane sulfofluoride gives trifluoromethanesulphonyl fluoride (321).

61    $CH_3-SO_2Cl$ $\xrightarrow[electrolysis]{HF}$ $CF_3-SO_2F$ $\xleftarrow[electrolysis]{HF}$ $CH_3-SO_2F$    [321]
                              87%    96%

Electrolytic fluorination of the homologous alkanesulfochlorides leads to the corresponding perfluoroalkanesulfofluorides in yields dropping with increasing number of carbon atoms (144, 322) (Table 15).

TABLE 15

*Yields of Perfluoroalkanesulphonyl Fluorides Obtained by the Electrochemical Fluorination Process (321, 322)*

| $C_nH_{2n+1}SO_2Cl$ $n =$ | 1 | 2 | 3 | 4 | 5 | 6 | 7 | 8 |
|---|---|---|---|---|---|---|---|---|
| $C_nF_{2n+1}SO_2F$ % | 87 | 79 | 68 | 58 | 45 | 36 | 31 | 25 |

Compounds containing divalent sulphur are transformed to perfluoroderivatives having sulphur in tetracovalent or hexacovalent state. Dimethylsulphide gives, in addition to large amounts of carbon tetrafluoride, mainly trifluoromethylsulphurpentafluoride and sulphurhexafluoride and only small amount of bis(trifluoromethyl)sulphurtetrafluoride (175).

62    $CH_3-S-CH_3$ $\xrightarrow[electrolysis]{HF}$ $(CF_3)_2SF_4$ + $CF_3SF_5$ + $CF_4$ + $SF_6$    [175]
                              2%        20%      58%    17%

Electrolytic fluorination of carbon disulphide has been mentioned elsewhere (p. 72).

Complete literature about the electrochemical fluorination process and its applications has been reviewed recently (*562, 966*).

## REPLACEMENT OF HALOGENS BY FLUORINE

The ease of the replacement of halogen atoms in organic molecules depends substantially upon its reactivity. Generally, iodine is replaced more easily than bromine and bromine again more easily than chlorine. The ease of the replacement of the same halogen depends on its position in the molecule, or on the kind of its bonding. Replacement of halogen atoms by fluorine is easy in reactive halogen derivatives such as sulphohalides, acid halides, α-halocarbonyl compounds, α-haloacids or esters, allylic halides, benzylic halides, and aromatic halogen derivatives with nitro groups in *o*- or *p*-positions to the halogen atom. Substitution of fluorine for a halogen in an aliphatic chain is more difficult.

The reactivity of the halogens in being replaced by fluorine determines the choice of the fluorinating reagent. Sulphohalides, phosphohalides, acid halides, and α-haloketones, acids and their derivatives are usually treated with potassium fluoride in order to accomplish the halogen exchange reaction. For the conversion of alkyl halides to alkyl fluorides, silver fluoride, mercurous fluoride and mercuric fluoride are used in addition to potassium fluoride. The replacement of one or two halogen atoms in geminal polyhalogen derivatives is best carried out by means of hydrogen fluoride, especially in the presence of catalysts such as antimony halides, by antimony trifluoride, simple or combined with a pentavalent antimony halide, and by mercury fluorides. Mercurous fluoride replaces usually only one halogen atom, whereas mercuric fluoride either one, or two geminal halogens, eventually even three, provided the last of the halogen atoms to be replaced is bromine or iodine.

Elemental fluorine, silver difluoride and cobalt trifluoride can also replace halogens by fluorine; their reactions with organic halogen derivatives are however difficult to keep within the required limits.

In addition to the nature of halogens to be replaced there are still other factors which influence the choice of the fluorinating reagent. In the laboratory, the work in glass apparatus is always preferred, consequently methods are chosen which do not preclude the use of glass (reaction with metal fluorides). On the other hand, in industry where metal apparatus is used anyway, anhydrous hydrogen fluoride is the most favoured fluorinating reagent, since it is available in quantities and its excess can usually be recovered.

## Replacement of Halogens by Fluorine by Means of Elemental Fluorine

Elemental fluorine can easily replace any halogen atom, irrespective of its mode of bonding. The reaction suffers, however, from the same disadvantages as the replacement of hydrogen atoms: It is difficult to limit it to the replacement of one definite halogen atom. Usually several atoms are replaced simultaneously, both halogens or hydrogens, so that a mixture of compounds is obtained from which the required product is sometimes isolated only with difficulties and in low yields. The preparation of fluorine derivatives by the reaction of organic halides with elemental fluorine is therefore convenient only in cases where a large number of fluorine atoms are to be introduced into the organic molecule. The advantage of this reaction over direct replacement of hydrogen atoms by fluorine is higher thermal and chemical resistance of the halogenated compounds and less decomposition.

In special equipment hexachloroethane was transformed by elemental fluorine to tetrachlorodifluoroethane in a yield of 20% (732), and hexachlorobenzene to a rich mixture of compounds with various content of fluorine in the molecules (293). Carbon tetrachloride was fluorinated to a mixture of carbon tetrafluoride and chlorotrifluoromethane by treatment with fluorine in the presence of arsenic (953), dichlorodifluoromethane to chlorotrifluoromethane by fluorine at 340–370°C in the presence of mercury (953).

**63**
$$CCl_4 \xrightarrow[70°]{F_2, As} CF_4 + CClF_3 \qquad\qquad [953]$$
$$74\% \quad 17\%$$

Generally speaking, replacement of halogens by means of elemental fluorine is impractical. The yields of the reaction are usually not very good, the fluorine is relatively expensive, and the replacement can be effected more conveniently by other reagents.

## Replacement of Halogens by Fluorine by Means of Hydrogen Fluoride

Hydrogen fluoride is a very general means for replacing halogen atoms by fluorine. It is only necessary to find suitable conditions which are dictated by the strength of the bonds and by the number of halogen atoms to be replaced. One of the most easily replaceable halogen atoms is a halogen attached to silicon. Alkylhalosilanes can be converted to alkylfluorosilanes by simple heating with aqueous -alcoholic solution of hydrogen fluoride (230).

**64**
$$(C_2H_5)_3SiCl \xrightarrow[70°]{HF, H_2O, C_2H_5OH} (C_2H_5)_3SiF \xleftarrow[75°]{HF, H_2O, C_2H_5OH} (C_2H_5)_3SiBr \quad [230]$$
$$85\% \qquad\qquad 80\%$$

In order to replace halogen atoms bound to carbon, anhydrous or at least almost anhydrous hydrogen fluoride has to be used. A relatively easy replacement of halogens occurs in the acyl chlorides (*954*):

65 $\qquad$ $COCl_2$ $\xrightarrow[\text{80°, 20 atm}]{\text{HF}}$ $COClF$ 25% $\qquad$ [*954*]

A very smooth replacement is that of halogens attached to a carbon alpha to a double bond or an aromatic ring. The transformation of benzotrichloride to benzotrifluoride was effected by means of anhydrous hydrogen fluoride over a very broad range of temperatures and with very good yields (*137, 955, 960*):

66 [*137, 955, 960*]

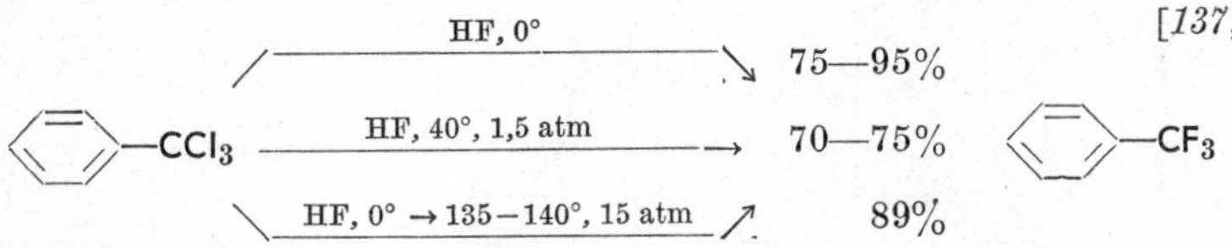

The same results were obtained with chlorinated homologues of pyridine (*638*):

67 45% [*638*]

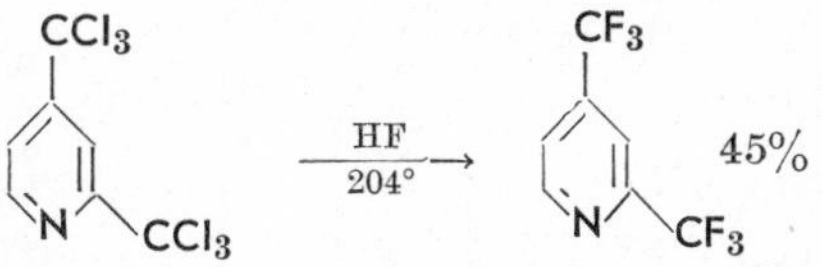

Hydrogen fluoride can exchange even relatively unreactive halogens for fluorine. During the addition of hydrogen fluoride to chlorinated olefins (Equation 19, p.65) partial replacement of chlorine by fluorine was very often observed, especially when the reaction was carried out at elevated temperatures (*452*). For complete replacement of all the halogens by fluorine in groups like $-CCl_2$ or $-CCl_3$, temperatures higher than 100°C are required (*1017*):

68 $\quad CCl_3-CH_2-CCl\begin{smallmatrix}CH_3\\CH_3\end{smallmatrix}$ $\xrightarrow[\text{130°}]{\text{HF}}$ $CF_3-CH_2-CCl\begin{smallmatrix}CH_3\\CH_3\end{smallmatrix}$ 40% [*1017*]

*Procedure 7.*

*Replacement of Chlorine by Fluorine by Means of Hydrogen Fluoride*
*Preparation of 1,1,1-Trifluoropropane from 1,1-Dichloro-1-propene\**

A five-litre steel rotating autoclave fitted with an electrically heated mantle, a thermocouple, a manometer reading up to 350 atm, and a needle valve is evacuated and

---

\* Carried out by the author in the laboratory of E. T. McBee.

charged with 700 g (6·3 moles) of 1,1-dichloro-1-propene and 920 g (46 moles) of liquid anhydrous hydrogen fluoride. The mixture is heated and stirred for 72 hours at 115–120°C, under autogenous pressure of 70–75 atm. The gaseous contents of the autoclave are then bled off through two fifteen-litre bottles containing altogether 20 litres of a 10% solution of sodium hydroxide. After being thus washed free from acids, the gas is dried by passage through a column of anhydrous calcium sulphate, and condensed in a receiver cooled with dry ice. Approximately 530 g (5·2 moles) of crude product is obtained whose distillation in a low temperature column yields round 450 g (4·6 moles) (73%) 1,1,1-trifluoropropane distilling over a range of −15°C to −11°C.

The replacement of chlorine by fluorine in carbon tetrachloride by means of anhydrous hydrogen fluoride requires considerably higher temperatures (637) (Table 16).

TABLE 16

*Effect of Variables on the Yield and Product Ratio in the Reaction of Carbon Tetrachloride with Anhydrous Hydrogen Fluoride at 70 atm. (637)*

| Temperature, °C | 400 | 435 | 460 | 490 |
|---|---|---|---|---|
| HF : CCl$_4$ Ratio ........................ | 1 | 1·9 | 2·3 | 4·9 |
| Total Yield % .......................... | | 95 | 85 | 59 |
| Product Composition CCl$_2$F$_2$ % ......... | 0 | 5 | 34 | 79 |
| CCl$_3$F % ......... | 52 | 77 | 59 | 21 |
| CCl$_4$ % ......... | 48 | 18 | 7 | 0 |

The replacement of the third and even fourth atom of chlorine in carbon tetrachloride cannot be effected by plain hydrogen fluoride.

The exchange reaction between the organic halogen derivatives and hydrogen fluoride can be promoted by catalysts, either in vapour phase, or in liquid phase.

*Catalytic replacement of chlorine by fluorine by means of anhydrous hydrogen fluoride in vapour phase* is described in a series of patents; however, the actual reaction conditions, under which the reaction is really successfully operated, are not disclosed. One of the best published descriptions pertains to the reaction in the presence of activated charcoal or activated charcoal containing ferric chloride. This particular catalyst gave very good results and high yields of dichlorodifluoromethane which is the coveted product of the reaction between carbon tetrachloride and hydrogen fluoride (637). The contrast brought by the catalysts is shown in Table 17.

TABLE 17

*Results of Non-catalytic and Catalytic Reaction of Carbon Tetrachloride with Anhydrous Hydrogen Fluoride in Vapour Phase (637)*

| Conditions | No Catalyst | Catalyst Activated Charcoal | Catalyst Activated Charcoal Containing 26% of FeCl$_3$ |
|---|---|---|---|
| | | Yields in % | |
| Optimum Pressure, atm. ............... | 70 | 70 | 7 |
| Optimum Temperature, °C .............. | 450 | 425 | 300 |
| Optimum HF : CCl$_4$ Ratio .............. | 2 | 2 | 1·8 |
| Optimum space velocity, ml./hr. ........ | 600 | 800 | 5000 |
| Conversion CCl$_4$ → (CCl$_3$F + CCl$_2$F$_2$), % ... | 85 | 95 | 95 |
| Exhaustion of HF, % .................. | 50 | 70 | 90 |
| Product Composition, %   CCl$_2$F$_2$ ........ | 30 | 50 | 75 |
| CCl$_3$F ........ | 60 | 50 | 20 |
| CCl$_4$ ........... | 10 | 0 | 5 |

For the *catalytic reaction of hydrogen fluoride* with organic halogen derivatives in a *liquid phase*, the best catalysts are antimony halides, which are used almost exclusively in the majority of laboratory and industrial conversions of polyhalogen derivatives to partly or totally fluorinated compounds. Other catalysts find use only in special cases.

The most frequently used antimony catalyst is *antimony trichloride* (*133*) together with a certain amount of bromine, chlorine or antimony pentachloride, which act as promoters and increase the efficiency of the catalyst. Still more advantageous is pure *antimony pentachloride* (*630*). All these catalysts support the exchange of the halogens for fluorine and permit the replacement to occur under much milder  conditions than with hydrogen fluoride alone. The exchange reaction probably takes place specifically between the organic polyhalogen derivative and the mixed chlorofluorides of pentavalent antimony, which are formed in the reaction vessel from the antimony trihalide or pentahalide and hydrogen fluoride:

69
$$SbCl_5 + 3\,HF \rightarrow SbF_3Cl_2 + 3\,HCl$$

$$R\!-\!CHCl_2 + SbF_3Cl_2 \rightarrow R\!-\!CHF_2 + SbFCl_4$$

$$SbFCl_4 + 2\,HF \rightarrow SbF_3Cl_2 + 2\,HCl$$

For better understanding an idealized scheme is outlined: the reaction course is, in reality, more complicated and non-stoichiometric. The exact mechanism is guess-work. As in the case of hydrogen fluoride alone, the catalytic reaction replaces most easily the halogen atoms in α-position to a double bond or to an aromatic ring. The —CCl₃ groups attached to aromatic nuclei are fluorinated to the full extent (630),

70

$$\text{(structure: benzene ring with Cl, CCl}_3\text{, CCl}_3\text{)} \xrightarrow[20^\circ]{\text{HF, SbCl}_5} \text{(structure: benzene ring with Cl, CF}_3\text{, CF}_3\text{)} \quad 91\% \qquad [630]$$

whereas if the same groups are part of a saturated chain, the replacement is smooth only to the stage of the difluoro-derivative. The exchange of the third halogen atom is sometimes difficult and gives only a low yield (133):

71

$$CCl_4 \xrightarrow[110^\circ,\ 30\ atm]{\text{HF, SbCl}_3,\ Cl_2} CCl_3F + CCl_2F_2 + CClF_3 \qquad [133]$$

$$\phantom{CCl_4 \xrightarrow{} } 9\% \qquad 90\% \qquad 0.5\%$$

If also the third halogen atom is to be replaced by fluorine by the action of anhydrous hydrogen fluoride, an especially active catalyst must be used. Such a reagent is suitably prepared by heating one mole of antimony penta-chloride with five moles of anhydrous hydrogen fluoride at 150°C in an auto-clave. The resulting product corresponds to a *molecular compound of an empirical formula of* SbF₃Cl₂·2HF (1081). It was succesfully used for the transformation of chloroform to fluoroform and of 1,1,1,2-tetrachloroethane to 1-chloro-2,2,2-trifluoroethane in very high yields (1081).

72

$$CH_2Cl—CCl_3 \xrightarrow[133^\circ,\ 46\ atm]{\text{SbF}_3Cl_2\cdot2HF} CH_2Cl—CF_3 \quad 95\% \qquad [1081]$$

When milder fluorinating action is required, the reaction between hydrogen fluoride and the organic halide is catalyzed by *stannic chloride* (1082).

73

$$CH_3—CHCl_2 \xrightarrow[70^\circ]{\text{HF, SnCl}_4} CH_3—CHClF + CH_3—CHF_2 \qquad [1082]$$

$$\phantom{CH_3—CHCl_2 \xrightarrow{}} 40\% \qquad 40\% \uparrow 82\%$$

$$\xrightarrow[50^\circ]{\text{HF, SbCl}_5}$$

It is interesting to note that hydrogen fluoride, alone or in the presence of catalysts, is unable to replace by fluorine a lone halogen atom such as the halogen in primary, secondary or tertiary halides (1017, 1081), or the halogen attached to an aromatic ring (630).

## Replacement of Halogens by Fluorine by Means of Antimony Fluorides

The preparation of fluorinated compounds by treatment of organic halogen compounds with antimony fluorides (*the Swarts reaction*) is one of the oldest and still most important laboratory methods for the production of organic fluorine derivatives. It resembles to a certain extent the reaction of organic halides with anhydrous hydrogen fluoride in the presence of antimony halides. Whereas in the latter reaction the necessary fluorine is supplied by hydrogen fluoride and the antimony halides act only as fluorine carriers and are therefore used only in catalytic amounts, the reaction of organic compounds with antimony fluorides requires at least an equivalent quantity of fluorine and more often an excess. This disadvantage is compensated by the circumstance that the spent fluorinating reagent can usually be regenerated by means of hydrogen fluoride. Another great advantage of the antimony fluorides is the possibility of working in glass apparatus, since the reactions very often take place at atmospheric pressure.

*Antimony trifluoride* alone is not very efficient as a fluorinating reagent. Its activity can be increased by transforming the antimony partly or totally from the trivalent to the pentavalent stage by the addition of chlorine, bromine, or antimony pentachloride. The more antimony is present in the pentavalent state, the more reactive is the fluorinating reagent. According to their relative reactivities, the antimony fluorides can be arranged into the following series (*698*).

74     $SbF_3$  <  $SbF_3 + SbCl_3$  <  $SbF_3 + SbCl_5$  <  $SbF_3Cl_2$  <  $SbF_5$     [*698*]

A common feature of all the antimony fluorinating reagents is the ability to react with $-CX_2$ and $-CX_3$ groupings and replace their halogens successively by fluorine. This circumstance is usually explained by an easy formation of complexes between the above mentioned types of compounds and antimony halides (*112*).

On the basis of this hypothesis compounds containing single halogen atoms should not react with antimony fluorides. Generally, this holds true, and antimony fluorides are not suitable for transforming the groups like $-CH_2Cl$, $>CHCl$, and $\geqslant CCl$ to the corresponding fluorinated derivatives. However, deviations from this rule should be quoted.

Antimony fluoride, alone or activated with a halogen or antimony pentachloride, substituted fluorine for the halogens in the following compounds: In a sulphonyl chloride (*416*),

75     $\begin{matrix} CH_3 \\ \quad \searrow \\ CH_3 \end{matrix}$ N—SO$_2$Cl  $\xrightarrow[\text{C}_6\text{H}_6]{\text{SbF}_3,\,\text{SbCl}_5}$  $\begin{matrix} CH_3 \\ \quad \searrow \\ CH_3 \end{matrix}$ N—SO$_2$F     78,2%     [*416*]

in an acid chloride (*912, 984*),

**76**          $CH_2F\!-\!COCl$ $\xrightarrow[\text{reflux}]{\text{SbF}_3}$ $CH_2F\!-\!COF$   64,4%           [*912*]

in chloroacetamide *(1053)*

**77**        $ClCH_2\!-\!CONH_2$ $\xrightarrow[\text{vacuum distillation}]{\text{SbF}_3}$ $FCH_2\!-\!CONH_2$   65%     [*1053*]

in α-iodothiophene (*1054*) (α-chloro- and α-bromothiophene do not react (*1054*),

**78**                            $\xrightarrow[\text{90—100°}]{\text{SbF}_3,\ \text{CH}_3\text{NO}_2}$                     10%      [*1054*]

and in *sym.*-trichlorotriazine (cyanuric chloride) (*579, 698*).

**79**                                             [*579*]

                                      $\xrightarrow[\text{160—180°, 24 hrs.}]{\text{SbF}_3,\ \text{SbCl}_5}$                        91%

All of the compounds mentioned contained a relatively reactive halogen. An isolated halogen atom in aliphatic chains is practically resistant toward the action of antimony fluorides (*368, 482*).

**80**    $CCl_3\!-\!CH_2\!-\!CH_2Cl$ $\xrightarrow[\substack{\text{C}_6\text{H}_5\text{CF}_3\\20° \to \text{reflux}}]{\text{SbF}_3,\ \text{SbF}_2\text{Cl}_2}$ $\begin{cases} CClF_2\!-\!CH_2\!-\!CH_2Cl & 10\% \\ CF_3\!-\!CH_2\!-\!CH_2Cl & 61\% \end{cases}$   [*368*]

**81**    $CCl_3\!-\!CH_2\!-\!CH_2Br$ $\xrightarrow{\text{SbF}_3}$ $CF_3\!-\!CH_2\!-\!CH_2Br$   40%     [*482*]

One of the few exceptions are fluorinations of 1,2,2-tribromoethane (*991*) and 2-bromo-1,4,4,4,-tetrachlorobutane (*1023*) during which the isolated bromine atom was exchanged for the fluorine atom, even if only to a small extent.

**82**                                           $\gg$               [*991*]

   $CHBr_2\!-\!CH_2Br$ $\xrightarrow[\text{180°}]{\text{SbF}_3,\ \text{Br}_2}$ $CHF_2\!-\!CH_2Br\ +\ CHBrF\!-\!CH_2Br\ +\ CHBr_2\!-\!CH_2F$

**83**                                                              [*1023*]

                                       $CFCl_2\!-\!CH_2\!-\!CHBr\!-\!CH_2Cl$   6.2%

$CCl_3\!-\!CH_2\!-\!CHBr\!-\!CH_2Cl$ $\xrightarrow[\text{40—50°}]{\text{SbF}_3,\ \text{SbF}_3\text{Cl}_2}$ $CF_2Cl\!-\!CH_2\!-\!CHBr\!-\!CH_2Cl$   51%

                                       $CF_2Cl\!-\!CH_2\!-\!CHF\!-\!CH_2Cl$   4.5%

Antimony trifluoride alone, without the addition of halogens or the antimony pentahalide, is used only for the exchange of sufficiently reactive halogen atoms, such as the halogens bound to silicon (*109*) or phosphorus (*695*),

84     $C_2H_5SiCl_3 \xrightarrow[35—40°]{SbF_3} C_2H_5SiCl_2F + C_2H_5SiClF_2 + C_2H_5SiF_3$     [*109*]

      215 g                15 ml       8 ml       70 ml

85            $CH_3OPCl_2 \xrightarrow[<50\,mm]{\substack{SbF_3 \\ <10°}} CH_3OPClF + CH_3OPF_2$     [*695*]

                                        8%        41%

the chlorine atoms in phosgene (*245*),

86     $\xleftarrow[\substack{SbF_3 \\ 200—280°,\ 150\ atm}]{} COCl_2 \xrightarrow[\substack{SbF_3,\ SbCl_5 \\ 135°,\ 32\ atm}]{}$     [*245*]

       $COF_2$                                          $COClF$

in a trichloromethyl group attached to sulphur (*1091*),

87     ⟨benzene⟩$-S-CCl_3 \xrightarrow[distillation]{SbF_3}$ ⟨benzene⟩$-S-CF_3$    70%     [*1091*]

or neighboring to a double bond (*488*)

                                 $CCl_2{=}CCl{-}CCl_2F$    13%

88     $CCl_2{=}CCl{-}CCl_3 \xrightarrow[150°]{SbF_3}$   $CCl_2{=}CCl{-}CClF_2$    28%     [*488*]

                                   $CCl_2{=}CCl{-}CF_3$      43%

                                         recovered     10%

or an aromatic nucleus (*996*):

89     ⟨benzene⟩$-CCl_3 \xrightarrow[125—140°]{SbF_3}$ ⟨benzene⟩$-CF_3$    47%     [*996*]

*Procedure 8.*

    *Replacement of Halogen by Fluorine by Means of Antimony Trifluoride*
    *Preparation of Benzotrifluoride from Benzotrichloride (110, 518)\**

The reaction of benzotrichloride with antimony trifluoride is to be carried out as fast as possible, since the prolonged contact of the reaction components at higher temperatures causes side reactions ending in resinification.

In a 50 ml. distilling flask fitted with a thermometer reaching almost to the bottom and connected to a water-cooled receiver, a mixture of 30 g (0·17 mole) of finely powdered antimony trifluoride and of 30 g (0·15 mole) of benzotrichloride is heated with

---

\* Carried out by the author.

a free flame to approximately 125°C. As soon as a spontaneous reaction sets in, the thermometer is shifted to a normal distillation position and the mixture is gently heated to maintain a lively reaction. The products distil at 108–115°C. The heating is stopped when the vapour temperature rises to 125°C.

The distillate in the receiver (13–15 g) is washed with an equal volume of 20% hydrochloric acid, with water, refluxed for 5–10 minutes with an equal volume of 5% solution of sodium hydroxide and steam-distilled. The organic layer in the distillate is separated, dried with calcium chloride, and rectified. The yield of the pure benzotrifluoride distilling over a range of 103–104·5°C varies from 58 to 66%.

Common aliphatic polyhalogen derivatives can be fluorinated by antimony trifluoride alone (458),

90     $CHCl_2{-}CCl_3 \xrightarrow{SbF_3} CHCl_2{-}CCl_2F \longrightarrow CHCl_2{-}CClF_2$     [458]

$$\downarrow$$

$$CHF_2{-}CClF_2 \longleftarrow CHClF{-}CClF_2$$

however, the addition of a halogen or antimony pentahalide promotes the reaction and leads to deeper fluorination and higher yields (1047):

91     $\left| \xrightarrow[140°]{SbF_3} \right.$    $CH_3{-}S{-}CCl_3 \xrightarrow[95°]{SbF_3,\ SbCl_5}$     [1047]

$$\downarrow \qquad\qquad\qquad\qquad\qquad\qquad\qquad\qquad \downarrow$$

$CH_3{-}S{-}CClF_2 + CH_3{-}S{-}CF_3$            $CH_3{-}S{-}CF_3$

    32·4%        37·8%               73%

*Antimony trifluoride "fortified" by the addition of chlorine, bromine or antimony pentafluoride* converts chloroform smoothly to chlorodifluoromethane (107) and bromoform to bromodifluoromethane (427):

92                 $CHCl_3 \xrightarrow[100°,\ 58\ atm]{SbF_3,\ SbCl_5} CHClF_2$                [107]

93                 $CHBr_3 \xrightarrow[4\ atm]{SbF_3,\ Br_2} CHBrF_2$    80%             [427]

The third chlorine atom resists fluorination even at 200°C and 80 atm (107).

Carbon tetrachloride and antimony trifluoride with a small amount of antimony pentachloride react to give dichlorodifluoromethane (708) as end product:

94            $CCl_4 \xrightarrow{SbF_3,\ SbCl_5} CCl_2F_2$    88—94%             [708]

In all the instances quoted so far, two halogen atoms attached to one carbon were replaced by fluorine at best. The more surprising is a recent communication about the reaction between 1,1,1-trichloroethane and a mixture of antimony trifluoride with antimony pentachloride leading to 1,1,1-trifluoroethane at a relatively low temperature and at atmospheric pressure (*688*).

95
$$CH_3-CCl_3 \xrightarrow{SbF_3 + SbCl_5} CH_3-CF_3 \qquad [688]$$

The mixture of antimony trifluoride and antimony pentachloride is used to substitute fluorine for halogen atoms in many types od aliphatic (see the above equations) or alicyclic halides (*605*), and

96 72·5% [*605*]

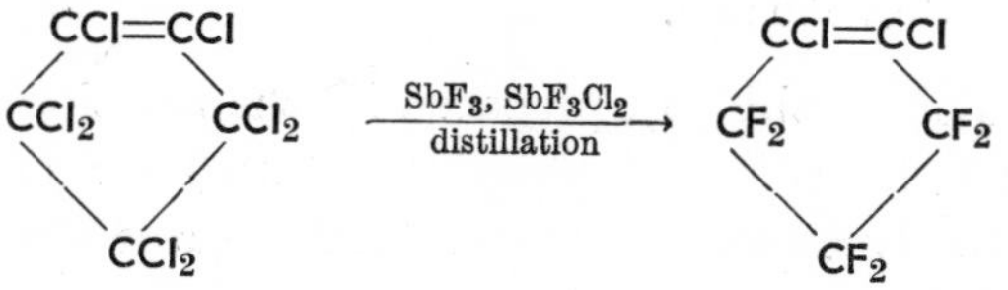

in aromatic (*650*) as well as heterocyclic halogen derivatives (*749*):

97 43% [*650*]

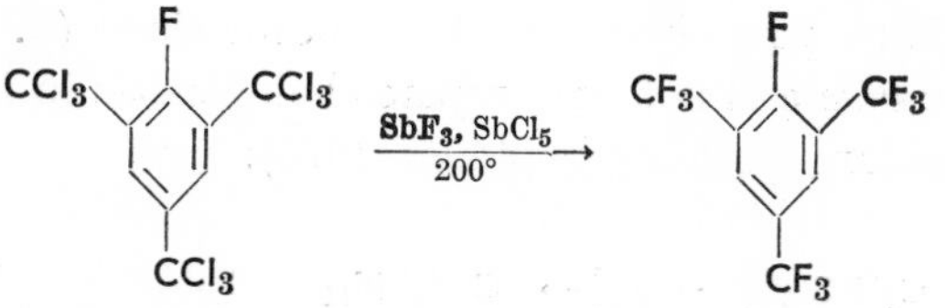

98 90% [*779*]

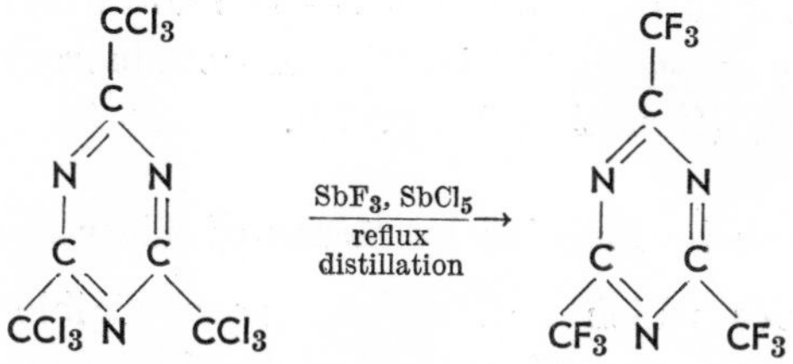

*Procedure 9.*

*Replacement of Chlorine by Fluorine by Means of Antimony Chlorofluoride*
*Preparation of 2,3-Dichlorohexafluoro-2-butene from Perchlorobutadiene (484)*\*

A two litre steel bomb (Fig. 2, p. 29) is charged with 1225 g (6·8 moles) of antimony trifluoride and closed by an assembly bearing a manometer for pressures up to 20 atm. and a needle valve. The bomb is inserted into the heating mantle of a rocking shaker and connected by means of a flexible copper tube to a chlorine cylinder. While rocking and heating the bomb at 135°C, chlorine is drawn in from a cylinder up to a weight increase of 480 g (6·8 moles). The copper tube is then disconnected, the excess of chlorine

---

\* Carried out by the author in the laboratory of A. L. Henne.

vented off through the valve, the bomb is opened by unscrewing the assembly 305 g (1·7 moles) of additional antimony trifluoride and 1044 g (4 moles) of perchlorobutadiene are added, the bomb is heated during 30 minutes in a steam-bath to 95–100°C, inserted into the heating mantle of the rocking shaker and heated while rocking two hours at 155°C. During the operations the pressure rises up to 6 atm. After cooling the bomb is opened, the assembly is replaced by another one consisting of a water-cooled descending condenser (Fig. 3. p. 29) connected to two 2-litre suction flasks half-filled with water and followed by a dry ice trap. The bomb is heated with a free flame up to 200°C while practically all of the organic material is distilled into the suction flasks. Their contents are combined, steam-distilled, the organic layer is separated, dried with calcium chloride (820 g, 3·7 moles) and distilled to give 680 g (2·9 moles) (73%) of pure 2,3-dichlorohexa-fluoro-2-butene, b.p. 66–68°C at 750 mm.

The individual compound $SbF_3Cl_2$, prepared by the reaction of equivalent amounts of antimony trifluoride and chlorine, is used for fluorinations of unreactive halogen derivatives such as tetrachloroethylene or hexachloroethane (621).

The complex compound $SbF_3Cl_2.2HF$ was dealt with in the paragraph on catalytic fluorination with hydrogen fluoride, since the hydrogen fluoride bound in the compound participates in the fluorinations (p. 92).

*Antimony pentafluoride* can not only replace halogens as in more profound fluorination of polychloropolyfluoroheptenes (675):

$$\textbf{99} \qquad C_7H_2Cl_6F_6 \xrightarrow[70°]{SbF_5} C_7H_2ClF_{11} \quad 83\% \qquad\qquad [675]$$

but—in contrast to other antimony fluorides — also add fluorine across double bonds, as in the preparation of 1,2-dichlorooctafluorocyclohexene from hexachlorobenzene (678) (Equation 33, p. 70).

## Replacement of Halogens by Fluorine by Means of Silver Fluorides

Primary monohalides which are, except for negligible instances, inert towards the fluorinating action of antimony fluorides can be converted to the corresponding fluoro-derivatives by means of *silver monofluoride* (998):

$$\textbf{100} \qquad C_8H_{17}I \xrightarrow[100°]{AgF, \ sand} C_8H_{17}F + C_6H_{13}CH{=}CH_2 \qquad\qquad [998]$$

Silver monofluoride forms molecular compounds with the other silver halides produced in the reaction with organic halides. Such mixed salts contain and immobilize one equivalent of silver fluoride, so that at least a 100% excess of silver fluoride over the stoichiometric amount is necessary (998). The results of this type of fluorination are not very encouraging. In older literature experimental details are lacking, and recent papers about the use of silver fluoride are very rare, since it has been displaced by more convenient reagents.

The reaction conditions for the exchange of halogen by means of silver fluoride are relatively mild, so that this reaction may be used even for sensitive compounds such as unsaturated halogenoesters (822) or halogenated ketones (1015):

**101**  Br—$CH_2$—CH=CH—$COOCH_3$  $\xrightarrow[\text{reflux}]{\text{AgF, (C}_2\text{H}_5)_2\text{O}}$  F—$CH_2$—CH=CH—$COOCH_3$  17% [822]

**102**

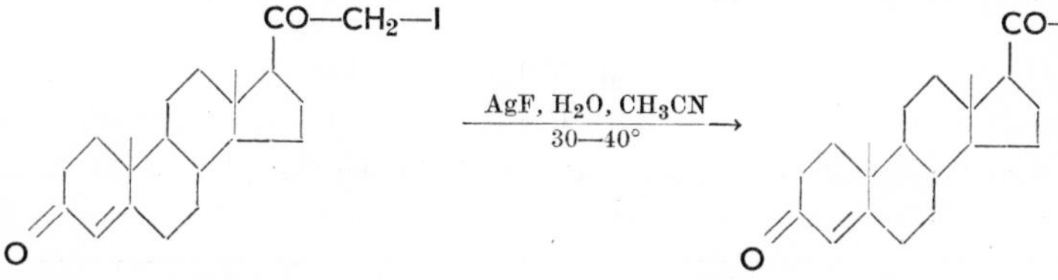

$\xrightarrow[30-40°]{\text{AgF, H}_2\text{O, CH}_3\text{CN}}$  63%  [1015]

Excellent yields were obtained with bis($\beta$-chloroethyl)sulphide  (687):

**103**  [687]

Cl—$CH_2$—$CH_2$—S—$CH_2$—$CH_2$—Cl  $\xrightarrow[40-50°,\ 2\ \text{hrs.}]{\text{AgF}}$  F—$CH_2$—$CH_2$—S—$CH_2$—$CH_2$—F  84%

The reaction of silver fluoride with $\alpha,\alpha,\alpha$-tribromo-$\alpha',\alpha',\alpha'$-trifluoroacetone gives, according to the reaction conditions, either dibromotetrafluoroacetone, or its mixture with bromopentafluoroacetone (937):

**104**  $\xleftarrow[\ \downarrow\ ]{\dfrac{\text{AgF}}{125°}}$  $CBr_3$—CO—$CF_3$  $\xrightarrow[\text{reflux 10 hrs.}\ \downarrow]{\text{AgF, acetone}}$  [937]

$CBr_2F$—CO—$CF_3$          $CBr_2F$—CO—$CF_3$ + $CBrF_2$—CO—$CF_3$

81%                 38%        46%

In 1,2-dibromo-1,1-difluoroethane the bromine in the group —$CH_2Br$ reacts preferentially with silver fluoride; the bromine atom bound to the carbon carrying the two fluorine atoms is not replaced by fluorine (993):

**105**       $CBrF_2$—$CH_2Br$  $\xrightarrow[120°]{\text{AgF}}$  $CBrF_2$—$CH_2F$  [993]

Contrary to the older literature data (163), tetrachloroethylene does not react with silver fluoride even when heated to 300°C in a bomb (108).

*Silver difluoride* is much more reactive than the monofluoride. It can replace any number of halogens located in any place in the organic molecule.

7*

It was successfully used for the conversion of chlorofluoroheptenes and chlorofluoroheptanes to perfluoroheptane and polyfluoroheptanes with a high content of fluorine (646).

$$
\textbf{106} \qquad \begin{matrix} C_7H_xCl_2F_{10} \\ x = 0 \text{ or } 2 \end{matrix} \xrightarrow[175-325°]{AgF_2} 90\% \left\{ \begin{matrix} C_7F_{16} & 15\% \\ C_7ClF_{15} & 45\% \\ C_7Cl_2F_{14} & 30\% \\ C_7Cl_3F_{13} & 5\% \end{matrix} \right. \qquad [646]
$$

In this respect silver difluoride is even more efficient than the similarly acting *cobalt trifluoride*, as can be seen from Table 18 (646).

TABLE 18

*Comparison of the Results of Replacement of Chlorine by Fluorine in the Reaction of Polychloropolyfluoroheptenes with Silver Difluoride and Cobalt Trifluoride (646)*

| Conditions and Results | Reaction with $AgF_2$ | Reaction with $CoF_3$ |
|---|---|---|
| Starting Material, Cl Content, % ................... | 38·7 | 38·7 |
| F Content, % ................... | 36·7 | 36·7 |
| Product, Cl Content, % ................... | 15·9 | 26·2 |
| F Content, % ................... | 63·9 | 56·1 |
| Yield, % ......................................... | 96 | 96 |
| Product Composition, % .......................... | | |
| $C_7F_{16}$ ................... | 8 | 5 |
| $C_7ClF_{15}$ ................. | 35 | 8 |
| $C_7Cl_2F_{14}$ ................. | 35 | 25 |
| $C_7Cl_3F_{13}$ ................. | 20 | 40 |

## Replacement of Halogens by Fluorine by Means of Mercury Fluorides

Mercury fluorides are among the most important fluorinating reagents for replacement of chlorine by fluorine.

*Mercurous fluoride* is more efficient for the replacement of halogens than silver fluoride. The reaction is promoted by the addition of iodine. Mercurous fluoride reacts with amyl bromide at 100°C to yield amyl fluoride and only a small amount of amylene, while the olefin prevails in the reaction with silver fluoride (998). Heptyl bromide was transformed to heptyl fluoride in a fair yield (998):

$$
\textbf{107} \qquad C_7H_{15}Br \xrightarrow[100°]{HgF, I_2} C_7H_{15}F \qquad 67\% \qquad [998]
$$

Secondary bromides such as 2-octyl bromide (*998*) or cyclohexylbromide (*1002*) give considerable amounts of olefins as by-products:

108    ⟨H⟩—Br $\xrightarrow[130-140°]{HgF}$ ⟨H⟩—F + ⟨⟩ + polymers    [*1002*]

                         30%      31%    12%

secondary bromine atoms seem to be replaced better than primaries (*1006*)

109                                                 [*1006*]

$CH_2Br—CHBr—CH_2—O—C_2H_5$ $\xrightarrow[200°, 1\,day]{\substack{HgF \\ 160-170°, 1\,day}}$ $CH_2Br—CHF—CH_2—O—C_2H_5$   18·5%

In geminal polybromides and polyiodides, only one halogen atom is usually replaced by fluorine (*475*).

The efficacy of mercurous fluoride is increased by the addition of large amounts of calcium fluoride. Heating a mixture of 39·4 parts (by weight) of iodoform, 87·6 parts of mercurous fluoride, and 131 parts of calcium fluoride gave 9·8% fluorodiiodomethane, 28·9% of difluoroiodomethane and 24·8% of fluoroform (*895*). A similar mixture with chloroform yielded 57% of dichlorofluoromethane at 100°C, 48% of chlorodifluoromethane at 250°C and 10% of fluoroform at 340°C (*895*).

Addition of an equivalent amount of halogen to mercurous fluoride produces *mercuric fluorochloride* or *fluoroiodide*, which are much more potent fluorinating reagents than mercurous fluoride (*475, 1004*). These reagents convert methyl iodide smoothly to methyl fluoride (*1004*):

110          $CH_3I$ $\xrightarrow{HgF,I_2}$ $CH_3F$   80%         [*1004*]

In geminal polybromoderivatives mercuric fluoroiodide can replace even both atoms of bromine by fluorine (*475*):

111          $CHBr_2—CH_2Br$ $\xrightarrow{HgF,I_2}$ $CHF_2—CH_2F$       [*475*]

A very important fluorinating agent is *mercuric fluoride*. It reacts with aliphatic monobromides at temperatures of 0–100°C in yields ranging from 80–90% (*461*):

112          $C_2H_5Br$ $\xrightarrow{HgF_2}$ $C_2H_5F$   quant.       [*461*]

Iododerivatives react with mercuric fluoride so vigorously that their dilution with chloroform or methylene chloride is advisable (*461*). In polybromoderivatives even three bromine atoms attached to the same carbon can all be replaced by fluorine with mercuric fluoride (427, 461, 472).

113    $CHBr_2-CHBr_2 \xrightarrow[150-160°]{HgF_2} CHBr_2-CHF_2$  quant.    [472]

114    $CHBr_3 \xrightarrow{HgF_2} CHF_3 \xleftarrow{HgF_2} CHI_3$    [427,461]

$HgF_2 \bigg| 50°, 7-8$ atm.

$CHBrF_2$

Polychloroderivatives are less reactive. Usually only two of the halogens are replaced (430, 461, 472):

115    $CH_2Cl-CHCl_2 \xrightarrow[140°]{HgF_2} CH_2Cl-CHClF + CH_2Cl-CHF_2$    [472]

50%            8—10%

116    $CHCl_3 \xrightarrow{HgF_2} CHClF_2$    [430]

Procedure 10.

Replacement of Halogen by Fluorine by Means of Mercuric Fluoride
Preparation of Methyl α-Fluoro-α,β-dibromopropionate from Methyl α,α,β-Tribromo-propionate (446)*

A half-litre three-necked flask fitted with a gas-tight stirrer, a thermometer and a water-jacketed descending condenser connected to a vacuum adapter and receiver, is charged with 225 g (0·72 mole) of methyl α,α,β-tribromopropinate and 70 g (0·29 mole) of mercuric fluoride. The apparatus is evacuated to 18–22 mm and the flask is heated in an oil bath with stirring as long as any organic material distils. The temperature in the flask is gradually raised from 140°C to 170°C. The distillate is diluted with an equal volume of ether, the solution is shaken with 100 ml. of 10% hydrochloric acid, washed with water, and the organic layer is dried and fractionated in a column in vacuo. The fraction distilling at 85–91°C at 11·5 mm (29 g, 0·115 mole) represents a 16% yield of methyl α-fluoro-α,β-dibromopropionate, $n_D^{20}$ 1·4892. In addition, 128 g (0·41 mole) (57%) of the starting methyl α,α,β-tribromopropionate is recovered. The net yield of the pure product is 37% based on the consumed starting material.

Practically the same results were obtained, using an equivalent amount of mercurous fluoride with half an equivalent of iodine added.

Evidently, a group of two halogens attached to one carbon atom reacts in preference to a single halogen atom (equation 115) (procedure 10) unless this

---

* Carried out by the author.

halogen atom is activated in some way, like in $\alpha$-haloethers whose $\alpha$-chlorine is replaced by fluorine under very mild conditions (*697*):

117    $CCl_3{-}CHCl{-}O{-}CH_3 \xrightarrow[10{-}20°]{HgF_2} CCl_3{-}CHF{-}O{-}CH_3$    42%    [*697*]

Instead of using ready-made mercuric fluoride, the preparation of which is somewhat tedious, methods have been worked out which prepare mercuric fluoride *in situ* from anhydrous hydrogen fluoride and mercuric chloride (*866*),

118

[*866*]

or mercuric oxide (*430, 443*):

119    $CCl_3{-}CF_2{-}CH_2Cl \xrightarrow[135°, \ 6{-}8 \ hrs.]{HgO, HF} CClF_2{-}CF_2{-}CH_2Cl$    50%    [*443*]

*Procedure 11.*

*Replacement of Halogen by Fluorine by Means of Mercuric Oxide and Hydrogen Fluoride*

*Preparation of 1-Chloro-2,2-difluoropropane from 1,2,2-Trichloropropane (443)*

A one-litre steel bomb is charged with 217 g (1 mole) of red or yellow mercuric oxide, cooled with dry ice, and filled with 200–240 g (10–12 moles) of anhydrous hydrogen fluoride. Thereafter 147·5 g (1 mole) of 1,2,2-trichloropropane cooled with dry ice is added, the bomb is closed with an assembly bearing a manometer and a needle valve, strapped to a rocking machine and allowed to come to room temperature while rocking. After about three hours the bomb is cooled with ice, the gaseous products are bled off through a dilute solution of sodium hydroxide, the bomb is opened, its contents poured over crushed ice, the organic layer is separated, washed with water and sodium hydroxide, and steam-distilled. The organic material is separated, dried and distilled to give 57 g (0·5 mole) (50%) of 1-chloro-2,2-difluoropropane, b.p. 55°C.

## Replacement of Halogens by Fluorine by Means of Halogen Fluorides

Replacement of halogen atoms by fluorine can be effected by means of chlorine trifluoride, bromine trifluoride, and iodine pentafluoride.

The action of *chlorine trifluoride* and bromine trifluoride is complicated by the ability of these compounds to add fluorine across double bonds and to replace hydrogen atoms by fluorine and chlorine or bromine, respectively. In the reaction of chlorine trifluoride with 1,2,3,4-tetrachloro-1,3-butadiene, both addition and replacement of hydrogen by fluorine and chlorine take place simultaneously (*756*).

The reaction of *bromine trifluoride* with hexachlorobenzene leads to a complicated mixture of halogenated products with various content of fluorine (*652*). On the other hand, the reaction of bromine trifluoride with carbon tetrabromide is clean-cut and gives excellent yields of bromotrifluoromethane (*30*):

120 $\qquad$ $CBr_4 \xrightarrow[\text{heating in autoclave}]{BrF_3} CBrF_3$ $\quad$ 94% $\qquad\qquad$ [*30*]

An analogous reaction takes place when *iodine pentafluoride* reacts with carbon tetraiodide (*30, 354*).

121 $\qquad$ $CI_4 \xrightarrow[90—100°]{IF_5} CIF_3$ $\quad$ 73—95% $\qquad\qquad$ [*30,354*]

Another example of the fluorinating ability of iodine pentafluoride is the conversion of tetrafluoro-1,2-diiodoethane to perfluoroiodoethane (*380*):

122 $\quad$ $CF_2I—CF_2I \xrightarrow[90—100°]{IF_5} CF_3—CF_2I \xleftarrow[0° \rightarrow 130°]{IF_5} CF_2=CF_2$ $\qquad$ [*30,380*]
$\qquad\qquad\qquad\qquad\qquad$ 85% $\quad$ 26%

The action of iodine pentafluoride is not simply fluorination. The reaction is far more complicated and probably iodine monofluoride is produced, as can be concluded from the reaction of iodine pentafluoride with tetrafluoroethylene (*30*) (Reaction 122).

## Replacement of Halogens by Fluorine by Means of Alkaline Fluorides

Of alkaline fluorides, potassium fluoride is most frequently used for the replacement of halogen atoms by fluorine. The other alkaline fluorides do not find so much use since they are less efficient and in some instances even less available. The efficiency of the alkali fluorides in respect to the replacement reaction drops in the sequence (*558, 1056, 1068*):

$$CsF > RbF > KF > NH_4F, NaF > LiF$$

The replacement reaction is easiest with a $-OSO_2-$group (p. 113). Next are halogens in the sequence $I > Br > Cl$, and finally organic groups bound by means of an oxygen atom such as $-OCOCH_3$, or $-OC_6H_5$. The ease of the replacement by fluorine decreases in the series: tert. > sec. > prim. bond (*558*).

*Potassium fluoride* has a very broad spectrum of applications. It is used for replacing reactive chlorine in sulphonyl chlorides (*1048*).

123 ⟨◯⟩$-CH=CH-SO_2-Cl \xrightarrow[\text{reflux}]{KF, \text{xylene}}$ ⟨◯⟩$-CH=CH-SO_2-F$ 51.4% [*1048*]

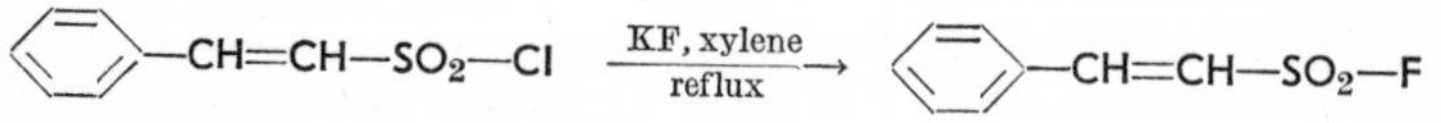

This reaction is so easy that it can be accomplished already by an aqueous (70%) solution of potassium fluoride (*201*):

124 $\qquad$ $C_2H_5—SO_2—Cl \xrightarrow[\text{reflux 20 min.}]{70\% \text{ KF aq.}} C_2H_5—SO_2—F$ $\quad$ 66.8% $\qquad$ [*201*]

Similarly acid fluorides are prepared from acid chlorides by heating with potassium fluoride. The starting material can be either ready-made acid chlorides (*767*),

125 $\qquad$ $CH_3$—CO—Cl $\xrightarrow[100°]{KF,CH_3COOH}$ $CH_3$—CO—F 76% $\qquad$ [*767*]

or free carboxylic acids, which are transformed to the corresponding chlorides in situ by heating with benzoyl chloride (*767, 788*). By this procedure even the fluoride of formic acid was successfully prepared, despite the fact that formyl chloride has not yet been isolated (*767, 788*):

126 $\qquad$ HCOOH + $C_6H_5$COCl $\xrightarrow[100°]{KF}$ HCO—F 16% $\qquad$ [*767*]

The reactive chlorine atom in chloroformates can also be replaced by the action of potassium fluoride (*788*):

127 $\qquad$ Cl—COOC$_2$H$_5$ $\xrightarrow[h\nu,\ 50-60°]{KF,CH_3COCH_2COCH_3}$ F—COOC$_2$H$_5$ 51·2% $\qquad$ [*788*]

The main interest has been concentrated on the preparation of derivatives of fluoroacetic acid from the corresponding halogen derivatives. The exchange reaction started with chloroacetonitrile (*559*), chloroacetamide (*25*),

128 $\qquad$ Cl—CH$_2$—CO—NH$_2$ $\qquad$ F—CH$_2$—CO—NH$_2$ $\qquad$ [*25*]

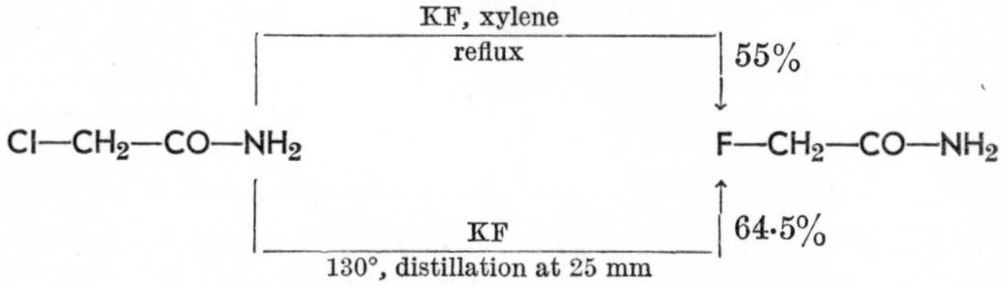

KF, xylene reflux — 55%

KF, 130°, distillation at 25 mm — 64·5%

ethyl bromoacetate (*25*), and most frequently with various esters of chloro-acetic acid (*84, 100, 334, 336, 777, 788, 912*).

129 $\qquad$ [*25, 84, 336*]

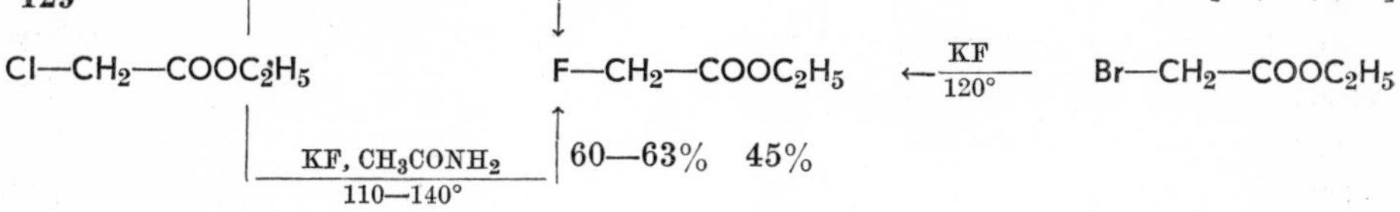

Cl—CH$_2$—COOC$_2$H$_5$ $\qquad$ F—CH$_2$—COOC$_2$H$_5$ $\xleftarrow[120°]{KF}$ Br—CH$_2$—COOC$_2$H$_5$

KF, 150—250° — 75%

KF, CH$_3$CONH$_2$, 110—140° — 60—63%   45%

130 $\qquad$ [*334*]

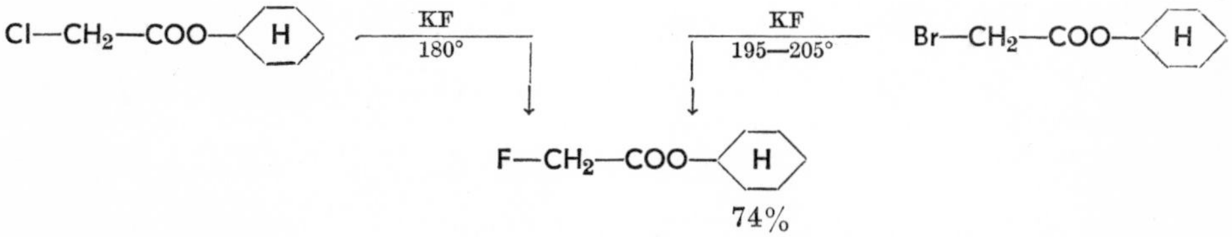

Cl—CH$_2$—COO—⟨ H ⟩ $\xrightarrow[180°]{KF}$ $\qquad$ $\xleftarrow[195-205°]{KF}$ Br—CH$_2$—COO—⟨ H ⟩

F—CH$_2$—COO—⟨ H ⟩

74%

*Procedure 12.*

*Replacement of Chlorine by Fluorine by Means of Potassium Fluoride*
*Preparation of Ethyl Fluoroacetate from Ethyl Chloroacetate (84)\**

A two-litre three-necked flask, fitted with a gas-tight stirrer and a column with a water-cooled reflux head, is charged with 475 g (8 moles) of acetamide and 720 g (5·9 moles) of ethyl chloroacetate, the flask is heated in a bath to 110°C, and 475 g (8·2 moles) of finely powdered dried potassium fluoride is added portionwise to the mixture while stirring vigorously. After the addition has been completed, the temperature is raised to 140°C. After about 15 minutes the crude product starts to distil at a reflux ratio set to 10:1. Two fractions are collected, the lower one distilling over the range of 76–115·5°C (245 g), and the pure ethyl fluoroacetate, boiling at 115·5–117·5°C (210 g). Redistillation of both fractions increases the yield of the pure compound, distilling over the range of 2°C, to 325 g (3 moles) (52%).

Potassium fluoride is further able to exchange halogen for fluorine in benzyl chloride in a yield of 20–28% (*558*) and in chlorobenzenes having nitro groups in *ortho-* and *para*-positions almost quantitatively (*276, 1057*):

The last reaction applies also to chloro-derivatives of pyridine in which the chlorine atom is activated by an *ortho-* or *para*-nitro group. Thus 2-chloro-3-nitropyridine and 2-chloro-5-nitropyridine give the corresponding 2-fluoro-3-nitropyridine (76%) and 2-fluoro-5-nitropyridine (78%) by heating with potassium fluoride in dimethylformamide (*280*). Unsubstituted α-halopyridines do not exchange their halogen for fluorine under comparable conditions (*280*).

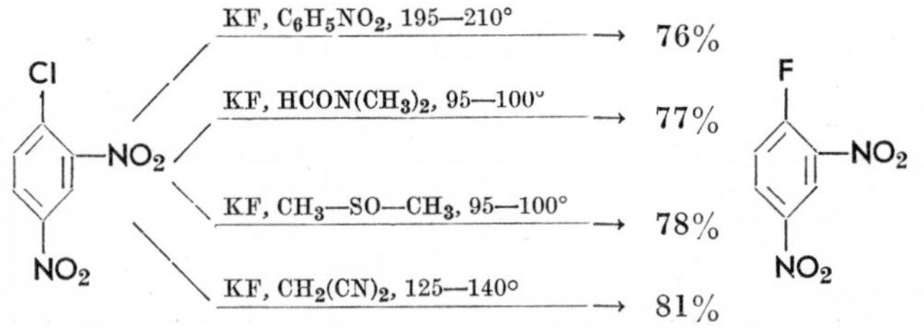

Heating of chloranil with potassium fluoride at higher temperatures results in successive replacement of the chlorine atoms by fluorine, so that fluoranil may be the final product (*916, 1067, 1068*):

**132**                                                                                 [*916, 1067*]

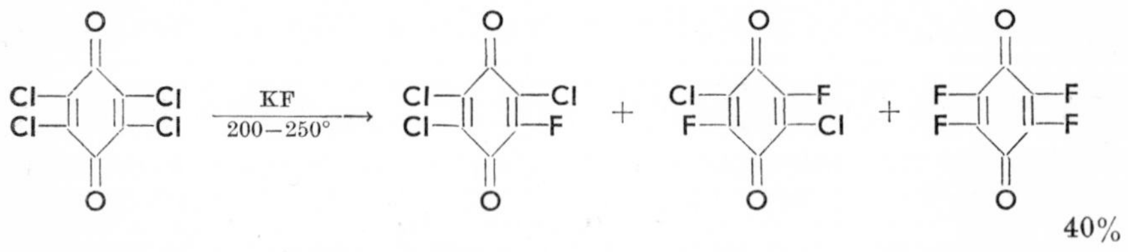

40%

---

\* Carried out by J. Salák.

Similar treatment of dichloromaleic anhydride leads to a mixture of chlorofluoromaleic anhydride with difluoromaleic anhydride (*1067*).

133
$$\underset{Cl-C-CO}{\overset{Cl-C-CO}{\Big\|}}\!\!\!\!>\!\!O \quad \xrightarrow[200-250°]{KF} \quad \underset{F-C-CO}{\overset{Cl-C-CO}{\Big\|}}\!\!\!\!>\!\!O \;+\; \underset{F-C-CO}{\overset{F-C-CO}{\Big\|}}\!\!\!\!>\!\!O \qquad [1067]$$

It is only in recent years that potassium fluoride has been successfully applied to replacing poorly reactive halogen atoms in aliphatic compounds. The clue to this problem is the application of suitable solvents such as acetamide (*84*), nitrobenzene (*276*), dimethylsulphoxide (*276*), ethylene glycol and diethylene glycol (*500, 558, 817*). Whereas the yields of the reaction with potasssium fluoride without the solvent seldom exceeded 20%, the use of ethylene glycol and other solvents raised the yields up to 30–70%, and sometimes more than that. By this improvement potassium fluoride competes successfully with silver monofluoride and mercuric fluoride (*819*) (p. 98,101).

134
$$Br(CH_2)_6COOC_2H_5 \quad \overset{\displaystyle \nearrow \; \overline{\underset{60°}{AgF}} \; \searrow}{\underset{\searrow \; \underset{130°,\,8\ hrs.}{KF,\ HOCH_2CH_2OH} \; \nearrow}{}} \quad \begin{matrix} 34\cdot1\% \\ F(CH_2)_6COOC_2H_5 \\ 40\cdot3\% \end{matrix} \qquad [819]$$

In order to obtain maximum yields, pure and absolutely dry chemicals must be used. The reaction is best carried out by heating one mole of the halogen derivative with a 100% excess of potassium fluoride and approximately 700 g of diethylene glycol at a temperature of 125±5°C with vigorous stirring. The addition of potassium iodide to the reaction mixture does not increase the yields (*817*).

135
$$C_6H_{13}Cl \quad \begin{matrix} \overline{\underset{200°}{KF}} \qquad 20\% \\ \big| \qquad\qquad \big| \\ C_6H_{13}F \\ \underline{\underset{175-185°}{KF,\ (CH_2OH)_2,\ (HOCH_2CH_2)_2O}} \quad 54\cdot1\% \end{matrix} \qquad [336,\ 500]$$

The reaction with potassium fluoride is not limited to the exchange of single halogen atoms. Recently several instances have been reported in which even two chlorine atoms bound to the same carbon were replaced by fluorine. Methyl dichloroacetate was thus converted to methyl difluoroacetate (*336*),

136
$$CHCl_2-COOCH_3 \quad \xrightarrow[220-230°,\ 25\ hrs.]{KF} \quad CHF_2-COOCH_3 \quad 18\% \qquad [336]$$

methylene chloride (296), chloroform (793) and carbon tetrachloride (793) to the derivatives containing up to two atoms of fluorine:

137    $CH_2Cl_2$    $\xrightarrow[\text{180—200°, 2 hrs.}]{\text{KF, NaF, HOCH}_2\text{CH}_2\text{OH}}$    $CH_2ClF + CH_2F_2$    [296]

　　　　　　　　　　　　　　　　　　　　　　　　　　　　19,7%　　17,2%

Even if the last quoted examples lack practical value, they demonstrate greater versatility of potassium fluoride than was believed.

Very important is the reaction of potassium fluoride with polyhalogen derivatives that have their halogens attached to several carbon atoms. Depending on the amount of the fluorinating reagent and the reaction conditions, α,ω-dihalogen paraffins can be transformed to mixed halogen-fluorides, or to difluorides (501, 502). The yields of such reactions are listed in Table 19.

TABLE 19

*The Results of the Reaction Between α,ω-Dihaloparaffins and Potassium Fluoride: a) Without Solvent (502), b) With Ethylene Glycol as Solvent (501)*

| Starting Material | Method | Temp. °C | Fluoro-halide % | Difluoride % |
|---|---|---|---|---|
| $Br(CH_2)_2Br$ | a | 90 | 24 | — |
|  | b | 150 | 23·7 | — |
| $Br(CH_2)_3Br$ | a | 80—100 | 31 | 10·2 |
|  | b | 150 | 6·2 | 23·3 |
| $Br(CH_2)_4Br$ | a | 100—110 | 19·6 | 23·4 |
| $Br(CH_2)_5Br$ | a | 100—110 | 31·4 | 25 |
| $Cl(CH_2)_4Cl$ | a | 110 | 36·6 | 15·9 |
|  | a | 125 | 34·2 | 29·3 |
|  | b | 170 | 6·6 | 44·2 |
| $Cl(CH_2)_5Cl$ | a | 125 | 22 | — |
|  | a | 130 | 38·5 | 14·1 |
|  | b | 180 | 9·4 | 36·6 |
| $Cl(CH_2)_6Cl$ | b | 185 | — | 30 |

By using an excess of potassium fluoride, all the four bromine atoms in tetrabromoneopentane were replaced by fluorine (335):

138    $\underset{BrCH_2}{\overset{BrCH_2}{>}}C\underset{CH_2Br}{\overset{CH_2Br}{<}}$    $\xrightarrow[\text{175—180°, 200—210°}]{\text{KF, (HOCH}_2\text{CH}_2)_2\text{O}}$    $\underset{FCH_2}{\overset{FCH_2}{>}}C\underset{CH_2F}{\overset{CH_2F}{<}}$    57—60%    [335]

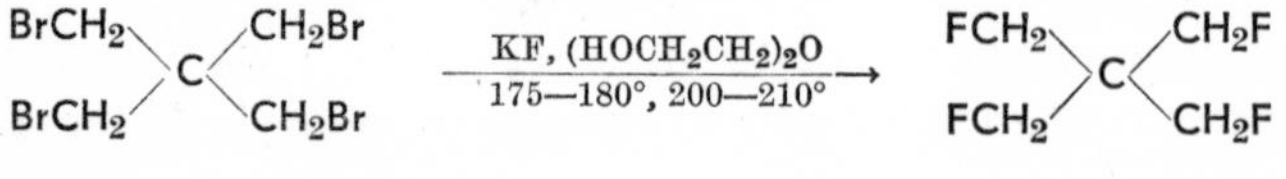

Fluorohydrins, important intermediates for the synthesis of fluorinated compounds, are also prepared by means of potassium fluoride. Ethylene fluorohydrin, which could not be prepared from ethylene bromohydrin and silver or mercuric fluoride (*995*), was obtained in fair yields by treatment of ethylene chlorohydrin with potassium fluoride (*560, 789, 914*):

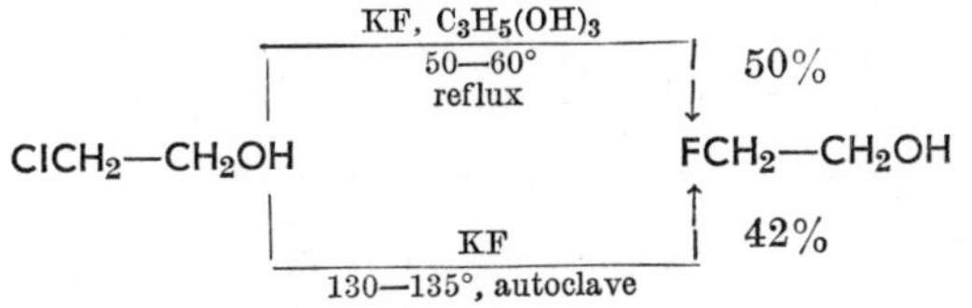

An interesting compound is formed from potassium fluoride and gaseous or liquid sulfur dioxide. It is formulated as $KF.0.85\ SO_2$ or sometimes $KSO_2F$ and is able to exchange its fluorine for the halogen atoms of phosphorus chlorides, arsenic chlorides, sulphochlorides, acid chlorides (*931*) and some other compounds with sufficiently reactive halogens (*326*).

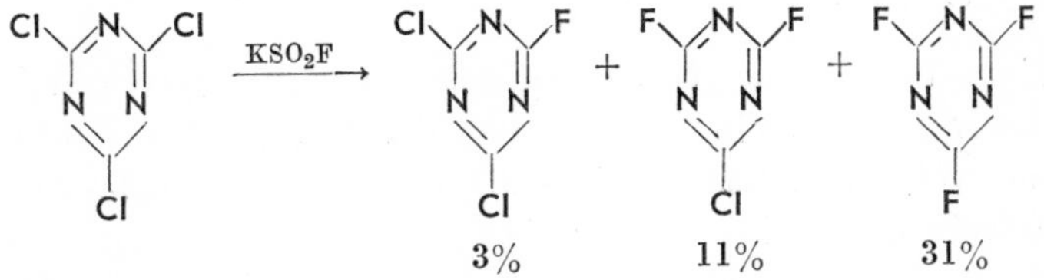

The use of *acid potassium fluoride* for fluorination is very limited. One of its applications is the conversion of sulphonyl chlorides to sulphonyl fluorides which is accomplished by an aqueous solution of the reagent (*321, 736*):

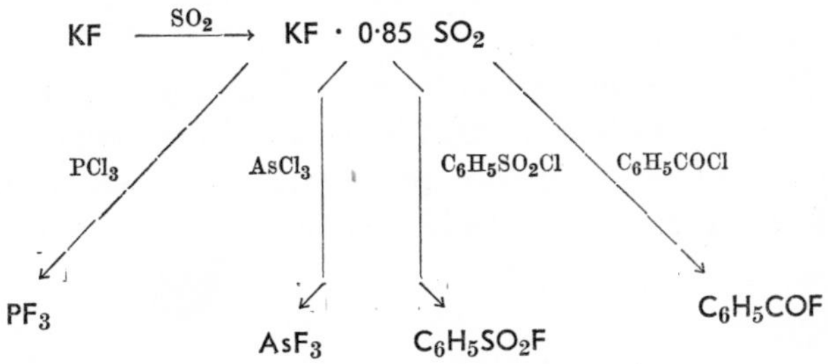

142    $FCH_2—CH_2—SO_2Cl$  $\xrightarrow[20° \rightarrow 70°]{KHF_2\ aq.}$  $FCH_2—CH_2—SO_2F$    45%    [*736*]

A somewhat more important application is the reaction between a carboxylic acid, benzoyl chloride and anhydrous acid potassium fluoride, which leads to acid fluorides (*786*). The yields of this reaction are listed in Table 20.

TABLE 20

*Yields of the Reaction of Carboxylic Acids RCOOH with Benzoyl Chloride and Acid Potassium Fluoride after Heating at 100°C for 1 hour (573)*

| R in RCOOH | H | CH$_3$ | C$_2$H$_5$ | C$_3$H$_7$ | iso-C$_3$H$_7$ | C$_4$H$_9$ | iso-C$_4$H$_9$ | C$_5$H$_{11}$ |
|---|---|---|---|---|---|---|---|---|
| Yield of RCOF, % | 35·4 | 81·6 | 81·0 | 80·0 | 79·5 | 66·5 | 64·5 | 67 |

The extent of applications for *sodium fluoride* is also very narrow. This compound is convenient for the preparation of ester fluorides of phosphoric acid (*911, 1040*),

**143**

but it failed entirely in the reaction with ethylene chlorohydrin (*914*) and with methyl chloroacetate (*912*).

One very peculiar reaction has been accomplished by means of sodium fluoride in formamide. It is the transformation of l-chloroperfluoro-2-propene to perfluoropropene (*720, 729*). The reaction is not a simple replacement of the halogen, but the result of a whole sequence of combined addition and elimination reactions:

**144**                                                                      [*729*]

$$CF_2{=}CF{-}CF_2Cl \xrightarrow[\substack{HCONH_2 \\ 35°, 24 \text{ hrs.}}]{NaF} [CF_3{-}\bar{C}F{-}CF_2Cl \xrightarrow{-[Cl]^{\ominus}} CF_3{-}\bar{C}F{-}\overset{\oplus}{C}F_2] \longrightarrow CF_3{-}CF{=}CF_2$$

*Ammonium fluoride* may be applied only to very reactive halides such as chlorophosfates (*1040*), silicon tetrachloride, phosphorus pentachloride or phosphorus oxychloride. Even with such compounds the replacement reaction affords relatively low yields (*1087*):

**145**          $POCl_3 \xrightarrow[\text{reflux 8 hrs.}]{NH_4F} POCl_2F + POClF_2 + POF_3$          [*1087*]

                                    12%        6%        38%

*Lithium fluoride* transformed glycerol chlorohydrin to the fluorohydrin in a fair yield (*792*):

**146**   $CH_2OH{-}CHOH{-}CH_2Cl \xrightarrow[140{-}160°, \, h\nu]{LiF} CH_2OH{-}CHOH{-}CH_2F$  52%   [*792*]

## Replacement of Halogens by Fluorine by Means of Other Fluorides

*Thallium fluoride* was used for the conversion of an acyl chloride to the acyl fluoride (*337*) and, with little success, to transform a bromoacetate to the fluoroacetate (*25*). However, it is important for the elegant preparation of alkyl fluorides, involving the decomposition of alkyl fluoroformates, which are obtained from the corresponding aliphatic or alicyclic chloroformates (*762*).

147 $\qquad$ Cl—COOR $\xrightarrow{\text{TlF}}$ F—COOR 51—67% $\qquad$ [*762*]

*Zinc fluoride* converted chlorosilanes to fluorosilanes under mild conditions and with fair yields (*244, 770*):

148 $\qquad$ $C_2H_5SiCl_3 \xrightarrow[55°]{ZnF_2} C_2H_5SiF_3$ 60% $\qquad$ [*244*]

It is, however, unsuitable for the replacement of chlorine bound to a carbon atom (*914, 928*).

*Cobalt trifluoride* can replace even very unreactive halogen atoms. In the successive preparation of polyfluoroheptanes, the mixed chlorofluoroheptenes were subjected to final drastic fluorination with silver difluoride or cobalt trifluoride, during which the remaining halogen atoms were largely replaced by fluorine (*646*) (Table 18, p. 100).

An interesting fluorination was carried out by means of *sodium fluoroborate*, which reacted with triphenylsilyl chloride to give the corresponding triphenyl-fluorosilane (*607*):

$(C_6H_5)_3SiCl$ + $NaBF_4$ + $CH_3COCH_3$ $\xrightarrow[\text{2 hrs.}]{20°}$ $(C_6H_5)_3SiF$ + NaCl + $BF_3\cdot CH_3COCH_3$

149 $\qquad\qquad\qquad\qquad\qquad\qquad$ 54—66% $\qquad$ 77% $\qquad\qquad$ [*607*]

Most recently another complex fluoride, *sodium fluorosilicate*, was tried as a fluorinating agent for the production of chlorofluoromethanes and ethanes from the corresponding chloro-derivatives (*191*).

150 $\qquad\qquad\qquad\qquad\qquad\qquad\qquad\qquad\qquad\qquad\qquad\qquad$ [*191*]

$CCl_3$—$CCl_3$ $\xrightarrow[\substack{280°, 35\ atm.\\ 2\ hrs.}]{Na_2SiF_6}$ $CCl_3$—$CCl_2F$ + $CCl_2F$—$CCl_2F$ + $CClF_2$—$CCl_2F$ + $CCl_2$=$CCl_2$

$\qquad\qquad\qquad\qquad\qquad\qquad$ 21·8% $\qquad\qquad$ 31% $\qquad\qquad$ 2·7% $\qquad\qquad$ 28·2%

The reaction of sodium fluorosilicate with benzotrichloride affords a mixture, in which the corresponding difluoro-derivative prevails over the mono-fluoro-derivative and benzotrifluoride (*191*). Phthaloyl chloride was converted by sodium fluorosilicate to phthaloyl fluoride in a good yield (*191*).

151 $\qquad\qquad\qquad\qquad\qquad\qquad\qquad\qquad\qquad\qquad\qquad\qquad\qquad$ [*191*]

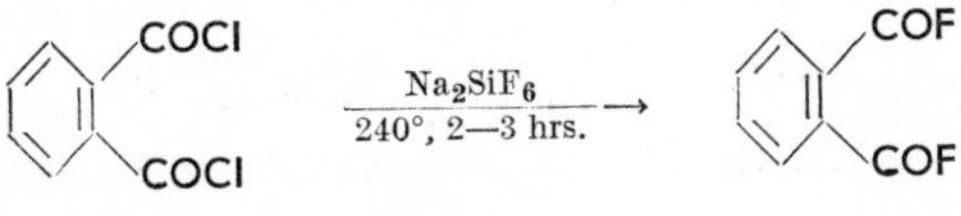

$\qquad\qquad\qquad\qquad$ $\xrightarrow[\text{240°, 2—3 hrs.}]{Na_2SiF_6}$ $\qquad\qquad\qquad$ 51·5%

Some organic fluorides are able to replace halogen atoms in other organic compounds by fluorine. *Benzotrifluoride* can substitute fluorine for chlorine in aliphatic alkyl chlorides (*1085*). Since benzotrifluoride is prepared very easily, this reaction may be of preparative importance.

152    $3 R—Cl +$ ⬡$—CF_3$ ⟶ $3 R—F +$ ⬡$—CCl_3$    [*1085*]

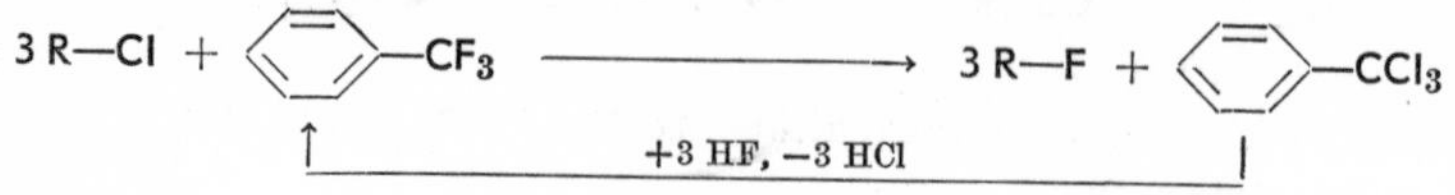

$+3 HF, -3 HCl$

Another organic fluoro-derivative, *trifluoromethyl hypofluorite*, converts chloroform to a mixture of dichlorodifluoromethane, chlorotrifluoromethane and carbon tetrafluoride with a trace of trichlorofluoromethane (*3*).

## REPLACEMENT OF OXYGEN BY FLUORINE

Compared with the substitution of fluorine for halogen, the substitution for oxygen is far less common. However, thanks to the discovery of new reactions in this field, the importance of this method for introducing fluorine into organic molecules is rapidly increasing.

Replacement of oxygen by two atoms of fluorine takes place in the interaction of *aromatic iodosoderivatives* with 40% hydrofluoric acid (*1071*). The iodide difluorides thus formed act themselves as fluorinating reagents, since fluorine is bound to the iodine atom very loosely (p. 84).

153    ⬡$—IO$ $\xrightarrow[\text{heating}]{40\% \text{ HF}}$ ⬡$—IF_2$    [*1071*]

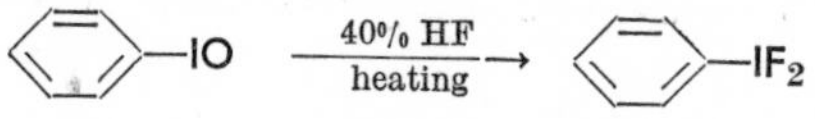

*Carbonyl oxygen* can be replaced by fluorine by means of a new fluorinating reagent, *sulphur tetrafluoride*. By virtue of this reaction aldehydes and ketones can be converted to geminal difluoroderivatives (*962, 963, 964*).

154    $CH_3—CO—CH_3$ $\xrightarrow[110°, 16 \text{ hrs.}]{SF_4}$ $CH_3—CF_2—CH_3$    63.5%    [*962*]

*The carboxyl* in carboxylic acids is replaced by trifluoromethyl group (*962, 964*):

155    $C_{11}H_{23}COOH$ $\xrightarrow[\substack{130°, \text{autogenous} \\ \text{pressure}}]{SF_4}$ $C_{11}H_{23}CF_3$    88%    [*964*]

Fluorine can substitute *alkoxy groups* in siloxanes. The reaction is accomplished by means of aqueous hydrofluoric acid (*691*).

156    $\begin{matrix} C_4H_9 \\ C_4H_9 \end{matrix}$Si$\begin{matrix} OC_2H_5 \\ OC_2H_5 \end{matrix}$ $\xrightarrow{48\% \text{ HF}}$ $\begin{matrix} C_4H_9 \\ C_4H_9 \end{matrix}$Si$\begin{matrix} F \\ F \end{matrix}$    [*691*]

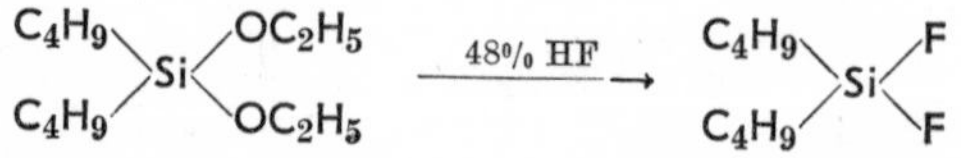

It is generally accepted that the direct conversion of *alcohols to fluorides* is not practical. However, some preparations were described as successful *(704)*. Treatment of 3,5-cyclocholestan-6β-ol with aqueous hydrofluoric acid produced—under a rearrangement—3β-fluoro-5-cholestene *(940)*.

157      $\xrightarrow[20°, 4 \text{ hrs.}]{40\% \text{ HF}}$      75%      *[940]*

In order to obtain fluorides from alcohols, the alcohols must be converted to the esters of sulphuric acid *(869)*, methanesulphonic acid *(820)*, benzenesulphonic acid *(869)* or *p*-toluenesulphonic acid *(231)*, and these esters react then with potassium fluoride alone, or dissolved in diethylene glycol or other suitable solvents *(87, 231, 820)*.

158      $(ClCH_2-CH_2-O)_2SO_2 \xrightarrow[175°]{KF} ClCH_2-CH_2F$    60%      *[869]*

159      $R-OSO_2-\langle\!\!\!\bigcirc\!\!\!\rangle-CH_3 \xrightarrow[250°, 500 \text{ mm}]{KF} R-F$      *[231]*

| R | R—F% |
|---|---|
| $CH_3$ | 89·4 |
| $C_2H_5$ | 85·6 |
| $C_3H_7$ | 63·3 |
| *iso*-$C_3H_7$ | 50·7 |

*Procedure 13.*

  *Replacement of Tosyl Group by Fluorine*
  *Preparation of Ethyl α-Fluoropropionate from Ethyl Tosyllactate and Potassium Fluoride (87)**

In a half-litre three-necked flask equipped with a thermometer, a gas-tight mechanical stirrer and a water-cooled condenser for distillation fitted with a vacuum receiver, a mixture of 133 g (0·49 mole) of ethyl tosyllactate, 42 g (0·73 mole) of finely powdered dry potassium fluoride and 100 ml. of diethylene glycol is heated 30 minutes at 100–120°C under a pressure of 25–30 mm and with vigorous stirring. The crude product which distils to the receiver gives by an additional distillation 31·6 g (0·26 mole) (54%) of ethyl α-fluoropropionate boiling over a range of 119–122°C.

The conversion of an alcohol to the sulphonic ester and that of the ester to the fluoride may be carried out in one step provided the alcohol is allowed

---

  * Carried out by J. Moural.

to react with a sulphonyl fluoride and potassium fluoride and the reaction mixture heated, after a certain time, with diethylene glycol to a high temperature (820):

**160**    $ClCH_2—CH_2OH + CH_3—SO_2F + KF$ $\xrightarrow[\substack{2.\ (HOCH_2CH_2)_2O \\ 100°,\ 300\ mm}]{1.\ 20°,\ 24\ hr}$ $ClCH_2—CH_2F$    53%    [820]

## REPLACEMENT OF THE AMINO GROUP BY FLUORINE

The replacement of an amino group by fluorine is of enormous importance in the aromatic series, since it is practically the only feasible way for preparing aromatic derivatives containing fluorine attached to the nucleus.

A rare application of this reaction in the aliphatic series is the conversion of ε-caprolactam to ε-fluorocaproic acid (776):

**161**                                                                                          [776]

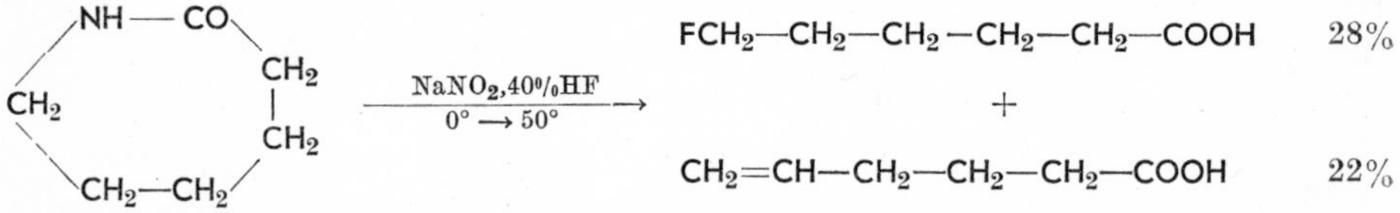

$FCH_2—CH_2—CH_2—CH_2—CH_2—COOH$    28%

$+$

$CH_2{=}CH—CH_2—CH_2—CH_2—COOH$    22%

Aromatic amines are first diazotized to diazonium salts and these are decomposed by a suitable method. The simplest modification, the diazotization of the amine in aqueous concentrated (60–70%) hydrofluoric acid, does not give good yields (331, 994). It is much more suitable to carry out this reaction by treating an amine with anhydrous sodium nitrite in anhydrous hydrogen fluoride, and then heating the resulting mixture (271).

**162**    $NH_2$—⟨ ⟩—$COOH$ $\xrightarrow[0°\ \to\ 100°]{NaNO_2,\ HF}$ $F$—⟨ ⟩—$COOH$    98%    [271]

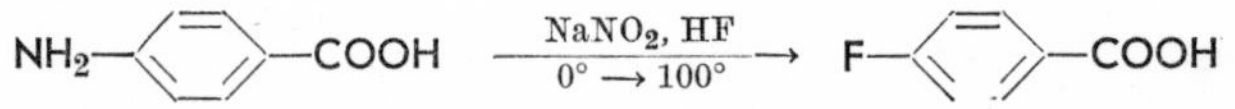

This modification is applicable also to sulphonamides but not to the amides of carboxylic acids (271).

**163**    $CH_3$—⟨ ⟩—$SO_2NH_2$ $\xrightarrow{NaNO_2,\ HF}$ $CH_3$—⟨ ⟩—$SO_2F$    [271]

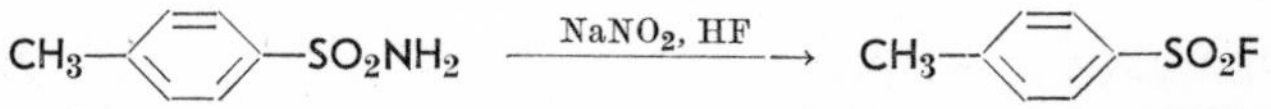

An obsolete way of carrying out the replacement of the amino group is stabilization of the diazonium salt by its reaction with piperidine, followed by the decomposition of the diazoaminocompound formed (1065, 1066).

**164**                                                                                          [1065,1066]

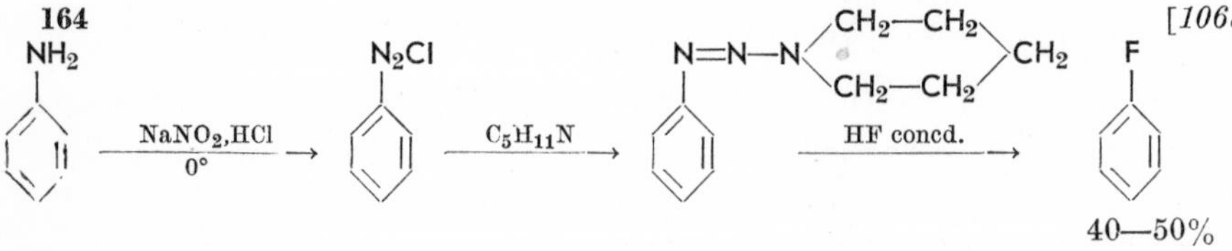

40—50%

The most generally used way of substituting fluorine for the amino group is the Schiemann reaction, which is based on the precipitation of slightly soluble diazonium fluoroborate, and on thermal decomposition of this dry diazonium salt. In the original modification, the diazonium fluoroborate was obtained by precipitation with 40% aqueous fluoroboric acid (*28*). Nowadays

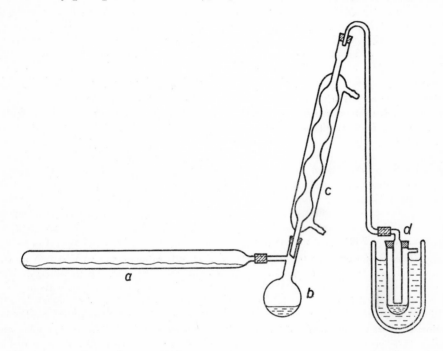

FIG. 17. Apparatus for the Schiemann reaction: a) reaction tube with diazonium fluoroborate, b) receiver, c) reflux condenser, d) dry-ice trap.

precipitation with sodium or ammonium fluoroborate is preferred (*884*). The precipitated diazonium fluoroborate is dried and decomposed by carefully heating in a suitable apparatus (Fig. 17). When diazonium fluoroborates of nitrocompounds are to be decomposed, it is advisable to dilute the dry salt with an inert material such as sand, sodium fluoride, or barium sulphate in order to prevent too vigorous a decomposition (*884*).

**165**

$NH_2 \xrightarrow[\text{2. } 40\% HBF_4]{\text{1. NaNO}_2,\ HCl} N_2BF_4 \xrightarrow{\text{heating}} F$ [*28*]

63%        97%

*Procedure 14.*

*The Schiemann Reaction—Replacement of Amino Group by Fluorine*

*Preparation of Fluorobenzene from Aniline\**

The preparation of fluorobenzene includes the preparation of sodium fluoroborate (a), benzene diazonium fluoroborate (b) and its thermal decomposition (c).

(a) Boric acid (125 g, 2 moles) is added in small portions and under continuous stirring with a copper rod to 400 g (8 moles) of 40% aqueous hydrofluoric acid placed in a copper dish cooled with an ice-salt mixture. The reaction mixture is allowed to stand overnight at room temperature and then neutralized to methyl orange with 5 N sodium hydroxide, while cooling with salt and ice.

(b) To 370 g (3·76 moles) of commercial hydrochloric acid placed in a 5-litre diazotization beaker cooled with an ice-salt mixture, is added 120 g (1·3 moles) of aniline in portions and with mechanical stirring. After the reaction mixture has been cooled to −5°C, a solution of 91 g (1·3 moles) of sodium nitrite in 360 ml. of water is added dropwise. During the diazotization the temperature should not exceed +5°C. After the diazotization is over the mixture is cooled to −10°C and treated all at once with the fore-prepared solution of sodium fluoroborate cooled to 0°C. The thick foam of the benzenediazonium fluoroborate is stirred for an additional 30 minutes, then the precipitate is filtered with suction and washed successively with ice-cold water and alcohol. The crystalline mass is crushed to small pieces, spread over a paper and allowed to dry thoroughly. The yield of the dry benzenediazonium fluoroborate, with decomposition point of 121°C, is approximately 230 g (1·2 moles) (92%).

(c) The decomposition of the dry salt is advantageously carried out in the apparatus shown in Fig. 17. The decomposition tube is filled with the fluoroborate, the reflux condenser supplied with cold water, and the Dewar flask charged with dry ice. The reaction vessel is heated locally to promote a spontaneous reaction, which is self-sustaining and propagates throughout the whole mass. After the spontaneous decomposition is over, the contents of the reaction tube are heated briefly to complete the reaction. The crude fluorobenzene collected in the distilling flask of the apparatus is washed with four-millilitre portions of 10% sodium hydroxide, then with water, dried with calcium chloride and distilled. Approximately 80 g (0·83 mole) (64%) of pure fluorobenzene distilling over the range of 84–85°C is obtained.

During the introduction of additional fluorine atoms into the molecule of fluorobenzene by the Schiemann reaction, replacement of fluorine by chlorine sometimes takes place, especially when the diazonium group to be exchanged for the fluorine atom is in *ortho*-position to any of the fluorine atom already bound to the ring (nucleophilic replacement of fluorine by chlorine from the reaction medium) *(277)*. To a lesser extent the chlorine derivative is formed during the thermal decomposition, owing to the presence of some sodium chloride which has been co-precipitated with the diazonium fluoroborate:

---

\* Carried out by E. Nušlovà.

The reaction of compound **166** (2,4-difluoroaniline):

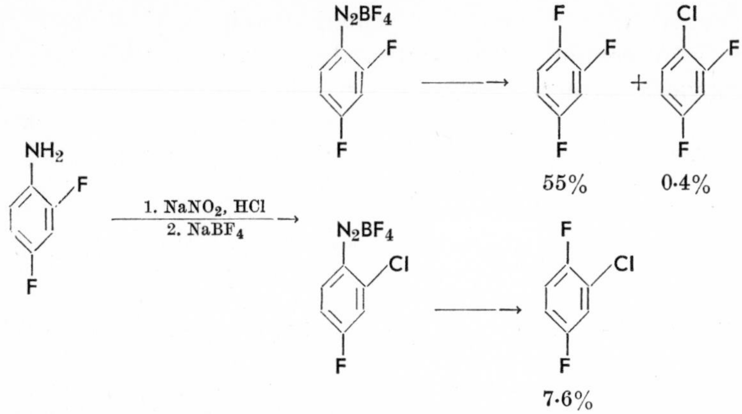

**166**

1. NaNO$_2$, HCl
2. NaBF$_4$

gives the diazonium fluoroborates (N$_2$BF$_4$) which decompose to fluorinated products:

55%   0.4%   [*277*]

7.6%

Some compounds are claimed to give better yields of the fluorinated derivatives, when the decomposition of the diazonium fluoroborates is carried out in liquid phase, in the medium of acetone and in the presence of catalytic amounts of copper powder or cuprous chloride (*83*):

**167**

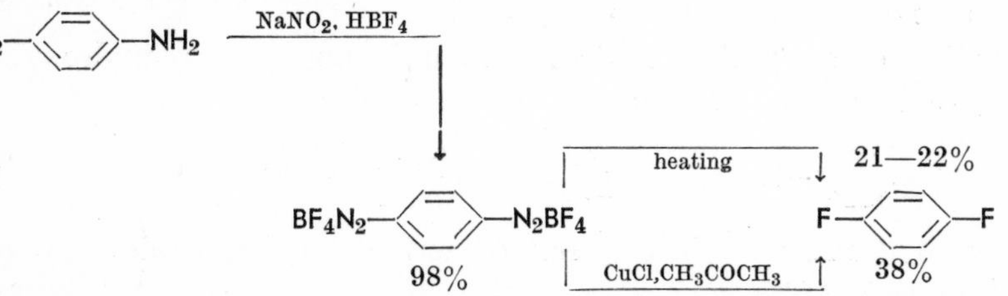

NH$_2$—C$_6$H$_4$—NH$_2$ $\xrightarrow{\text{NaNO}_2,\ \text{HBF}_4}$ BF$_4$N$_2$—C$_6$H$_4$—N$_2$BF$_4$  98%

heating → 21—22%   [*83*]

CuCl, CH$_3$COCH$_3$ → F—C$_6$H$_4$—F  38%

On the other hand, some results obtained by this modification, could not be duplicated (*34*). Thus the decomposition of *o*-nitrobenzenediazonium fluoroborate in acetone in the presence of cuprous chloride gave only nitrobenzene in a yield of 85%, and the decomposition of *p*-carboxybenzenediazonium fluoroborate yielded, under comparable conditions, a mixture of 33% of benzoic acid and 16% of *p*-chlorobenzoic acid (*34*).

Another modification of the method is the immediate decomposition of the diazonium fluoroborate in the reaction mixture in which it has been formed (*82*):

**168**                                                                 [*82*]

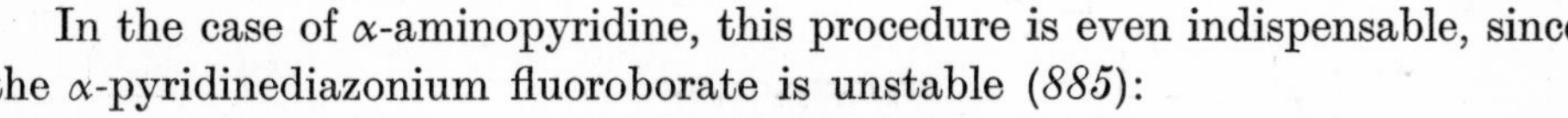

CH$_3$—CO—NH—C$_6$H$_4$—NH$_2$ $\xrightarrow[\text{2. Cu powder}]{\text{1. NaNO}_2,\ \text{HBF}_4}$ CH$_3$—CO—NH—C$_6$H$_4$—F   82%

In the case of $\alpha$-aminopyridine, this procedure is even indispensable, since the $\alpha$-pyridinediazonium fluoroborate is unstable (*885*):

**169**

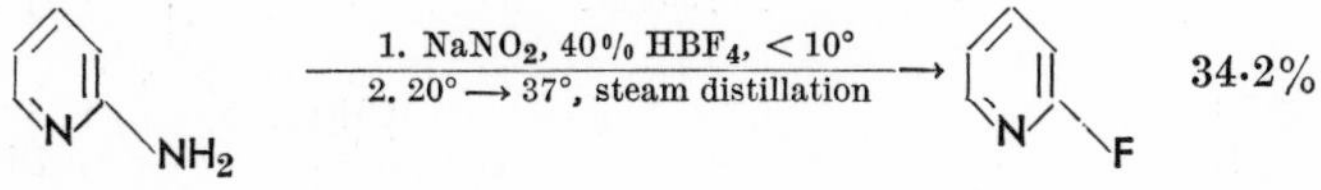

$\xrightarrow[\text{2. }20° \rightarrow 37°,\ \text{steam distillation}]{\text{1. NaNO}_2,\ 40\%\ \text{HBF}_4,\ <10°}$   34.2%   [*885*]

An alternative way of preparing aromatic fluorine derivatives is the decomposition of diazonium fluorosilicates. However, cases where this process gives better yields than the Schiemann reaction are exceptional (53, 169):

**170**    [169]

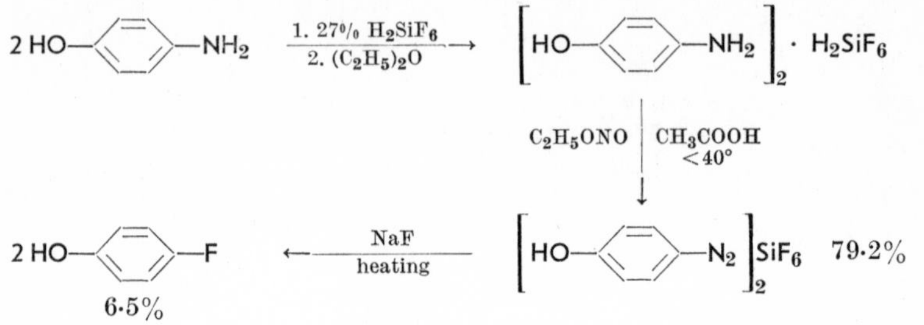

6·5%

## OTHER METHODS FOR THE PREPARATION OF ORGANIC FLUORINE DERIVATIVES

As special methods for preparing fluorinated compounds, only the following ones will be mentioned in this paragraph:

The action of anhydrous hydrogen fluoride upon phenylhydroxylamine produces p-fluoroaniline (1045):

**171**       45%    [1045]

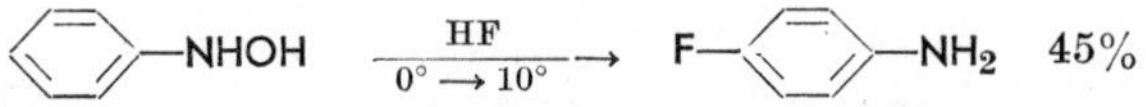

Aqueous hydrofluoric acid converts trimethylbenzylammonium hydroxide to benzyl fluoride (527).

**172**    60%    [527]

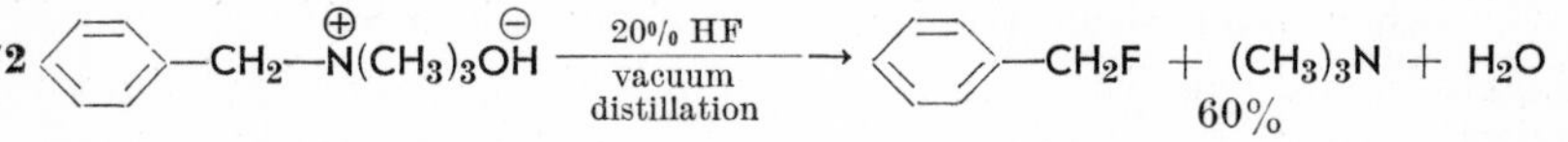

The three-membered ring in ethylene oxides (oxiranes) is broken by anhydrous hydrogen fluoride and ethylene fluorohydrins are formed in fair yields (571). This reaction is of great importance in the steroid field. Another method for opening the oxirane rings is a reaction of an epoxide with the complex formed by boron trifluoride and ether, $BF_3 \cdot (C_2H_5)_2O$, in benzene and at room temperature (125, 424). 3β-acetoxy-5β, 6β-oxidocholestane was thus converted to 3β-acetoxy-5α-fluoro-6β-cholestanol in 62% yield (424).

**173**    [424]

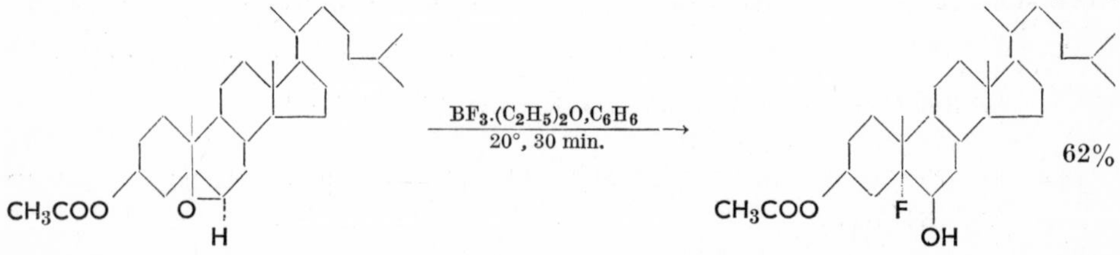

62%

Anhydrous hydrogen fluoride also opens the three-membered ring of 1,1-dichlorocyclopropane and splits it to a mixture of 1-chloro-1,1-difluoropropane and 1,1,1-trifluoropropane (*639*);

174
$$\underset{CH_2}{\overset{CCl_2-CH_2}{\diagdown\diagup}} \xrightarrow[130°,\ 56\ atm]{HF} CClF_2-CH_2-CH_3 + CF_3-CH_2-CH_3 \qquad [639]$$
$$\qquad\qquad\qquad\qquad\qquad\qquad\qquad 25\% \qquad\qquad 47\%$$

Chlorination of carbon in the presence of hydrogen fluoride leads to a mixture of all possible halogen derivatives of methane (*752*),

175  C $\xrightarrow[\substack{500°,\ 3\ hrs \\ autogenous \\ pressure}]{Cl_2,\ HF}$ $CCl_3F$ + $CCl_2F_2$ + $CClF_3$ + $CF_4$ + haloethanes [*752*]

8 parts                6·1%       18·3%      47%      12·8%        4·6%

7 parts

the reaction of silicon carbide with fluorine leads to a mixture of carbon tetrafluoride and silicon tetrafluoride (*853*). Treatment of various carbides with elemental fluorine or cobalt trifluoride produces varying proportions of various fluorocarbons, among which carbon tetrafluoride always prevails (*927*) (Table 21);

TABLE 21

*Results of Fluorination of Metal Carbides with Fluorine at 20°C for 90 minutes, or with Cobalt Trifluoride at 440° for 9 hours (927)*

| Metal Carbide | $CF_4$, % | $C_2F_6$, % | $C_3F_8$, % | $C_4F_{10}$, % |
|---|---|---|---|---|
| SiC | 68 | 16 | 14 | 2·2 |
| $B_4C$ | 88 | 9·2 | 2·4 | 0·5 |
| TiC | 86 | 11 | 2·9 | 0·4 |
| WC | 67 | 26 | 7 | 0·6 |
| $Fe_3C$ | 86 | 9·5 | 3·5 | 0·7 |
| $Cr_2C_3$ | 91 | 6·4 | 2·4 | 0·5 |
| $Al_4C_3$ | 88 | 9·2 | 2·9 | — |
| $CaC_2$ | 75 | 19 | 5·5 | 0·6 |

High-temperature fluorination of activated charcoal, in the presence of mercury as catalyst, gives a mixture of fluorocarbons containing one to six carbon atoms (*951*).

From the preparative point of view, an elegant method for the production of alkyl fluorides is based on the decomposition of alkyl fluoroformates (p.111). The decomposition is effected either by heating of the alkyl fluoroformates in pyridine at 100–110°C (762), or in the presence of boron trifluoride-ether complex at 0–50°C (763).

**176**     R—OCOF

$$\begin{array}{c} \overset{C_5H_5N,\ 100—110°}{\nearrow} \quad 58—75\% \\ \underset{\underset{0—50°}{BF_3 \cdot (C_2H_5)_2O}}{\searrow} \quad 36—86\% \end{array} \qquad R—F + CO_2 \qquad [762,763]$$

The results of the preparation of alkyl fluorides by this method are shown in Table 22.

TABLE 22

*Preparation of Alkyl Fluorides by the Decomposition of Alkyl Fluoroformates Obtained from Alkyl Chloroformate and Thallium Fluoride (762, 763)*

| Alkyl | $C_2H_5$ | iso-$C_3H_7$ | sec-$C_4H_9$ | cyclo-$C_6H_{11}$ |
|---|---|---|---|---|
| Yield of Alkyl Fluoroformate, % | 58 | 51 | 64 | 67 |
| Yield of Alkyl Fluoride by Decomposition in the Presence of: | | | | |
| Pyridine............ | 63 | 75 | 61 | 58 |
| $BF_3 \cdot (C_2H_5)_2O$ ........ | 86 | 82 | — | 36 |

*Procedure 15.*

*Preparation of Alkyl Fluorides by the Decomposition of Alkyl Fluoroformates (762)*

(a) *Preparation of Alkyl Fluoroformates.* A chloroformate is treated at room temperature with an equivalent amount of thallium fluoride added with stirring in small portions over several hours. The stirring is continued for an additional 12 hrs., the mixture is then heated for 30 min. on the steam bath, the product is vacuum distilled from the mixture and purified by redistillation. The pure alkyl fluoroformates are obtained in yields of 50–65%.

(b) *Decomposition of the Alkyl Fluoroformates.* The decomposition of the alkyl fluoroformates is carried out in a flask fitted with a reflux condenser, which is connected, by means of two tubes containing soda lime, to a dry ice trap. In a nitrogen atmosphere the alkyl fluoroformate is diluted with an equivalent amount of dry pyridine at room temperature. The white precipitate which is formed at the start, redissolves readily and the solution becomes red and finally brown. The flask is slowly heated in an oil bath to 100–110°C and kept at this temperature for several hours as long as some product collects in the dry ice receiver. The content of the receiver is washed with water, dried with calcium chloride, and redistilled to give 60–75% of the alkyl fluoride.

Of a similar nature is the reaction by which bis($\beta$-fluoroethyl)sulphide was obtained by treatment of thiodiglycol chloroformate with hydrogen fluoride (554):

$$S(CH_2—CH_2OH)_2 \xrightarrow{COCl_2} S(CH_2—CH_2—OCOCl)_2$$

**177**

$$HF \downarrow \begin{array}{c} -80° \rightarrow\ 0° \\ 20° \rightarrow 75° \end{array} \qquad\qquad\qquad [554]$$

$$S(CH_2—CH_2F)_2 + FCH_2—CH_2—S—CH_2—CH_2Cl$$

# V

# PREPARATION OF ORGANIC COMPOUNDS OF FLUORINE

In the preceding chapter, covering the methods for introducing fluorine into organic molecules, attention was directed toward the experimental conditions, scope and limitation of various fluorinating reactions. The purpose of the present chapter will be to select the fluorinating processes which are most suitable for the preparation of individual types. Exceptions and peculiarities will be omitted, and only such practical procedures, which are actually in use in the laboratory or in industry, will be stressed. Usually only the first members of the homologous series will be dealt with in detail and the higher members will be mentioned only when their preparations deviate from the regular one, or when they are of particular interest.

In the practical preparation of fluorinated organic compounds two types of methods will be represented: introduction of fluorine into a non-fluorinated molecule, and conversion of a fluorinated compound to another fluorinated compound. Only the former will be described in detail. The latter will be mentioned only briefly here and reference will be made to the next chapter on "Reactions of Organic Fluorine Compounds" where such preparations will be discussed more thoroughly (p. 157).

The individual types of organic compounds will be arranged in the following order: fluoroparaffins, fluoroolefins, fluorocycloparaffins and fluorocyclo-olefins, fluoroacetylenes, fluorinated aromatic systems, fluorinated hydroxy-derivatives and ethers, fluorinated aldehydes and ketones, fluorinated acids and their derivatives, fluorinated compounds of sulfur, fluorinated amines and other nitrogen derivatives, and fluorinated derivatives of silicon, phosphorus and other elements. In all these types the corresponding halogen derivatives, i.e. compounds containing chlorine, bromine and iodine, will also be considered

## PREPARATION OF FLUOROPARAFFINS AND THEIR HALOGEN DERIVATIVES

All fluorinated derivatives of methane can be prepared by fluorination of methane with elemental fluorine (*340*); in practice, however, each derivative s produced in a specific way. *Methyl fluoride* was obtained by treatment of methyl iodide with mercurous fluoride and iodine (*1004*) (Equation 110, p.101) or by the reaction of methyl *p*-toluenesulphonate with potassium fluoride (*231*) (equation 159, p.113). The same reaction was used for the preparation of *trideuteromethyl fluoride* (*232*):

$$\textbf{178} \qquad CD_3OH \xrightarrow{\text{p-C}_7\text{H}_7\text{SO}_2\text{Cl}} CD_3OSO_2\!\!-\!\!\langle\!=\!\rangle\!\!-\!\!CH_3 \xrightarrow{\text{KF}} CD_3F \qquad [232]$$

$$\qquad\qquad\qquad\qquad\qquad\qquad\qquad\quad 82\% \qquad\qquad\qquad\qquad\quad 80\%$$

*Methylene fluoride* results from the action of antimony fluorides upon any dihalogeno-derivative of methane (*428*). *Fluoroform* was obtained by a two-step fluorination of bromoform, first with antimony fluoride to the stage of bromodifluoromethane, then with mercuric fluoride (*427*) (Equations 93 and 114, p. 96,102). The same compound was prepared in one step by heating chloroform with a complex $SbF_3Cl_2.2\,HF$ (*1081*). *Carbon tetrafluoride* appears in better than 50% yield in the fluorination of activated charcoal (*951*), of carbon tetrachloride (*953*) (Equation 63, p. 88), or in the fluorination of various carbides (*927*), especially silicon carbide (*853*), where the silicon tetrafluoride formed simultaneously is washed out with water. As final product of exhaustive fluorination, carbon tetrafluoride often accompanies other fluorinated compounds in most fluorination processes (*175, 294, 376,*) (Equations 39, 62, 246, p. 72,86,144).

Fluorinated halogen derivatives of methane are important refrigerants called *Freons, Frigens* etc. (p. 23). Their manufacturing is based on the successive replacement of chlorine atoms in chloroform or carbon tetrachloride. *Dichlorofluoromethane* and *chlorodifluoromethane* are prepared from chloroform and antimony trifluoride, activated by an addition of antimony pentachloride, (*107*) (Equation 92, p. 96), or more frequently by treatment of chloroform with anhydrous hydrogen fluoride with antimony chlorides as catalysts (*198*). Both procedures can be applied to the conversion of carbon tetrachloride to *trichlorofluoromethane* and *dichlorodifluoromethane* (*35, 133, 198, 199, 200, 539, 708*). The fluorination can be carried out in liquid as well as in vapour phase, by passing a mixture of carbon tetrachloride and anhydrous hydrogen fluoride over activated charcoal impregnated with catalysts such as copper chloride (*199*) or ferric chloride (*637*) (Table 16, 17, p. 90,91). The ratio of monofluoro- and difluoroderivatives during the fluorinations of chloroform and carbon tetrachloride is easily influenced by the amount of the fluorinating reagent and by the reaction conditions. By recycling the monofluoro-derivatives, practically complete conversion to difluoro-derivatives can be effected. Under the conditions mentioned the third halogen atom is replaced only to a small extent.

In addition to these classical methods for the preparation of fluoro- and difluorohalogenomethanes, some modern processes are under investigation, especially for the production of dichlorodifluoromethane. These are for example simultaneous action of chlorine and anhydrous hydrogen fluoride upon methane (*35*), the action of hydrogen fluoride upon phosgene (*374*),

179      $COCl_2 \xrightarrow[\text{425°}]{\text{HF, FeCl}_3\text{/C}} CCl_2F_2 + CClF_3 + CF_4$      [*374*]

or chlorinolysis of 1,1-difluoroethane obtained by the addition of hydrogen fluoride to acetylene (*35*).

180      $CH{\equiv}CH \xrightarrow{\text{2 HF}} CH_3{-}CHF_2 \xrightarrow[\text{550°}]{\text{Cl}_2} CCl_2F_2$      [*35*]

In spite of many new inventions in this field the old reaction, liquid phase fluorination by means of hydrogen fluoride and antimony fluorides, remains still the most convenient method for an industrial scale, especially when carried out under ordinary pressure and at relatively low temperatures (*200, 877*).

*Chlorotrifluoromethane* was prepared in a yield of 95% by treating carbon tetrafluoride with a complex compound $SbF_3Cl_2.2$ HF at 160°C and 70 atm. (*1081*).

Practical procedures for the preparation of fluorohalogenoderivatives of methane are given in the chapter on applications of these compounds as refrigerants (p. 335, 336).

As to brominated fluoromethanes, *bromodifluoromethane* was prepared by the reaction of bromoform with antimony trifluoride and bromine (*427*) (Equation 93, p. 96).

Similar treatment of carbon tetrabromide at 120–130°C afforded a 65–75% yield of *tribromofluoromethane* (*97*), an important intermediate for the synthesis of hexafluorobenzene (equation 232, p. 139).

*Bromotrifluoromethane* was obtained by bromination of fluoroform with bromine (*35*) or fluorination of carbon tetrabromide with bromine trifluoride (*30*) (Equation 120, p. 104) and *dibromodifluoromethane* either by fluorination of carbon tetrabromide with hydrogen fluoride, or by bromination of difluoromethane with bromine (*35*). The latter two compounds are produced on a large scale as fire extinguishers.

The preparation of *difluoroiodomethane* from iodoform (*953*) and *trifluoroiodomethane* from carbon tetraiodide (*354*), can be achieved by means of iodine pentafluoride:

181          $CI_4 \xrightarrow{\text{JF}_5} CIF_3$     95%      [*354*]

Fluorinated halogen derivatives of methane, containing varying number of various halogens, can be conveniently prepared by the reaction of halogens with silver salts of halogenated fluoroacetic acids (*364*). The following compounds were prepared in this way:

182      $CClBrF_2$, $CClBr_2F$, $CClIF_2$, $CCl_2BrF$, $CCl_2IF$, $CHClIF$, $CHBrIF$      [*364*]

Fluorinated derivatives of ethane and higher paraffins can be obtained not only by the replacement of hydrogen or halogen atoms by fluorine but also by the addition of hydrogen fluoride to olefins and acetylenes and addition of fluorine to some olefins.

*Ethyl fluoride* was prepared by treatment of ethyl bromide with mercuric fluoride (*461*) (Equation 112, p. 101), or by heating ethyl *p*-toluenesulphonate with potassium fluoride (*231*) (Equation 159, p. 113), or by the reaction of ethyl chloroformate with thallium fluoride and decomposition of the resulting fluoroformate (*762, 763*) (Equation 147, 176, p. 111, 120).

*Ethylidene fluoride* can be obtained either by the addition of hydrogen fluoride to acetylene (*328, 769*), or by the reaction of ethylidene chloride or ethylidene bromide with antimony trifluoride (*473*) or mercuric fluoride (*461, 473, 475*). *1,1,1-Trifluoroethane* was prepared by treatment of 1,1,1-trichloroethane with a mixture of antimony trifluoride and pentachloride (*688*) (Equation 95, p. 97), or by the reaction of anhydrous hydrogen fluoride with 1,1,1-trichloroethane or vinylidene chloride (*636*):

183    $CCl_3—CH_3 \xrightarrow[\substack{350—420\ atm. \\ 16—20\ hrs.}]{HF,\ 225°} CF_3—CH_3 \xrightarrow[\substack{350—420\ atm. \\ 16—20\ hrs.}]{HF,\ 225°} CCl_2{=}CH_2$    [*636*]
                                                              80%   90%

The reaction of ethylene bromide with potassium fluoride led only to *1-bromo-2-fluoroethane* (*336, 501, 502*) (Table 19, p. 108) whereas the reaction with mercuric fluoride gave a low yield of *1,2-difluoroethane* (*461*). The same compound was obtained in high yields by treatment of ethylene glycol bis-(*p*-toluenesulphonate) with potassium fluoride (*231*).

The starting material for the preparation of *1,1,2-trifluoroethane* is 1,1,2-tribromoethane. Since its reaction with antimony trifluoride (*991*) or mercuric fluoride (*461*) leads only to monofluoro- or geminal difluoro compound, the tribromoethane is first transformed to 1,1-dibromo-2-iodoethane by means of calcium iodide; the iodocompound undergoes full fluorination when heated with mercuric fluoride (*461*):

          $CH_2Br—CHBr_2 \xrightarrow[180°]{SbF_3,\ Br_2} CH_2Br—CHF_2$

184       $CaJ_2 \downarrow$                                    [*461,991*]

          $CH_2I—CHBr_2 \xrightarrow[160°]{HgF_2} CH_2F—CHF_2$
          60%                              60%

*1,1,2,2-Tetrafluoroethane* is obtained by successive replacement of halogen atoms by fluorine according to the scheme (*461*),

185
$CHBr_2—CHBr_2 \xrightarrow{HgF_2}$   $CHBr_2—CHBrF \xrightarrow[\substack{heating, \\ pressure}]{HgF_2}$   $CHBrF—CHF_2$   10%   [*461*]

                                    $CHBr_2—CHF_2$                  $CHF_2—CHF_2$   40—50%

*pentafluoroethane* by direct fluorination of ethane with elemental fluorine (*1098*).

186          $C_2H_6 + F_2 + N_2$ (1 : 4 : 16—80)  →  $CHF_2$—$CF_3$     30%          [*1098*]
volumes

The most convenient preparation of *hexafluoroethane* seems to be the reaction of ethane with elemental fluorine in a specially designed reactor (Fig. 4, p. 31); yields up to 83% of hexafluoroethane have been reached (*1051*).

Mixed fluorohalogenoderivatives of ethane are obtained by fluorination of the corresponding halogenated compounds (*461, 1035, 1081*),

187          $CHCl_2$—$CHCl_2$  $\xrightarrow[\text{90—170, 30 atm.}]{\text{HF/SbCl}_5}$  $CF_3$—$CH_2Cl$     80%          [*1035*]

or by halogenation of fluoroethanes (*640*),

188     $CF_3$—$CH_3$ $\xrightarrow[500°]{\text{Br}_2}$ $CF_3$—$CH_2Br$ + $CF_3$—$CHBr_2$ + $CF_3$—$CBr_3$ + $CF_2Br$—$CH_2Br$     [*640*]
                                      50%          7%          traces          traces

or by the addition of hydrogen fluoride to chlorinated ethylenes (*470*)

189          $CH_2$=$CCl_2$  $\xrightarrow[65°]{\text{HF}}$  $CH_3$—$CFCl_2$     50%          [*470*]

which may by accompanied by substitution of fluorine for chlorine, especially when antimony fluoride is present (*290*):

190          $CCl_2$=$CCl_2$  $\xrightarrow[150°]{\text{HF, SbF}_3\text{Cl}_2}$  $CHCl_2$—$CCl_2F$ + $CHCl_2$—$CClF_2$     [*290*]
                                      54%          35%

*1,1,1-Trifluoro-2-bromo-2-chloroethane* which is made either by chlorination of 1,1,1-trifluoro-2-bromoethane, or by bromination of 1,1,1-trifluoro-2-chloroethane (*976, 977*), is a newly favoured inhalation narcotic (*976*) (p. 356).

191          $CF_3$—$CH_2Cl$ $\xrightarrow[465°]{\text{Br}_2}$ $CF_3$—$CHClBr$ $\xleftarrow[427°]{\text{Cl}_2}$ $CF_3$—$CH_2Br$     [*976,977*]

*Tetrachloro-1,2-difluoroethane* is easily produced by the addition of fluorine to tetrachloroethylene. The use of elemental fluorine is avoided by treating the compound with anhydrous hydrogen fluoride and lead dioxide (*485*) (Equation 29, Procedure 3, p. 69).

A whole series of chlorofluoroethanes which do not contain hydrogen atoms results from the reaction of hexachloroethane with antimony fluorides (*111, 621*) or with anhydrous hydrogen fluoride in the presence of antimony pentachloride (*200, 538*). Instead of solid high-melting hexachloroethane, a mixture of liquid tetrachloroethylene and chlorine is used to avoid handling difficulties

(*200, 877*). Depending on the amount of fluorinating reagent and the experimental conditions, monofluoro-, difluoro-, trifluoro- and tetrafluoro-derivatives can be prepared. A more thorough discussion of this reaction will be made in chapter X (p. 337).

Another route to fluorinated halogen derivatives of ethane is the addition of halogens or hydrogen halides to fluorinated ethylenes. The procedure can be illustrated by the preparation of *1,2-diiodotetrafluoroethane* which can be further converted to pentafluoroiodoethane by treatment with iodine pentafluoride (*380*):

192        $CF_2{=}CF_2$ $\xrightarrow[150°, \text{ autoclave}]{I_2}$ $CF_2I{-}CF_2I$ $\xrightarrow[90-110°]{IF_5}$ $CF_3{-}CF_2I$        [*380*]

          76%                        85%

Fluorinated derivatives of propane and higher paraffins are prepared in ways similar to those used for fluorinated ethanes. Monofluoro-derivatives are obtained by the reaction of halogenoparaffins with mercurous (*998*) or better still mercuric fluoride (*461*), with potassium fluoride (*336*) (Equation 135, p. 107), by the reaction of alkyl *p*-toluenesulfonates with potassium fluoride (*231*) (Equation 159, p. 107), or by reaction of alkyl chloroformates with thallium fluoride followed by decomposition of the alkyl fluoroformates thus formed (*762, 763*) (Equations 147 and 176, p. 111, 120, Table 22, p. 120).

For the preparation of *geminal difluoro-derivatives* the following ways are good: the addition of hydrogen fluoride to acetylenes (*328, 469*) (Equation 23, Procedure 2, p. 67), the addition of hydrogen fluoride to halogen derivatives carrying one halogen atom attached to the double bond, and the simultaneous replacement of this halogen by fluorine (*440, 452*) (Equation 19, p. 65), the action of hydrogen fluoride or antimony fluorides upon geminal halogen derivatives, conveniently obtained from ketones (*440*):

193                                                                      [*440*]

$CH_3{-}CH_2{-}CO{-}CH_2{-}CH_3$ $\xrightarrow[12° \to 100°]{PCl_5}$ 
$\begin{array}{ll} CH_3{-}CH_2{-}CCl_2{-}CH_2{-}CH_3 & 14\% \\ CH_3{-}CH_2{-}CCl{=}CH{-}CH_3 & 73\% \end{array}$ $\longrightarrow$

$\xrightarrow[\to 60°]{HF}$ $CH_3{-}CH_2{-}CF_2{-}CH_2{-}CH_3$

                67—73%

and finally by direct action of sulphur tetrafluoride on aldehydes or ketones (*962*) (Equation 154, p. 112).

*Vicinal difluoro-derivatives* or fluorohalo-derivatives are produced by treating with mercuric fluoride vicinal halogenoparaffins (*461*).

*α,ω-Fluorohalo-derivatives* and *α,ω-difluoro-derivatives* can be prepared from the corresponding α,ω-dihalogenoparaffins by treatment with potassium fluoride (*501, 502*). α,ω-Difluoroparaffins containing an even number of methylene groups are obtained in the electrolysis of $\omega$-fluorocarboxylic acids in sodium methoxide (*823*):

**194** [*823*]

$$F(CH_2)_nCOOH \xrightarrow[\text{electrolysis}]{\text{Na, CH}_3\text{OH}} F(CH_2)_{2n}F$$

| $n$ | 4 | 5 | 6 | 9 |
|-----|-----|-----|--------|-----|
| Yield | 45% | 45% | 57,5% | 69% |

*Derivatives containing a terminal* $-CF_3$ *group* are produced either by the action of antimony fluorides upon 1,1,1-trichloroparaffins (*368, 482*) (Equations 80 and 81, p. 94), or by treatment of 1,1-dichloro-1-olefins with hydrogen fluoride (*639*),

**195** $\quad CCl_2=CH-CH_3 \xrightarrow[90-100°]{\text{HF}} CClF_2-CH_2-CH_3 + CF_3-CH_2-CH_3$ [*639*]
$$2·2\% \qquad\qquad 88·5\%$$

or by fluorination of a $-CCl_3$ group in 1,1,1-trichloro-2-olefins with a subsequent saturation of the double bond (*450*);

**196** $\quad CCl_2=CF-CCl_3 \xrightarrow[125°]{\text{SbF}_3}$
$$CCl_2=CF-CClF_2 \quad 27\%$$
$$CCl_2=CF-CF_3 \quad 37\% \qquad\qquad\qquad [450]$$
$$\Big\downarrow \xrightarrow[h\nu]{\text{Cl}_2} CCl_3-CFCl-CF_3 \quad \text{quant.}$$

or by the reaction of carboxylic acids with sulphur tetrafluoride (*962, 964*) (Equation 155, p. 112), or finally by the addition of trifluoroiodomethane to olefins (*370, 379*) (Equations 613, 614, 616, p. 253, 254).

There is no exact recipe for preparing fluorinated paraffins and haloparaffins with a definite number of fluorine atoms in appropriate positions. An individual path is to be followed and combinations of additions of hydrogen fluoride with halogen atoms exchange and halogenating and dehalogenating reactions are utilized as can be seen in the following example (*486*):

**197** [*486*]

$$CH_3-CH=CH_2 \xrightarrow{\text{Cl}_2} CH_3-CHCl-CH_2Cl \xrightarrow[\text{Fe}]{\text{Cl}_2} CH_3-CHCl-CHCl_2$$
$$\downarrow \text{KOH} \mid C_2H_5OH$$

$$CH_3-CH_2-CF_3 \xleftarrow{\text{HF, HgO}} \begin{array}{l} CH_3-CH_2-CCl_2F \quad 12\% \\ CH_3-CH_2-CClF_2 \quad 60\% \end{array} \xleftarrow[100°, 20\,\text{atm.}]{\text{HF}} CH_3-CH=CCl_2$$
$$88\%$$

$$\Big\downarrow \text{Cl}_2$$

$$CCl_3-CH_2-CF_3 \xrightarrow{\text{HF, HgO}} CF_2Cl-CH_2-CF_3 + CF_3-CH_2-CF_3$$
$$5\% \qquad\qquad 84\%$$

It is obvious that the preparation of well-defined fluorinated paraffins and their halogen derivatives is often tedious.

For the preparation of higher perfluoroparaffins (fluorocarbons), elemental fluorine is almost generally required. Various ways for obtaining fluorocarbons will be illustrated by the preparation of *perfluoroheptane*. The direct conversion of heptane to perfluoroheptane is achieved by catalytic fluorination with elemental fluorine over a silver carrier (*152, 759*) or by treatment with cobalt trifluoride (*287*). Both reactions give approximately the same yields, the reaction with cobalt trifluoride is, however, considered more convenient (Table 11, p. 76, and Table 13, p. 81).

A special mention should be made of the preparation of *perfluoroalkyl halides* by Hunsdiecker's method. Silver salts of perfluorocarboxylic acids are treated with elemental halogens, chlorine, bromine, or iodine (*406, 409, 410*) (Equations 366, 367, p. 183, 184). This method is of great importance for the preparation of perfluoroalkyl iodides and the Grignard reagents therefrom (p. 217).

## PREPARATION OF FLUOROOLEFINS AND THEIR HALOGEN DERIVATIVES

Fluorinated olefins and their halogen derivatives are most frequently prepared by dehalogenation of fluorinated vicinal halogenoparaffins with zinc or by dehydrohalogenation of fluorinated halogenoparaffins with alkalis. Controlled addition of hydrogen fluoride to acetylenes is less common. Recently two methods have been adopted, viz. addition of fluorinated halogenoparaffins to acetylenes and decomposition of sodium salts of perfluorocarboxylic acids.

The simplest fluoroolefin, *vinyl fluoride*, was obtained by dehalogenation of 1,1,2-fluorodihalogenoethanes (*986*),

$$198 \quad CH_2Br—CHBr_2 \xrightarrow[100°]{SbF_3, Br_2} \begin{matrix} CH_2Br—CHBrF \\ CH_2Br—CHF_2 \end{matrix} \xrightarrow{Zn, C_2H_5OH} CH_2{=}CHF \quad [986]$$

however, the addition of hydrogen fluoride to acetylene in the presence of mercuric chloride is nowadays the only practical way (*769*) (Equation 24, p. 67). *Vinylidene fluoride* is prepared by dehalogenation of 1,1-difluoro-1,2-dichloroethane (*644*) obtained from hydrogen fluoride and trichloroethylene (*349*),

199                                                                                          [349,644]

$$CHCl{=}CCl_2 \xrightarrow[199—204°, 9.5 \text{ hrs.}]{HF} \underset{35\%}{CH_2Cl—CClF_2} \xrightarrow[\substack{CH_3CONH_2 \\ C_4H_9\overset{|}{C}HCH_2OH \\ C_2H_5 \ 145°}]{Zn, NaI} CH_2{=}CF_2 \quad \begin{matrix} 76\% \quad \text{conversion} \\ 97\% \quad \text{yield} \end{matrix}$$

or by following series of reactions starting with acetylene *(35)*:

**200** $CH{\equiv}CH \xrightarrow{2\ HF} CHF_2{-}CH_3 \xrightarrow{Cl_2} CClF_2{-}CH_3 \xrightarrow{OH^{\ominus}} CF_2{=}CH_2$    *[35]*

*Trifluoroethylene* is best prepared by adding hydrogen bromide to chlorotrifluoroethylene, then treating with zinc *(803)*:

**201**    $CF_2{=}CFCl \xrightarrow[100°]{HBr} CF_2Br{-}CHFCl \longrightarrow CF_2{=}CHF$   85%    *[803]*

                       quant.

*Tetrafluoroethylene* was obtained by dehalogenation of 1,2-dichlorotetrafluoroethane *(621)*, prepared from hexachloroethane *(111)*.

**202**                $CF_2Cl{-}CF_2Cl \xrightarrow[60\ atm.]{Zn,\ C_2H_5OH} CF_2{=}CF_2$          *[621]*

Low yields when the reaction was carried out in ethanol were improved by using acetic anhydride as the solvent *(434)*. Industrial production of this compound is based on the pyrolysis of chlorodifluoromethane *(799)*, which will be described more thoroughly in chapters VI and X (p. 268, 343).

Other methods for laboratory-scale preparation of tetrafluoroethylene are thermal decomposition of sodium perfluoropropionate *(609)* (Equation 682, p. 271), heating of sodium trifluoroacetate with sodium hydroxide *(609)* (p. 271) and pyrolytic depolymerization polytetrafluoroethylene *(618)* (p. 275).

Pyrolysis and pyrolytic or catalytic hydrogenation of difluorodihalogenomethanes produces usually tetrafluoroethylene in low yield and in complex mixtures with other fluorinated compounds from which it is separated only with difficulty *(80, 227)*. The high-temperature fluorinations of carbon with fluorine *(269)*, phosphorus trifluoride *(267)* or phosphorus tribromide *(268)* are also for the time being without practical value.

Among the fluorinated halogen derivatives of ethylene, *chlorotrifluoroethylene* is definitely the most important one. The most convenient and to date also the only large-scale procedure for its preparation is dehalogenation of 1,1,2-trichlorotrifluoroethane obtained by the reaction of hexachloroethane with antimony fluorides *(111)* or hydrogen fluoride in the presence of antimony pentachloride *(200, 538)* (p. 337). The dehalogenation is carried out by means of zinc in methyl alcohol or ethyl alcohol *(65, 621)* (Equation 732, p. 343). Dehalogenation with sodium amalgam is not suitable for industrial uses *(710)*. On the other hand, the catalytic dehalogenation by means of hydrogen gives promising results *(171, 690)* (Table 49 p. 265).

In addition to this classical method for producing chlorotrifluoroethylene, pyrolytic dehydrohalogenation of 2,2-dichloro-1,2,2-trifluoroethane (*798*) and a series of reactions starting with trichloroethylene (*1090*) also lead to this important compound:

$$CHCl=CCl_2 \xrightarrow[20—100°]{Br_2} CHClBr—CCl_2Br$$

**203**       $\Big\downarrow\begin{matrix} HF \\ SbCl_5 \end{matrix}\Big|\begin{matrix} 60—100° \\ 5\ atm \end{matrix}$       [*1090*]

$$CHClF—CClF_2 \xrightarrow[25—50°]{KOH} CClF=CF_2$$

*1-Fluoro-2-propene—allyl fluoride*—was prepared by the reaction of allyl bromide with silver fluoride (*703*) or potassium fluoride (*647*). Of special interest in the laboratory as well as in industry are the propylene derivatives with all three hydrogen atoms in the methyl group replaced by fluorine. *1,1,1-Trifluoropropene* (*368*) was prepared according to the equation:

**204**       [*368*]

$$CCl_3—CHBr—CH_2Br \xrightarrow[90°]{SbF_3,\ SbF_3Cl_2} CF_3—CHBr—CH_2Br \xrightarrow{Zn,\ C_2H_5OH} CF_3—CH=CH_2$$
$$\phantom{CCl_3—CHBr—CH_2Br \xrightarrow[90°]{SbF_3,\ SbF_3Cl_2}} 53\% \phantom{CF_3—CHBr—CH_2Br \xrightarrow{Zn,\ C_2H_5OH}} 90\%$$

or by the following sequence of reactions (*1092*):

**205**   $CCl_4 + CH_2=CH_2 \xrightarrow{\hspace{2cm}} CCl_3—CH_2—CH_2Cl$       [*1092*]

$$\Big\downarrow SbF_3$$

$$CF_3—CH_2—CH_2Cl \xrightarrow{\hspace{2cm}} CF_3—CH=CH_2$$
$$58—63\% \phantom{CF_3—CH_2—CH_2Cl \xrightarrow{\hspace{2cm}}} 98\%$$

*1,1,1-trifluorotrichloropropene* by the reaction of perchloropropene with antimony trifluoride (Equations 88, p. 95).

*Perfluoropropene*, prepared originally by a series of reactions (*465*),

**206**       [*465*]

$$CH_3—CFCl—CH_2Cl \xrightarrow[hv,\ 1\ week]{Cl_2} 53\%CCl_3—CFCl—CCl_3 \xrightarrow[225°]{SbF_3,Cl_2} 70\%CF_2Cl—CFCl—CF_2Cl$$

$$\phantom{xxxxxxxxxxxxxxxxxxxxxxxxxxxxxxxxxxxxxxxxxxxxxxxxxxxxxx} Zn\ \Big|\ C_2H_5OH\ \downarrow$$

$$CFCl_2—CFCl—CF_3 \xleftarrow{Cl_2} 82\%CFCl=CF—CF_3 \xleftarrow[125°]{SbF_3,\ Cl_2} 86\%CF_2Cl—CF=CF_2$$

$$\Big\downarrow SbF_3$$

$$CF_2Cl—CFCl—CF_3 \xrightarrow{Zn} CF_2=CF—CF_3$$

is now being produced either by thermal decomposition of sodium perfluoro-butyrate (*609*) (table 50, p. 273), or, on the industrial scale, by pyrolysis of perfluorocyclobutane (*827*) or tetrafluoroethylene (*618*) (p. 274–276).

The propylene derivatives containing the $CF_3$ group are utilized for the production of trifluoroacetic acid (p. 172). Perfluoropropene is of increasing importance as a monomer for copolymerization to plastics (p. 348).

*1,1,1-Trifluoro-3-iodo-2-propene* can be obtained by the addition of trifluoro-roiodomethane to acetylene (360):

**207**  $CH{\equiv}CH + CF_3I \xrightarrow{200-220°} CF_3{-}CH{=}CHI$  70—80%  [*360*]

Similarly, halogenated olefins containing various halogen atoms in various positions are prepared from the corresponding fluorohalogenomethanes (p.251). Another general method for preparing 1,1,1-trifluoroolefins is dehydration of trifluoromethylcarbinols (p. 270) which are produced by the Grignard synthesis using perfluoromethyliodide (p. 221) or by the reduction of trifluoromethyl ketones obtained by the Claisen condensation (p. 229).

One of the most important fluorinated derivatives of butene is *2,3-dichloro-hexafluoro-2-butene* which is the starting material for many other fluorinated compounds. It results from the reaction of hexachlorobutadiene (perchloro-butadiene) with antimony chlorofluoride; chlorine is first added to the 1,4-positions and the octachloro-2-butene thus formed is readily fluorinated in both its trichloromethyl groups (*484*). The main product is accompanied by smaller amount of 2-chloroheptafluoro-2-butene:

**208**  [*484*]

$CCl_2{=}CCl{-}CCl{=}CCl_2 \xrightarrow{SbF_3Cl_2} [CCl_3{-}CCl{=}CCl{-}CCl_3] \xrightarrow{SbF_3,Cl_2}$  $\begin{array}{l} CF_3{-}CCl{=}CCl{-}CF_3 \\ CF_3{-}CF{=}CCl{-}CF_3 \end{array}$

2,3-Dichlorohexafluoro-2-butene can be converted, by means of hydrogen fluoride and lead dioxide, to 2,3-dichlorooctafluorobutane whose dehalogenation gives *perfluoro-2-butene* (465):

**209**  [*465*]

$CF_3{-}CCl{=}CCl{-}CF_3 \xrightarrow{HF,PbO_2} CF_3{-}CClF{-}CClF{-}CF_3 \xrightarrow[reflux]{Zn, C_2H_5OH} CF_3{-}CF{=}CF{-}CF_3$

conversion 26%
recovered 63%

Isomeric perfluorobutylenes, *1-perfluorobutene, 2-perfluorobutene and per-fluoroisobutylene,* are conveniently prepared by thermal decomposition of sodium or potassium salts of perfluorocarboxylic acids having one more carbon atom (130) (p. 271). Another method for preparing perfluoroisobutylene is illustrated by the following equation (397):

$$CF_3-CH=CF_2 + CF_3I \xrightarrow[20°, \text{ 4 days}]{h\nu} CF_3-CH-CF_2I \qquad 80\%$$
$$| $$
$$CF_3$$

210                                                                                                    [397]

$$Cl_2 \Big| h\nu$$
$$\downarrow$$

$$CF_3-CCl-CF_2Cl \xrightarrow[70°]{Zn, \text{ dioxane}} CF_3-C=CF_2$$
$$| \qquad\qquad\qquad\qquad\qquad | $$
$$CF_3 \qquad\quad 90\% \qquad\qquad\qquad CF_3 \quad 88\%$$

For industrial purposes the best preparation seems to be pyrolysis of perfluorocyclobutane (827) (p. 274) or polytetrafluoroethylene (751) (p. 276).

The lowest possible diolefin, *perfluoroallene*, was prepared only recently by the following series of reactions (531, 532):

$$CH_2=CF_2 + CF_2Br_2 \rightarrow CF_2Br-CH_2-CF_2Br$$

211                    KOH $\Big|$ aq.                                               [531]
$$\downarrow$$

$$CF_2=CH-CF_2Br \xrightarrow{KOH(s)} CF_2=C=CF_2$$

A fluorinated analogue of chloroprene, *fluoroprene*, was obtained by the addition of hydrogen fluoride to vinylacetylene (743).

212    $CH\equiv C-CH=CH_2 \xrightarrow[50-100°]{HF; \text{ C, Hg}} CH_2=C-CH=CH_2$    conversion 40%    [743]
$$| $$
$$F \qquad\qquad\qquad \text{yield } 50\text{-}70\%$$

Other fluorinated butadienes have been synthesized in many ways, e.g. *1,1,2-trifluoro-1,3-butadiene* from chlorotrifluoroethylene and ethylene by the following procedure (1016):

$$CF_2=CFCl \xrightarrow{Br_2} CF_2Br-CFClBr \qquad\qquad CH_2=CH_2$$
$$\Big| \underline{\qquad\qquad\qquad\qquad\qquad\qquad}\Big| $$

213                                                                                                    [1016]
$$\downarrow$$

$$CF_2=CF-CH=CH_2 \xleftarrow[\text{2. KOH}]{\text{1. Zn}} CF_2Br-CFCl-CH_2-CH_2Br$$

Another fluorinated butadiene important as a monomer for copolymerizations is *1,1,2,3-tetrafluoro-1,3-butadiene* prepared as shown below *(530)*:

$$CF_2Cl—CFCl—CF_2—COOH \xrightarrow{\text{LiAlH}_4} CF_2Cl—CFCl—CF_2—CH_2OH$$

**214**     $\left. \begin{array}{l} 1. \\ 2. \end{array} \right| \begin{array}{l} CH_3C_6H_4SO_2Cl \\ NaI \end{array}$     *[530]*

$$CF_2Cl—CFCl—CF_2—CH_2I \xrightarrow{Zn} CF_2=CF—CF=CH_2$$

*Perfluoro-1,3-butadiene* which, in contrast to its chlorine analogue, is readily polymerizable was prepared by several methods: by pyrolysis of perfluorocyclobutene *(385)* (p. 274), by thermal decomposition of sodium perfluoroadipate *(373)* (Equation 683, p. 272), or by a synthesis starting with chlorotrifluoroethylene *(365, 471)*, or with 1,2-dichlorodifluoroethylene *(898)*:

**215**     *[365, 471]*

$$CF_2=CFCl \xrightarrow[\text{25—40°}]{ICl} CF_2Cl—CFClI$$     97%

$\nearrow \xrightarrow[\text{ice cooling, 3 hrs.}]{Zn, (CH_3CO)_2O, CH_2Cl_2} CF_2Cl—CFCl—CFCl—CF_2Cl$     51%     $Zn \left| C_2H_5OH \right.$

$\searrow \xrightarrow[\text{100°}]{Zn, dioxane} CF_2=CF—CF=CF_2$     98%

**216**     *[898]*

$$CClF=CClF \xrightarrow[\text{10 hrs.}]{275°} CCl_2F—CClF—CF=CClF \xrightarrow[hV]{Cl_2} CCl_2F—CClF—CClF—CCl_2F$$     99.4%

conversion 77%
yield 87%

$SbF_5 \left| \text{ or } SbF_3Cl_2 \right.$
$250° \left| \text{5 hrs.} \right.$

$$CF_2=CF—CF=CF_2 \longleftarrow \qquad\qquad CClF_2—CClF—CClF—CClF_2$$

93.7%     86.6%

The simplest triene, *1,1,4,4-tetrafluoro-1,2,3-butatriene*, has recently been prepared from 1.1,4,4-tetrafluoro-1,3-butadiene *(696)*.

**217**     *[696]*

$$CF_2=CH—CH=CF_2 \xrightarrow{Br_2} CBrF_2—CH=CH—CF_2Br \xrightarrow[\text{110—150°}]{85\% \ KOH} CF_2=C=C=CF_2$$

## PREPARATION OF FLUOROCYCLOPARAFFINS AND FLUOROCYCLOOLEFINS AND THEIR HALOGEN DERIVATIVES

Fluorinated derivatives of cycloparaffins are generally prepared by the same methods as fluorinated paraffins. *Fluorocyclohexane* was obtained as a product of the addition of hydrogen fluoride to cyclohexene *(679)* (Equation 18, p. 65), or by heating bromocyclohexane with mercurous fluoride *(1002)* (Equation 108, p. 101). The treatment of hexachloro-1,3-cyclopentadiene *(678)* and octachlorocyclopentene *(605)* with antimony fluorides gives *1,2-dichlorohexafluoro-1-cyclopentene* (Equation 96, p. 97), a useful intermediate for more complicated fluorinated derivatives.

These general methods can be supplemented by some new ones which are specific for the fluorinated compounds. In the first place, dimerization of fluorinated haloolefins should be mentioned which leads, in contrast to the non-fluorinated analogues, to cyclic compounds, mostly the derivatives of cyclobutane *(477)* (Equations 629, 630, 631, p. 257). It is probably by this mechanism that the important Freon C318, *perfluorocyclobutane*, is formed in the depolymerization of polytetrafluoroethylene *(722)* (Table 53, p. 275).

Another important method suitable for six membered rings is the addition of fluorine to aromatic systems and fluorination of aromatic hydrocarbons. Thus the reaction of antimony pentafluoride with hexachlorobenzene yields *1,2-dichlorooctafluoro-1-cyclohexene (617,678)* (Equation 33, p. 70), and the reaction of cobalt trifluoride with benzene gives monohydro-, dihydro- and trihydroperfluorocyclohexanes *(37, 263, 317, 961)* in addition to *perfluorocyclohexane.*

**218**                                                                 [*37, 263, 317, 961*]

*Perfluorocycloparaffins* with six membered rings are the final products of the action of elemental fluorine *(390, 759)* (Equation 44, 46, p. 77, 78) or cobalt trifluoride *(391)* (Equation 51, p. 82) upon aromatic as well as hydroaromatic hydrocarbons.

Very often partly fluorinated derivatives prepared without the action of elemental fluorine are used as starting materials for perfluorinated cycloparaffins. Such compounds resist the energetic action of fluorine better, and in addition do not consume such amounts of elemental fluorine as the non-fluorinated material. Thus *perfluorodimethylcyclohexanes* were prepared from xylenes by the following series of reactions *(69, 287, 641, 757)*. For this particular case the consumption of elemental fluorine is reduced almost to one half.

**219**                                                                 [*69, 641, 757*]

Some heterocyclic aromatic compounds such as benzothiophene *(748)* or α-methylindol *(748)* lose the heteroatom during the fluorination and yield perfluorocycloparaffins.

Fluorinated cycloolefins with six-membered rings are further obtainable by the diene synthesis from fluorinated olefins and aliphatic, alicyclic and aromatic dienes (*647*) (Equation 638, p. 259).

**220**

$$
\begin{array}{ccc}
\overset{\displaystyle CH_2}{\underset{\displaystyle CH}{\overset{\displaystyle |}{\underset{\displaystyle CH}{\overset{\displaystyle |}{\underset{\displaystyle CH_2}{\diagdown}}}}}} & \overset{\displaystyle CF_2}{\underset{\displaystyle CF-CF_3}{\|}} & \xrightarrow[\text{24 hrs.}]{180°} \quad \overset{\displaystyle CH_2}{\underset{\displaystyle CH_2}{\overset{\displaystyle CH}{\underset{\displaystyle CH}{\|}}}} \quad \overset{\displaystyle CF_2}{\underset{\displaystyle CF}{\overset{\displaystyle |}{\underset{\displaystyle \diagdown CF_3}{}}}} \qquad 64\% \qquad [647]
\end{array}
$$

Fluorinated cycloolefins with large-membered rings are exceptional. One such example is perfluorooctamethylcyclooctatetraene which is formed by tetramerization of *perfluoro-2-butyne* (*136, 233*).

## PREPARATION OF FLUORINATED ACETYLENES

The preparation of *fluoroacetylene* and *difluoroacetylene* was accomplished only recently by thermal decomposition of monofluoromaleic anhydride (*707*) and difluoromaleic anhydride (*706, 862*), respectively. The reaction seems to proceed stepwise since fluoropropiolyl fluoride was mentioned as an intermediate (*706*).

**221**

$$
\begin{array}{c}
\overset{\displaystyle CH-CO}{\underset{\displaystyle CF-CO}{\|\qquad\qquad}}\diagdown O \quad \xrightarrow[1-2mm]{650°} \quad \overset{\displaystyle CH}{\underset{\displaystyle CF}{\||\|}} \; + \; \overset{\displaystyle CO}{\underset{\displaystyle CO_2}{}} \qquad [707]
\end{array}
$$

**222**

$$
\begin{array}{c}
\overset{\displaystyle CF-CO}{\underset{\displaystyle CF-CO}{\|\qquad\qquad}}\diagdown O \quad \xrightarrow[5-7mm]{650°} \quad \overset{\displaystyle CF}{\underset{\displaystyle CF}{\||\|}} \; + \; \overset{\displaystyle CO}{\underset{\displaystyle CO_2}{}} \qquad [706]
\end{array}
$$

For the synthesis of higher fluorinated acetylenes, two ways are available. The one is based on the preparation of a halogenated fluoro-derivative whose dehalogenation or dehydrohalogenation yields finally the required acetylenic compound (*441, 462*):

**223** $CF_3-CCl=CCl-CF_3 \xrightarrow[40°]{Zn, C_2H_5OH} CF_3-C\equiv C-CF_3$ conversion 49%  yield 54.4%  [441]

**224** [462]

$CCl_3-CH_2-CH_2Cl \xrightarrow[-78°\,\rightarrow\,20°]{SbF_3,\,SbF_3Cl_2} CF_3-CH_2-CH_2Cl \xrightarrow[2.\,Br_2]{1.\,KOH} CF_3-CHBr-CH_2Br$ 92,5%

KOH, C₂H₅OH $\Big\downarrow$ 0—5°

$90\% \; CF_3-CBr=CHBr \xleftarrow[0-5°]{KOH,\,C_2H_5OH} CF_3-CBr_2-CH_2Br \xleftarrow{Br_2} CF_3-CBr=CH_2$ 90%

Zn, C₂H₅OH $\Big\downarrow$ reflux $\qquad\qquad$ quant.

$CF_3-C\equiv CH$ 97%

The other route to fluorinated acetylenes is the addition of a fluoroalkyl iodide to an acetylene, followed by the recovery of the triple bond by splitting off the hydrogen iodide (*360, 381*).

**225** [381]

$$CH_3—C{\equiv}CH + CF_3I \xrightarrow{h\nu} CH_3—CI{=}CH—CF_3 \xrightarrow{Br_2} CH_3—CIBr—CHBr—CF_3$$

$$
\begin{array}{l}
39\% \ CH_3—C{\equiv}C—CF_3 \\
36—42\% \ CH{\equiv}C—CH_2—CF_3
\end{array}
\xleftarrow{\ Zn,\ C_2H_5OH\ }
\left\{
\begin{array}{ll}
CH_3—CBr{=}CBr—CF_3 & 76\% \\
CH_3—CBr{=}CH—CF_3 & 16\%
\end{array}
\right.
$$

## PREPARATION OF FLUORINATED AROMATIC COMPOUNDS

Aromatic hydrocarbons and their derivatives can have fluorine atoms bound to the aromatic nucleus, the side chain, or both. The preparation of the individual types differ substantially.

For introducing fluorine atoms into the side chains of benzene homologs, the same methods can generally be utilized as for the fluorinated paraffins. However, direct fluorination must be avoided, since both elemental fluorine and silver difluoride or cobalt trifluoride add fluorine to the double bonds of the aromatic nuclei, and hydroaromatic polyfluoro- or perfluoro-derivatives would result (p. 77). The most frequently employed method is the exchange reaction between the halogen derivatives and hydrogen fluoride or metal fluorides.

The easiest to prepare are the derivatives with fluorine atoms on the carbon adjacent to the ring, (i.e. in $\alpha$-positions). Heating of benzyl chloride with potassium fluoride afforded *benzyl fluoride* in a yield of 20–28% (*558*). Better results were obtained by treatment of benzyl bromide with mercuric fluoride (*93*) or by the thermal decomposition of trimethylbenzylammonium fluoride (*93, 527*) (Equation 172, p. 118). Yields up to 60% were reached in both instances. *Benzal fluoride* is available from benzal chloride and antimony trifluoride (*508*). Of numerous methods offered for the preparation of the important *benzotrifluoride*, the best yields are described for a treatment of benzotrichloride with anhydrous hydrogen fluoride (*137, 955, 960*) (Equation 66, p. 89). However, the reaction of benzotrichloride with antimony trifluoride is more convenient for the laboratory scale preparation (*110, 996*) (Equation 89, p. 95). Similar reactions are applied to the preparation of derivatives with two or more trifluoromethyl groups attached to the rings (*641, 650, 757*) (Equation 219, p. 134).

The preparation of compounds with fluorine atoms in more distant positions is much more tedious. It requires laborious replacements of halogens by fluorine, and halogenation and dehalogenation reactions as shown in the following examples for fluorinated styrenes (*184, 856*):

**226** [*184*]

$$\text{C}_6\text{H}_5\text{—CO—CH}_3 \xrightarrow[55–60°]{\text{Cl}_2} \text{C}_6\text{H}_5\text{—CO–CCl}_3 \xrightarrow[\text{reflux}]{\text{HF, AgF}_2} \text{C}_6\text{H}_5\text{—CO—CCl}_2\text{F} \quad 54\%$$

$$\downarrow \text{PCl}_5$$

$$\text{C}_6\text{H}_5\text{—CF=CClF} \xleftarrow{\text{Zn, C}_2\text{H}_5\text{OH}} \text{C}_6\text{H}_5\text{—CClF—CCl}_2\text{F} \xleftarrow[125–130°]{\text{SbF}_3, \text{Br}_2} \text{C}_6\text{H}_5\text{—CCl}_2\text{—CCl}_2\text{F}$$
$$48\% \qquad\qquad 46\% \qquad\qquad 89\%$$

**227** [*856*]

$$\text{C}_6\text{H}_5\text{—CO—CHF}_2 \xrightarrow{\text{PCl}_5} \text{C}_6\text{H}_5\text{—CCl}_2\text{—CHF}_2 \xrightarrow[180–195°]{\text{SbF}_3} \text{C}_6\text{H}_5\text{—CClF—CHF}_2$$
$$91\% \qquad\qquad 67\%$$

$$\text{Zn} \qquad\qquad \text{NaOH}$$

$$\text{C}_6\text{H}_5\text{—CF=CHF} \qquad\qquad \text{C}_6\text{H}_5\text{—CF=CF}_2$$
$$52\% \qquad\qquad 17\%$$

The Friedel–Crafts synthesis of *α,β,β-trifluorostyrene* from benzene and trifluorochloroethylene does not give satisfactory yields (*856*) (Equation 468, p. 211).

An entirely different method is used to attach fluorine directly to the aromatic ring. Electrophilic substitution commonly used for other halogens does not apply to fluorine because it does not form the fluorine cation required for electrophilic attack. Replacement of halogens by fluorine is limited to halogen atoms which are activated by nitro groups in *ortho-* or *para-*positions (p. 106).

The only general method for introducing fluorine into aromatic rings is the replacement of the amino group via the diazonium salts. Aromatic amines are diazotized and the diazonium group replaced by fluorine by means of hydrogen fluoride (*271*) or by the Schiemann reaction (*28, 884*) (p. 115). By this method *fluorobenzene* (*28*) (Equation 165, p. 115) and fluorinated benzene homologues were prepared in very high yields (*884*). In order to prepare difluoro- and polyfluoro-derivatives, the fluorinated aromatic compounds are nitrated, the nitrocompounds reduced and the amines converted to the fluoro-derivatives in the described way. Thus isomeric *difluorobenzenes* (*884,*

*920*), *trifluorobenzenes* (*279, 920*), and *tetrafluorobenzenes* (*278*) have been prepared:

[*920*]

228

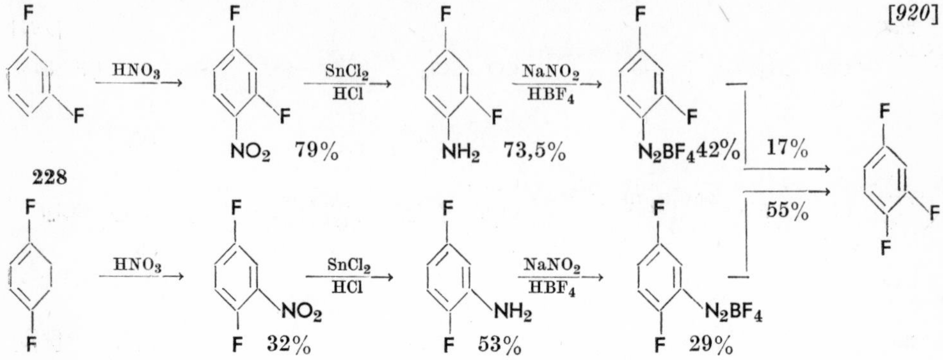


229 [*278*]

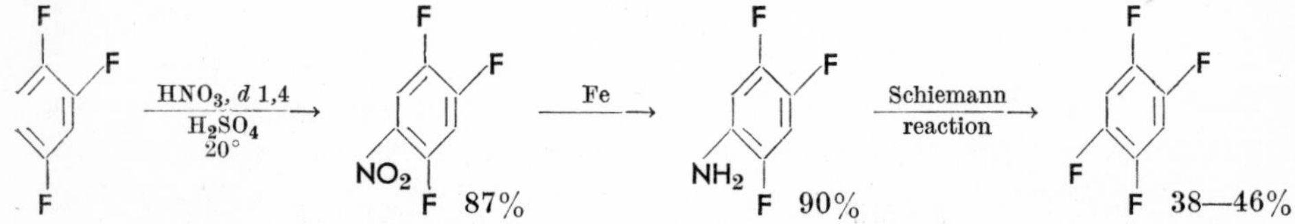


So far, this method of successive fluorination has failed to give pentafluorobenzene and hexafluorobenzene. These two interesting compounds have, however, been prepared by special methods. *Pentafluorobenzene* resulted from splitting off of three moles of hydrogen fluoride from 1,2,4,5-tetrahydrooctafluorocyclohexane (*774, 967*), or from Raney nickel desulphuration of pentafluorothiophenol (*1028*).

230 [*774, 967, 1028*]

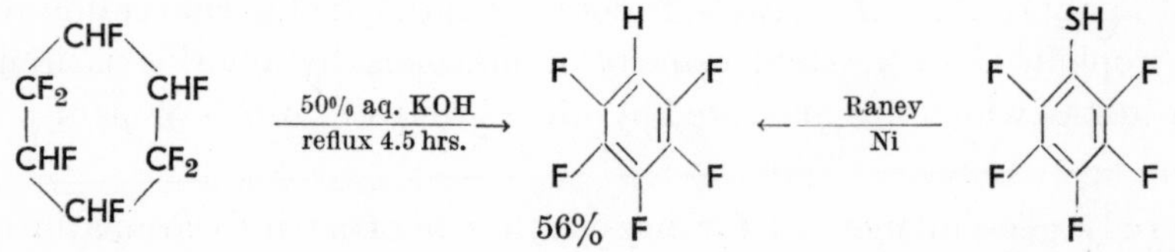


The oldest preparation of *hexafluorobenzene* from hexachlorobenzene (*652*)

231 [*652*]

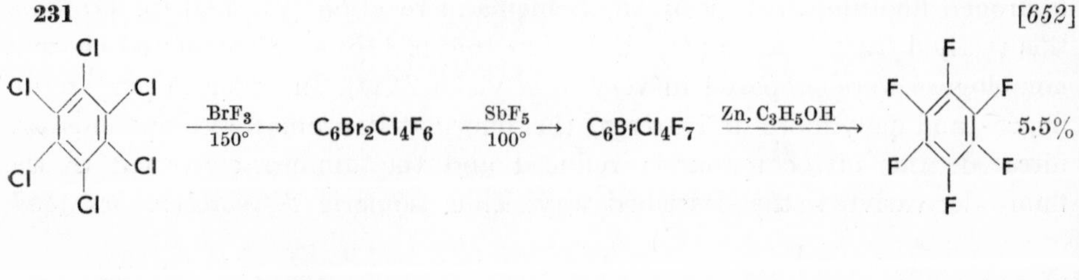

gave way to more modern methods, the pyrolysis of tribromofluoromethane (*97, 208, 209*),

**232**  [209]

$$CFBr_3 \xrightarrow[630-640°]{\text{Pt tube}}$$

(structure) + (structure) + $C_2F_2Br_4$

45%    6%    2%

dehydrofluorination of 1,2,4-trihydrononafluorocyclohexane (*316*) and the high-temperature catalytic defluorination of perfluorocyclohexane (*1028*).

**233**  [316, 1028]

(structure) $\xrightarrow{\text{aq. KOH}}$ (structure) $\xleftarrow[550°]{\text{Fe or Ni}}$ (structure)

13·2%

The latter reaction is general for the conversion of perfluorohydroaromatic compounds to perfluoroaromatic compounds. For instance *perfluoronaphthalene* was obtained in a yield of approximately 50% from perfluorodecalin (*303, 1028*).

Both pentafluorobenzene and hexafluorobenzene show many interesting reactions which will be discussed in the proper paragraphs.

Fluorinated compounds with fluorine both in the ring and the side chain are usually prepared by first attaching fluorine to the ring, then halogenating the side chain, and finally replacing the side-chain halogens by fluorine (*650*) (Equation 97, p. 97).

The described methods are applicable not only to the hydrocarbons but also to derivatives with substituents or functions resistant to the subsequent halogenation and fluorination reactions. Otherwise special methods must be chosen.

Some heterocyclic compounds can be treated like aromatics. The replacement reaction has introduced fluorine in the side chains of the pyridine homologues (*638*) (Equation 67, p. 89). The introduction of fluorine into the pyridine ring has so far been limited to the α- and β-derivatives. The pyridinediazonium fluoroborates are unstable and decompose already at low temperatures, sometimes—as in the case of the γ-isomer—to give products other than expected (*885*). Decomposition of α-pyridinediazonium fluoroborate in the reaction mixture without isolation gave 34% of *α-fluoropyridine* (*885*) (Equation 169, p. 117). *β-Fluoropyridine* was obtained in a yield of 50% by diazotization of β-aminopyridine with ethyl nitrite in an alcoholic solution, by precipitation of the diazonium fluoroborate with ether, and by decomposition of the salt in petroleum ether at 15–25°C (*885*). The decomposition at higher temperatures is uncontrollable, and in dry state it is explosive. Relatively good yields (42% and 36%, respectively) were obtained by the decomposition of α- and β-pyridinediazonium fluorosilicates (*53*). Only recently *perfluoropyridine* has been prepared by catalytic defluorination of *perfluoropiperidine* (*145*).

Among other heterocycles, the preparation of *α-fluorothiophene* has been mentioned in the paragraph on fluorinations with antimony trifluoride (*1054*) (Equation 78, p. 94). More important are the fluorinated derivatives of sym.-triazines. *2,4,6-Trifluoro-sym.-triazine* (*cyanuric fluoride*) arose from the action of antimony trifluoride on cyanuric chloride (*698*) (Equation 79, p. 94), *2,4,6-tris(trifluoromethyl)-sym.-triazine* by similar treatment of 2,4,6-tris(trichloromethyl)-sym.-triazine (*655, 779*) (Equation 251, p. 147).

## PREPARATION OF FLUORINATED HYDROXY-DERIVATIVES AND THEIR ESTERS AND ETHERS

The simplest fluorinated hydroxy-derivative, *fluoromethanol*, resulted from the reduction of formyl fluoride (*790*) or ethyl fluoroformate (*790*) with lithium aluminium hydride (Equation 293, p. 162). An interesting derivative of methanol, *trifluoromethyl hypofluorite*, arises from the fluorination of methylalcohol (*552*) or carbon monoxide (*154*), or from treatment of some organic fluoro-derivatives with oxygen difluoride (*192*) (Equation 38, p. 71). The fluorinating properties of this compound have been described elsewhere (p. 84).

Higher fluorinated alcohols unless prepared by the reduction of more oxidized derivatives (*311, 454, 1005*) (Equation 277, 278, 279, p. 158, 154) are obtained by the reaction of halogenohydrins with metal fluorides. *Ethylene fluorohydrin* which failed to arise from the reaction of the bromohydrin or iodohydrin with silver fluoride or mercuric fluoride (*995*), was isolated in fair yields from the reaction of ethylene chlorohydrin with potassium fluoride (*500, 560, 789, 914*).

**234**                               $[500, 789, 914]$

$$\text{ClCH}_2\text{—CH}_2\text{OH} \begin{cases} \xrightarrow[\text{130—135°, pressure}]{\text{KF}} & 42\% \\[4pt] \xrightarrow[\text{170—180°}]{\text{KF, (CH}_2\text{OH)}_2} \text{FCH}_2\text{—CH}_2\text{OH} & 42.5\% \\[4pt] \xrightarrow[\text{reflux 2 hrs.}]{\text{KF, C}_3\text{H}_5\text{(OH)}_3, \ h\nu} & 50\% \end{cases}$$

Glycerol α-monochlorohydrin was transformed to the *glycerol α-fluorohydrin* by heating with lithium fluoride (*792*) (Equation 146, p. 110).

*Polyfluoroalcohols* $C_nF_{2n+1}CH_2OH$ are easily made by reduction of perfluoro-carboxylic acids or their derivatives (*159, 523*) (Equations 291, 292, p. 161), *secondary and tertiary polyfluorinated alcohols* by the Grignard synthesis (*215, 332, 479*) (Equations 473, 474, p. 214). *Fluorinated phenols* are produced by the Schiemann reaction (*884*) or by the decomposition of diazonium fluoro-silicates (*169*) (Equation 170, p. 118). *Pentafluorophenol* was prepared by treat-ment of hexafluorobenzene with potassium hydroxide (*97, 283*) (Equation 430, p. 202).

*Esters of fluorinated alcohols* can be obtained either by replacement of halogens by fluorine in the corresponding esters of halogenated alcohols (*995*),

$$\mathbf{235} \qquad BrCH_2\text{—}CH_2OCOCH_3 \quad \xrightarrow[140°]{AgF} \quad FCH_2\text{—}CH_2OCOCH_3 \qquad\qquad [995]$$

or by esterification of ready-made fluorinated alcohols. This reaction is rather difficult with polyfluorinated alcohols which must be treated in some special ways, such as a reaction with acid chlorides (*177*) (Equation 419, p. 198), or esterification in trifluoroacetic anhydride (*1, 177*) (Equation 700, p. 280).

For the production of *fluorinated ethers* there is a number of possibilities. The ether bond is stable enough to stand treatment even with the most ener-getic fluorinating reagents.

Frequently halogenated ethers are transformed to the fluorinated ones by heating with potassium fluoride (*336*):

$$\mathbf{236} \quad CH_3\text{—}CH_2\text{—}O\text{—}CH_2\text{—}CH_2X \quad \xrightarrow[240-250°]{KF} \quad CH_3\text{—}CH_2\text{—}O\text{—}CH_2\text{—}CH_2F \quad 40\% \quad [336]$$

This replacement reaction in usually easy with α-haloethers whose α-halogen is very mobile (*697*) (Equation 117, p. 103).

The most important methods for the preparation of fluorinated ethers are direct etherification of fluorinated alcohols (*476*), the Williamson reaction of alcoholates with fluorinated alkyl halides (*480, 629, 1101*) (Equations 552, 553, p. 238), the addition of alcohols (or phenols) to fluorinated olefins (*258, 480, 1020, 1101*) (Equation 554, p. 238), and the addition of fluorinated alcohols to olefins or acetylenes (*2*):

$$CF_3\text{—}CH_2OH + CH\equiv CH \quad \xrightarrow[<50°,\ 7.5\ hrs.]{HgO,BF_3.(C_2H_5)_2O} \quad CF_3\text{—}CH_2\text{—}O\text{—}CH_2{=}CH_2$$

$$Na \quad \Big| \quad CF_3\text{—}CH_2OH$$

$$\mathbf{237} \qquad\qquad\qquad\qquad\qquad\qquad\qquad\qquad\qquad\qquad\qquad\qquad\qquad\quad \downarrow \qquad\qquad\qquad\qquad [2]$$

$$CF_3\text{—}CH_2\text{—}O\text{—}CH{=}CH_2 \quad \xleftarrow[200-400°]{clay} \quad \begin{matrix} CF_3\text{—}CH_2O \\ CF_3\text{—}CH_2O \end{matrix}\!\!>\!CH\text{—}CH_3$$

An example for direct fluorination is the conversion of diphenyl ether to the perfluoro-derivative by means of cobalt trifluoride (287). The reaction is accompanied by fluorolysis with formation of perfluorocyclohexane:

[287]

238

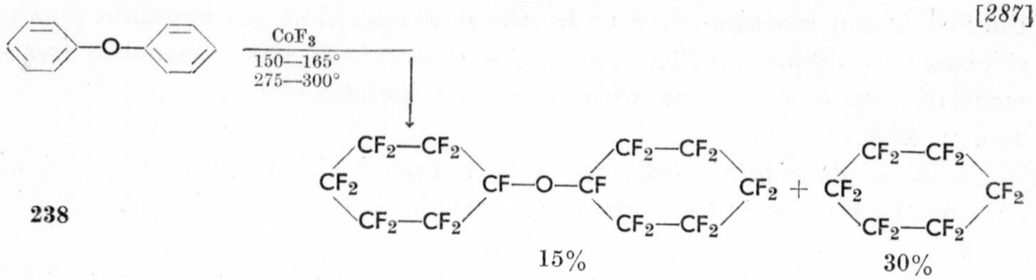

15%    30%

By combining various methods for the preparation of fluorinated ethers, *perfluorotetrahydrofuran* and *perfluorotetrahydropyran* have been prepared (476).

239

[476]

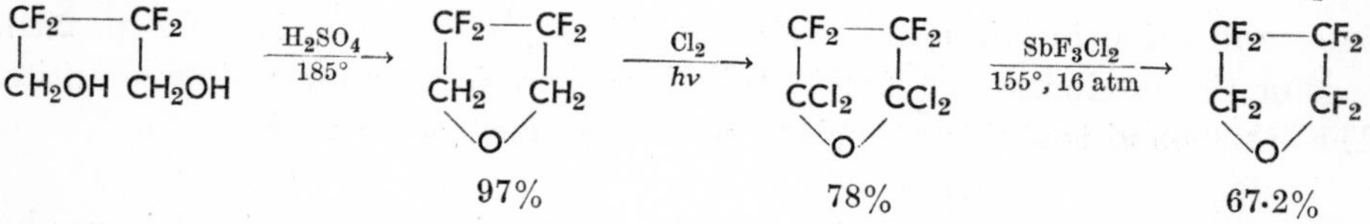

97%    78%    67·2%

Both these perfluorinated ethers are very stable and do not have basic properties (p. 310).

## PREPARATION OF FLUORINATED ALDEHYDES AND KETONES

*Fluorinated aldehydes* are most generally prepared by oxidation or reduction of other fluorinated compounds (1089):

240

[1089]

$F(CH_2)_9CH{=}CH_2 \xrightarrow{HCO_3H} F(CH_2)_9CH{-}CH_2 \xrightarrow{NaIO_4} F(CH_2)_9CHO$

OH  OH    39·5%

75%

241    $F(CH_2)_7COCl \xrightarrow[xylene]{H_2, Pd(S)} F(CH_2)_7CHO$    75—85%    [1089]

242    [1089]

$FCH_2{-}CH_2{-}CH_2{-}CH_2{-}NO_2 \xrightarrow[\text{2. } H_2SO_4,\text{ ice}]{\text{1. NaOH aq.}} FCH_2{-}CH_2{-}CH_2{-}CHO$    73%

*Fluoroacetaldehyde* was obtained in a yield of 6% by oxidation of ethylene fluorohydrin (*914*), in a yield of 24.5% by treatment of formyl fluoride with diazomethane and decomposition of the diazoacetaldehyde with anhydrous hydrogen fluoride (*783*) (Equation 581, p. 244), and in a yield of 80% by the reaction of glycerol monofluorohydrin with periodic acid (*792*) (Equation 333, p. 174). *Trifluoroacetaldehyde*, too, was prepared by several methods: by a controlled reduction of trifluoroacetonitrile (*468*) (Equation 297, p. 163), by oxidation of 1,1,1-trifluoropropane (*936*) (Equation 319, p.169), by the Rosenmund hydrogenation of trifluoroacetyl chloride (*134*), and by the reduction of trifluoroacetic acid with lithium aluminium hydride (*127*) (Equation 291, p. 161).

*Fluorinated aromatic aldehydes* are produced by the reduction of the corresponding acid chlorides according to Rosenmund (*939*), or, more frequently, by oxidation of fluorinated toluenes. In addition to the direct conversion of the fluorotoluenes to the aldehydes by means of chromyl chloride (*590*), two indirect methods are available: chlorination or bromination to the corresponding fluorinated benzyl halides followed by oxidative hydrolysis with copper or lead nitrate (*518*), or halogenation of the fluorotoluenes to the fluorobenzal halides and their hydrolysis with calcium carbonate (*586*):

243                                                                 [*518,586,590,939*]

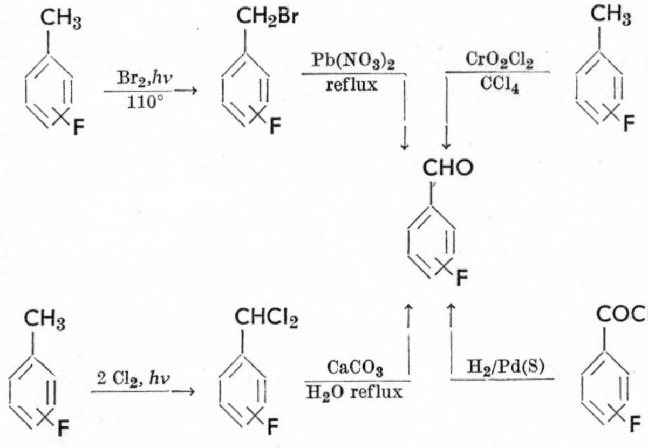

*Procedure 16.*

*Preparation of p-Fluorobenzaldehyde from p-Fluorotoluene (518)\**

A 250 ml. three-necked flask fitted with a reflux condenser, a thermometer reaching to the bottom, and a dropping funnel, is charged with 55 g (0·5 mole) of p-fluorotoluene and heated to 105–110°C. While illuminating the flask with a 500 W bulb from a distance

---

\* Carried out by the author.

of 5–15 cm, 29 ml. (0·53 mole) of bromine is added at such a rate that the bromine colour does not persist more than a few seconds. After the addition has been completed (20–30 min.) the mixture is rapidly heated up to 165°C, then cooled to room temperature, and the hydrogen bromide is expelled by replacing the dropping funnel by a tube reaching to the bottom and passing a mild stream of air through the liquid.

The crude *p*-fluorobenzyl bromide thus formed is placed in a 1-litre flask fitted with a reflux condenser and refluxed with a solution of 88 g (0·38 mole) of lead nitrate in 625 ml. of water for 6–7 hours with occasional shaking. Steam-distillation of the reaction mixture gives 47 g (0·38 mole) (76%) of a crude product, which is transformed to its bisulphite compound by shaking with 100 ml. of a 5 N solution of sodium bisulphite. The precipitate (78 g, 0·34 mole) is filtered with suction, washed with ethanol, decomposed by a solution of 48 g (0·45 mole) of sodium carbonate in 300 ml. of water, and the mixture is subjected to steam-distillation: 36 g (0·29 mole) (58%) of pure *p*-fluorobenzaldehyde is obtained after drying with calcium chloride and distillation over the range of 176–178°C; $n_D^{20}$ 1·5210.

*Fluorinated ketones* are prepared both by introducing fluorine into ketones and by oxidative and synthetic reactions of fluorinated compounds.

*Fluoroacetone* was obtained from chloroacetone (*868*), bromoacetone (*867*) or diazoacetone (*85, 783*),

**244**                                                                                          [*783,867*]

$$BrCH_2{-}CO{-}CH_3 \xrightarrow{TlF,(C_2H_5)_2O} FCH_2{-}CO{-}CH_3 \xleftarrow[anhyd.]{HF} N_2CH{-}CO{-}CH_3$$

$$66\% \qquad\qquad\qquad 84\%$$

and in a mixture with other compounds, by the reaction of acetone with elemental fluorine (*294*) (Equation 246). Fluorinated acetones containing almost any possible number of fluorine atoms are readily accessible by the ketonic fission of various fluorinated acetoacetic esters (*207, 660*) (Equation 524, p. 230, Table 40, p. 231).

*α,α'-Dichlorotetrafluoroacetone* was the main product of the action of antimony trifluoride upon hexachloroacetone (*715, 716*).

**245**    $CCl_3{-}CO{-}CCl_3 \xrightarrow[120°{-}140°,\,5\,hrs.]{SbF_3,\,SbCl_5} CClF_2{-}CO{-}CClF_2$    66%    [*715*]

Fluorination of *α,α,α*-tribromo-*α',α',α'*-trifluoroacetone with silver fluoride produced *bromopentafluoroacetone* and *α,α-dibromotetrafluoroacetone* (*937*) (Equation 104, p. 99). *Hexafluoroacetone* was one of the products of the action of elemental fluorine upon acetone (*294*),

**246**    $CH_3{-}CO{-}CH_3 \xrightarrow[60°]{F_2,N_2}$    $CH_3{-}CO{-}CH_2F + CF_3{-}CO{-}CF_3$

$+ CF_3{-}COF + (COF)_2 + COF_2 + CF_4$    [*294*]

however, oxidation of fluorinated isobutylenes seems to be a more suitable means of preparation (*370, 751, 827*) (Equation 326, p. 172).

Of the homologous ketones, methyl $\beta$-fluoroethyl ketone was obtained in a low yield by fluorination of methyl ethyl ketone with elemental fluorine (*1031*) (Equation 43, p. 73). A fruitful way to higher polyfluorinated and perfluorinated ketones is the interaction of perfluoroalkylmetal halides with derivatives of carboxylic acids. *Methyl perfluoropropyl ketone,* made from per-fluoropropylzinc iodide and acetyl chloride (*400*) (Equation 503, p. 225), and *perfluoropropyl trifluoromethyl ketone,* made from perfluoropropylmag-nesium iodide and trifluoroacetyl chloride (*367*), illustrate this method of pre-paration. The action of trifluoroacetyl chloride on dihexylcadmium resulted in the formation of hexyl trifluoromethyl ketone (*159*) (Equation 506, p. 226). *Bis(perfluoropropyl) ketone* was obtained by the action of sodium upon ethyl perfluorobutyrate (*403*) (Equation 529, p. 232).

The simplest unsaturated fluorinated ketone, *difluoroketene,* was prepared recently by dehalogenating the bromide of difluorochloroacetic acid (*1095*).

**247**                                                            [*1095*]

$$CCl_3-COOCH_3 \xrightarrow{SbF_3} CClF_2-COOCH_3 \quad 19\%$$

$$\downarrow Ba(OH)_2 \ | \ reflux$$

$$CClF_2-COO\frac{Ba}{2} \xrightarrow[160°, 3\,hrs.]{PBr_3} CClF_2-COBr \xrightarrow[(C_2H_5)_2O]{Zn/Cu} CF_2{=}C{=}O$$

$$47.5\% \qquad\qquad 87\%$$

Aliphatic-aromatic ketones carrying fluorine atoms in the aliphatic chain can be produced by the replacement of halogens by fluorine in the appropriate halogenated derivatives. This way was successful with *ω-fluoroacetophenone* (*867*)

**248**    ⟨◯⟩—CO—CH$_2$Br $\xrightarrow{TlF, C_2H_5OH}$ ⟨◯⟩—CO—CH$_2$F    79%    [*867*]

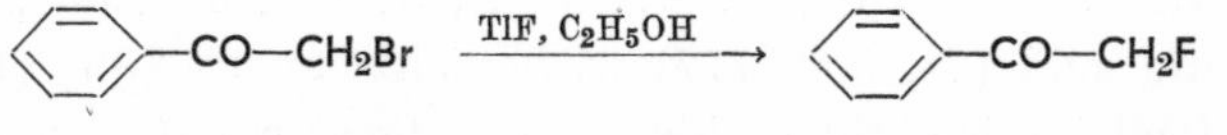

and *dichlorofluoroacetophenone* (*184*) (Equation 226, p. 137).

More frequently, such ketones are prepared by the Friedel–Crafts synthesis starting with chlorides of the fluorinated acids (*90, 183, 337, 856, 958*). This method was utilized for the preparation of *ω-mono-, ω,ω-di-* and *ω,ω,ω-tri-fluoroacetophenone* (Equation 469, p. 212).

Aromatic ketones with fluorine on the ring may also be obtained by the Friedel-Crafts reaction (*285*):

**249**    F—⟨◯⟩ + (CH$_3$CO)$_2$O $\xrightarrow[reflux]{AlCl_3, CS_2}$ F—⟨◯⟩—CO—CH$_3$    55%    [*285*]

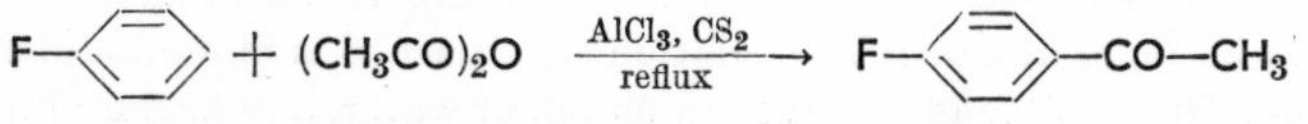

In addition, they can be prepared by the Schiemann reaction from the corre-sponding amino ketones (*884*) (p. 115).

*Fluorinated diketones,* both aliphatic and aliphatic-aromatic, are conveniently prepared by the Claisen condensation of fluorinated acetates with ketones or fluorinated ketones (*466, 780*) (Equations 520, 521, 522, 523, p. 229, 230).

## PREPARATION OF FLUORINATED ACIDS AND THEIR DERIVATIVES

The preparation of fluorinated acids and their derivatives has been worked out very thoroughly, since these compounds are useful intermediates for the synthesis of a number of other fluorinated compounds. They are obtained by introducing fluorine into acids or their derivatives, or else by oxidation or other reaction of fluorinated compounds.

The first member of the homologous series of fluorinated acids, *fluoroformic acid,* is known only in the form of its derivatives. *Fluoroformates* are produced by treatment of chloroformates with potassium fluoride (*788*) (Equation 127, p. 105) or thallium fluoride (*762*) (Equation 147, p. 111). Chloride and fluoride of fluoroformic acid, FCOCl and FCOF, respectively, are formed by the reaction of phosgene with antimony fluorides (*245*) (Equation 86, p. 95).

The greatest amount of attention has been directed to *fluorinated acetic acids,* their halogenated derivatives and their functional derivatives. They are obtained by the conversion of the corresponding halogenated derivatives to the fluoroderivatives, or by oxidation of fluorinated olefins or aromatic compounds.

The halogen atom in the derivatives of monochloro- and monobromoacetic acid (the free acid is not used for this purpose) is very reactive and therefore easily replaced by fluorine. This exchange reaction has been accomplished with chloroacetonitrile (*559*), chloroacetamide (*25*) (Equation 128, p. 105) or most frequently with esters of chloracetic acid (*84, 100, 334, 336, 337, 777, 912*) or bromoacetic acid (*25, 334, 777*) with yields up to 85% (Equations 129, 130, p. 105). Hydrolysis of the amide (839) or the esters (*337, 910*) affords free fluoroacetic acid.

*Procedure 17.*

*Preparation of Fluoroacetic Acid by the Hydrolysis of Ethyl Fluoroacetate (912)\**

To a mixture of 53 g (0·5 mole) of ethyl fluoroacetate and 100 ml. of water, containing a few drops of an alcoholic solution of phenolphthalein, finely powdered barium hydroxide octahydrate is added in small portions while stirring the mixture vigorously. Approximately 80–100 g of barium hydroxide (according to its quality) is consumed before the reaction becomes alkaline. The solution thus formed is filtered and evaporated to dryness *in vacuo.* The residue is dissolved in 50 ml. of water, and barium fluoroacetate

---

\* Carried out by the author.

is precipitated by the addition of 400 ml. of ethyl alcohol. The precipitate weighing 63 g (0·22 mole) (86·5%) is added in portions to 133 g (1·35 mole) of stirred 100% sulphuric acid. The mixture is then slowly heated in a distillation apparatus *in vacuo* (15–20 mm) up to 200°C. Approximately 32 g of crude fluoroacetic acid is collected in the receiver. Its redistillation yields 24 g (0·31 mole) (61·5%) of pure fluoroacetic acid distilling over the range of 164–170°C and solidifying in the receiver.

*Difluoroacetic acid* results from the oxidation of 2,2-difluoroethanol (*988*) or more conveniently, of 1,1-dichloro-3,3-difluoro-1-propene (*436*):

**250** $\qquad$ $CHF_2{-}CH{=}CCl_2 \xrightarrow{\text{KMnO}_4} CHF_2{-}CO\,OH$ $\quad$ 86% $\qquad$ [*436*]

A more recent way of preparation from tetrafluoroethylene (*467*) by way of tris(difluoromethyl)-*sym*-triazine is described in the paragraph on addition reactions (Equation 589, p. 246).

Much interest was and still is attached to the production of *trifluoroacetic acid*. One of the methods is the hydrolysis of trifluoroacetyl fluoride obtained by the action of elemental fluorine upon acetone (*294, 413*), or better by the electrolysis of a solution of acetic anhydride in anhydrous hydrogen fluoride (*546*) (Equation 60, p. 86). More frequently trifluoroacetic acid is produced by oxidation of compounds with a trifluoromethyl group in their molecules. Oxidative degradation of *m*-aminobenzotrifluoride (*412, 1000, 1060*) gives trifluoroacetic acid in yields up to 95% (Equation 325, p. 171), oxidation of 1,1,1-trifluoro-2-propene in a yield of 80% (*1092*), that of 1,1,2-trichlorotrifluoro-1-propene in a yield of 90% (*436*) and oxidation of 2,3-dichlorohexafluoro-2-butene in a yield of 83% (*484*) (Equation 327, p. 172). A somewhat more complicated synthesis starting from easily available raw materials is displayed in the following equation (*655, 779*):

$$CCl_3{-}COOC_2H_5 \xrightarrow{NH_3,(C_2H_5)_2O} CCl_3{-}CONH_2 \xrightarrow{P_2O_5} CCl_3{-}CN \xleftarrow{HCl,Cl_2} CH_3{-}CN$$

$$\qquad\qquad\qquad 91\% \qquad\qquad\qquad 85\%$$

$$HBr \;\Big|\; AlBr_3$$

**251**

[*655, 779*]

$$CF_3{-}COOC_2H_5 \xrightarrow[\text{2. H}_2\text{SO}_4]{\text{1. NaOH}} CF_3{-}COOH$$

$$92.5\% \qquad\qquad 91\%$$

10*

*Procedure 18.*

*Preparation of Trifluoroacetic Acid (655, 779)\**

(a) *Ethyl Trichloroacetate.* Azeotropic esterification of 2500 g (15·3 moles) of trichloroacetic acid with 4500 ml. (75 moles) of 96% ethanol in the presence of 10 ml. of hydrochloric acid and 4000 ml. of benzene gives 2340 g (12·2 moles) (80%) of ethyl trichloroacetate distilling at 162–165·5°C.

(b) *Trichloroacetamide.* Ethyl trichloroacetate (2340 g, 12·2 moles) is dissolved in 2000 ml. of ether and the solution is saturated with gaseous ammonia for 20 hrs. while cooling with ice. The crystals are filtered by suction, the filtrate is evaporated to a small volume, and the second crystals separated and added to the first crop, yielding 1760 g (10·8 moles) (89%) of trichloroacetamide, m.p. 141°C.

(c) *Trichloroacetonitrile.* Trichloroacetamide (605 g, 3·7 mole) is thoroughly mixed with 620 g (4·35 moles) of phosphorus pentoxide and the mixture is slowly heated to 180°C (with an oil-bath) in a distilling apparatus with a descending condenser. The crude trichloroacetonitrile distilling at 86–95°C (482 g, 3·35 moles) is mixed with 50 g of phosphorus pentoxide and redistilled to give 450 g (3·1 moles) (83%) of pure trichloroacetonitrile boiling over a range of two degrees (84–86°C).

(d) *1,3,5-Tris(trichloromethyl)-sym-triazine.* Trichloroacetonitrile (560 g, 3·9 moles) is mixed with 12 g (0·045 mole) of aluminium bromide and the mixture is saturated with gaseous hydrogen bromide. In 45 minutes the temperature rises from 20°C to 65°C. The content of the flask solidifies overnight. It is then melted on a steam-bath with an equal volume of water, the mixture is cooled, the crystals are filtered by suction, and recrystallized from 750 ml. of ethanol. 1,3,5-Tris(trichloromethyl)-*sym*-triazine, m. 90–92°C, is obtained in a yield of 470 g (3·26 moles) (83·5%).

When the same amount of trichloroacetonitrile is treated with half the amount of aluminium chloride (in place of aluminum bromide) and saturated with hydrogen bromide, the trimerization reaches a yield of 95%.

(e) *1,3,5-Tris(trifluoromethyl)-sym-triazine.* A three-necked two-litre flask fitted with a gas-tight stirrer and a column with a dephlegmator is charged with 800 g (4·5 moles) of antimony trifluoride, 105 p. (1·4 mole) of antimony trichloride, and chlorine is passed through the mixture at 150°C until the weight increase has reached 100 g (1·4 mole). Into the heated and stirred mixture is added on the whole 500 g (1·15 moles) of 1,3,5-tris(trichloromethyl)-sym-triazine at half-hour intervals in portions of 22·5, 22·5, 45, 45, 90, 90, and 95 g. After the first half-hour of boiling, the distillate is collected in the column head under the reflux ratio of 10 : 1 or higher. Two fractions are obtained: one (43 g) boiling at 98–105°C , and the other (153 g) distilling from 105 to 125°C. Repeating of the whole procedure with the higher boiling fraction and 500 g of antimony trifluoride and 105 g of antimony trichloride saturated at 150°C with chlorine gave 95 g of a distillate boiling at 98–100°C, and a smaller portion of the higher boiling liquid whose subsequent fluorination yielded still 22 g of the product distilling at 98–100°C. The total yield is approximately 160 g (0·56 mole) (49%) of 1,3,5-tris(trifluoromethyl)-*sym*-triazine, boiling at 98–100°C (notably less than described in the literature).

(f) *Ethyl Trifluoroacetate.* A mixture of 227 g (0·8 mole) of 1,3,5-tris(trifluoromethyl)-*sym*-triazine, 270 ml. (4·5 moles) of 95% ethanol, 370 ml. of 32% hydrochloric acid, and 300 ml. of water is refluxed under a column for 40 min. during which time the temper-

---

\* Carried out by the author and by J. Salák.

ature in the column head drops from 60°C to 52°C. At a reflux ratio of 10:1,330 g of a fraction distilling at 52·5–55°C, and 50 g of a fraction boiling over the range of 55–76°C are collected. The first fraction, the azeotropic mixture of ethyl trifluoroacetate, ethanol, and water, is mixed under cooling with 105 g of 100% sulphuric acid and distilled in a column to yield 280 g (1·95 moles) (81%) of ethyl trifluoroacetate, boiling at 61–62°C.

(*g*) *Sodium Trifluoroacetate.* In a three-necked-litre one flask fitted with a gas-tight stirrer, a separating funnel, and a reflux condenser, 216 ml. of the azeotropic mixture of ethyl trifluoroacetate and water (the distillate in the range of 52·5–55°C from the previous preparation) is added dropwise to a solution of 72 g (1·8 moles) of sodium hydroxide in 270 ml. of water. The reaction is carried out by stirring and heating the mixture at 90°C for two hours. Thereafter the solution is evaporated to dryness *in vacuo* to yield 230 g (1·7 moles) of sodium trifluoroacetate, contamined with sodium hydroxide.

(*h*) *Trifluoroacetic Acid.* A one-litre three-necked flask fitted with a gas-tight stirrer and a column with a dephlegmator is charged with 540 g (5·5 moles) of 100% sulphuric acid to which 223 g (1·64 moles) of sodium trifluoroacetate is added portionwise with efficient stirring over a period of 45 minutes. After stirring the mixture for two additional hours, distillation is started and crude trifluoroacetic acid distilling at 72·5–78°C is collected (133 g, 0·98 mole). Redistillation of this product after the addition of 16 g of phosphorus pentoxide afforded 95 g (0·7 mole) of pure trifluoroacetic acid boiling over a range of 1 degree (72–73°C).

*Halogenated fluoroacetic acids* are prepared by oxidation of the appropriate halogenated fluoroolefins (*436*),

252 $\qquad$ $CClF_2$—$CCl{=}CCl_2$ $\xrightarrow{KMnO_4}$ $CClF_2$—$COOH$ $\quad$ 70% $\qquad$ [*436*]

or by halogenation of fluoroacetic acids (*988, 990*),

253 $\qquad$ $\overset{\displaystyle \xleftarrow{\;Cl_2.\,h\nu\;}}{\Big\downarrow}$ $CHF_2$—$COOH$ $\overset{\displaystyle \xrightarrow[160°]{Br_2,\,Fe}}{\Big\downarrow}$ $\qquad$ [*988, 990*]

$\qquad$ $CClF_2$—$COOH$ $\qquad\qquad\qquad\qquad$ $CBrF_2$—$COOH$

or by the addition of alcohols (*258*) or amines (*1097*) to fluorinated halogeno-ethylenes and subsequent hydrolysis of the addition products (Equations 597 and 603, p. 248 and 250, respectively):

254 $\qquad\qquad\qquad\qquad\qquad\qquad\qquad\qquad\qquad\qquad\qquad\qquad\qquad$ [*1097*]

$CClF{=}CF_2 + NH(C_2H_5)_2$ $\xrightarrow[\text{2. } H_2O]{\text{1. } 0°}$ $CHClF$—$CON(C_2H_5)_2$ $\xrightarrow{H_2SO_4}$ $CHClF$—$COOH$

$\qquad\qquad\qquad\qquad\qquad\qquad\qquad\qquad\qquad\qquad\qquad\qquad\qquad\qquad\qquad\qquad\qquad$ 59·5%

Fluorinated homologous acids are obtained by oxidation of suitably fluo-rinated compounds, or by replacement of halogen in halogenated acids by fluorine, or by introducing a carboxyl into a fluorinated molecule. As an example, preparation of some *ω-fluorocarboxylic acids* may be quoted (*816, 819*):

**255**

$$Br(CH_2)_6COOC_2H_5 \begin{array}{c} \overset{KF,\ HOCH_2CH_2OH}{\underset{130°,\ 8\ hrs.}{\nearrow}} \quad \searrow 40\% \\[4pt] \underset{60°}{\overset{AgF}{\searrow}} \quad \nearrow 34\% \end{array} F(CH_2)_6COOC_2H_5 \qquad [819]$$

**256**   $F(CH_2)_6Cl \xrightarrow[\text{2. }CO_2]{\text{1. Mg, }(C_2H_5)_2O} F(CH_2)_6COOH \quad 64\% \qquad [816]$

A general method for the production of *perfluorocarboxylic acids* is the electrolysis of acids or their anhydrides in anhydrous hydrogen fluoride and the subsequent hydrolysis of the perfluoroacyl fluorides thus formed (*546, 957*) (Equation 60, p. 86). Another route to the perfluorocarboxylic acids starts with the addition of perfluoroalkyl iodides to acetylene and oxidation of the resulting olefin (*380*), or with the addition of perfluoroalkyl iodides to olefins, dehydrohalogenation, and oxidation of the resulting fluorinated olefins (*397*). Both ways lead to acids containing more carbon atoms than the starting perfluoroalkyl halide. A considerable disadvantage of this method is the fact that the nesessary perfluoroalkyl iodides are best prepared from the perfluoro-carboxylic acids.

**257**                                                                                 [*380, 397*]

$$C_2F_5I + CH\equiv CH \xrightarrow{240°} \underset{72\%}{C_2F_5-CH=CHI} \xrightarrow{KMnO_4} 22-30\% \quad C_2F_5-COOH$$

$$\phantom{C_2F_5I + CH\equiv CH} \qquad\qquad\qquad\qquad\qquad\qquad\qquad\qquad\qquad 55\%\ \uparrow KMnO_4$$

$$C_2F_5I + CH_2=CF_2 \xrightarrow{220°} \underset{95\%}{C_2F_5-CH_2-CF_2I} \xrightarrow{KOH,\ C_2H_5OH} \underset{72\%}{C_2F_5-CH=CF_2}$$

*Fluorinated unsaturated aliphatic acids* result either from the action of fluo-rinating agents upon unsaturated halogenoacids (*822*) (Equation 101, p. 99), or from the partial replacement of halogens in polyhalogenoacids by fluorine followed by dehalogenation or dehydrohalogenation (*446*) (Procedure 10, p. 102).

**258**                                                                                 [*446*]

$$CH_2Br-CBr_2-COOCH_3 \xrightarrow{HgF_2} \underset{24\%}{CH_2Br-CBrF-COOCH_3} \xrightarrow{Zn,\ C_2H_5OH} CH_2=CF-COOCH_3$$

As an example for a combination of various reactions, the preparation of perfluoroacrylic acid nitrile is shown in the following scheme (*164*):

259                                                                    [*164*]

$$CF_2Cl-CF{=}CCl \xrightarrow[35-50°]{O_2,\ h\nu} CF_2Cl-\underset{\underset{O}{\diagdown\diagup}}{CF}-CCl_2 \rightarrow \begin{array}{l} CF_2Cl-CClF-COCl \\ CF_2Cl-CCl_2-COF \end{array} \Big\}$$

$$\downarrow NH_3 \mid (C_2H_5)_2O$$

$$CF_2{=}CF-CN \xleftarrow[70-125°]{Zn,\ (CH_3CO)_2O} \begin{array}{l} CF_2Cl-CFCl-CONH_2 \\ CF_2Cl-CCl_2-CONH_2 \end{array}$$

*Fluorinated dicarboxylic aliphatic acids* are produced almost exclusively by oxidation of fluorinated diolefins or cycloolefins. *Difluoromalonic acid* (stable up to 160°C, in contrast to malonic acid) was obtained by oxidation of 3,3-difluoro-1,4-pentadiene (*440*) and of fluorinated cycloolefins (*146, 263*), perfluorodicarboxylic acids containing four, five and six carbon atoms in a straight chain were obtained by oxidation of the appropriate fluorinated cycloolefins, hexafluorocyclobutene (*492*) or 1,2-dichlorotetrafluorocyclobutene (*489, 492*) (Equation 329, p. 173), 1,2-dichlorohexafluorocyclopentene (*491, 678*) (Equation 331, p. 173), and perfluorocyclohexene (*146*) or 1,2-dichlorooctafluorocyclohexene (*678*), respectively:

260                                                                    [*146, 678*

83%          75%

A convenient method for the preparation of *fluorinated aromatic acids* having fluorine atom attached to the nucleus is the Schiemann reaction (*884*).

Functional derivatives of fluorinated acids are prepared either by modifying the carboxyl group, or by introducing fluorine into the appropriate acid derivative.

*Fluorinated acid halides* result from the reaction of free fluorinated acids with benzoyl chloride, thionyl chloride or phosphorus halides. Fluorinated acid fluorides are conveniently obtained by heating the acid chlorides with potassium fluoride (*767, 788*) or the free acids with benzoyl chloride and potassium fluoride (*767, 788*) (Equations 125, 126, p. 105), or with acid potassium fluoride (*786*) (Table 20, p. 110).

*Anhydrides* of the fluorinated acids are formed by heating the acids with phosphorus pentoxide (*118*) (p. 281). *Trifluoroacetic anhydride* is a very important esterification and acylation reagent and is used for converting carboxylic acids to the mixed anhydrides which act as strong acylating agents (p. 279).

*Esters of fluorinated acids* result from the reaction of the esters of halogenated acids with fluorinating reagents (*822*) (Equations 101, 127, 129, 130, p. 99, 105), or they are produced by esterification of free fluorinated acids (*436, 872*) (Equation 418, p. 197).

Both ways of preparation hold for the preparation of amides and nitriles of fluorinated acids: (*25, 559*) (Equation 128, p. 105), (*142, 309, 655*) (Equation 577, p. 243). Amides of $\alpha$-H-perfluorocarboxylic acids are formed by the addition of primary or secondary amines to perfluoroolefins (*180, 564*) (Equation 603, p. 250), the corresponding nitriles by similar addition of ammonia (*608*) (Equation 587, p. 245). *N-bromo-derivatives of fluorinated acetamides* (*804*), *perfluorosuccinimide* (*489*) and *perfluoroglutarimide* (*489*) are prepared by the action of bromine and silver oxide on the fluorinated amides or imides (Equation 716, p. 285) and function as special halogenating agents (p. 286).

Among the derivatives of fluorinated acids mention should be made of *peroxy acids. Trifluoroperacetic acid* arises from treating highly concentrated hydrogen peroxide with trifluoroacetic acid (*247, 249*) or better still with trifluoroacetic anhydride (*246, 247, 250, 252, 253, 902*) (p. 284) and acts as a special oxidizing reagent (p. 282). A compound resulting from the action of elemental fluorine upon trifluoroacetic acid can be considered as the fluoride of trifluoroperacetic acid (*153*):

261          $CF_3-COOH \xrightarrow[25°]{F_2,\ N_2,\ H_2O} CF_3-COOF$  25%          [*153*]

## PREPARATION OF FLUORINATED SULPHUR DERIVATIVES

Sulphur derivatives containing fluorine are not very common. *Fluorinated sulphides* are obtained from the chlorinated derivatives by replacement of halogens (*687*) (Equation 103, p. 99). The reaction of $\alpha,\alpha,\alpha,\alpha'$-tetrachlorodimethyl sulphide with antimony fluorides gives chlorofluorosulphides with varying proportions of fluorine atoms (*1047*):

262    $CH_2Cl-S-CCl_3 \xrightarrow[80°]{SbF_3,\ SbCl_5} CH_2Cl-S-CClF_2 + CH_2Cl-S-CF_3$    [*1047*]
                                                    9%                 66%

Electrolytic fluorination of dimethylsulphide produces perfluoro-derivatives containing hexacovalent sulphur (*175*) (Equation 62, p. 86). Similar compounds are obtained by the fluorination of methylmercaptan (*944*) or carbon disulphide (*944*) with fluorine or cobalt trifluoride (Equation 39, p. 72) or by electrolysis in anhydrous hydrogen fluoride (*945*) (Equation 40, p. 72). Fluorination of carbon disulphide with mercuric fluoride (*753*) or iodine penta-fluoride (*376*) leads to perfluorodimethyl disulphide (Equation 39, p. 72).

Fluorinated sulphides are normally oxidized to *fluorinated sulphones* (*1047*) (Equations 323, 324, p. 171). Oxidation by chlorine of fluorinated isothio-cyanates (*736*), prepared by alkylation of potassium thiocyanate (*510*) (Equa-tion 335, p. 174), or of isothiuronium salts (*736*) results in the formation of *fluorinated sulphonyl chlorides*:

**263**                                                                 [*736*]

$$FCH_2\text{—}CH_2OH \xrightarrow[\substack{2.\ NaOH \\ 3.\ CS(NH_2)_2}]{1.\ CH_3C_6H_4SO_2Cl} FCH_2\text{—}CH_2\text{—}S\text{—}C\diagup^{NH}_{\diagdown NH_2.CH_3C_6H_4SO_3H} \quad 73\%$$

$$\xrightarrow{Cl_2} FCH_2\text{—}CH_2\text{—}SO_2Cl \quad 51\%$$

*Polyfluorosulphonic acids* arise from the addition of alkali sulphites to poly-fluoroolefins (*180, 589*) (Equations 414, 415, p. 196). Perfluorosulphonic acids are conveniently prepared by hydrolysis of perfluorosulphonyl fluorides pro-duced by the electrolytic fluorination process from sulphonyl chlorides (*321, 322*) (Equation 61, p. 86). They are acids of an unusual strength (p. 309). Fluorinated sulphonic acids of the aromatic series can be prepared by the Schiemann reaction (*884*), or by sulphonation of aromatic fluoro-derivatives (*504, 774*) (Equations 416, 417, p. 196).

## PREPARATION OF FLUORINATED AMINES AND OTHER NITROGEN DERIVATIVES

*Fluorinated aminoderivatives* are accessible by several methods: by alkyla-tion of ammonia (*989*) (Equation 558, p. 239) or primary or secondary amines (*212, 598*) (Equations 559, 560, p. 239), by addition of ammonia (*563, 1064*) (Equations 585, 586, p. 245) or amines (*858*) (Equation 602, p. 249) to fluo-rinated unsaturated compounds, by reduction of aliphatic (*818*) and aromatic nitrocompounds (*66*) (Equation 282, p. 159), by reduction of amides (*684*) and nitriles of fluorinated acids (*678*) (Equation 294, p. 162), by reduction of oximes (*543*) (Equation 281, p. 159) and by the Curtius degradation (*482*) (Equation 544, p. 235).

Entirely new and interesting compounds originate from the reaction of amines with elemental fluorine (*1037*), cobalt trifluoride (*355, 356*) or by electrolytical fluorination of amines (*547*). Tertiary aliphatic amines afford *perfluoroalkylamines* (*547*) (Equation 59, p. 85), secondary ones give *perfluoro-dialkylfluoroamines* (*1037*)

264        $CH_3-NH-CH_3$ $\xrightarrow[\substack{CoF_3 \\ 130-220°}]{F_2, N_2}$ $CF_3-NF-CF_3$ 40—70%        [*1037*]

and primary amines *perfluoroalkyldifluoroamines* (*356*):

265        $CH_3-NH_2$ $\xrightarrow[190°]{CoF_3}$ $CF_3-NF_2$        [*356*]

Cyclic perfluorinated amines are obtained in poor yields by fluorination of pyridine and its homologues with fluorine or cobalt trifluoride (*355, 356, 392*). These compounds do not possess any basic properties and are very stable towards acids and alkalis (*1037*).

266        [*355*]

Fluoroamines were also isolated from the reaction of acetonitrile with fluorine (*190*), and with mercuric fluoride (*515, 765*) (Equation 35, p. 71). Fluorination of malonic dinitrile produced among other products an interesting cyclic compound, *perfluoropyrazolidine* (*23*) (Equation 36, p. 71).

Pyrolysis of *perfluoromethyloxazetidine* obtained by the addition of trifluoronitrosomethane to tetrafluoroethylene (*43*), and pyrolysis of *perfluoro-dimethylcarbamyl fluoride* obtained by electrolysis of dimethylcarbamyl chloride in anhydrous hydrogen fluoride (*1099*) afford a new interesting compound, *perfluoro(methylmethylene)imine* (Equations 689, 690, p. 277).

*Fluorinated isocyanates* are formed by the reaction of fluorinated primary amines with phosgene (*739, 874*) (Equation 579, p. 243), by the Curtius degradation of fluorinated acid azides (*46*) (Equation 545, p. 235), and by careful hydrolysis of perfluoro(alkylalkyleneimines) (*46*) (Equation 451, p. 208).

Fluorinated nitro-, nitroso-, azo- and diazo-compounds are usually obtained by introducing the corresponding functions into molecules which already contain fluorine. Thus *fluorinated aliphatic nitrocompounds* result in good yields from treatment of fluorohalogen compounds with silver nitrite (*815*) (Equation 404, p. 192). *1-Nitroperfluoroalkanes* have been prepared by the action of dinitrogen tetroxide upon perfluoroalkyl iodides, or better still by oxidation of nitrosoperfluoroalkanes obtained by treatment of perfluoroalkyl iodides with nitric oxide (*45, 536*) (Equation 405, p. 192). A very convenient way to aliphatic nitrocompounds is addition of dinitrogen tetroxide to fluorinated olefins. The primary products, nitro-nitrites (*566*), undergo further transformations to fluorinated nitro-acids which possess significant stability (*563, 567*).

267 $\begin{array}{c} CF_3 \\ \diagdown \\ CF_3 \diagup \end{array} C{=}CF_2 \xrightarrow{N_2O_4} \begin{array}{c} CF_3 \\ \diagdown \\ CF_3 \diagup \end{array} \underset{NO_2\ ONO}{C{-}CF_2} \xrightarrow{H_2O} \begin{array}{c} CF_3 \\ \diagdown \\ CF_3 \diagup \end{array} \underset{NO_2}{C{-}COOH}$ [563]

*Aromatic fluorinated nitrocompounds* are accessible by nitration of fluorohydrocarbons (*979, 1102*) (Equations 406, 407, 408, p. 193, 194). Aromatic nitrocompounds having fluorine in *ortho-* or *para*-positions to the nitro group can also be obtained from the corresponding chloro-derivatives (*276*).

268

$\xrightarrow[\text{var}]{KF,\ C_6H_5NO_2}$  76%  $\xleftarrow[\substack{H_2SO_4\ (d\ 1,84) \\ 20°\rightarrow100°}]{HNO_3\ (d\ 1,52)}$  87%  [276,1102]

The reaction of trifluoroiodomethane with nitric oxide leads to *trifluoronitrosomethane* (*43*), which disproportionates to *trifluoronitromethane* and *hexafluoroazoxymethane* (*537*).

269    $CF_3{-}NO \xrightarrow[\text{7,5 \% NaOH}]{C,\ 100°} CF_3{-}N{=}N{-}CF_3$  47%    [537]
          $\underset{O}{|}$

*Hexafluoroazomethane* results from the fluorination of cyanogen iodide with iodine pentafluoride (*854*), or from a reaction of cyanogen chloride with silver difluoride (*315*) (Equation 37, p. 71).

*Aromatic diazonium compounds* are readily accessible by diazotization of aromatic amines (*978*) (Equation 412, p. 195). In this way, perfluoroaniline yields a diazonium compound which reacts with another molecule of perfluoroaniline to give *perfluorodiazoaminobenzene* (*284*).

270                                                                    [284]

Owing to the strong stabilizing effect of the perfluoroalkyl group, perfluoroalkylmethylamines afford by diazotization perfectly stable *1H, 1-diazoalkanes (309)* (Equation 410, p. 194).

## PREPARATION OF FLUORINATED DERIVATIVES OF SILICON, PHOSPHORUS AND OTHER ELEMENTS

Halogen atoms attached to silicon and phosphorus are much more reactive than halogens bound to carbon. Therefore fluorinated compounds of silicon and phosphorus are readily made from other halogen derivatives of these elements by mild fluorinating agents (*109, 230, 695*), even those which are unable to replace halogen attached to carbon (*244, 770, 911, 928, 1087*) (Equations 64, 84, 85, 143, 145, 148, p. 88, 95, 110). *Fluoro-derivatives of silicon* can also be prepared by the action of hydrogen fluoride upon the appropriate silicon alkoxyderivatives (*691*) (Equation 156, p. 112), and by the addition of some silicon compounds to fluorinated olefins (*305, 672*) (Equation 594, 595, p. 247).

One of the commercially important fluorinated derivatives of phosphorus is *diisopropyl fluorophosphate*, which is prepared from phosphorus trichloride, isopropyl alcohol, chlorine, and sodium fluoride in one reaction vessel (*911*).

[*911*]

PCl$_3$ + 2 (CH$_3$)$_2$CHOH

$$\xrightarrow[55-60°]{CCl_4} [(CH_3)_2CHO]_2POH \xrightarrow[CCl_4]{Cl_2} [(CH_3)_2CHO]_2\,P{<}^{O}_{Cl}$$

**271**    $\xrightarrow[\text{reflux}]{\text{NaF, CCl}_4} [(CH_3)_2CHO]_2P{<}^{O}_{F}$   75%

Similar one-step reaction between phosphorus oxychloride, dimethylamine and sodium fluoride gives *tetramethyldiamidofluorophosphate (905)*.

**272**   POCl$_3$ + 2 (CH$_3$)$_2$NH  $\longrightarrow$  $[(CH_3)_2N]_2\,P{<}^{O}_{Cl}$  $\xrightarrow{\text{NaF} \atop C_6H_6}$  $[(CH_3)_2N]_2\,P{<}^{O}_{F}$  [*905*]

*Perfluoroalkylderivatives of sulphur (404), phosphorus (147) and arsenic (240)* result from the reaction of perfluoroalkyl iodides with the appropriate elements (Equation 557, 652, p. 239, 263). The interesting *iodide fluorides*, obtainable by treatment of iodoso-derivatives with aqueous hydrofluoric acid (*1071*) (Equation 153, p. 112), have been mentioned in the preceding chapter.

Fluorinated derivatives of magnesium, lithium, zinc and mercury will be dealt with later (p. 213).

# VI

# REACTIONS OF ORGANIC FLUORINE COMPOUNDS

FLUORINATED derivatives include almost every type of organic compounds and there are many reactions to which they are subjected. In most reactions fluorinated compounds behave like the non-fluorinated ones; some reactions are, however, specific. Preparative reactions will be fully described and side-reactions will be briefly mentioned.

The individual reactions will be classified as unambiguously as possible, and any remaining overlap will be taken care of by cross-references. The sequence will be: reduction, oxidation, halogenation, nitration, nitrosation, sulphonation, esterification, hydrolysis, Friedel–Crafts reaction, reactions with organometallic compounds, basic condensation, molecular rearrangements, alkylation, arylation and acylation, addition, elimination, and pyroreactions.

## REDUCTION

The paragraph on reduction will include catalytic hydrogenation, reduction with complex metal hydrides, with metals, with inorganic compounds and with organic compounds; the reduced functions will include: carbon-carbon double bond and triple bond, aromatic nuclei, carbonyl, carboxyl, nitrile and nitro groups, and the carbon-halogen bond.

Generally speaking, fluorine, either as a single atom or as perfluorinated clusters, exerts little influence on the reduction results. Many fluorinated compounds have been reduced under standard conditions with current yields.

The carbon-fluorine bond resists the action of most but not all reducing reagents. Replacement of fluorine by hydrogen was observed during some catalytic hydrogenations (115, 573, 601, 860, 997), exceptionally during reduction with lithium aluminium hydride (193, 194, 232), sodium (659, 700), zinc (435, 519), aluminium amalgam (519), during electroreduction (237), and some radical-type hydrogenations (534).

Some of the reductions are important as general preparative methods: the reduction of perfluorocarboxylic acids to perfluoroalkylcarbinols and perfluoroaldehydes, the reduction of perfluorinated amides and nitriles to perfluoroalkylamines.

## Catalytic Hydrogenation

In most cases the fluorinated compounds are hydrogenated normally, sometimes more slowly and less easily than their non-fluorinated models but usually with the expected results.

Hydrogenation of a carbon-carbon double bond was effected on palladium (573), nickel (455, 573) or Raney nickel (150):

273          $CF_3—CF{=}CF_2$  $\xrightarrow[20°]{H_2/Pd}$  $CF_3—CHF—CHF_2$   96%          [573]

274     $CF_3—CH{=}CH_2$  $\xrightarrow[125-200°,\ 250-300\ atm]{H_2/Ni\ (Kieselguhr),\ C_2H_5OH}$  $CF_3—CH_2—CH_3$  quant.     [455]

275          $CF_3—\underset{\underset{CH_2}{\|}}{C}—COOH$  $\xrightarrow[20°,\ 1\ atm]{H_2/Ni\ (Raney)}$  $CF_3—\underset{\underset{CH_3}{|}}{CH}—COOH$          [150]

Hydrogenation of fluorinated alkynes leads to fluorinated olefins and paraffins (441):

276                                                                                          [441]

$CF_3—C{\equiv}C—CF_3$  $\xrightarrow[C_2H_5OH,\ 100\ atm]{H_2/Ni\ (Raney)}$  $CF_3—CH{=}CH—CF_3$ + $CF_3—CH_2—CH_2—CF_3$

                                                            34%                         21%

Ketones are easily hydrogenated on platinum (1005), platinum oxide (311, 543), palladium (543) or Raney nickel (454).

277          $CF_3—CO—CH_3$  $\begin{array}{c} \xrightarrow[50°,\ 48\ atm]{H_2/Pt} 94\% \\[6pt] \xrightarrow[20°,\ 54\ atm]{H_2/PtO_2,\ H_2O} 90\% \end{array}$  $CF_3—\underset{\underset{OH}{|}}{CH}—CH_3$     [311, 1005]

Hydrogenation of the carboxyl group is mentioned in only one instance: conversion of trifluoroacetic anhydride to a mixture of trifluoroethyl trifluoroacetate, 2,2,2-trifluoroethanol and trifluoroacetic acid (1009). Ethyl and butyl trifluoroacetates, on the other hand, resisted the action of hydrogen on copper chromite at 250°C and 245 atm. and were recovered (311).

The Rosenmund reduction is a suitable way to obtain fluorinated aldehydes. Applied to perfluorobutyryl chloride, it gives a mixture of perfluorobutyraldehyde (hydrate) and 1H,1H-heptafluorobutyl alcohol (523).

278                                                                                          [523]

$CF_3—CF_2—CF_2—COCl$  $\xrightarrow[\text{S-quinoline}]{H_2/Pd\ (C)}$

$+\ \begin{array}{l} CF_3—CF_2—CF_2—CH_2OH \\ CF_3—CF_2—CF_2—CH(OH)_2 \end{array}$  $\xrightarrow{H_2SO_4}$  $CF_3—CF_2—CF_2—CHO$  32%

By hydrogenation over platinum, trifluoroacetamide yielded 2,2,2-trifluoroethanol (*311*),

279 $\quad$ CF₃—CONH₂ $\xrightarrow[\text{90°, 105 atm}]{\text{H}_2/\text{PtO}_2,\,(\text{C}_2\text{H}_5)_2\text{O}}$ CF₃—CH₂OH $\quad$ 76.5% $\qquad$ [*311*]

trifluoroacetonitrile gave 2,2,2-trifluoroethylamine (*309*).

280 $\quad$ CF₃—C≡N $\xrightarrow[\text{55—60°, 105 atm}]{\text{H}_2/\text{PtO}_2,\,(\text{C}_2\text{H}_5)_2\text{O}}$ CF₃—CH₂—NH₂ $\quad$ 80% $\qquad$ [*309*]

Hydrogenation of a fluorinated oxime over palladium afforded a primary amine in high yield (*543*).

281 $\quad$ ⟨⟩—CH₂—C—CF₃ (‖ NOH) $\xrightarrow[\text{150°, 100 atm}]{\text{H}_2/\text{Pd(C)},\,(\text{C}_2\text{H}_5)_2\text{O}}$ ⟨⟩—CH₂—CH—CF₃ (| NH₂) $\quad$ 87% $\quad$ [*543*]

As an example for hydrogenation of fluorinated nitrocompounds, conversion of *p*-fluoronitrobenzene to *p*-fluoroaniline can be quoted (*66*). The same compound was also obtained by hydrogenation of nitrobenzene in an excess of anhydrous hydrogen fluoride.

282 $\quad$ F—⟨⟩—NO₂ $\xrightarrow[\text{85°, 85 atm}]{\text{H}_2/\text{Ni(Raney)}}$ F—⟨⟩—NH₂ $\xleftarrow[\text{HF}]{\text{H}_2/}$ ⟨⟩—NO₂ $\quad$ [*66*]

The behaviour of the carbon-halogen including the carbon-fluorine bond during hydrogenations will be described more thoroughly. In chlofluoro-derivatives, chlorine is replaced by hydrogen (*80, 573, 712*).

283 $\quad$ CCl₂F₂ $\xrightarrow[\text{685°}]{\text{H}_2}$ CHClF₂ + CH₂F₂ + CHF₂—CHF₂ + CF₂=CF₂ $\qquad$ [*80*]
$\qquad\qquad\qquad\qquad$ 36% $\qquad$ 19% $\qquad$ 14% $\qquad\quad$ 13%

284 $\quad$ CF₂=CCIF $\xrightarrow[\text{20°}]{\text{H}_2/\text{Pd}}$ CF₂=CHF + CHF₂—CH₂F $\qquad\qquad$ [*573*]
$\qquad\qquad\qquad\qquad\qquad$ 60% $\qquad\quad$ 25%

Fluorine atoms stand the majority of hydrogenations without change (*455, 822, 997, 1003*). However there are quite a few examples for the fission of the carbon-fluorine bond. Curiously enough *p*-fluorobenzoic acid lost its fluorine during hydrogenation over platinum black (*997*), whereas hydrogenation over palladium on calcium carbonate in the presence of alkali, which are the conditions suitable for eliminating other halogens, left the *p*-fluorobenzoic acid unchanged (*520*).

285 $\quad$ F—⟨⟩—COOH $\xrightarrow[\text{H}_2\text{O, NaOH}]{\text{H}_2/\text{Pt black}}$ ⟨⟩—COOH → ⟨⟩—COOH $\quad$ [*997*]

Propyl fluoride and isopropyl fluoride resist hydrogenation in the presence of palladium on activated charcoal at low temperatures. At 110°C hydrogenation of isopropyl fluoride to propane amounts to 5–10%, and at 155–235°C it is practically quantitative (601). Hydrogenation of ethyl fluoride and methyl fluoride is still more difficult and reaches only 10% at 240°C (601).

During hydrogenation of $\alpha,\alpha$-difluorodesoxybenzoin the two fluorine atoms are replaced by hydrogen (115).

**286**               [115, 700]

$$C_6H_5-CF_2-CH_2-C_6H_5 \xrightarrow[C_2H_5OH]{H_2/Pd} \Big\downarrow \quad \Big\downarrow \xrightarrow{\ Na\ }_{CH_3OH} \quad C_6H_5-CF=CF-C_6H_5$$

$$C_6H_5-CH_2-CH_2-C_6H_5$$

Quite unexpectedly, even perfluoro-derivatives show some tendency to split out fluorine during hydrogenation. Whereas hydrogenation of tetrafluoroethylene (573) and perfluoroisobutylene (573) over palladium ends by saturating the double bonds, hydrogenation of these compounds in the presence of nickel at 90–100°C leads to partial replacement of fluorine by hydrogen (573).

**287**               [573]

$$\xleftarrow[20°]{H_2/Pd} \quad (CF_3)_2C=CF_2 \quad \xrightarrow[100°]{H_2/Ni}$$

$$(CF_3)_2CH-CHF_2 \qquad\qquad (CF_3)_2CH-CHF_2 \; + \; (CF_3)_2CH-CH_3$$

$$96\% \qquad\qquad\qquad\qquad 10\% \qquad\qquad\qquad 75\%$$

Perfluorobenzene is converted to pentafluorobenzene by treatment with hydrogen on palladium or platinum on activated charcoal (281, 860). The same compound is obtained by desulphuration of pentafluorothiophenol with Raney nickel (1028).

**288**               [860, 1028]

$$C_6F_6 \xrightarrow{H_2/Pt\ or\ Pd(C)} C_6F_5H \xleftarrow{Ni\ (Raney)} C_6F_5SH$$

Whereas perfluorocyclohexane stands treatment with hydrogen at 830–850°C without change (534), aliphatic perfluoroparaffins are split in the carbon chain (534):

**289**    $CF_3-CF_2-CF_3 \xrightarrow[775-840°]{H_2} CHF_3 \; + \; CHF_2-CF_3 \; + \; CH_2F_2 \; + \; C$    [534]

                                     73%        34.5%        traces     5.5%

Perfluorinated derivatives of arsenic are relatively very resistant to hydrogenation. The action of hydrogen at 220–240°C for 93 hours broke tris(trifluoromethyl)arsine to only 2% of fluoroform and 89% of the starting material was recovered (*242*).

## Reduction with Complex Metal Hydrides

Reduction with complex metal hydrides and especially with lithium aluminium hydride is a very convenient preparative method. Since the carbon-fluorine bond is only exceptionally affected, lithium aluminium hydride is frequently used for reducing oxygen and nitrogen functions and for reductive eliminations of other halogen atoms.

Fluorinated ketones are smoothly reduced to fluorinated alcohols (*90, 658*),

290      $CF_3-CO-CHFBr \xrightarrow{\text{LiAlH}_4} CF_3-\underset{\underset{OH}{|}}{CH}-CHFBr$    91%      [*658*]

Free acids give aldehydes or alcohols, depending on the conditions used; usually both types are formed together (*127, 523*).

291                                                               [*127*]

$CF_3-COOH \xrightarrow[\substack{(C_2H_5)_2O \\ -5°--0°}]{\text{LiAlH}_4} CF_3-CH_2OH + CF_3-CH(OH)_2 \xrightarrow[P_2O_5]{H_2SO_4} CF_3-CHO$

                                                  21%                                          77·5%

Esters of fluorinated acids are converted to alcohols (*159*), but under special conditions, especially low temperature and progressive addition of the hydride, to fluorinated aldehydes (*840*).

292                                                                               [*159,840*]

$$CF_3-CHO \xleftarrow[\substack{C_2H_5)_2O \\ -80°}]{\substack{\text{LiAlH}_4 \text{ equiv.}}} CF_3-COOR \xrightarrow[\substack{(C_2H_5)_2O \\ \text{reflux}}]{\text{LiAlH}_4 \text{ excess}} CF_3-CH_2OH$$

CF$_3$—CHO   71%                                                  CF$_3$—CH$_2$OH   76%

*Procedure 19.*

    *Reduction of Esters of Perfluorocarboxylic Acids with Lithium Aluminium Hydride*
    *Preparation of Perfluoroaldehydes and Pseudoperfluoroalcohols**

    (*a*) *Preparation of perfluoroaldehydes (840).* A half-litre three-necked flask fitted with a gas-tight stirrer, a reflux condenser carrying a calcium chloride tube, and a nitrogen inlet is charged with 300 ml. of anhydrous ether into which 10 g (0·26 mole) of finely powdered lithium aluminium hydride has been added. The flask content is stirred in a nitrogen atmosphere for 1–2 hours, and allowed to settle.

---

   * Pseudoperfluoroalcohols are compounds of the general formula $C_nF_{2n+1}CH_2OH$.

11

In one-litre three-necked flask equipped with gas-tight stirrer, a reflux condenser, and a separating funnel fitted with a pressure equalizer, one mole of a dry ester of per-fluorocarboxylic acid is dissolved in one and a half of volume of absolute ether, the flask is cooled with dry ice, and the separating funnel is filled with the solution of lithium aluminium hydride in ether decanted from the first flask. The sediment is stirred with a new portion of ether and the solution is decanted to the separating funnel. While passing nitrogen through the apparatus, the content of the separating funnel is added over a period of 3 hours into the cooled ethereal solution of the ester. Thereafter 25 ml. of 95% ethanol is dropped in from the separating funnel, the mixture is allowed to warm to room temperature and poured into a two-litre beaker containing a mixture of ice and approximately 75 ml. of concentrated sulphuric acid. The aqueous layer is separated and extracted twice with ether. The combined ethereal solutions are evaporated, the residue is treated with an equal volume of concentrated sulphuric acid, and distillation is continued, using a properly cooled receiver. Redistillation of the distillate yields 70–80% of pure perfluoroaldehyde.

(b) *Preparation of 2,2,2-Trifluoroethanol (159)*. In a one-litre three-necked flask fitted with a gas-tight stirrer, a reflux condenser, and a separating funnel, 116 g (0·68 mole) of butyl trifluoroacetate is added in the course of two hours to a stirred solution of 17·8 g (0·47 mole) of lithium aluminium hydride in 500 ml. of absolute ether. The heat of reaction brings the mixture to boiling. After completing the addition the mixture is refluxed for 15 min., the excess of the lithium aluminium hydride is destroyed by the addition of water, and the content of the flask is poured on a mixture of ice, dilute sulphuric acid and ether. The aqueous layer is separated, extracted with two 100 ml. portions of ether, the combined ethereal solutions are evaporated, and the residue is distilled with phosphorus pentoxide in order to remove water from the product. The yield o 2,2,2-trifluoroethanol amounts to 52 g (0·52 mole) (76%).

Lithium aluminium hydride reduction of ethyl fluoroformate afforded fluoromethanol (790). The same compound was also prepared from formyl fluoride (790).

293 $\quad$ $FCOOC_2H_5$ $\xrightarrow[(C_2H_5)_2O]{LiAlH_4}$ $FCH_2OH$ $\xleftarrow[(C_2H_5)_2O]{LiAlH_4}$ $HCOF$ $\qquad$ [790]

$\qquad\qquad\qquad$ 50% $\qquad$ 2·5%

Reduction of acid halides with lithium aluminium hydride affords alcohols (437). Fluorinated amides are reduced mainly to amines (99, 523, 684);

294 $\qquad$ $CF_3—CONH_2$ $\xrightarrow[reflux]{LiAlH_4,\ (C_2H_5)O_2}$ $CF_3—CH_2—NH_2$ $\quad$ 75% $\qquad$ [684]

substituted amides give secondary or tertiary amines (99, 692). The reduction can also be performed by sodium borohydride with an equivalent of aluminium chloride or boron trifluoride etherate. Sodium borohydride itself does not effect the reduction (99).

295 $\ CF_3—CON(CH_3)_2$ 
$\qquad$ $\nearrow$ $\overline{\quad LiAlH_4 \quad}$ $\searrow$ $\quad$ 11–32%
$\qquad$ $\overline{NaBH_4.BF_3.(C_2H_5)_2O}$ $\rightarrow$ 31–64·7% $\quad CF_3—CH_2N(CH_3)_2$ $\quad$ [99]
$\qquad$ $\searrow$ $\overline{\quad NaBH_4.AlCl_3 \quad}$ $\nearrow$ $\quad$ 16·4–47·9%

In special instances aldehydes can be obtained by reducing properly substituted amides (*1063*):

**296** [*1063*]

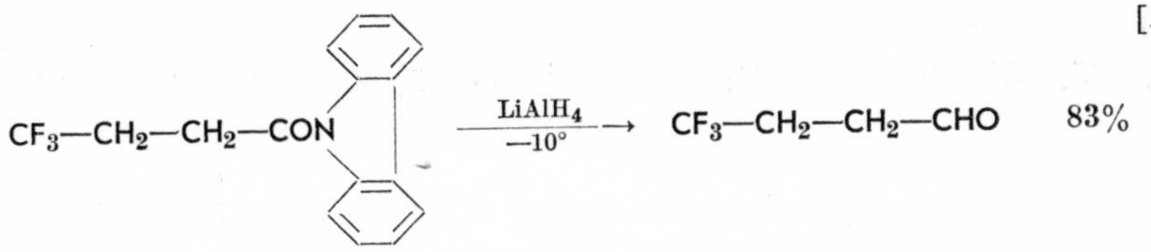

$$CF_3-CH_2-CH_2-CON\Big< \xrightarrow[-10°]{LiAlH_4} CF_3-CH_2-CH_2-CHO \quad 83\%$$

Fluorinated nitriles yield generally amines, but under special conditions the intermediate aldimine can be intercepted which gives aldehyde on hydrolysis (*468*):

**297** $\qquad CF_3-CN \xrightarrow[-80 \to 20]{LiAlH_4,(C_2H_5)_2O} CF_3-CHO \quad 46\%$ [*468*]

Lithium aluminium hydride cleaves readily the carbon-halogen bond. Even in halogenoperfluoro-derivatives, which are very stable toward reduction, the halogens are replaced by hydrogen when exposed to lithium aluminium hydride (*151, 386, 890*).

**298** [*386*]

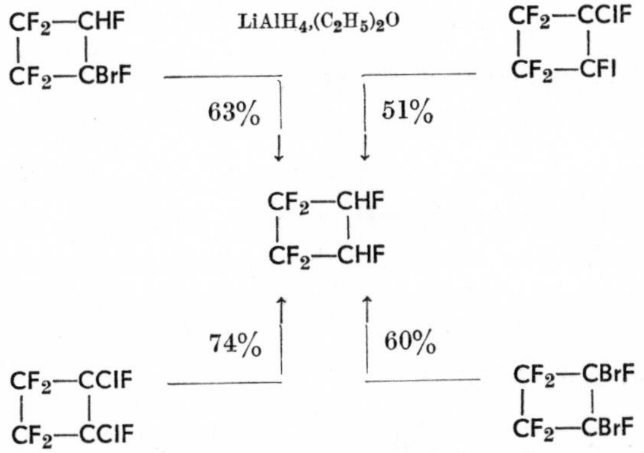

$$
\begin{array}{ccc}
\underset{\begin{subarray}{l}|\quad\ |\\ CF_2-CBrF\end{subarray}}{CF_2-CHF} & \xrightarrow{LiAlH_4,(C_2H_5)_2O} & \underset{\begin{subarray}{l}|\quad\ |\\ CF_2-CFI\end{subarray}}{CF_2-CClF} \\
& 63\% \quad\quad 51\% & \\
& \downarrow\qquad\downarrow & \\
& \underset{\begin{subarray}{l}|\quad\ |\\ CF_2-CHF\end{subarray}}{CF_2-CHF} & \\
& 74\%\ \uparrow\qquad\uparrow\ 60\% & \\
\underset{\begin{subarray}{l}|\quad\ |\\ CF_2-CClF\end{subarray}}{CF_2-CClF} & & \underset{\begin{subarray}{l}|\quad\ |\\ CF_2-CBrF\end{subarray}}{CF_2-CBrF}
\end{array}
$$

The carbon–fluorine bond is very resistant to the attack of lithium aluminium hydride. There are, however, a few instances recorded in which replacement of fluorine by hydrogen has occurred. Partial removal of fluorine was observed in the reduction with lithium aluminium hydride of tribromofluoromethane (*232*).

**299** $\qquad CBr_3F \xrightarrow{LiAlH_4} CH_3F + CH_4$ [*232*]

In the reduction of perfluoroalkylurethanes (*193*) and perfluoroalkyliso-cyanates (*194*) with lithium aluminium hydride, the fluorine atoms, bound to the carbon adjacent to nitrogen, are replaced by hydrogen:

**300**                                                                                    [*193, 194*]

$$CF_3-CF_2-CF_2-NH-COOC_2H_5 \xrightarrow{\text{LiAlH}_4} CF_3-CF_2-CH_2-NHCOOC_2H_5 \quad 63\%$$

$$\xrightarrow{\text{LiAlH}_4,\text{ excess}}$$

$$\downarrow 60\%$$

$$CF_3-CF_2-CF_2-NCO \xrightarrow{\text{LiAlH}_4} CF_3-CF_2-CH_2-NH-CH_3 \quad 73.5\%$$

## Reduction with Metals

Reductions of fluorinated compounds with metals are not very frequent. Treatment of $\alpha$-chloro-$\alpha,\alpha$-difluorotoluene with sodium amalgam resulted in the replacement by hydrogen of the chlorine atom while the fluorine atoms stayed unattacked (*985*). The $-CF_2-$ group in the chain of ethyl difluoro-caprate resists the action of sodium in ethanol (*772*). On the other hand reduction with metals easily occurs in the vicinity of electron attracting groups. Thus treatment with sodium in methanol converted $\alpha,\beta$-difluorostilbene to 1,2-diphenylethane (*700*) (Equation 286, p. 160), and ethyl pentafluoroaceto-acetate was reduced to ethyl $\gamma,\gamma,\gamma$-trifluoroacetoacetate by sodium in ether (*659*).

**301**                                                                                    [*659*]

$$CF_3-CO-CF_2-COOC_2H_5 \xrightarrow{\text{Na, (C}_2\text{H}_5)_2\text{O}} \begin{array}{l} CF_3-CO-CF_2-COOC_2H_5 \\ CF_3-CO-CH_2-COOC_2H_5 \end{array} \Big\} \quad 12\text{ g}$$

$$23\text{ g}$$

Zinc is a favourite reagent for reductive cleavage of carbon-halogen bonds. In paraffinic chains chlorine is rather resistant to the reduction with zinc, while bromine splits out readily. Dibromofluoromethane treated with zinc in 70% ethanol affords a mixture of bromofluoromethane and fluoromethane (*992*). Another example is removal of bromine from 2-bromo-4-chloro-4,4-di-fluoro-2-methylbutane (*1025*):

**302**
$$\begin{array}{c} CH_3 \\ \phantom{x} \diagdown \\ CH_3 \diagup \end{array} \begin{array}{c} C-CH_2-CCIF_2 \\ | \\ Br \end{array} \xrightarrow[\text{reflux}]{\text{Zn, HCl}} \begin{array}{c} CH_3 \\ \phantom{x} \diagdown \\ CH_3 \diagup \end{array} CH-CH_2-CCIF_2 \quad 57\% \quad [1025]$$

In vinylic systems flanked by electron-attracting groups, chlorine (*435*) and iodine (*366*) are smoothly replaced by hydrogen.

**303**     $$CF_3-CCl=CCl-CF_3 \xrightarrow{\text{Zn or Fe}}_{\text{HCONH}_2} CF_3-CH=CH-CF_3 \quad (trans) \quad [435]$$

**304**     $$CCl_3-CH=CI-CF_3 \xrightarrow{\text{Zn, CHI}} CCl_3-CH=CH-CF_3 \quad 63\% \quad [366]$$

Fluorine atom seems to be stable toward zinc in alkaline (*338*), neutral (*992*) or acidic medium (*366, 1025*). Interestingly ethyl fluorooxalacetate gave ethyl malate when treated with zinc in acetic acid (*519*) or with amalgamated aluminium (*519*) while treatment with sodium borohydride afforded ethyl fluoromalate (*519*).

**305** $\xrightarrow[\text{CH}_3\text{COOH}]{\text{Zn}}$ $\text{C}_2\text{H}_5\text{OCO}-\text{CHF}-\text{CO}-\text{COOC}_2\text{H}_5$ [*519*]

$\text{AlHg}_x \downarrow (\text{C}_2\text{H}_5)_2\text{O}$ $\qquad\qquad$ $\text{NaBH}_4 \downarrow \text{C}_2\text{H}_5\text{OH}$

$\text{C}_2\text{H}_5\text{OCO}-\text{CH}_2-\underset{\text{OH}}{\text{CH}}-\text{COOC}_2\text{H}_5$ $\qquad\qquad$ $\text{C}_2\text{H}_5\text{OCO}-\text{CHF}-\underset{\text{OH}}{\text{CH}}-\text{COOC}_2\text{H}_5$

$\quad$ 20.4% $\quad$ 26% $\qquad\qquad\qquad\qquad$ 26%

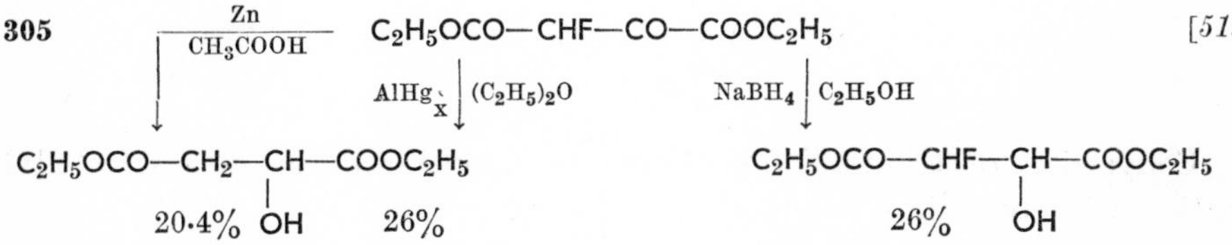

Fluorinated sulphonylchlorides are reduced with zinc to the corresponding sulphinic acids (*378*).

**306** $\qquad$ $\text{CF}_3-\text{SO}_2\text{Cl}$ $\xrightarrow[\text{2. Na}_2\text{CO}_3]{\text{1. Zn, H}_2\text{O, 25°}}$ $\text{CF}_3-\text{SO}_2\text{Na.H}_2\text{O}$ $\quad$ 38% $\qquad$ [*378*]

## Reduction with Inorganic Compounds

Stannous chloride was used for the reduction of *o*-fluorobenzenediazonium chloride to *o*-fluorophenylhydrazine in a yield of 55% (*978*); sodium hydrogen sulphide failed in this particular case (*978*). Reduction of di-*p*-difluorobenzohydrol with hydriodic acid afforded 75% of di-*p*-difluorodiphenylmethane (*338*). The action of hydroiodic acid on perfluorocyclobutene resulted in the addition of hydrogen across the double bond and splitting of the ring (*385*):

**307** $\qquad$ $\begin{matrix} \text{CF}_2-\text{CF} \\ | \quad\; || \\ \text{CF}_2-\text{CF} \end{matrix}$ $\xrightarrow[275°]{\text{HI}}$ $\text{CHF}_2-\text{CHF}-\text{CHF}-\text{CHF}_2$ $\quad$ 90% $\qquad$ [*385*]

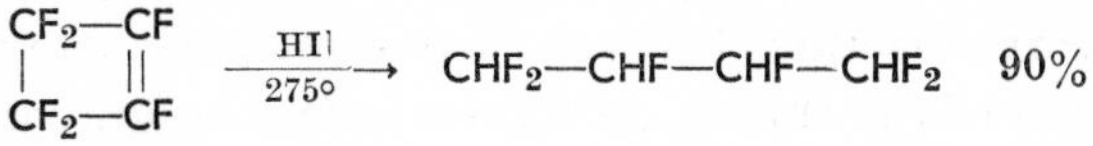

An interesting reduction was recorded during polarographic investigation of *ω*-fluoroacetophenone in which first the fluorine atom is replaced by hydrogen and then the keto-group reduced to an alcoholic function (*237*).

**308** [*237*]

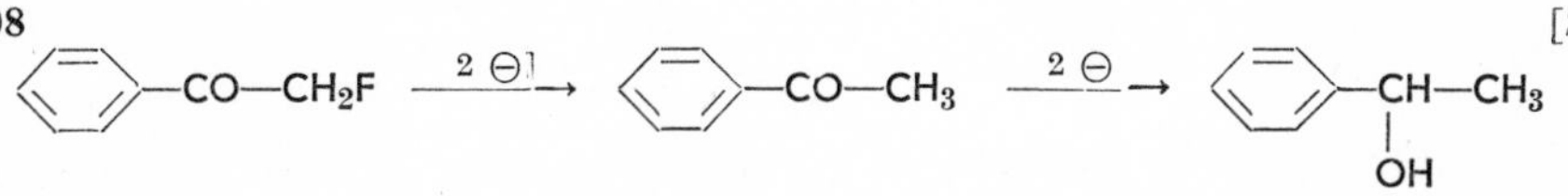

When trifluoroiodomethane is allowed to stand with potassium hydroxide for 15 days at room temperature, or heated for 30 hours at 90°C, 9% and 7%, respectively, of fluoroform is found in the reaction products, in addition to 88% of recovered trifluoroiodomethane and some potassium iodate. A similar behaviour was noted with iodoperfluoroethane which yielded 7% of penta-fluoroethane after a 7 day contact with potassium hydroxide at room temperature (*33*).

**309**     $CF_3I$

- KOH, 20°, 15 days → 9%
- $+C_6H_{14}$, 205°, 24 hrs / $-C_6H_{12}$ (35%) → 13% $CHF_3$          [*32,33*]
- $+C_2H_5OH$, 180°, 24 hrs / $-CH_3CHO$ → 30%

Conversion of trifluoroiodomethane to fluoroform was also effected by heating with organic compounds (see later).

### Reduction with Organic Compounds

An interesting kind of reduction occurs when fluorinated compounds able to split into radicals are heated with organic compounds apt to furnish hydrogen atoms. Heating of trifluoroiodomethane with hexane gives fluoroform and hexane (*32*), with ethyl alcohol fluoroform and acetaldehyde (*32*) (Equation 309. Similarly heating of 1,1,1-trifluoro-3-iodopropane with hexane yields 1,1,1-trifluoropropane (*361*).

**310**     $CF_3—CH_2—CH_2I$

- 1. Mg / 2. $H_2O$ → 75%     $CF_3—CH_2—CH_3$          [*361*]
- $C_6H_{14}$ / 280—300° → 52%

A high-temperature reaction between di(perfluorocyclohexyl) and toluene results in the fission of the carbon-carbon bond in the perfluorocompound and in the transformation of the latter to hydroperfluorocyclohexane while toluene forms dibenzyl (*40*).

**311**          Φ—Φ  +  —$CH_3$          [*40*]

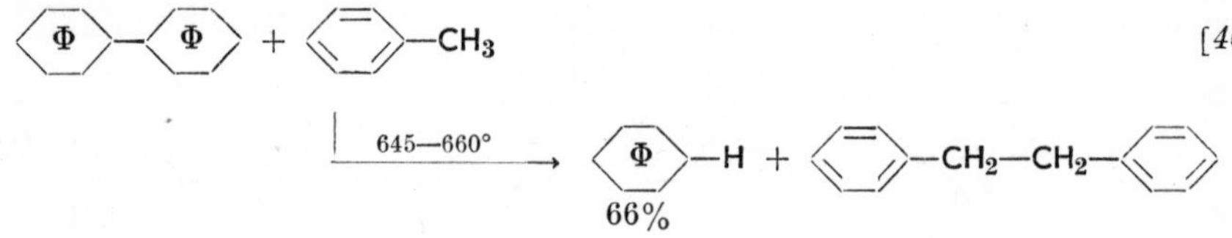

645—660° →  Φ—H  +  —$CH_2$—$CH_2$—

66%

Reduction occurs frequently during the syntheses with organometallic compounds. Though not of preparative importance, this kind of reduction may sometimes prevail over the normal products. Such cases happen especially when the Grignard reagent contains a hydrogen atom in *beta*-position, which is readily detachable with formation of an olefin. During the reaction of tri-fluoroacetaldehyde and pentafluoropropionaldehyde with Grignard reagents the following ratios between the normal addition reaction and abnormal reduction were observed *(657)* (Table 23).

TABLE 23

*Results of the Reaction of Various Grignard Reagents with Trifluoroacetaldehyde and Per-fluoropropionaldehyde (657)*

| Perfluoroaldehyde | $CF_3CHO$ | | $C_2F_5CHO$ | |
|---|---|---|---|---|
| Alkyl of the Grignard Reagent | $CF_3CHROH$ | $CF_3CH_2OH$ | $C_2F_5CHROH$ | $C_2F_5CH_2OH$ |
| R = $CH_3$ | 67% | 0% | 87% | 0% |
| $C_2H_5$ | 60% | 20% | 33·6% | 55·5% |
| $(CH_3)_2CH$ | 0% | 87% | 0% | 90% |
| $(CH_3)_3C$ | 7% | 84% | 14·3% | 76·2% |
| $C_6H_5$ | 88% | 0% | 86% | 0% |
| $C_6H_5CH_2$ | 81% | 0% | 83% | 0% |

A large extent (up to 68%) of reduction was also noted in the reactions of Grignard reagents with perfluoroketones *(657)*, perfluoroesters *(657, 844)*, and perfluoronitriles *(663)*. Sometimes conditions have some influence upon the course and results of the reaction. Thus adding isopropylmagnesium bromide to perfluorobutyronitrile yields mainly perfluoropropyl isopropyl ketone, whereas the reverse order of mixing the reagents gives more perfluorobutyr-aldehyde *(663)*.

$$C_3F_7{-}CN + CH_3{-}CH{-}CH_3$$

**312** [*663*]

$$C_3F_7{-}CO{-}CH\begin{smallmatrix}CH_3\\CH_3\end{smallmatrix}$$
44%

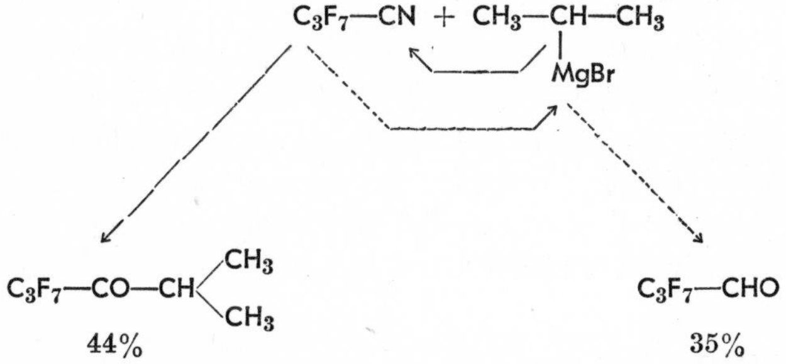

$$C_3F_7{-}CHO$$
35%

A small extent of reduction occurs also during the syntheses using perfluoro-alkylmagnesium halides *(367)* (p. 220). Similar reduction was observed in the reaction of perfluoropropyl lithium with ethyl perfluorobutyrate. In addition to the expected di(perfluoropropyl)ketone the corresponding reduced product, di(perfluoropropyl)carbinol was obtained. The hydrogen necessary for the reduction was abstracted from the solvent (diethyl ether) (668).

**313**
$$C_3F_7-COOC_2H_5 + C_3F_7Li \xrightarrow[-35° \text{ to } -50°]{(C_2H_5)_2O}$$

$$\begin{array}{ll} C_3F_7-CO-C_3F_7 & 8.7\% \\[4pt] C_3F_7-\underset{\underset{\displaystyle OH}{|}}{CH}-C_3F_7 & 42\% \\[10pt] CF_3-CF{=}CF_2 & 25\% \end{array}$$

[668]

The Meerwein-Ponndorf method of reduction applied to fluorinated compounds has given, so far, high yields without by-products (339, 581).

**314**

$$F-\langle \rangle-CO-CH_3 \xrightarrow[(CH_3)_2CHO \frac{Al}{3}]{(CH_3)_2CHOH} F-\langle \rangle-\underset{\underset{\displaystyle OH}{|}}{CH}-CH_3 \quad 75-79\%$$

581

During the reaction of chlorofluoroparaffins with alkaline alkoxides or phenoxides, replacement of chlorine by hydrogen was sometimes observed (629) (Equation 553, p. 238).

## OXIDATION

Oxidative reactions of fluorinated compounds have great preparative importance. Oxidations by potassium permanganate represent a general route to polyfluorinated and perfluorinated ketones and acids from fluorinated olefins.

In addition to permanganate, chromic oxide and dichromates find use in the oxidation of alcohols and in oxidative degradation of the benzene ring, especially when its stability has been weakened by suitable substitution.

Oxidation of trifluoroacetic acid to trifluoroperacetic acid furnishes a new important selective oxidation reagent. An increased effort is devoted to the air oxidation of fluorinated olefins as this oxidation represents a simple route to substituted fluoroacetic acids.

### Oxidation by Oxygen

Fluorinated derivatives of ethylene react readily with oxygen and yield fluorinated acetyl halides. *sym*-Dibromodifluoroethylene gives dibromofluoroacetyl fluoride (982), chlorotrifluoroethylene chlorodifluoroacetyl fluoride

(*384, 522*), probably via chlorotrifluoroethylene oxide as an intermediate (*384*), which can be isolated at low temperatures (*754*).

315 

$$\text{CCIF}{=}\text{CF}_2 \xrightarrow[-80°,\ 10\ \text{hrs.}]{O_2} \quad \text{CCIF}{=}\text{CF}_2 \xrightarrow[25-50°,\ 7-21\ \text{atm.}]{O_2}$$

[*384, 522, 754*]

$$\underset{25\%}{\text{CCIF}{-}\text{CF}_2 \diagdown_O\diagup} \xrightarrow{\quad \text{probably}\quad} \underset{43\%}{\text{CCIF}_2{-}\text{COF}}$$

Energetic oxidation of tetrafluoroethylene by oxygen leads to carbon tetrafluoride and carbon dioxide (*79*).

Oxidation of 1,1,2,2,3-pentafluoropropane with oxygen gives 1,1,2,2,3-pentafluorotrimethylene oxide (*426*).

316 

$$\text{CHF}_2{-}\text{CF}_2{-}\text{CH}_2\text{F} \xrightarrow[500°]{O_2} \underset{O\ -\ CHF}{\overset{CF_2{-}CF_2}{|\qquad|}} \quad 85\%$$

[*426*]

Oxidation of 2,3-dichlorohexafluoro-2-butene by oxygen in the presence of chlorine yields trifluoroacetylchloride (*434*). Similarly, 1,2-dichloroperfluorocycloalkenes give perfluorodicarboxylic acid chlorides (*434*).

317 

$$\text{CF}_3{-}\text{CCl}{=}\text{CCl}{-}\text{CF}_3 \xrightarrow[h\nu]{O_2} \quad 2\ \text{CF}_3{-}\text{COCl}$$

[*434*]

## Oxidation by Hydrogen Peroxide

Hydrogen peroxide forms with trifluoroacetic acid or trifluoroacetic anhydride trifluoroperacetic acid (*247*) (p. 282). Another important reaction of hydrogen peroxide is oxidation of mercury trifluoromethyl mercaptide to trifluoromethanesulphonic acid (*377*):

318 

$$\text{CF}_3{-}\text{SHgS}{-}\text{CF}_3 \xrightarrow[2.\ H_2SO_4]{1.\ 35\%\ H_2O_2,\ 92-105°} \quad \text{HgO} + \text{CF}_3{-}\text{SO}_3\text{H} \quad 64\%$$

[*377*]

## Oxidation by Nitrogen Oxides and Nitric Acid

Nitrogen tetroxide and oxygen convert 1,1,1-trifluoropropane to a mixture of trifluoroacetaldehyde and 1,1,1-trifluoro-3-nitropropane (*936*).

319 

$$\text{CF}_3{-}\text{CH}_2{-}\text{CH}_3 \xrightarrow[437-462°]{N_2O_4,\ O_2} \quad \underset{16\%}{\text{CF}_3{-}\text{CH}_2{-}\text{CH}_2{-}\text{NO}_2} + \underset{20-24\%}{\text{CF}_3{-}\text{CHO}}$$

[*936*]

A rather interesting reaction results from the treatment of perfluoroalkane-sulphonic acid fluorides with nitrogen oxides. Perfluorooctanesulphonyl fluoride and nitrogen tetroxide or nitric oxide give, at 550°C, perfluoroenanthoic acid fluoride accompanied by its lower homologue (935).

320 $\xrightarrow[550°]{\text{NO}}$ $C_8F_{17}SO_2F$ $\xrightarrow[\text{2. } H_2SO_4 \text{ dil.}]{\text{1. } N_2O_4, 550°}$ [935]

$C_7F_{15}COF$        $C_7F_{15}COOH$ + $C_6F_{13}COOH$
37%        54%       13%

1-Hydroperfluoroalkanes and a mixture of nitrogen tetroxide with chlorine (2 : 1) produce perfluorocarboxylic acids containing an equal number of carbon atoms (935). The same compounds are formed by heating perfluoroalkyl bromides or iodides with nitrogen tetroxide at 550–600°C and hydrolysing the resulting perfluoroacid halides (935).

321    $C_7F_{15}H$ $\xrightarrow[600°]{NO_2, Cl_2}$ $C_6F_{13}COOH$ $\xleftarrow[\text{2. } H_2O]{\text{1. } N_2O_4, 550°}$ $C_7F_{15}I$    [935]
                      45.6% 59%

Nitric acid ($d$ 1·15) oxidized trifluoromethylcyclohexane at 135°C to a mixture of trifluoroacetic, succinic and trifluoromethyladipic acids (999). Bis($\beta$-fluoroethyl)sulphide was converted to the corresponding sulphoxide by nitric acid (687) (Equation 323, p. 171).

Copper nitrate and better still lead nitrate are suitable for the conversion of benzylic halides to aromatic aldehydes (518) (Equation 243, p. 143).

## Oxidation with Chromium Compounds

Chromium trioxide and potassium dichromate are frequently employed for oxidizing fluorinated secondary alcohols to ketones (339) and primary alcohols to acids, when the alcohols are easily available (333).

322    $FCH_2-CH_2-CH_2OH$ $\xrightarrow{K_2Cr_2O_7, H_2SO_4}$ $FCH_2-CH_2-COOH$ 80%    [333]

Perfluorinated alcohols are quite resistant toward potassium dichromate-sulphuric acid mixtures. Di(perfluoropropyl)carbinol was recovered in 75% yield after such a treatment (668). Oxidation of trifluoromethyl-N,N'-diacetyl-p-phenylene diamine yielded trifluoromethyl-p-benzoquinone in 35% yield by dichromate oxidation (419).

Potassium dichromate effected the oxidation of bis($\beta$-fluoroethyl) sulphide to the corresponding sulphone whereas nitric acid oxidation led to the sulphoxide stage (*687*).

**323**    $\xrightarrow{\text{HNO}_3}$    $FCH_2-CH_2-S-CH_2-CH_2F$    $\xrightarrow{\text{K}_2\text{Cr}_2\text{O}_7}$    [*687*]

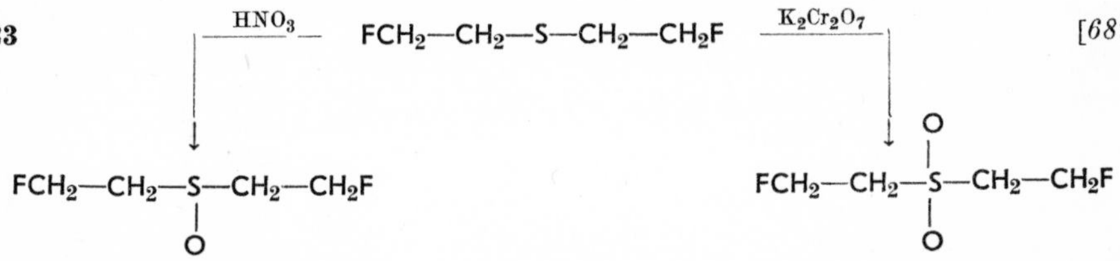

$FCH_2-CH_2-S-CH_2-CH_2F$
                    |
                    O

$FCH_2-CH_2-\overset{\displaystyle O}{\underset{\displaystyle O}{S}}-CH_2-CH_2F$

Similarly polyfluorodimethyl sulphides such as methyl trifluoromethyl sulphide, $\alpha$-chloro-$\alpha,\alpha$-difluoro-dimethyl sulphide, and $\alpha'$-chloro-$\alpha,\alpha,\alpha$-trifluoro-dimethyl sulphide were converted to the sulphones. On the other hand, pentafluorodimethyl sulphide resisted oxidation entirely (*1047*).

**324**    $CF_3-S-CH_2Cl$    $\xrightarrow[95°]{\text{CrO}_3,\ \text{CH}_3\text{COOH}}$    $CF_3-\overset{\displaystyle O}{\underset{\displaystyle O}{S}}-CH_2Cl$  46%    [*1047*]

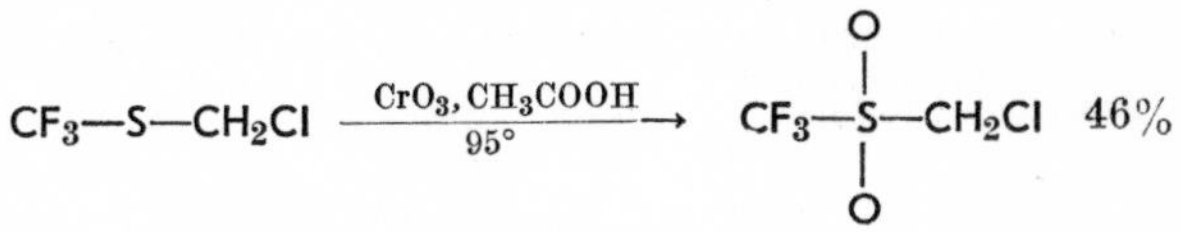

For preparative purpose, oxidation of *m*-aminobenzotrifluoride by potassium dichromate to trifluoroacetic acid is still of considerable importance. Though more economical processes have been developed since, this reaction has the advantage of an easily available starting material (*1000, 1060*):

**325**

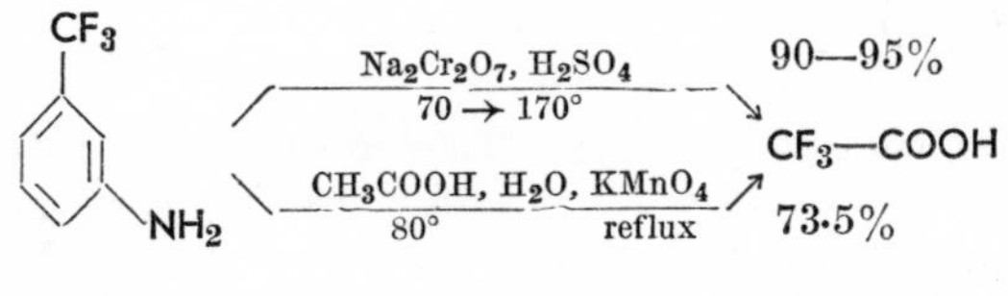

(structure of *m*-aminobenzotrifluoride with CF$_3$ and NH$_2$ groups)

$\xrightarrow[70 \to 170°]{\text{Na}_2\text{Cr}_2\text{O}_7,\ \text{H}_2\text{SO}_4}$  90—95%

$CF_3-COOH$    [*412, 1060*]

$\xrightarrow[80° \quad \text{reflux}]{\text{CH}_3\text{COOH},\ \text{H}_2\text{O},\ \text{KMnO}_4}$  73.5%

Oxidation with chromyl chloride (Etard reaction) does not seem to possess the advantages claimed in many textbooks. In the case of fluorinated aromatic aldehydes at least, other methods proved to give better results (*518*) (Equation 243, p. §43).

## Oxidation by Potassium Permanganate

Although the spectrum of oxidation reactions effected by potassium permanganate is very broad, its applications in the field of fluorinated compounds are limited mostly to the fission of double bonds and production of fluorinated ketones or acids. Other oxidative reactions of practical value such as the preparation of trifluoroacetic acid from *m*-aminobenzotrifluoride are not very frequent (*412*).

Polyfluoroolefins, with the double bonds deficient of electrons, are easily attacked by permanganate anion.

Oxidation of properly fluorinated isobutylenes is a source of perfluoro-acetone (*370, 751, 827*).

**326**                                                                  [*370, 751, 827*]

Olefins containing one —CF₃ group attached to the double bond represent convenient materials for the production of trifluoroacetic acid. 1,1,1-Trifluoro-2-propene (*1092*), and 1,1,2-trichlorotrifluoro-1-propene (*436*) give generally high yields of trifluoroacetic acid. However, the most advantageous reaction is the oxidation of 2,3-dichlorohexafluoro-2-butene, since no carbon atom is lost during the process (*484*).

**327**                                                                  [*436, 484, 1092*]

*Procedure 20*

*Oxidation of Fluoroolefins*
*Preparation of Trifluoroacetic Acid from 2,3-Dichlorohexafluoro-2-butene (484)\**

In a five-litre three-necked flask fitted with a gas-tight stirrer, a separating funnel and an efficient reflux condenser connected to a dry ice trap, 460 g (2·9 moles) of potassium permanganate and 315 g (5·6 moles) of potassium hydroxide are dissolved in 3500 ml. of water while heating the mixture at 60°C. To the warm solution thus formed, 450 g (1·93 moles) of 2,3-dichlorohexafluoro-2-butene is added at a rate such that the reflux condenser should not be flooded. After the addition has been completed, the reaction mixture is heated until its temperature reaches 95°C (approximately 10 hours). Thereafter the flask is cooled to 40°C, a current of sulphur dioxide is passed through the liquid cooled below 60°C until the permanganate colour disappears; the solution is then neutralized with 50% sulphuric acid using Congo Red as indicator, and sulphur dioxide is again introduced until the mixture becomes colourless. A continuous extraction with ether lasting approximately 60 hours and evaporation of the ether from the extract through a column leaves about 370 g of a fraction distilling at 100–104·5°C and representing an azeotropic mixture of trifluoroacetic acid and water in the ratio 80:20. The yield of trifluoroacetic acid contained therein is 2·6 moles (68%).

---

\* Carried out by the author in the laboratory of A. L. Henne.

For the preparation of higher perfluorocarboxylic acids, oxidation by potassium permanganate is applied to 1-perfluoroalkyl-2-iodoethylenes which are easily obtained by the addition of perfluoroalkyl iodides to acetylene (*357*).

**328**   $C_3F_7—CH=CHI$   $\xrightarrow[\text{KOH}]{\text{KMnO}_4}$   $C_3F_7—COOH$   63%   [*357*]

A very important application of potassium permanganate is the oxidation of fluorinated cycloolefins which leads to fluorinated dicarboxylic acids. Oxidation of hexafluorocyclobutene (*492*) and of 1,2-dichlorotetrafluoro-1-cyclo-butene (*489, 492*) affords perfluorosuccinic acid, oxidation of 1-ethoxypenta-fluoro-1-cyclobutene its acidic ester (*864*).

**329**   [*489, 492*]

$$
\begin{array}{ccc}
CF_2—CCl & & CF_2—COOH \\
|\quad\quad \| & & | \\
CF_2—CCl & & CF_2—COOH
\end{array}
$$

KMnO₄ / H₂O  65%
KMnO₄ / CH₃COCH₃  74%

$\xleftarrow{\text{KMnO}_4,\ \text{KOH}\quad 5\ \text{atm, 10 hrs.}}$   25%

$$
\begin{array}{c}
CF_2—CF \\
|\quad\quad \| \\
CF_2—CF
\end{array}
$$

**330**   
$$
\begin{array}{c}
CF_2—C—OC_2H_5 \\
|\quad\quad \| \\
CF_2—CF
\end{array}
$$
$\xrightarrow[\substack{\text{H}_2\text{O, CH}_3\text{COCH}_3 \\ 0—10°}]{\text{KMnO}_4}$
$$
\begin{array}{c}
CF_2—COOC_2H_5 \\
| \\
CF_2—COOH
\end{array}
$$
49%   [*864*]

The starting cyclobutene derivatives are readily obtained by dimerization of fluorinated ethylenes and subsequent reactions (p. 257).

Similarly, oxidation of 1,2-dichlorohexafluoro-1-cyclopentene gives per-fluoroglutaric acid (*146, 491, 678*),

**331**   
$$
\begin{array}{c}
\diagup CF_2—CCl \\
CF_2 \quad\quad \| \\
\diagdown CF_2—CCl
\end{array}
$$
$\xrightarrow[\text{reflux 12 hrs.}]{\text{KMnO}_4\ \text{aq.}}$
$$
\begin{array}{c}
\diagup CF_2—COOH \\
CF_2 \\
\diagdown CF_2—COOH
\end{array}
$$
86%   [*491*]

oxidation of 1,2-dichlorooctafluoro-1-cyclohexene (*678*) and of perfluorocyclo-hexene (*146*) analogously perfluoroadipic acid (Equation 260, p. 151). Oxidation of 3- and 4-H-nonafluoro-1-cyclohexenes yields heptafluoroadipic acid (*264*), that of perfluoro-1,4-cyclohexadiene difluoromalonic acid (*146*).

**332**   
$$
\begin{array}{c}
\diagup CF \diagdown \\
CF \quad\ CF_2 \\
|\quad\quad\ | \\
CF_2 \quad\ CF \\
\diagdown CF \diagup
\end{array}
$$
$\xrightarrow[20°,\ 30\ \text{min.}]{\text{KMnO}_4,\ \text{CH}_3\text{COCH}_3}$
$$
\begin{array}{c}
\diagup COOH \\
CF_2 \\
\diagdown COOH
\end{array}
$$
60%   [*146*]

The last-mentioned oxidation can be conveniently carried out by a solution of potassium permanganate in anhydrous acetone. This reagent is very efficient and causes an exothermic reaction lasting for a few seconds. The reaction is carried out at room temperature by using practically an equivalent amount (5% excess) of potassium permanganate required by the double bond, and gives satisfactory yields (146).

## Oxidation by Other Reagents

Of other oxidative reactions, mention will be made of the oxidation by manganese dioxide (914) and periodic acid (792), both applied to fluorohydrins and yielding fluoroacetaldehyde.

$$333 \qquad CH_2F-CH_2OH \xrightarrow[98°]{MnO_2, H_2SO_4} CH_2F-CHO \xleftarrow[20°]{H_5IO_6} \begin{array}{c} CH_2OH \\ | \\ CHOH \\ | \\ CH_2F \end{array} \qquad [792, 914]$$

$$6\% \qquad\qquad 80\%$$

Chlorine was used for converting alcohols to aldehydes or ketones (661)

334                                                                                                    [661]

$$CF_3-CF_2-CF_2-CH_2-OH \xrightarrow[hv,\ 60°]{Cl_2,\ CCl_4} CF_3-CF_2-CF_2-CHO$$

conversion 24%

yield 80%

and for oxidizing isothiocyanates (736, 914) or isothiuronium salts to sulphonyl chlorides (736).

335                                                                                              [736, 914]

$$FCH_2-CH_2-SCN \xrightarrow[0°]{Cl_2,\ H_2O} \qquad \xleftarrow{Cl_2,\ H_2O} FCH_2-CH_2-S-C\begin{array}{c} \diagup NH \\ \diagdown NH_2 \cdot HO_3SC_6H_4CH_3 \end{array}$$

$$60\cdot6\% \downarrow \qquad \downarrow 51\%$$

$$FCH_2-CH_2-SO_2Cl$$

The paragraph on oxidations should also include the Kolbe electrolytic doubling of alkali salts of ω-fluorocarboxylic acid esters which yield α,ω-difluoroparaffins (823) (Equation 194, p. 127).

## HALOGENATION

The paragraph on halogenation will cover not only the reactions with the halogens themselves but also other preparations such as treatment of organic compounds with hydrogen halides and halides of metals or non-metals. For convenience the term "halogen" will be applied to chlorine, bromine and iodine only. These reactions which involve fluorine, hydrogen fluoride and metal fluorides have already been described in the chapter dealing with preparations of organic fluorine compounds.

The presence of fluorine, specially when concentrated in one place, affects behaviour of the organic molecules toward halogenating reagents. Special attention will be given to the effect fluorine exerts upon the replacement of hydrogen by halogens, upon the ease and direction of addition of hydrogen halides across double and triple bonds, and upon deviations from usual rules due to the strong inductive and electromeric effects of fluorine.

Reactions with halogens, reactions with hydrogen halides, and reactions with non-metal halides will be treated in sequence.

## Reactions with Halogens

Elemental halogens cause both addition and substitution reactions. Besides replacement of hydrogen, other elements or whole groups can be exchanged for halogen atoms and, in rare instances, the carbon chain is cleaved.

*Addition reactions of halogens.* Addition of halogens occurs with fluorinated olefins, acetylenes, aromatic systems, and compounds of phosphorus or other metalloids which are able to change from trivalent to pentavalent stage.

Addition of chlorine and bromine to fluorinated olefins is practically quantitative and represents a valuable aid in the isolation and purification of fluorinated compounds, and very often an intermediate stage in the synthesis of more complicated fluorinated derivatives (*462, 465*) (Equations 206, 224, p. 130, 135). Addition is conveniently carried out under illumination (*450*).

**336** $\quad\quad CH_2{=}CF{-}CF_3 \xrightarrow[h\nu]{Cl_2} CH_2Cl{-}CFCl{-}CF_3$ quant. $\quad\quad$ [*450*]

Under these conditions, chlorine was succesfully added even to perfluoro-olefins (*132*).

**337**

$$
\begin{array}{c}
CF{=}CF \\
CF_2 \qquad CF_2 \\
CF_2{-}CF_2
\end{array}
\xrightarrow[h\nu]{Cl_2}
\begin{array}{c}
CFCl{-}CFCl \\
CF_2 \qquad CF_2 \\
CF_2{-}\!\!-CF_2
\end{array}
\;62\% \qquad [132]
$$

Addition of iodine to tetrafluoroethylene gives good yields at 150°C, which is from the thermodynamic point of view rather surprising, as vicinal diiodides usually show instability at higher temperatures (*380*).

**338** $\quad\quad CF_2{=}CF_2 \xrightarrow[150°, 24\,hrs.]{I_2} CF_2I{-}CF_2I \;\; 76\% \quad\quad [380]$

Iodine monochloride adds to chlorotrifluoroethylene with iodine linking to the carbon atom holding chlorine (*365*). This is a consequence of unambigous polarization of the double bond due to the strong electromeric effect of fluorine:

A similar evidence for the electromeric effect of fluorine is apparent from the reaction of trifluoroethylene with bromine in acetic acid (575).

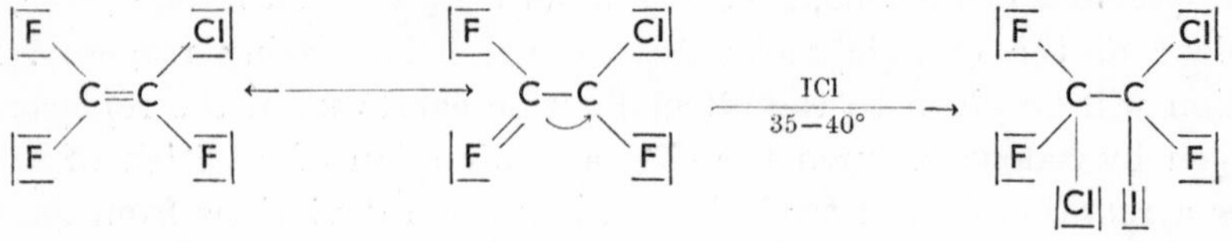

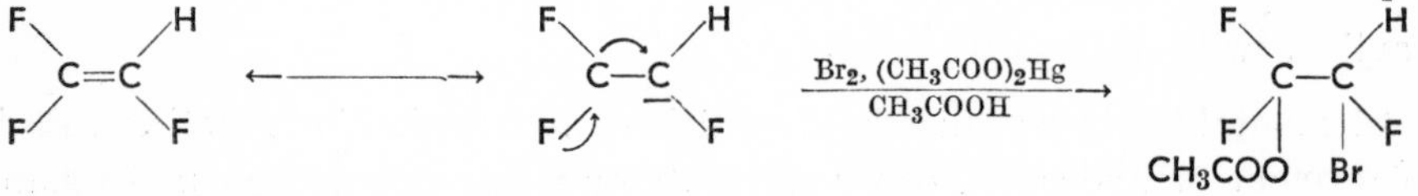

The strong inductive effect exhibited by fluorine is visible on the course of addition of bromine in acetic acid to 1,1,1-trifluoro-2-propene (575). Bromine cation as attacking particle joins the carbon atom adjacent to the trifluoromethyl group, which increases its electron density, while the more distant carbon atom accepts the acetate anion.

**341**                                                                 [575]
$$CF_3-CH=CH_2 \longleftrightarrow CF_3-\overset{\ominus}{\underset{}{CH}}-\overset{\oplus}{CH_2} \xrightarrow[\text{CH}_3\text{COOH}]{\text{Br}_2,\ (\text{CH}_3\text{COO})_2\text{Hg}} CF_3-\underset{\underset{Br}{|}}{CH}-\underset{\underset{OCOCH_3}{|}}{CH_2}$$

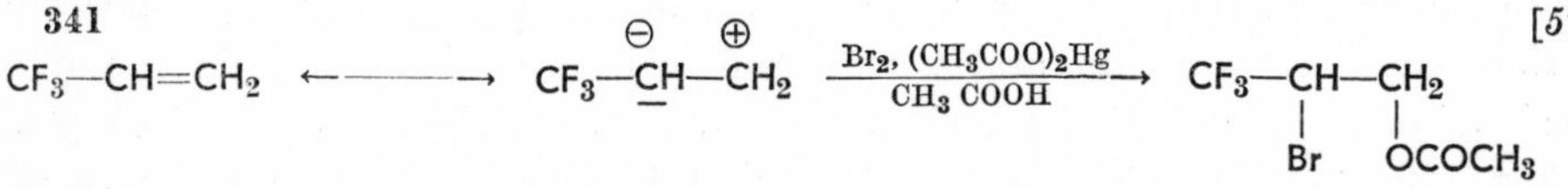

Addition of chlorine across a triple bond is very vigorous under ultraviolet illumination (462).

Fluorinated benzene derivatives accept halogens and form halogenated derivatives of cyclohexane. Pentafluorobenzene is thus converted to 1-H-1,2,3,4,5,6-hexachloropentafluorocyclohexane (774, 967). Sometimes, the addition is accompanied by replacement of hydrogen by the halogen (301).

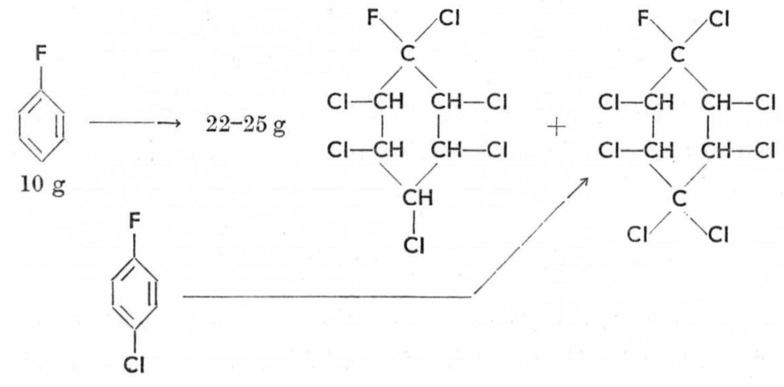

Another addition reaction is the conversion of tris(trifluoromethyl)phosphine to its dichloride (*71*):

343 $\qquad$ (CF$_3$)$_3$P $\xrightarrow{\text{Cl}_2}$ (CF$_3$)$_3$PCl$_2$ 56% $\qquad$ [*71*]

*Replacement of hydrogen by halogens.* The replacement of hydrogen by halogens in paraffins is a radical reaction which requires homolytic cleavage of the carbon–hydrogen bond. The ease of such substitution depends therefore on the nature of the C—H bond: the less polar it is, the easier the attack by chlorine atoms. The presence of fluorine in a paraffinic chain exhibits a strong polar influence on the neighbouring bonds, especially when more fluorine atoms are concentrated at one carbon. The most polar carbon–hydrogen bond in a fluorinated paraffin is that of the carbon atom adjacent to a —CF$_2$— or —CF$_3$ group. Curiously enough (at least for people not acquainted with quantum-mechanical backgrounds of modern chemistry), the carbon–hydrogen bond in groups like —CHF$_2$ and CHF$_3$ seems to be entirely homopolar. The polarizing effect of a fluorine cluster fades off along the chain, so that the carbon–hydrogen bond of the more distant carbons is again non-polar (*426*). These deductions can be illustrated on the following series of fluorinated paraffins arranged according to the increasing polarity of the carbon–hydrogen bond (or acidity of bold-face hydrogen):

344 $\qquad\qquad\qquad\qquad\qquad\qquad\qquad\qquad\qquad\qquad\qquad$ [*426*]

CHF$_2$—CH$_3$ $<$ CHF$_2$—CHF$_2$ $<$ CHF$_2$—CF$_3$ $<$ CH$_2$F—CF$_3$ $<$ CF$_3$—CH$_2$—CF$_3$

As the acidity of the hydrogen atoms increases, their chance of being replaced by chlorine atoms decreases. The validity of these observations can be proved by some examples.

The hydrogen atom in fluoroform can be replaced by chlorine (*427*) or bromine (*320*), though under rather severe conditions.

345 $\qquad\qquad$ $\xrightarrow[\substack{h\nu}]{\text{Cl}_2}$ CHF$_3$ $\xrightarrow[655-680°]{\text{Br}_2}$ $\qquad\qquad$ [*320, 427*]

$\qquad\qquad\qquad$ CClF$_3$ $\qquad\qquad\qquad\qquad\qquad\qquad\qquad$ CBrF$_3$

In ethylidene fluoride the preponderance of chlorination at the carbon carrying the fluorine atoms is evident from the composition of the reaction product (*453*):

346 $\qquad$ CHF$_2$—CH$_3$ $\xrightarrow{\text{Cl}_2}$ CCl$_2$—CH$_3$ F + CClF$_2$—CH$_2$Cl $\qquad$ [*453*]
$\qquad\qquad\qquad\qquad\qquad\qquad\qquad$ 70% $\qquad\qquad$ 6%

The relative resistance of the hydrogen atoms bonded to the carbon adjacent to a trifluoromethyl group is further apparent from the chlorination of

1,1,1-trifluoropropane. Hydrogen atoms on the end carbon are replaced preferentially (*361, 487, 670, 677*).

**347**                                                                                                                  [*487*]

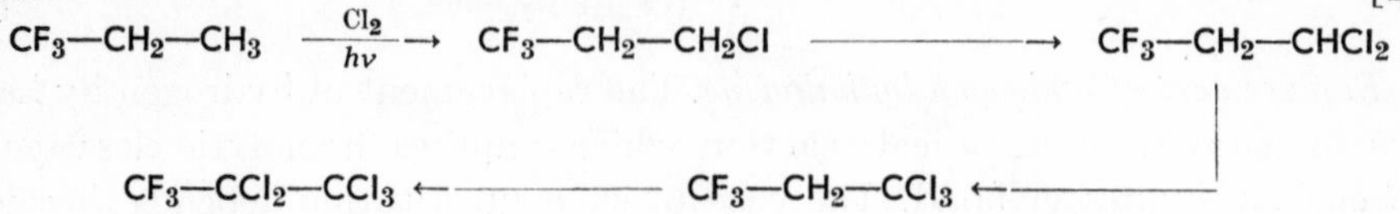

$$CF_3{-}CH_2{-}CH_3 \xrightarrow[hv]{Cl_2} CF_3{-}CH_2{-}CH_2Cl \longrightarrow CF_3{-}CH_2{-}CHCl_2$$

$$CF_3{-}CCl_2{-}CCl_3 \longleftarrow CF_3{-}CH_2{-}CCl_3 \longleftarrow$$

TABLE 24

*Chlorination of 1,1,1-Trifluoropropane under Various Conditions (361, 670)*

| Conditions of Chlorination | Composition of the Product, % | | | |
|---|---|---|---|---|
| | $CF_3CH_2CH_2Cl$ | $CF_3CH_2CHCl_2$ | $CF_3CH_2CCl_3$ | $CF_3CHClCH_3$ |
| $Cl_2$, $-14{-}45°C$ | 65 | 21 | | 1·5 |
| $Cl_2$, $hv$, $80{-}100°C$ | 55 | 17 | 2 | 18 |
| $Cl_2$, $420°C$ | 36 | 18 | 8 | 3 |

In 1H,1H,3H-pentafluoropropane, both chlorination and bromination occur with more ease in the $-CHF_2$ group than in the $-CH_2F$ group (*426*).

$$CHF_2{-}CF_2{-}CH_2F$$

**348**   $\xrightarrow[250°]{Cl_2}$                                $\xrightarrow[250°]{Br_2}$        [*426*]

| $CClF_2{-}CF_2{-}CH_2F$ | 53% | $CBrF_2{-}CF_2{-}CH_2F$ |
| $CHF_2{-}CF_2{-}CHClF$ | 47% | $CHF_2{-}CF_2{-}CHBrF$ |

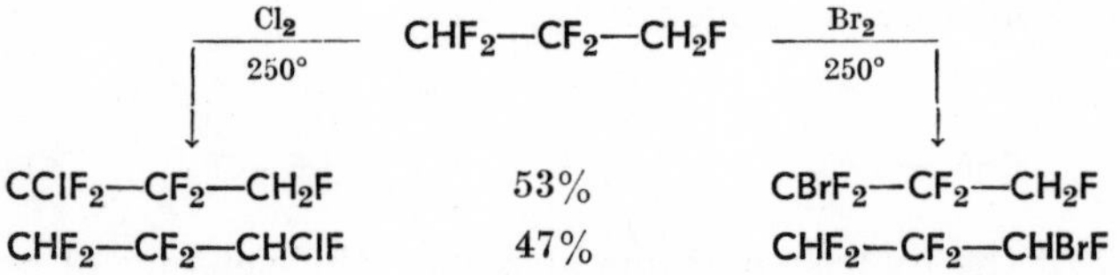

The strong polar effect of fluorine clusters in 2H,2H-hexafluoropropane upon the $-CH_2-$ group prevents chlorination entirely (*453*).

In 1,1,1-trifluorobutane chlorination does not occur at the adjacent methylene group (*452*).

**349**                               $CF_3{-}CH_2{-}CH_2{-}CH_3$                               [*452*]

$$\begin{array}{cc} Cl_2 & H_2O \\ hv & \end{array}$$

$CF_3{-}CH_2{-}CHCl{-}CH_3$   4 parts                    $CF_3{-}CH_2{-}CH_2{-}CH_2Cl$   5 parts

$Cl_2$ | 6 parts        1 part        $Cl_2$

$CF_3{-}CH_2{-}CCl_2{-}CH_3$ 8 parts                     $CF_3{-}CH_2{-}CH_2{-}CHCl_2$   2 parts

$CF_3{-}CH_2{-}CHCl{-}CH_2Cl$

$Cl_2$

$CF_3{-}CH_2{-}CH_2{-}CCl_3$

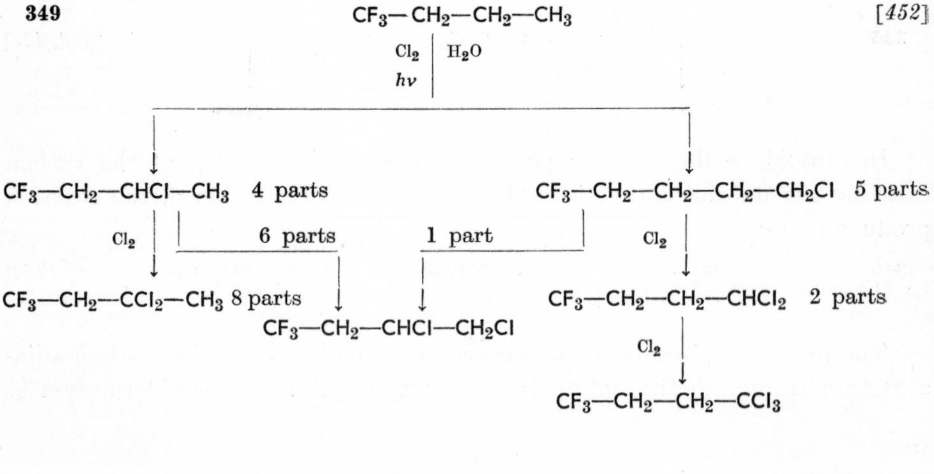

Chlorination of 2,2-difluorobutane does not take place at the methyl adjacent to the $-CF_2-$ group *(451)*.

3,3-Difluoropentane is attacked preferentially at the end carbons *(440)*.

$CH_3-CH_2-CF_2-CH_2-CH_3 \xrightarrow[H_2O]{Cl_2}$

**350**

$CH_3-CH_2-CF_2-CH_2-CH_2Cl$
70%

$\xrightarrow[H_2O]{Cl_2} ClCH_2-CH_2-CF_2-CH_2-CH_2Cl$
60%

*[440]*

The results of the chlorination of paraffins containing both fluorine and chlorine atoms are hard to predict and deductions about directing effects are extremely difficult. The following examples illustrate the course of some chlorinations *(402, 418, 458)*.

**351** $\quad CHFCl-CH_2Cl \xrightarrow[hv]{Cl_2} CHFCl-CHCl_2 + CFCl_2-CH_2Cl \qquad$ *[402]*
1·7 parts $\qquad$ 1 part

**352** $\quad CHF_2-CH_2Cl \xrightarrow[hv]{Cl_2} CHF_2-CHCl_2 + CClF_2-CH_2Cl \qquad$ *[402]*
4 parts $\qquad$ 1 part

**353** $\quad CHF_2-CHCl_2 \xrightarrow[60-70°]{Cl_2, hv} CClF_2-CHCl_2 + CHF-CCl_3 \qquad$ *[418]*
56% $\qquad$ 44%

In contrast to paraffins, the halogenation of aromatic rings has a polar mechanism and follows the rules for electrophilic substitution. The presence of fluorine on the benzene ring has a strong *ortho-para* directing effect in catalyzed chlorinations and brominations *(795)*.

**354**

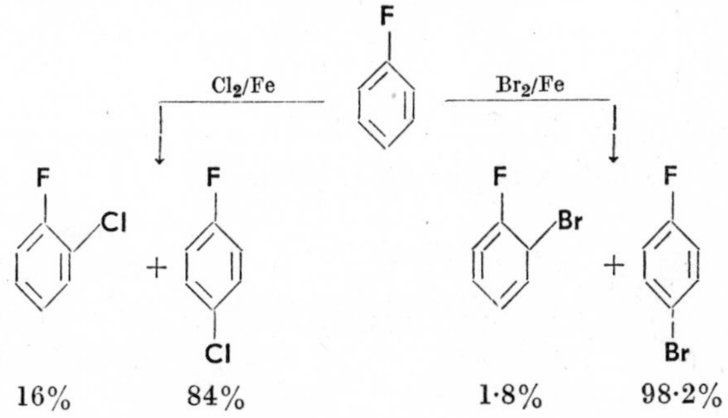

*[795]*

16%      84%              1·8%      98·2%

In the above reactions, *meta*-substitution does not exceed 1%. On the other hand, elevated temperature applied to the halogenation of fluorobenzene shifts the distribution of the isomers toward *meta*-substitution, as is apparent from Table 25 (*795*).

TABLE 25

*Composition of the Products of High Temperature Halogenation of Fluorobenzene (795)*

| Temperature of Halogenation, °C for: | | Composition of $C_6H_4ClF$ | | | Composition of $C_6H_4BrF$ | | |
|---|---|---|---|---|---|---|---|
| $C_6H_5Cl$ | $C_6H_5Br$ | % *ortho* | % *meta* | % *para* | % *ortho* | % *meta* | % *para* |
| 260 | 260 | 1·7 | 32·0 | 66·3 | 1·8 | 6·4 | 91·8 |
| 345 | 345 | 2·7 | 47·2 | 50·1 | 1·9 | 5·6 | 92·5 |
| 450 | 420 | 9·5 | 57·5 | 33·0 | 3·7 | 8·0 | 88·3 |
| 520 | 500 | 9·0 | 58·0 | 33·0 | 5·2 | 31·0 | 63·8 |
| 600 | 600 | 11·5 | 56·5 | 32·0 | 11·5 | 60·5 | 28·0 |
| 680 | 680 | 10·0 | 58·0 | 32·0 | 13·8 | 60·2 | 26·0 |

In addition to the *ortho-para* directing effect, fluorine exhibits quite a strong activating influence upon electrophilic substitutions. Surprisingly its activating effect is stronger than that of hydrogen as can be seen from a study of halogenation of durene and its 3-halogen derivatives. If the position *3* is occupied by fluorine, position *6* is much more sensitive to bromination than in the unsubstituted durene. Relative reactivities of durene and its 3-halogen derivatives toward bromine dissolved in nitromethane are expressed by the following relative data (*526*):

355

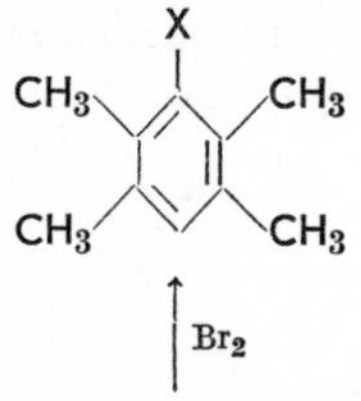

| X | = | H | F | Cl | Br | I | [526] |
|---|---|---|---|---|---|---|---|
| rel. reactivity | | 1000 | 2310 | 72·6 | 30·9 | 40·0 | |

However strongly fluorinated derivatives of benzene require severe conditions to undergo halogenation (*421*, *422*).

356                       $\xrightarrow[50-60°]{Br_2/AlBr_3,\ 60\%\ oleum}$            78%       [421]

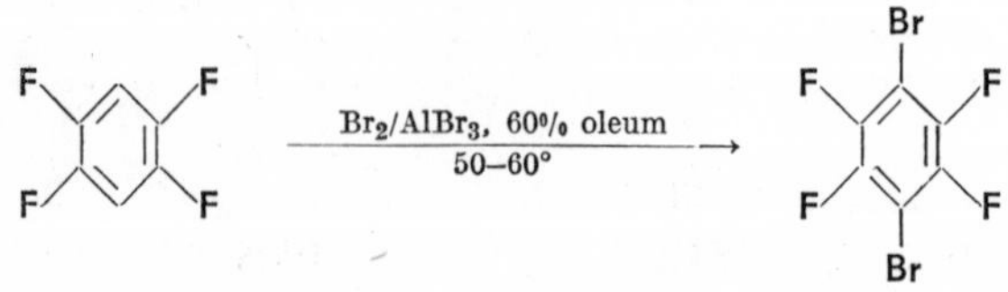

**357**

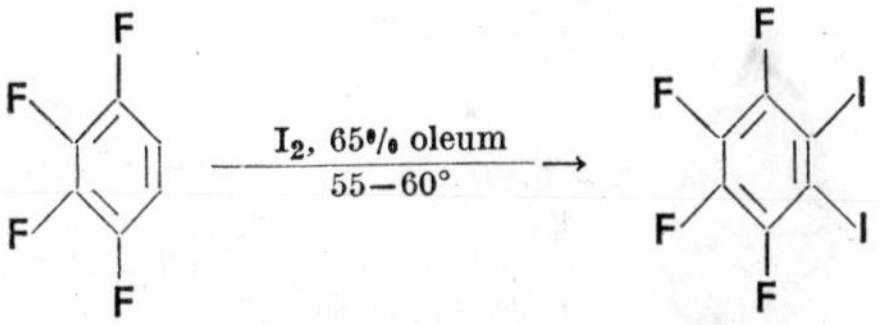

$$\xrightarrow[\text{55--60°}]{\text{I}_2,\ 65\%\ \text{oleum}}$$

82%     [*422*]

In this way even pentafluorobenzene was successfully converted to pentafluorohalobenzenes (*774*).

**358**

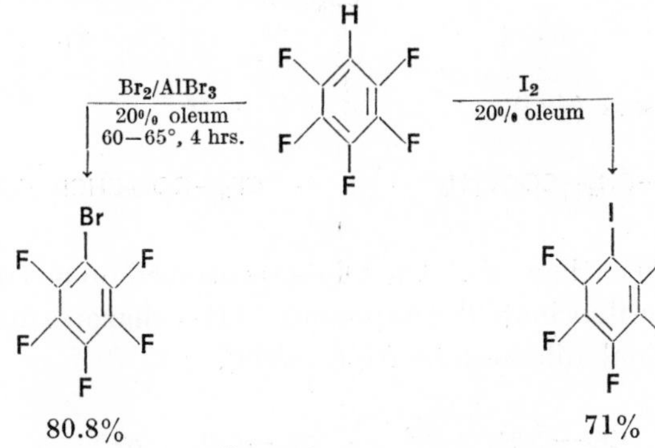

$$\xrightarrow[\substack{20\%\ \text{oleum}\\60-65°,\ 4\ \text{hrs.}}]{\text{Br}_2/\text{AlBr}_3}$$

$$\xrightarrow[\text{20\% oleum}]{\text{I}_2}$$

80.8%          71%       [*774*]

A trifluoromethyl side-chain directs the entering halogen to the *meta*-position (*958*):

**359**

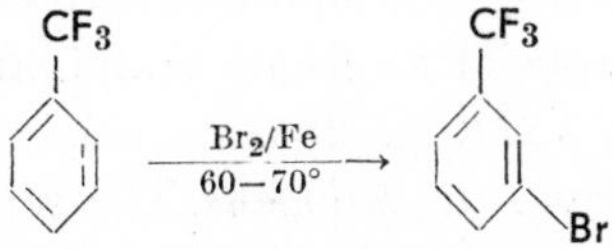

$$\xrightarrow[\text{60--70°}]{\text{Br}_2/\text{Fe}}$$

52%     [*958*]

Trifluoroderivatives with the halogen atom in *ortho*-positions must be prepared indirectly (Equation 373, p. 185).

As to action of halogens upon other organic fluorides, mention will be made of the chlorination of fluorinated ethers (*808, 811*),

**360**                $CH_3-CH_2-O-CF_2-CHF_2$               [*811*]

$Cl_2\downarrow$

$CH_3-CHCl-O-CF_2-CHF_2$   81.5%            $CH_2Cl-CH_2-O-CF_2-CHF_2$   11.5%

$Cl_2\ |$   1 part            1 part  $\downarrow$        $|\ Cl_2$

$CH_3-CCl_2-O-CF_2-CHF_2$               $CH_2Cl-CHCl-O-CF_2-CHF_2$

$Cl_2\ |$         $\downarrow$   2 parts          1 part  $|\ Cl_2$

$CH_2Cl-CCl_2-O-CF_2-CHF_2$   81%         $CHCl_2-CHCl-O-CF_2-CHF_2$

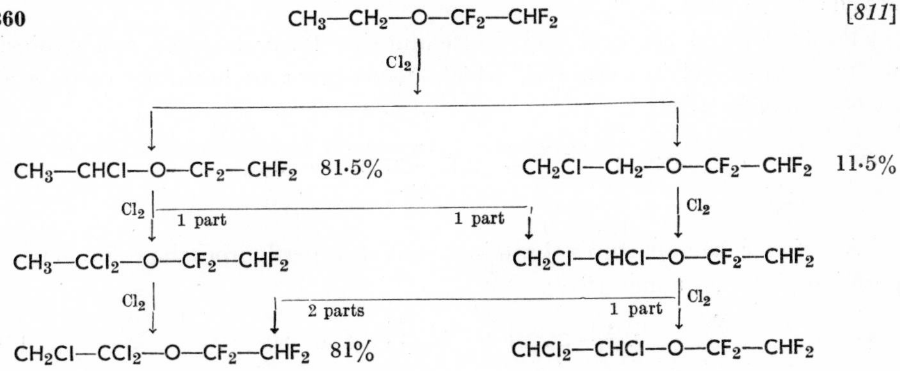

fluorinated ketones (*63, 631*),

$$
\textbf{361} \quad CF_3\text{—}CO\text{—}CH_3 \quad
\begin{array}{c}
\xrightarrow[20°]{Br_2,\ H_2SO_4} \\[4pt]
\xrightarrow[20°]{Br_2,\ CH_3COOH,\ CH_3COONa}
\end{array}
\quad
\begin{array}{c}
80\text{—}87.6\% \\[4pt]
CF_3\text{—}CO\text{—}CH_2Br \\[4pt]
19.2\%
\end{array}
\quad [\textit{631}]
$$

$$
\downarrow Br_2 \mid H_2SO_4
$$

$$
CF_3\text{—}CO\text{—}CHBr_2 \xrightarrow[CH_3COOH]{Br_2,\ CH_3COONa} CF_3\text{—}CO\text{—}CBr_3
$$

$$
86.9\% \qquad\qquad\qquad 81.1\%
$$

and fluorinated esters (*495*).

**362**    $CF_3\text{—}CO\text{—}CH_2\text{—}COOC_2H_5 \xrightarrow[20°]{Cl_2} CF_3\text{—}CO\text{—}CHCl\text{—}COO_2H_5$    [*495*]

Generally, halogenations of all the types mentioned do not deviate substantially from the non-fluorinated compounds. 1H-1-diazoperfluoroalkanes and halogens give geminal dihalo-derivatives (*309*).

**363**    $CF_3\text{—}CHN_2 \xrightarrow{I_2} CF_3\text{—}CHI_2$    87%    [*309*]

Bromination of trifluoroacetamide (*804*), perfluorosuccinimide (*489*) and perfluoroglutarimide (*489*) yields N-bromo-derivatives which themselves act as halogenating reagents in special kinds of bromination. Their applications will be discussed in Chapter VIII, p. 285.

*Replacement of Carboxyl groups by halogens.* The replacement of a carboxyl group by halogen atoms is of unusual significance, in the chemistry of organic fluorine compounds since it represents an easy route from readily available perfluorocarboxylic acids to perfluoroalkyl halides, which again are important intermediates in the synthesis of more complicated fluorinated derivatives, e.g. by way of the Grignard synthesis (p. 217), or the addition to olefins and acetylenes (p. 250).

Perfluoropropionic acid and dilute fluorine form unstable and explosive perfluoropropionyl hypofluorite, which decomposes to hexafluoroethane and carbon dioxide (*701*).

**364**    $C_2F_5\text{—}COOH \xrightarrow{F_2,\ N_2} C_2F_5\text{—}COOF \rightarrow C_2F_6 + CO_2$    [*701*]
    9.5%

At higher temperatures bromine converts perfluorocarboxylic acids to perfluoroalkyl bromides (*611*).

**365**    $C_2F_5\text{—}COOH \xrightarrow[350°]{Br_2} C_2F_5Br$    80%    [*611*]

with higher perfluoroalkyl bromides are not so good and this
ing perfluoroalkyl bromides cannot compete with the reaction
with silver salts of perfluorocarboxylic acids, the Hunsdieckers'

e action of halogens upon silver salts of perfluorocarboxylic acids pro-
es, according to the conditions used, either perfluoroacyl hypohalites
(p. 285), or perfluoroalkyl halides. The silver salts must be perfectly dry. The
reaction was originally carried out by heating a mixture of the salt and iodine
in a thin layer in a horizontal tube (*442*); later on an elaborate apparatus was
designed (Fig. 18) in which the dry salt is exposed to iodine vapors (*189*).

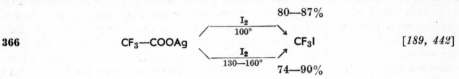

366   $CF_3$—$COOAg$   $CF_3I$   [*189, 442*]

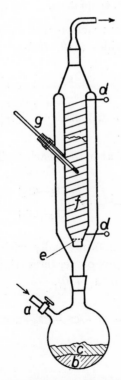

FIG. 18. Apparatus for the Hunsdiekers reaction:
a) air inlet, b) iodine layer, c) phosphorus pentoxide layer, d) heating wire leads, e) perforated
bottom, f) reactor with silver salt, g) thermometer.

*Procedure 21*

   *The Hunsdieckers Reaction with Perfluorocarboxylic Acids*
   *Preparation of Trifluoroiodomethane from Silver Trifluoroacetate and Iodine (359).*
   (a) *Preparation of Silver Trifluoroacetate.* Sodium trifluoroacetate dissolved in a minimum amount of water is added to a concentrated aqueous solution of silver nitrate. The solution is filtered, extracted with ether, the ethereal extract is evaporated to dryness on the steam bath, and the residue is powdered and dried in a vacuum desiccator over silica gel.

   Another way to silver trifluoroacetate is to dissolve silver oxide in a 50% aqueous trifluoroacetic acid, then to evaporate to dryness in vacuo.

   (b) *Reaction of Silver Trifluoroacetate with Iodine.* An intimate mixture of 100 g (0·45 mole) of silver trifluoroacetate and 300 g (1·18 moles) of iodine is placed in a two-litre flask fitted with a wide air and water-cooled reflux condenser, connected to two dry-ice traps. The reaction is started by gently heating the mixture with a free flame. Excessive heating is disadvantageous. The product, trifluoroiodomethane, passes through the reflux condensers and liquefies in the dry-ice traps together with the stripped iodine. After scrubbing with dilute sodium hydroxide approximately 80 g (0·41 mole) (91%) of trifluoroiodomethane, b.p. −22°, is obtained.

   Later on, the Hunsdieckers reaction was successfully applied to the preparation of bromides and chlorides (*359, 405, 409*).

$$
\textbf{367} \qquad C_5F_{11}\text{—COOAg}
\begin{cases}
\xrightarrow{\ Cl_2,\ 100°\ } & C_5F_{11}Cl \quad 71·2\% \\
\xrightarrow{\ Br_2,\ 80—90°\ } & C_5F_{11}Br \quad 82·5\% \\
\xrightarrow{\ I_2,\ 100°\ } & C_5F_{11}I \quad\ 73·9\%
\end{cases}
\qquad [409]
$$

   The reaction is applicable also to dicarboxylic acids (*410*).

$$
\textbf{368} \qquad
\begin{array}{l} CF_2\text{—}CF_2\text{—COOAg} \\ \ \ | \\ CF_2\text{—}CF_2\text{—COOAg} \end{array}
\xrightarrow{\ I_2\ }
\begin{array}{l} CF_2\text{—}CF_2I \\ \ \ | \\ CF_2\text{—}CF_2I \end{array}
\quad 64·2\% \qquad [410]
$$

   Silver perfluoroglutarate reacts normally with chlorine or bromine (*409*). With iodine only a small amount of perfluorotrimethylene diiodide is formed and the main product is a lactone (*410*).

$$
\textbf{369}
$$

$$
\xleftarrow{\ \frac{Cl_2}{100°}\ } \quad AgOCO\text{—}CF_2\text{—}CF_2\text{—}CF_2\text{—COOAg} \quad \xrightarrow{\ \frac{Br_2}{80—90°}\ } \quad [409,\ 410]
$$

$$
\downarrow \qquad\qquad\qquad\qquad I_2\ \Big|\ 120° \qquad\qquad\qquad\qquad \downarrow
$$

$$
ClCF_2\text{—}CF_2\text{—}CF_2Cl \qquad\qquad \downarrow \qquad\qquad BrCF_2\text{—}CF_2\text{—}CF_2Br
$$

$$
64·5\% \qquad\qquad ICF_2\text{—}CF_2\text{—}CF_2I \ +\
\begin{array}{c} CF_2\text{—}CF_2 \\ | \qquad | \\ CF_2 \quad CO \\ \ \ \diagdown O \diagup \end{array}
\qquad\qquad 80·3\%
$$

$$
\qquad\qquad 10—18·2\% \qquad\qquad\qquad 44—56\%
$$

The method is not limited to aliphatic fluorinated acids, and is satisfactory in the alicyclic series (*132*).

$$
\underset{370}{\quad} \xrightarrow[\;60-70°\;]{\text{Br}_2} \quad
\begin{array}{c} \text{COOAg} \\ | \\ \text{CF} \\ \text{CF}_2 \quad \text{CF}_2 \\ \text{CF}_2 \quad \text{CF}_2 \\ \text{CF}_2 \end{array}
\quad \xrightarrow[\;170°\;]{\text{I}_2} \quad \text{[132]}
$$

$$
\begin{array}{c} \text{Br} \\ | \\ \text{CF} \\ \text{CF}_2 \quad \text{CF}_2 \\ \text{CF}_2 \quad \text{CF}_2 \\ \text{CF}_2 \end{array} \quad 54\%
\qquad
\begin{array}{c} \text{I} \\ | \\ \text{CF} \\ \text{CF}_2 \quad \text{CF}_2 \\ \text{CF}_2 \quad \text{CF}_2 \\ \text{CF}_2 \end{array} \quad 63\%
$$

Sodium, barium, lead, and mercury salts generally give lower yields than the silver salts (*359*).

*Other reactions of fluorinated compounds with halogens.* Halogens can replace not only the carboxyl group but also a sulphonic group (*323*) or a nitro group (*1058*). However the importance of these two examples does not match that of the Hunsdieckers reaction.

371 $\qquad 2\,CF_3{-}SO_3Ag \xrightarrow{\;I_2\;} (CF_3{-}SO_2)_2O + CF_3I$ $\qquad$ [*323*]

372 $\qquad$ (NO$_2$, F substituted aromatic) $\xrightarrow[190-200°]{Cl_2}$ (Cl, F substituted aromatic) $\quad 90\%$ $\qquad$ [*1058*]

The reaction between a halogen and an organometallic compound is used for the introduction of the halogen into the position *ortho* to a trifluoromethyl group. Whereas the direct bromination of benzotrifluoride yields the *meta*-derivative (Equation 359, p. 181), *o*-bromobenzotrifluoride is prepared by the reaction of *o*-trifluoromethylphenyl lithium with bromine (*68*).

373 $\qquad$ (CF$_3$ benzene) $\xrightarrow[(C_2H_5)_2O]{C_4H_9Li}$ (CF$_3$, Li benzene) $\xrightarrow{Br_2}$ (CF$_3$, Br benzene) $\quad 28\%$ $\qquad$ [*68*]

At higher temperatures and under special conditions, replacement of fluorine by chlorine or other halogens was observed (969). Thus both partial and total substitution of chlorine for fluorine takes place during the chlorination of 1-chloro-1,1-difluoroethane in an iron reactor (49).

**374**                                                                     [49]

$$CH_3-CCIF_2 \xrightarrow[200°]{Cl_2} CH_2Cl-CCIF_2 + CCl_3-CCIF_2 + CCl_3-CCl_2F + CCl_3-CCl_3$$

                                       30%             15%            10%            40%

At temperatures higher than 700°C, chlorine cleaves the carbon chain in fluorinated paraffins. Tetrachloro-1,1-difluoroethane is converted by chlorine to a mixture of carbon tetrachloride and dichlorodifluoromethane in yields satisfactory for commercial production (434).

**375**                               $CF_2Cl-CCl_3 \xrightarrow[700°]{Cl_2} CF_2Cl_2 + CCl_4$                   [434]

Hexafluoroethane with chlorine yields chlorotrifluoromethane, with bromine bromotrifluoromethane (131).

**376**                               $\xrightarrow[900°]{Cl_2}$    $CF_3-CF_3$   $\xrightarrow[900°]{Br_2}$                           [131]

                       CF_3Cl                                      CF_3Br

This reaction takes place even with higher fluorocarbons (40). As a side reaction, replacement of some fluorine atoms by chlorine or other halogens can occur (131).

**377**                                                                         [40]

Φ—Cl                                                    Φ—Br

          56·8%                                         52·8%

## Reaction with Hydrogen Halides

As to reactions with hydrogen halides, the main attention will be given to additions which are interesting in theory and important in practice. Other reactions of hydrogen halides such as replacement of hydroxyl group or diazo group are far less significant.

*Addition reactions of hydrogen halides.* In additions of hydrogen halides to fluorinated olefins, distinction should be made as to whether the fluorine atoms are in vinylic or allylic positions. Under normal conditions the halide anion joins the carbon atom, which carries fluorine whose electromeric effect causes an unambiguous polarization of the double bond according to Equation 339, (p. 176). Polyfluorinated ethylenes accept hydrogen halides, so that the halide anion becomes attached to the carbon which holds the more fluorine (compare footnote on p. 66) (*396, 807*):

378 $\qquad$ $CF_2{=}CCl_2 \xrightarrow{\text{HBr/CaSO}_4\text{(C)}} CF_2Br{-}CHCl_2$ $\qquad$ [*807*]

379 $\qquad$ $CF_2{=}CFCl \xrightarrow[hv]{\text{HBr}} CF_2Br{-}CHFCl$ 92·5% $\qquad$ [*396*]

During the addition of hydrogen halides to tetrafluoroethylene polymerization of this reactive olefin takes place especially under conditions favouring the free radical reactions (*396*).

380 $\quad$ $CF_2{=}CF_2 \xrightarrow[hv]{\text{HBr}} CF_2Br{-}CHF_2 + CF_2Br(CF_2)_3H + CF_2Br(CF_2)_5H$ $\quad$ [*396*]
$\qquad\qquad\qquad\qquad\quad$ 18% $\qquad\qquad$ 3·2% $\qquad\qquad$ 0·9%

Addition reactions to fluorinated propylenes suffer certain limitation in the case of hydrogen fluoride. It was observed that hydrogen fluoride does not add to propylenes which carry fluorine on the central carbon atom (*486*). Addition therefore takes place with:

381 $\qquad\qquad$ $CH_3{-}CH{=}CF_2 \quad CF_3{-}CH{=}CF_2$ $\qquad$ [*486*]

whereas it does not occur with the compounds:]

382 $\qquad$ $CCl_3{-}CF{=}CHCl \quad CCl_2F{-}CF{=}CH_2$
$\qquad\qquad$ $CClF_2{-}CF{=}CH_2 \qquad CF_3{-}CF{=}CH_2$ $\qquad$ [*486*]

Of theoretical interest is the addition of hydrogen halides to olefins with allylic fluorine carrying a trifluoromethyl group on one of the doubly bonded carbons (*455, 807, 842*). This group is strongly electron-attracting, i.e. it concentrates electrons on the same carbon atom on which it is bonded, and furthermore decreases the ease of addition. The addition of hydrogen chloride to 1,1,1-trifluoropropene takes place only with the strongest catalysts such as aluminium halides. Boron trifluoride, ferric chloride, bismuth trichloride and zinc chloride are ineffective (*455*). The halide anion occupies the terminal carbon atom.

383 $CF_3{-}CH{=}CH_2$
$\qquad\qquad$ $\xrightarrow[100°]{\text{HCl/AlCl}_3} CF_3{-}CH_2{-}CH_2Cl$ $\quad$ conversion 28% yield 82%
$\qquad\qquad$ $\xrightarrow[100°]{\text{HBr/AlBr}_3} CF_3{-}CH_2{-}CH_2Br$ $\quad$ conversion 35% yield 46% $\qquad$ [*455*]

The addition of hydrogen halides to perfluoroalkylethylenes takes the same course and also requires strong catalysis (*842*).

**384**          $C_3F_7-CH=CH_2$  $\xrightarrow[120-140°]{HBr/CaSO_4(C)}$  $C_3F_7-CH_2-CH_2Br$          [*842*]

Compared with the directing effect of a carboxyl group, that of the trifluoromethyl group is somewhat weaker, as can be seen from the result of the addition of hydrogen bromide to ethyl $\gamma,\gamma,\gamma$-trifluorocrotonate (*1064*).

**385**          [*1064*]

$CF_3-CH=CH-COOC_2H_5$  $\xrightarrow[100° \text{ sealed}]{HBr/C_2H_5Br}$  $CF_3-CHBr-CH_2-COOC_2H_5$  80%

A similar directing effect is exerted by the trifluoromethyl group in fluorinated acetylenes. In contrast, the addition to 1,1,1-trifluoropropyne occurs readily even without catalysts (*363, 464*).

**386**          [*363*]

$CF_3-CH=CHF$  $\xleftarrow[60°]{HF}$     $\xrightarrow[20°]{HCl}$  $CF_3-CH=CHCl$

92%                              100%

$CF_3-C\equiv CH$

$CF_3-CH=CHBr$  $\xleftarrow[0°]{HBr}$     $\xrightarrow[100°]{HI}$  $CF_3-CH=CHI$

100%                              65%

*Substitution reactions of hydrogen halides.* Replacement of the hydroxy group by iodine in the reaction of trifluoroethanol with iodine and phosphorus (or hydrogen iodide) resulted only in a 5% conversion to 1,1,1-trifluoro-2-iodoethane (*310*). Much better yields were obtained by the reaction of hydrogen iodide with 1,1,1-trifluorodiazoethane (*310*).

**387**          $CF_3-CHN_2$  $\xrightarrow[-75°]{HI}$  $CF_3-CH_2I$  $\xleftarrow{P,I_2}$  $CF_3-CH_2OH$          [*310*]

77%        5%

Similarly 1H,1H,1-iodoheptafluorobutane was prepared in a single reaction (without the isolation of the intermediate diazo compound) from 1H,1H-heptafluorobutylamine (*595*).

**388**   $C_3F_7-CH_2-NH_2$  $\xrightarrow[2.\ HI\ (g)]{1.\ NaNO_2,\ HCl(g),\ (ClCH_2CH_2)_2O}$  $C_3F_7-CH_2I$  58%  [*595*]

## Reactions with Non-metal and Metal Halides

In the following paragraph some addition reactions of inorganic halides to fluorinated olefins, replacement of hydroxyl groups by halogens, and halogen exchange will be described.

*Addition reactions.* As in the additions of hydrogen halides, the additions of mixed halogen halides to fluorinated olefins follow the polarity of the double bond which again is determined by the number and position of fluorine atoms bound to its carbon atoms. The more positive particle of iodine mono-chloride (*365*), and of iodine monobromide, iodine (*395*), joins the carbon atom with higher electron density whereas the chlorine and bromine anions, respectively, become attached to the other end of the double bond (compare with Equation 339, p. 176).

389    $CF_2{=}CH_2$    $\xrightarrow[20°,\ dark]{IBr}$    $CF_2Br{-}CH_2I$    95%    [*395*]

In the additions of hypochlorous and hypobromous acid to 1,1,1-trifluoro--2-methylpropene, the halogen cations become fixed to the carbon adjacent to the trifluoromethyl group, whereas the negative part of the molecule, the hydroxyl, adds to the more distant end of the double bond (*211*).

390

$$\xrightarrow[0°\to20°]{HOCl}\quad CF_3{-}\underset{\underset{CH_3}{|}}{C}{=}CH_2 \quad \xrightarrow{HOBr}$$

$$CF_3{-}\underset{\underset{CH_3}{|}}{CCl}{-}CH_2OH \quad 58.7\% \qquad\qquad CF_3{-}\underset{\underset{CH_3}{|}}{CBr}{-}CH_2OH$$

[*211*]

In the case of nitrosyl chloride, the direction of addition is also dictated by the polarity of the double bond and by the scission of nitrosyl chloride. Since chlorine is released as an anion, it adds to the more positive end of the double bond (*1093*).

391    $CHF{=}CHCl$    $\xrightarrow[20°]{NOCl}$    $CHFCl{-}CHClNO \gtrless CHFCl{-}CCl{=}NOH$    45%    [*1093*]

392    [*1093*]

$CHF{=}CH_2$    $\xrightarrow[20°]{NOCl}$    $CHFCl{-}CH_2Cl$    +    $[CHFCl{-}CH_2NO]$    $\longrightarrow$    $CHFCl{-}CH_2NO_2$
 
　　　　　　　　　　　　　　　　34%　　　　　　　　　　　　　　　　　　　　54%

However, the additions of nitrosyl chloride are usually accompanied by side reactions, such as chlorination and oxidation, because both polar and homo-lytic fissions are possible.

*Substitution reactions.* Alkali hypobromites sometimes react anomalously with the perfluorocarboxylic amides and afford, in some instances, rather good yields of perfluoroalkylbromides (*524, 525*) (Equation 546, p. 236).

The hydroxyl groups in carboxylic acids can be conveniently replaced by halogens, by treatment with phosphorus oxychloride (*183*), and phosphorus trihalides (*958*) and pentahalides (*273*):

**393**

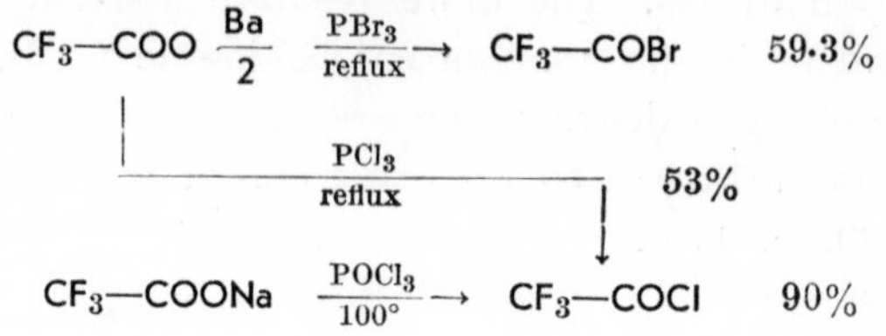

$$CF_3-COO\frac{Ba}{2} \xrightarrow[\text{reflux}]{PBr_3} CF_3-COBr \quad 59.3\%$$

$$\xrightarrow[\text{reflux}]{PCl_3} \quad 53\% \qquad [183,958]$$

$$CF_3-COONa \xrightarrow[100°]{POCl_3} CF_3-COCl \quad 90\%$$

**394**      $C_nF_{2n+1}COOH \xrightarrow{PCl_5} C_nF_{2n+1}COCl \quad 59-90\% \qquad [273]$

The same reactions can be effected by using thionyl chloride (*51*).

**395**

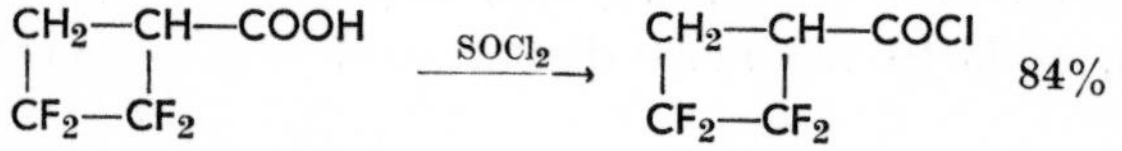

$$\begin{array}{c} CH_2-CH-COOH \\ | \quad\quad | \\ CF_2-CF_2 \end{array} \xrightarrow{SOCl_2} \begin{array}{c} CH_2-CH-COCl \\ | \quad\quad | \\ CF_2-CF_2 \end{array} \quad 84\% \qquad [51]$$

whereas sulphuryl chloride is able to replace a hydrogen atom by chlorine (*91*)

**396**      ⟨◯⟩—CO—CH₂F $\xrightarrow[\text{reflux}]{SO_2Cl_2}$ ⟨◯⟩—CO—CHClF    68%      [91]

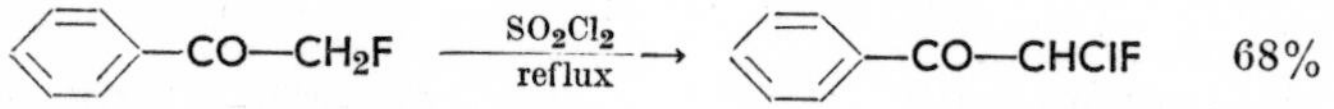

Metal halides were used for the preparation of perfluoroacyl halides from perfluoroacyl fluorides or anhydrides (*612*),

**397**      $C_3F_7-COF \xrightarrow[60°]{AlBr_3/act.C} C_3F_7-COBr \xleftarrow[\text{reflux}]{NaBr} (C_3F_7CO)_2O \qquad [612]$

and for the preparation of fluorinated alkylhalides from fluorinated alkyl benzenesulphonates (*633*).

**398**

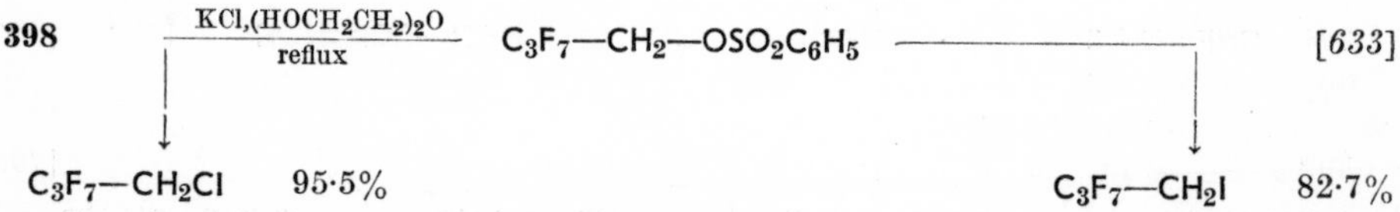

$$\xleftarrow[\text{reflux}]{KCl,(HOCH_2CH_2)_2O} C_3F_7-CH_2-OSO_2C_6H_5 \xrightarrow{\quad\quad\quad} \qquad [633]$$

$$C_3F_7-CH_2Cl \quad 95.5\% \qquad\qquad C_3F_7-CH_2I \quad 82.7\%$$

Fluorine atoms are often replaced by chlorine in the reactions of fluorinated compounds with aluminium chloride. Thus treatment of perfluoropropene with aluminium chloride at a moderate temperature affords complex mixture of almost all possible chlorofluoropropenes, together with perchloropropene (*801, 969*).

**399**      $CF_3-CF=CF_2 \xrightarrow[50-60°]{AlCl_3}$

$$\begin{array}{l} CF_3-CF=CFCl + CF_3-CF=CCl_2 \\ CF_2Cl-CF=CCl_2 + CFCl_2-CF=CCl_2 \\ CCl_3-CF=CCl_2 + CCl_3-CCl=CCl_2 \end{array} \qquad [801]$$

Especially sensitive to exchange with aluminium chloride are the fluorine atoms in the $\alpha$-positions of perfluoroethers (*1041*).

400

$$
\underset{\begin{array}{c} \\ \\ C_4F_9-\underset{\displaystyle \diagdown_O\diagup}{CF}\;\;\overset{|}{\underset{|}{CF_2}}'\end{array}}{\overset{\begin{array}{c}CF_2-CF_2\\ |\quad\quad|\end{array}}{}} \xrightarrow[200-220°]{AlCl_3\ (1.5\ mole)} \underset{\begin{array}{c}C_4F_9-\underset{\displaystyle \diagdown_O\diagup}{CCl}\;\;\overset{|}{\underset{|}{CCl_2}}\end{array}}{\overset{\begin{array}{c}CF_2-CF_2\\ |\quad\quad|\end{array}}{}} \quad 51\% \qquad [\textit{1041}]
$$

Perfluoroethers, which do not have any side chains in their $\alpha$-positions, split right away to perfluorocarboxylic acid chlorides and 1,1,1-trichloroperfluoro-alkanes (*1042*),

401 $\qquad C_4F_9{-}O{-}C_4F_9 \xrightarrow[175°,\ autoclave]{AlCl_3} C_3F_7{-}COCl + C_3H_7{-}CCl_3 \qquad [\textit{1042}]$

$\qquad\qquad\qquad\qquad\qquad\qquad\qquad\qquad 21\% \qquad\qquad 21\%$

while cyclic perfluoroethers afford, by the same token, chlorides of $\omega,\omega,\omega$-tri-chloroperfluorocarboxylic acids (*1043*).

402

$$
\underset{\begin{array}{c}CF_2 \quad CF_2\\ |\qquad\ |\\ CF_2 \quad CF_2\\ \diagdown_O\diagup\end{array}}{\overset{\diagup CF_2\diagdown}{}} \xrightarrow[180°]{AlCl_3} \underset{\begin{array}{c}CF_2 \quad CF_2\\ |\qquad\ |\\ CCl_3 \quad COCl\end{array}}{\overset{\diagup CF_2\diagdown}{}} \quad 38\% \qquad [\textit{1043}]
$$

This reaction belongs to those which do not seem to have parallels among the non-fluorinated compounds.

## NITRATION

Examples of the nitration of fluorinated compounds are not abundant. As in halogenation, the differences in the nitration results from the non-fluo-rinated compounds are due to polar effect of fluorine atoms present in the molecules to be nitrated.

Fluorinated olefins accept nitrogen tetroxide in such a way that both a dinitrocompound and a nitro-nitrite are formed (*566*).

403 $\hfill [\textit{566}]$

$$
CF_2{=}CF_2 \xrightarrow[CCl_4\ or\ CHCl_3]{N_2O_4} \underset{\begin{array}{c}|\quad\ |\\ NO_2\ NO_2\end{array}}{CF_2{-}CF_2} + \underset{\begin{array}{c}|\quad\ |\\ NO_2\ ONO\end{array}}{CF_2{-}CF_2} + O_2N{-}(CF_2{-}CF_2)_n{-}NO_2 \ \ 90\%\ \text{altogether}
$$

Higher perfluorinated olefins require higher temperatures. Thus with perfluo-ropropene the addition occurs at 100°C, with perfluoroisobutylene at 180°C (*566*).

Substitution of a nitro group for a hydrogen atom was effected by treatment of 1,1,1-trifluoropropane with nitrogen oxides (936). The nitro group joined the more distant carbon atom, since its carbon-hydrogen bonds are the more likely to split homolytically (compare p. 177) (Equation 319, p. 169).

A series of $\omega$-fluoronitroparaffins was prepared by the interconversion of $\omega$-fluoroalkyl halides with silver nitrite (815).

**404**       $F(CH_2)_5X \xrightarrow[\text{reflux 24 hrs.}]{\text{AgNO}_2,\ (C_2H_5)_2O} F(CH_2)_5NO_2$   73%       [815]

Perfluoronitroparaffins are formed in the reaction of perfluoroalkyl iodides with nitrogen tetroxide (45), or better still by oxidation of perfluoronitrosoparaffins, which are obtained from the same starting material by treatment with nitric oxide (31, 43, 45, 536).

**405**       $\xleftarrow[hv]{N_2O_4} C_3F_7I \xrightarrow[\text{Hg. } hv]{NO}$       [45]

$C_3F_7{-}NO_2$ + $C_2F_5{-}NO_2$ + $CF_3{-}NO_2$       $C_3F_7{-}NO \xrightarrow[70°]{O_2,\ hv} C_3F_7{-}NO_2$

38%    7%    17%       76%       53·5%

In the aromatic series, fluorine acts as a strong *ortho-para* directing substituent. The results of nitration of fluorobenzene by various nitrating agents are shown in Table 26 (794).

TABLE 26

*Results of Nitration of Fluorobenzene by Various Reagents (794)*

| Nitrating Reagent | Temperature °C | Total Yield % | *ortho-* % | *meta-* % | *para*-Nitrofluorobenzene % |
|---|---|---|---|---|---|
| $HNO_3$, $H_2SO_4$ | −10 | 84·5 | 8 | traces | 90 |
| $NaNO_3$, $H_2SO_4$ | 60−70 | 65 | 8 | traces | 90 |
| $CH_3COONO_2$ | 0 | 71·2 | 4 | 0 | 96 |
| $C_6H_5COONO_2$ | Cooling | 60·2 | 0 | 0 | 100 |
| $N_2O_5$, $CCl_4$ | Cooling | 93·8 | 28 | 0 | 72 |
| $N_2O_4$(1) | 20 | 13·2 | 16 | 6 | 78 |
| $N_2O_4$(g) | 130 | | 90 | 10 | 0 |

*Procedure 22.*

### Nitration of Fluorobenzene
### Preparation of 2,4-Dinitrofluorobenzene from Fluorobenzene (1102)*

In a 100 ml. flask fitted with a gas-tight stirrer, a thermometer and a reflux condenser, the mixed acid is prepared by adding 20 g (0·32 mole) of fuming nitric acid
(d 1·52) to a mixture of 20 g of concentrated sulphuric acid and 15 g of fuming sulphuric
acid containing 20% of free sulphur trioxide. While stirring the content of the flask,
9·6 g (0·1 mole) of fluorobenzene is slowly added from a dropping funnel through the
reflux condenser at such a rate that the temperature of the reaction mixture stays
below 55°C. After the addition has been completed the flask is heated for two hours on
the steam bath. The cooled content of the flask is poured over ice, the organic layer is
separated, the aqueous extracted with two 15 ml. portions of benzene, the benzene solution of the product is washed with water to the loss of acidity, dried with anhydrous
calcium chloride, freed from the solvent, and distilled under reduced pressure. Approximately 10 g (0·054 mole) (54%) of 2,4-dinitrofluorobenzene distilling at 156–158°C at
12 mm is thus obtained. The distillate solidifies almost entirely to a light yellow crystalline mass. Decanting of a small amount of a liquid portion yields crystals with
m.p. 25°C.

Nitration of *o*-fluorotoluene affords preponderantly 5-nitro-2-fluorotoluene.
This demonstrates not only a strong *ortho-para* directing effect but also the
activating influence of the fluorine atom which prevails over that of the methyl
group (*979*).

**406**                                                                    [*979*]

A trifluoromethyl group attached to the aromatic nucleus directs the
entering nitro group to the *meta*-position. Benzotrifluoride thus yields *m*-nitro-
benzotrifluoride (*983*), an important intermediate in the classical preparation
of trifluoroacetic acid (Equation 325, p. 179).

**407**                                                                    [*983*]

---

* Carried out by E. Nušlová.

A benzene ring carrying two trifluoromethyl groups is very resistant to nitration. Even under very severe conditions, only a low yield of the nitro-derivative was obtained (*889*).

408   CF$_3$—⟨ ⟩—CF$_3$   $\xrightarrow[\substack{90-100°; \quad 100-105° \\ 1 \text{ hrs.} \quad 2 \text{ hrs.}}]{\substack{100\% \text{ HNO}_3, 24\% \text{ oleum}}}$   CF$_3$—⟨ ⟩—CF$_3$ (NO$_2$)   35%   [*889*]

## NITROSATION

The paragraph on nitrosation will deal both with the preparation of nitro-so-derivatives and with the action of nitrous acid upon primary amines to form diazo compounds.

An interesting analogy of the Hunsdieckers reaction is the reaction be-tween silver trifluoroacetate and nitrosyl chloride, which leads to trifluoro-nitrosomethane (*375*). The same compound is formed from trifluoroiodometha-ne and nitric oxide (*31, 43, 536*).

409          CF$_3$—COOAg $\xrightarrow[-10° \text{ to } -5°]{\text{NOCl}}$ CF$_3$—NO $\xleftarrow[hv\ 20°]{\text{NO, Hg,}}$ CF$_3$I          [*375, 536*]

                                    16%          89%

Aliphatic amines with their primary amino group attached to a carbon atom adjacent to a perfluoroalkyl group are converted by nitrous acid to stable diazo compounds similar to those originating from α-amino ketones or α-amino esters. 2,2,2-Trifluoroethylamine yields thus 2,2,2-trifluorodiazo-ethane (*309*), 1H,1H,1-aminoheptafluorobutane 1H,1-diazoheptafluorobutane (*595*).

410          CF$_3$—CH$_2$—NH$_2$·HCl $\xrightarrow{\text{NaNO}_2,\ \text{H}_2\text{SO}_4}$ CF$_3$—CHN$_2$   50—70%          [*309*]

*Procedure 23*

*Nitrosation*

*Preparation of 2,2,2-Trifluorodiazoethane from 2,2,2-Trifluoroethylamine (309)*

In a 250 ml. thick-walled flask, 75 ml. of dibutyl ether and 7·5 g (0·11 mole) of sodium nitrite are added to a solution of 13·5 g (0·1 mole) of 2,2,2-trifluoroethylamine hydrochloride in 50 ml. of water. The flask is stoppered and agitated for 5 minutes. After cooling in an ice-salt mixture, the ethereal layer is separated and the aqueous layer allowed to stand with 75 ml. of dibutyl ether for 1 hour at room temperature with occa-sional shaking. After cooling and separating the ether layer, the extraction procedure is re-peated once more with 75 ml. of dibutyl ether. The combined ethereal solutions (225 ml.) are dried with anhydrous calcium chloride and distilled in a distilling column, the head of which is connected to three receivers of which the first is cooled with an ice-salt mix-ture and the other two are immersed in a dry-ice bath. The last receiver is attached to an aspirator. After evacuating the apparatus to approximately 200 mm, the connection to the pump is shut off and the distillation is started by heating the flask to 80°C in a bath. 2,2,2-Trifluorodiazoethane distils and liquefies in the traps. From time to time the apparatus is re-evacuated so as to maintain the pressure at about 200 mm. When the dibutyl ether solution in the distilling flask has become decolorized, the distillation is finished. The yellow product collected in the second receiver weighs 7·4 g (0·067 mole) which represents a 67% yield of 2,2,2-trifluorodiazoethane, b.p. 13–13,5°C.

If the pure 2,2,2-trifluorodiazoethane does not have to be isolated and its ethereal solution is satisfactory, diethyl ether can be substituted for the dibutyl ether and the solution is, after drying, distilled at ordinary pressure to a receiver kept at −15°C.

The conversion of primary amines to the diazo compounds is limited to such cases where the amino group is bound to the same carbon as the perfluoroalkyl group. If it is more distant, the normal reaction, replacement of the amino group by a hydroxyl, takes place. Thus 2,2,3,3-tetrafluorocyclobutylmethylamine treated with nitrous acid gives 2,2,3,3-tetrafluorocyclobutylcarbinol, curiously enough without rearrangement (27).

411
$$\begin{array}{c} CF_2\text{---}CF_2 \\ | \qquad | \\ CH_2\text{---}CH\text{---}CH_2\text{---}NH_2 \end{array} \xrightarrow[\text{reflux 4 hrs.}]{NaNO_2,\ NaH_2PO_4} \begin{array}{c} CF_2\text{---}CF_2 \\ | \qquad | \\ CH_2\text{---}CH\text{---}CH_2OH \end{array} \qquad 65\% \qquad [27]$$

Fluorinated aromatic amines are diazotized normally on treatment with nitrous acid (978).

412
978]

This statement applies even to such a heavily substituted amine as pentafluoroaniline, which affords pentafluorodiazonium salt coupling with the unreacted pentafluoroaniline to decafluorodiazoaminobenzene (284) (Equation 270, p. 155).

The amines containing trifluoromethyl groups can also be converted to the corresponding diazonium salts which again are useful intermediates for the synthesis of more complicated trifluoromethyl derivatives (126, 837).

413                                                                 [126, 837]

## SULPHONATION

Aliphatic sulphonic acids or their derivatives are prepared by indirect methods such as oxidation of isothiocyanates to sulphonyl chlorides (*736*) (Equation 335, p. 174), addition of bisulphites to fluorinated olefins (*180, 589*),

414    $CF_3—CF{=}CF_2$ $\xrightarrow[110-120°]{NaHSO_3,\ Na_2B_4O_7}$ $CF_3—CHF—CF_2—SO_3Na$  75%      [*589*]

and addition of sulphur trioxide to fluorinated olefins (*217, 256*).

415          $CF_2{=}CF_2$ $\xrightarrow{SO_3}$ $\begin{matrix} CF_2—CF_2 \\ |\qquad\ | \\ SO_2—O \end{matrix}$ $\xrightarrow{(C_2H_5)_3N}$ $\begin{matrix} CF_2—CO \\ |\qquad\ | \\ SO_2F\ \ F \end{matrix}$          [*217,256*]

True sulphonation takes place in the aromatic series according to the rules of electrophilic substitution (*504, 778*).

416          F—⟨benzene⟩ $\xrightarrow[70°]{10\%\ oleum}$ F—⟨benzene⟩—$SO_3H$          [*504*]

Surprisingly the sulphonation of pentafluorobenzene takes place under rather mild conditions (*774*).

417    $\xrightarrow[15°,\ 48\ hrs.]{20\%\ oleum}$ 88%          [*774*]

## ESTERIFICATION

The paragraph on esterification will cover methods for preparing fluorinated esters of organic acids not only by direct esterification or transesterification but also by other ways from various starting materials.

The reaction between fluorinated acids and alcohols is carried out as with the non-fluorinated compounds. Although some of the fluorinated acids possess enormous acidity sufficient for the esterification, mineral acids, most frequently sulphuric acid, are generally used as catalysts. A classical example is the esterification of trifluoroacetic acid. Since the anhydrous material is not readily available in many laboratories, the esterification can be started either with azeotropic (80/20) mixture of trifluoroacetic acid and water (*436*), or with sodium or potassium trifluoroacetate, from which the free acid is liberated *in situ* by means of concentrated sulphuric acid (*309, 872*). Distillation of such mixtures yields a ternary system ethyl trifluoroacetate—ethanol—water, which is freed from water and alcohol by distillation with concentrated sulphuric acid (*779*) or phosphorus pentoxide (*309*).

$$CF_3-COOH + C_2H_5OH \xrightarrow[\text{distillation}]{H_2SO_4} 82\%$$

**418**
$$CF_3-COOC_2H_5 \qquad [436, 872]$$

$$CF_3-COOK + C_2H_5OH \xrightarrow[\text{distillation}]{H_2SO_4} 73\%$$

*Procedure 24*

*Esterification of Perfluorocarboxylic Acids*
*Preparation of Ethyl Trifluoroacetate*

(a) *Preparation from Anhydrous Trifluoroacetic Acid.** Esterification of anhydrous trifluoroacetic acid with absolute ethanol proceeds without catalysts by mere heating of the components. A mixture of 570 g (5 moles) of trifluoroacetic acid and 460 g (10 moles) of ethanol is heated at 80–90°C in a 1·5 litre flask fitted with a distilling column of about 5 theoretical plates. At a reflux ratio of 5:1, an azeotropic mixture of ethyl trifluoroacetate-ethanol-water distils at 53·3–53·6°C in the course of 2–2·5 hours. The distillate weighing 840 g is shaken with four 300 ml. portions of a saturated calcium chloride. The ester layer from which thus ethanol has been removed is dried with calcium chloride solution, then with phosphorus pentoxide, filtered, and distilled in the distilling column. The pure ester distils at 59·8–60·1°C at 725 ml. in a yield of 570 g (4 moles) (80%). Considerable losses occur during drying and filtration.

(b) *Preparation from the Azeotropic Mixture of Trifluoroacetic acid and Water (436).*** A mixture of 316 g (6·5 moles) of 95% ethanol and 5 ml. of concentrated sulphuric acid is added to 372 g (2·6 moles) of the azeotrope containing 80% of trifluoroacetic acid and 20% of water, and the mixture is distilled in a fractionating column. After separating diethyl ether boiling at 34–35°C, a fraction boiling between 52·5 and 55·5°C representing a ternary mixture of ethyl trifluoroacetate, ethanol and water, is collected. This distillate separates into two layers, heavier ester layer weighing 367 g and lighter aqueous layer. Redistillation of the aqueous layer yields about 5 g of the heavier fraction which is added to the originally separated layer of the ester. This is washed with 900 ml. of a saturated solution of calcium chloride, dried with anhydrous calcium chloride and distilled in a fractionating column to give 303 g (2·1 moles) (82%) of ethyl trifluoroacetate, boiling at 60–61·5°C.

(c) *Preparation from Sodium Trifluoroacetate (309).**** A mixture of 170 g (1·25 moles) of anhydrous sodium trifluoroacetate, 140 g (3·0 moles) of absolute ethanol, and 160 g of concentrated sulphuric acid is allowed to stand for 3 hours at room temperature and then distilled from a steam bath until the temperature of the vapours of the distilling ester reaches 68°C. The distillate collected in an ice-cooled receiver is washed with 100 ml. of a cold 5% solution of sodium carbonate, dried with 25 g of anhydrous calcium chloride at 0°C, mixed with 15 g of phosphorus pentoxide and distilled to give 160 g (1·13 moles) (90%) of ethyl trifluoroacetate boiling at 60–62°C.

---

* Carried out by M. Hořický.

** Carried out by the author in the laboratory of A. L. Henne.

*** Carried out by the author.

Polyfluorinated alcohols having considerable acidity give but low yields in their esterification by acids; a better way to their esters is reaction with acid halides (*177*), or esterification in the presence of trifluoroacetic anhydride (*177*).

**419**                                                                                              [*177*]

$$C_5F_{11}-CH_2OH + CH_2=CH-COCl \xrightarrow[67°]{quinoline} C_5F_{11}-CH_2OCO-CH=CH_2 \quad 82.5\%$$

The reaction of alcohols with fluorinated acyl halides is, in addition to direct esterification, the most frequently used method of preparation (*273*).

**420**                                                                                              [*273*]

$$2 C_7F_{15}COCl + HO(CH_2)_5OH \xrightarrow[reflux]{} C_7F_{15}-COO(CH_2)_5OCOC_7F_{15} \quad 59\%$$

Fluorinated acid anhydrides can be used instead of acid chlorides (*172*):

**421**     $\langle \rangle$—OH + (C$_2$F$_5$CO)$_2$O $\xrightarrow{120°}$ $\langle \rangle$—OCOC$_2$F$_5$  94%     [*172*]

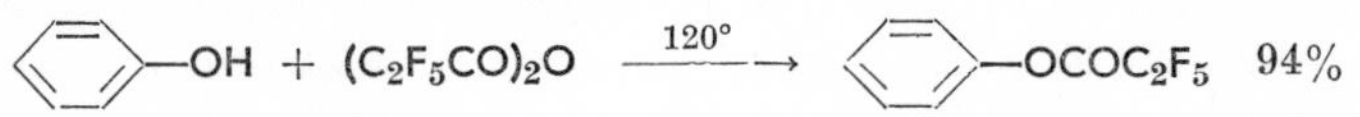

A reversal of this reaction, viz. the reaction between an alkyl halide and a salt of a fluorinated acid, has also been accomplished (*273*).

**422**      $2 C_3F_7COOAg + I(CH_2)_6I \rightarrow C_3F_7COO(CH_2)_6OCOC_3F_7$  88%     [*273*]

Methyl esters of fluorinated acids are conveniently prepared by treatment of free acids with diazomethane (*489*).

**423**
$$\begin{array}{l} CF_2-COOH \\ | \\ CF_2-COOH \end{array} + 2 CH_2N_2 \xrightarrow[(C_2H_5)_2O]{} \begin{array}{l} CF_2-COOCH_3 \\ | \\ CF_2-COOCH_3 \end{array}$$
[*489*]

Occasionally other acid derivatives can supply esters, e.g. fluoroacetamide gives a fair yield of fluoroacetate (*25*),

**424**     $FCH_2-CONH_2 \xrightarrow[20°]{C_2H_5OH,\ HCl\ (g)} FCH_2-COOC_2H_5$  66%     [*25*]

trimeric trifluoroacetonitrile, tris(trifluoromethyl)-*sym*-triazine, an excellent yield of ethyl trifluoroacetate (*779*) (p. 148). Treatment of 2H,1,2-dichloro-pentafluoropropane with alcoholic potassium hydroxide, or of 2-chloropenta-fluoropropene with sodium ethoxide, yields the same product, α-chloro-β,β,β-trifluoropropionate (*634*).

**425**                                                                                              [*634*]

$$CF_3-CHCl-CClF_2 \xrightarrow[reflux\ 6\ hrs.]{KOH,\ C_2H_5OH} \quad \xleftarrow[20°,\ 5\ hrs]{Na,\ C_2H_5OH} CF_3-CCl=CF_2$$

64%    $CF_3-CHCl-COOC_2H_5$    35%

The reaction between acetic anhydride and fluorinated aldehydes yields the aldehyde diacetates (*523*).

426       $C_3F_7$—CHO $\xrightarrow[\text{H}_2\text{SO}_4]{\text{(CH}_3\text{CO)}_2\text{O}}$ $C_3F_7$—CH$\Big\langle{}^{OCOCH_3}_{OCOCH_3}$   75%      [*523*]

Transesterification of vinyl acetate by fluorinated alcohols gives fluorinated vinyl ethers (*176*).

427                                                              [*176*]

$C_3F_7$—$CH_2OH$ + $CH_2$=CH—OCO—$CH_3$ $\xrightarrow[\text{Hg(OCOCH}_3)_2]{\text{H}_2\text{SO}_4}$ $C_3F_7$—$CH_2OCH$=$CH_2$ + $CH_3$—COOH

                                                    32%

## HYDROLYSIS

Of the many hydrolytic reactions undergone by fluorinated compounds, those which suffer splitting out of fluorine will receive main attention. The stability of a single fluorine atom bonded in various ways will be examined first. Then attention will be directed to the behaviour of a group of two fluorine atoms, and the stability of a trifluoromethyl group will be considered last. Mention will also be made of the resistance of perfluoro-derivatives to hydrolysis, and examples for hydrolytic cleavage of a carbon–carbon bond will be offered.

### Hydrolysis of Non-fluorinated Parts of the Molecules

The presence of fluorine atoms in esters does not affect their rate of hydrolysis too seriously. In ethyl acetate, the introduction of the first fluorine atom decreases, the presence of two and three fluorine atoms increases the rate of hydrolysis by 0,1 N hydrochloric acid in 70% aqueous acetone as compared with that of the unsubstituted ester (*319*). The effect of the accumulation of fluorine atoms in the molecules of esters of organic acids upon their rate of hydrolysis is shown in Table 27 (*744*).

TABLE 27

*Velocity Constants for the Hydrolysis of Esters in 70% Aqueous Acetone at 24·9°C (744)*

| Ester | $CH_3COOCH_3$ | $CF_3COOCH_3$ | $CF_3COOC_2H_5$ | $CF_3COOC_3H_7$ | $C_2F_5COOC_2H_5$ | $C_3F_7COOC_2H_5$ |
|---|---|---|---|---|---|---|
| $K \times 10^5$ | 5·49* | 12·8 | 3·99 | 2·69 | 0·272 | 0·131 |

\* At 25°C in 60% aqueous acetone.

Primary nitrofluoroparaffins were converted to fluorinated aldehydes by treatment with sodium hydroxide, followed by acidification with sulphuric acid. They were isolated as 2,4-dinitrophenylhydrazones and recovered by subsequent hydrolysis of these hydrazones (*1089*).

**428**                                                                    [*1089*]

$$FCH_2\text{--}CH_2\text{---}CH_2\text{---}CH_2\text{---}NO_2 \xrightarrow{\text{NaOH}} FCH_2\text{---}CH_2\text{---}CH_2\text{---}CH\text{=}N\overset{O}{\underset{ONa}{<}}$$

$$\xrightarrow{\text{H}_2\text{SO}_4} FCH_2\text{---}CH_2\text{---}CH_2\text{---}CHO \quad 22\%$$

## Hydrolysis of Monofluoro-derivatives

A single fluorine atom in an aliphatic chain is notably more resistant to alkaline hydrolysis than other halogens bound in like fashion (*501*). A kinetic study of hydrolytic removal of the halogens from polymethylene chlorofluorides shows that, with the exception of 1-chloro-3-fluoropropane, the differences in resistance to alkaline hydrolysis between the chlorine and fluorine are very substantial (*501*) (Table 28).

TABLE 28

*Hydrolysis of ω,ω'-Chlorofluoroparaffins in Alkaline Solutions (501)*

| $Cl(CH_2)_nF$ | Reflux (30 min.) of 0·01 mole of the chlorofluoride with 25 ml. of 1N $C_2H_5ONa$ in $C_2H_5OH$ | | Reflux (30 min.) of 0·01 mole of the chlorofluoride with 25 ml. of 1N KOH in 70% $C_2H_5OH$ | |
|---|---|---|---|---|
| | Split off, % | | Split off, % | |
| n | Cl | F | Cl | F |
| 2 | 49·5 | — | 42·2 | — |
| 3 | 57 | 11·6 | 26·3 | 0·9 |
| 4 | 38·3 | 0·2 | 20·4 | 0·2 |
| 5 | 31·4 | 0·1 | 15·4 | 0·2 |
| 6 | 28·7 | 0·2 | 12·1 | 0·2 |
| $C_6H_{13}X$ | 21·2 | Traces | 6·7 | 0·2 |

A similar resistance was noticed with fluorine bonded to a cyclohexane ring (*104*) or benzene ring (*1046*), provided it is not activated by nitro groups (see later). Low reactivity of fluorine toward the action of bases in contrast to the reactivities of other halogens can be demonstrated in Table 29 (*1046*).

TABLE 29

*Relative Reactivities of Isoamyl Halides and Halobenzenes toward Piperidine or Sodium Methoxide at 18° (1046)*

| $\begin{array}{c}CH_3\\ >CH-CH_2-CH_2X\\ CH_3\end{array}$ | F | Cl | Br | J |
|---|---|---|---|---|
| $C_5H_{11}N$ ............. | 1 | 68·5 | 17 800 | 50 500 |
| $CH_3ONa$ ............ | 1 | 71 | 3 550 | 4 500 |

| ⬡—X | F | Cl | Br | J |
|---|---|---|---|---|
| $C_5H_{11}N$ ............. | 1 | 1·9 | 74·5 | 132 |
| $CH_3ONa$ ............ | 1 | 1·8 | 4·4 | 35·6 |

Hydrolysis of fluorine is easier if its bond to carbon is weakened by a special position in the molecule. Benzylic fluorine is removed hydrolytically though less easily than chlorine bound in the same manner (731). α-Fluoro-α,α-dichlorotoluene was completely hydrolysed by boiling water (985). A similar arrangement of atoms in a paraffinic chain is much more stable. Dichlorofluoromethane did not show any signs of hydrolysis after 18 days of contact with water (107), A rather surprising reaction is the conversion of ethylene fluoride to ethylene glycol with water at room temperature (460).

Fluorine attached to a carbon atom adjacent to a carboxyl group possesses a pronounced tendency to hydrolytic removal. Much care was devoted to the study of the behaviour of fluorine in fluoroacetates. Methyl fluoroacetate dissolved in distilled water to form a 15% solution splits off, after 60 hours at 23°C, only 2·5% of its total fluorine (850). When hydrolysis of the fluoroacetate was attempted with calcium hydroxide in the cold (337) or by 5 minute contact with a 20% alcoholic potassium hydroxide (912), no hydrolytic cleavage of fluorine was observed. Prolonged action of a 20% potassium hydroxide solution upon methyl fluoroacetate removed 50% of the fluorine (912). No sodium fluoride was produced by refluxing methyl fluoroacetate for 1 hour with 10% sodium hydroxide (741). In an excess of 30% aqueous sodium hydroxide, an almost quantitative removal of fluorine was noted after 15 to 60 minutes of refluxing (741).

Aromatic fluoro-derivatives having nitro groups is *ortho-* and/or *para*-position readily exchange their fluorine for a hydroxyl group (*318*).

**429**

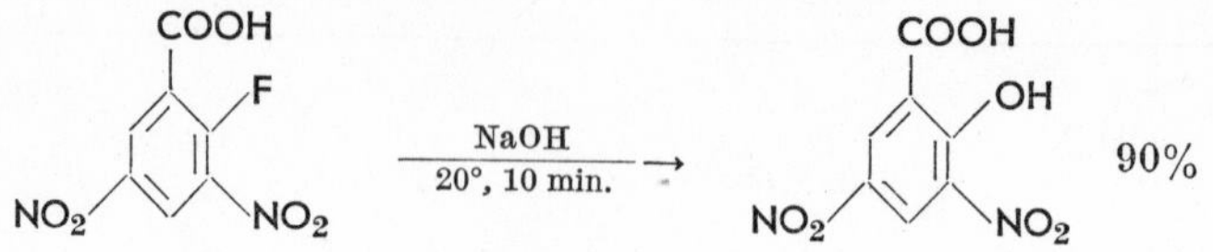

COOH / F ring with NO₂, NO₂ substituents → NaOH, 20°, 10 min. → COOH / OH ring with NO₂, NO₂ substituents, 90%          [*318*]

The easy reaction of this type of fluorine atom with alkalis is caused by the decrease of the electron density at the carbons in *ortho-* and *para*-positions to the nitro groups. The low electron density is probably also the reason for the relatively easy replacement of fluorine by hydroxyl in the reaction of hexafluorobenzene with potassium hydroxide in *tert.*-butyl alcohol (*97*).

**430**

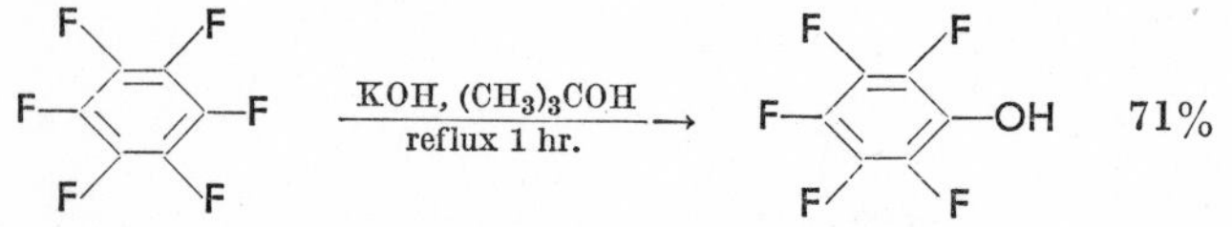

hexafluorobenzene (F₆ ring) → KOH, $(CH_3)_3COH$, reflux 1 hr. → pentafluorophenol (F₅–OH ring) 71%          [*97*]

Hydrolysis of acid fluorides is relatively easy. Carbonyl fluoride loses its fluorine by mere contact with water (*245*). Fluorine in sulphonyl fluorides undergoes hydrolysis in an alkaline solution easier than in acid medium (*321*).

**431**

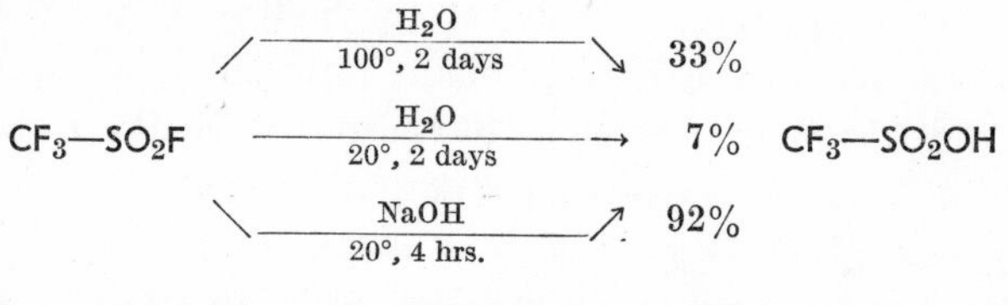

$$CF_3-SO_2F$$

- $H_2O$, 100°, 2 days — 33%
- $H_2O$, 20°, 2 days — 7%   $CF_3-SO_2OH$   [*321*]
- NaOH, 20°, 4 hrs. — 92%

A comparative rate of hydrolysis between an acid fluoride and a sulphonyl fluoride can be seen in the reaction of fluorosulphonyldifluoroacetyl fluoride with water and alkali (*256*):

**432**

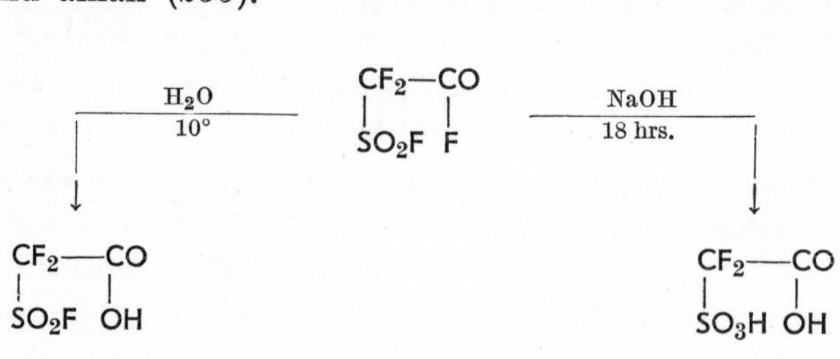

$$\begin{array}{c} CF_2-CO \\ | \quad | \\ SO_2F \ F \end{array}$$

$H_2O$, 10° →

$$\begin{array}{c} CF_2-CO \\ | \quad | \\ SO_2F \ OH \end{array}$$

NaOH, 18 hrs. →

$$\begin{array}{c} CF_2-CO \\ | \quad | \\ SO_3H \ OH \end{array}$$          [*256*]

Some sulphonyl fluorides can even be purified by steam-distillation (*1048*).

Fluorine attached to silicon (*244*) or phosphorus (*911*) is removed very easily. Dialkyl fluorophosphates are hydrolysed with water in the cold which splits off the fluorine without affecting the alkyl groups (*911*).

On the other hand, fluorine bonded to nitrogen in bis(trifluoromethyl)fluoroamine is stable at 20°C toward water or a 50% potassium hydroxide solution (*1037*).

## Hydrolysis of Geminal Difluoro Derivatives

The stability of a —CF$_2$— group toward hydrolysis strongly depends on its position in the molecule and on the conditions used. It is primarily transformed to a keto group and subsequent alterations leading to other derivatives are due to the nature of the adjacent substituents.

When 1,1-difluoro-2-bromoethane is heated with water and mercuric oxide at 150–160°C, only bromine is hydrolysed; fluorine needs temperatures as high as 180–200°C for hydrolytic removal (*460, 987*).

The vicinity of a double bond notably increases the hydrolysability of a —CF$_2$— group. Thus conversion of such groups to keto-groups was observed in many cyclobutene derivatives. An acidic medium seems to favour the hydrolysis and generally alkaline reagents do not attack such compounds (*102, 182, 942, 943*):

The presence of another substituent attached to the carbon carrying a $-CF_2-$ group also assists hydrolysis. Thus $\alpha$-chloro-$\alpha,\alpha$-difluorotoluene yields benzoic acid by mere contact with water at room temperature (985). Chlorodifluoromethane and bromodifluoromethane give alkaline formate on hydrolysis with an alkaline hydroxide (992). Treatment of chlorodifluoroacetate with sodium hydroxide at an elevated temperature yields oxalate (990). 1,1,3-Trichloro-2,3,3-trifluoro-1-propene affords $\beta,\beta$-dichloro-$\alpha$-fluoroacrylic acid on heating with concentrated sulphuric acid (973).

437        $CCl_2{=}CF{-}CClF_2 \xrightarrow[85-135°]{98\% \ H_2SO_4} CCl_2{=}CF{-}COOH$        88%        [973]

Hydrolysis is still easier if the substituent bonded to the $-CF_2-$ group is an oxygen or nitrogen atom. 2-Nitrotetrafluoroethyl nitrite readily hydrolyses to nitrodifluoroacetic acid (567):

438                                                                                      [567]

$O_2N{-}CF_2{-}CF_2{-}ONO \xrightarrow{H_2O} [O_2N{-}CF_2{-}CF_2OH] \longrightarrow O_2N{-}CF_2{-}COOH$

If the substituent attached to a difluoromethylene group is an alkoxy- or aryloxy group, an ester or an acid results in acidic hydrolysis (576, 1020, 1100, 1101):

$CHClF{-}CF_2OC_2H_5 \xrightarrow[0-10°]{96\% \ H_2SO_4} CHClF{-}COOC_2H_5$        83%

439                                                         NaOH $\downarrow$                    [1020,1100]

$CHClF{-}CF_2OC_6H_5 \xrightarrow{96\% \ H_2SO_4} CHClF{-}COOH$        50%

*Procedure 25*

*Hydrolysis of a Difluoromethylene Group*
*Preparation of Ethyl Chlorofluoroacetate (258)\**

A one-litre three-necked flask fitted with a gas-tight stirrer, a separating funnel, and a T-tube allowing the insertion of a thermometer and the escape of the hydrogen fluoride formed in the reaction, is charged with 137 g (0·85 mole) of $\beta$-chloro-$\alpha,\alpha,\beta$-trifluoroethyl ethyl ether. While cooling the flask with an ice bath to a temperature of 5—10°C, 170 g (93·5 ml., 1·65 moles) of concentrated sulphuric acid is added dropwise to the ether in the course of 30 minutes. The mixture is stirred for 2 more hours, poured onto a mixture of 400 g of ice and 200 ml. of water, the lower layer is separated, washed with a solution of sodium bicarbonate and with water, is dried with magnesium sulphate and distilled. Approximately 83 g (0·59 mole) (70%) of ethyl chlorofluoroacetate is obtained, distilling at 127–130°C.

---

\* Carried out by the author.

The hydrogen fluoride formed in the reaction attacks the flask, the thermometer and the stirrer. The most profound corrosion occurs at the level of the liquid in the flask. It is therefore advisable to protect the stem of the thermometer and the stirrer by a paraffin layer.

Another example of the activation of a difluoromethylene group for hydrolysis is the hydrolysis of perfluoro-$\gamma$-butyrolacetone which affords perfluorosuccinic acid (*410*).

440
$$
\begin{array}{cc}
CF_2-CF_2 \\
| \quad\quad | \\
CF_2 \quad CO \\
\diagdown O \diagup
\end{array}
\xrightarrow{\ H_2O\ }
\begin{array}{cc}
CF_2-\!-\!-CF_2 \\
| \quad\quad\quad | \\
COOH\ \ COOH
\end{array}
\qquad [410]
$$

If the substituent attached to the carbon atom with the $-CF_2-$ group is nitrogen (carrying alkyl or acyl groups), acid amides are the results of hydrolysis (*195, 576*).

441 $\quad CF_3-CHF-CF_2-N(C_2H_5)_2 \xrightarrow{\ H_2O\ } CF_3-CHF-CO-N(C_2H_5)_2 \qquad [576]$

A complex example of the hydrolysis of a difluoromethyl group is the conversion of 2,2,3,3-tetrafluoro-1-cyanocyclobutane to $\alpha,\alpha$-difluoroglutaric acid. This reaction seems to proceed through dehydrofluorination, hydration of the $\alpha,\beta$-double bond thus formed, and hydrolytic cleavage of the $\beta$-keto-acid (*41*).

442
$$
\begin{array}{cc}
CF_2-CF_2 \\
| \quad\quad | \\
CH_2-CH-CN
\end{array}
\xrightarrow[\text{reflux}]{25\%\ NaOH}
\begin{array}{cc}
CF_2-COOH \\
| \\
CH_2-CH_2-COOH \\
30\%
\end{array}
\leftarrow
\begin{array}{cc}
CF_2-CFCl \\
| \quad\quad | \\
CH_2-CH-CN
\end{array}
\qquad [41]
$$

## Hydrolysis of Trifluoromethyl Group

The behaviour of a trifluoromethyl group in hydrolysis depends on its position in the organic molecule, and on whether the hydrolysis is carried out with alkaline or acidic reagents.

The aliphatic trifluoromethyl group generally resists hydrolysis, either in acidic or in alkaline medium. The reported hydrolysis of fluoroform to potassium formate by means of potassium hydroxide (*702*) has been contradicted by more recent experiments (*427*).

The trifluoromethyl group in 3-bromo-1,1,1,3-tetrafluoro-2-propanol remained unchanged after heating with sodium hydroxide at 95–100°C (*658*). Similarly bis(trifluoromethyl) sulphide was recovered in a 97% yield after heating with 15% sodium hydroxide for 24 hours (*128*). Sometimes even the trifluoro-

methyl group adjacent to a double bond resists hydrolysis while a trichloromethyl group is converted to a carboxyl under the same conditions (*366*).

**443**                                                                      [*366*]

$$CF_3—CHI—CH_2—CCl_3 \xrightarrow[\text{2. } H_2SO_4, 80°]{\text{1. 10\% KOH, } C_2H_5OH} \underset{51\%}{\Big\downarrow} \quad \underset{68\%}{\Big\downarrow} \xrightarrow{H_2SO_4 \text{ concd.}} CF_3—CH{=}CH—CCl_3$$

$$CF_3—CH{=}CH—COOH$$

In 2-bromo-1,1,1-trifluoroethane, the trifluoromethyl group withstood heating
with sulphuric acid at 140°C (*1009*), in other trifluoromethyl-derivatives at
120–130°C (*150*).

**444**                                                                      [*150*]

$$\underset{\overset{|}{CH_2OH}}{CF_3—CH—CONH_2} \xrightarrow[120°]{H_2SO_4} \underset{\overset{|}{CH_2OH}}{CF_3—CH—COOH} \xleftarrow[120-130°]{H_2SO_4} \underset{\overset{||}{CH_2}}{CF_3—C—CN}$$

On the other hand, the trifluoromethyl group in $\beta,\beta,\beta$-trifluoroisobutyric acid
is easily hydrolysed in alkaline medium to methylmalonic acid (*150*).

**445**       $$\underset{\overset{|}{CF_3}}{CH_3—CH—COOH} \xrightarrow[100°]{2N\text{-}NaOH} \underset{\overset{|}{COOH}}{CH_3—CH—COOH} \quad 62\%$$       [*150*]

This example betrays the fact that the hydrolysis is probably preceded
by dehydrofluorination to $\alpha,\beta$-difluoromethacrylic acid which readily accepts
water, and the $\beta,\beta$-difluoro-$\beta$-hydroxyisobutyric acid thus formed undergoes
further hydrolysis. A similar mechanism could apply to the alkaline hydrolysis
of 1,1-bis(*p*-chlorophenyl)-2,2,2-trifluoroethane which affords bis(*p*-chloro-
phenyl)ketene as an intermediate (*699*).

**446**                                                                      [*699*]

The trifluoromethyl group attached to an aromatic nucleus is rather
resistant to the action of water and alkaline reagents (*983*), but undergoes

a quantitative hydrolysis to carboxyl in sulphuric acid (*616, 665, 956, 958, 996*).

**447**

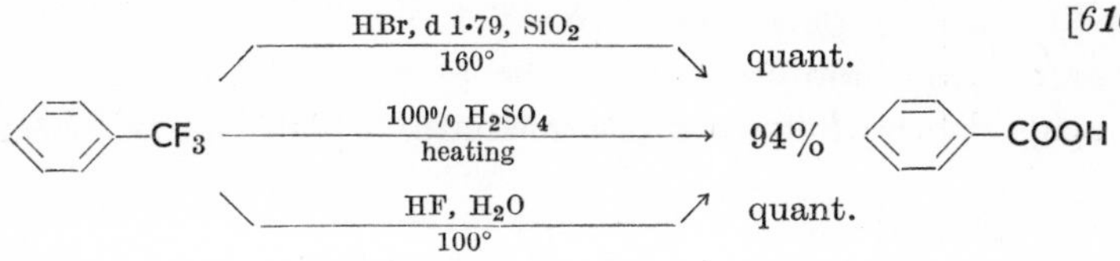

*Procedure 26*

*Hydrolysis of a Trifluoromethyl Group*
*Hydrolysis of Benzotrifluoride (616)*

A mixture of 35·6 g (0·25 mole) of benzotrifluoride and 28 g (0·28 mole) of 100% sulphuric acid is carefully heated until hydrogen fluoride begins to evolve, and the reaction is kept going by intermittent heating as long as hydrogen fluoride escapes. Pouring of the mixture into 1 litre of ice-cold water and filtration with suction of the crystals yields 28·6 g (0·235 mole) (94%) of benzoic acid.

It is interesting to note the occurrence of hydroxyterephthalic acid in the Kolbe synthesis of 4-trifluoromethylsalicylic acid, which is carried out in an alkaline medium (*408*) (p. 237).

## Hydrolysis of Perfluoro-derivatives

Saturated perfluoro-derivatives show remarkable resistance to hydrolysis. Traces of fluoride ions found after alternatively heating of perfluoroheptane with a 1% solution of sodium bicarbonate and a 20% solution of sodium hydroxide for four hours at 100°C could probably come from negligible impurities contaminating the fluorocarbons (*288*). Unsaturated and aromatic perfluoro-derivatives are less stable. Perfluorotoluene is hydrolysed under drastic conditions to perfluorobenzoic acid (*665*).

**448**

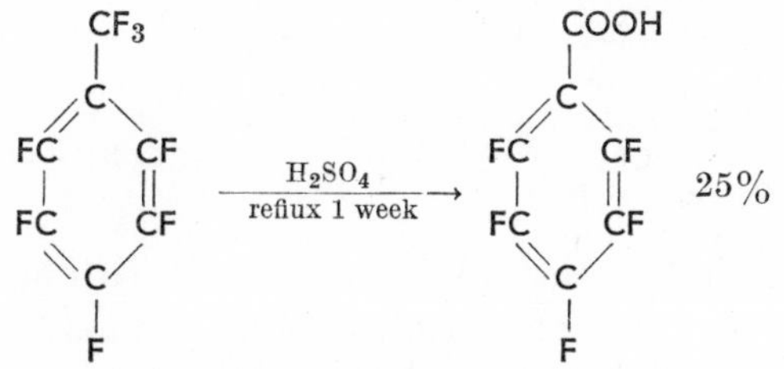

Double bonds surrounded by fluorine atoms have a decreased electron density and are therefore sensitive to nucleophilic attack by hydroxyl ions. An example for such a reaction is the conversion of perfluorobenzene to perfluorophenol (*97*) (Equation 430, p. 202).

By the same token, an olefinic double bond carrying fluorine atoms can easily accept anions and thus form intermediates much more sensitive to hydrolysis than the original perfluoroolefins.

For example, perfluoropropene can be hydrolysed to $\alpha$-hydrotetrafluoropropionic acid via the following intermediates which can be isolated (740).

**449**                                                                                    [740]

$$CF_3-CF{=}CF_2 \xrightarrow{\text{10\% KOH, CH}_3\text{OH}} CF_3-CHF-CF_2-OCH_3$$

$$\downarrow \substack{\text{96\% H}_2\text{SO}_4 \\ <\,10-20°}$$

$$CF_3-CHF-COOCH_3 \xrightarrow{\text{H}_3\text{PO}_4} CF_3-CHF-COOH$$

Similarly perfluoroacrylonitrile affords fluoromalonic acid on hydrolysis with sulphuric acid (613).

**450**     $CF_2{=}CF-CN \xrightarrow[\text{70°, 72 hrs.}]{\text{70\% H}_2\text{SO}_4\text{/Cu, quinol}} COOH-CHF-COOH$   45%     [613]

Partial hydrolysis of perfluoro (methyl-methyleneimine) carried out with an insufficient amount of water gives trifluoromethylisocyanate (46).

**451**     $CF_3-N{=}CF_2 \xrightarrow{\text{H}_2\text{O}} CF_3-N{=}C{=}O$     conversion 36%     [46]
                                                                       yield 50%

## The Fluoroform Reaction

Hydrolytic fission of a carbon chain which occurs with $\alpha,\alpha,\alpha$-trihalocarbonyl compounds and which affords haloforms was observed also with the fluorinated analogues. The fluoroform reaction has been described for trifluoroacetaldehyde (936), trifluoromethyl alkyl ketones (1012), trifluoroacetophenone (958) and for 2,4-dihydroxytrifluoroacetophenone (1080).

**452**     ⬡—CO—CF$_3$ $\xrightarrow{\text{10\% KOH}}$ ⬡—COOH + CHF$_3$     [958]

The same carbon-carbon bond fission was noted with higher perfluoroalkyl aldehydes and ketones. Thus the hydrate of heptafluorobutyraldehyde (523) or heptafluoropropyl methyl ketone (1012) yield 1H-heptafluoropropane and formic or acetic acid, respectively.

**453**     $CF_3-CF_2-CF_2-CH\begin{smallmatrix}OH\\OH\end{smallmatrix}$ $\xrightarrow[\text{reflux 90 min.}]{\text{10\% KOH}}$ $CF_3-CF_2-CF_2H$   88,5%     [523]

A similar cleavage was observed during the reaction of ethyl trifluoroacetate, perfluoropropionate and perfluorobutyrate with sodium ethoxide,
and 1H-perfluoroalkanes were isolated as the main products (*81*).

454 $\quad C_3F_7COOC_2H_5 \xrightarrow[\text{reflux}]{C_2H_5ONa,\ C_2H_5OH} C_3F_7H + CO(OC_2H_5)_2$ [*81*]
$\qquad\qquad\qquad\qquad\qquad\qquad\qquad\quad 71\% \qquad 30\%$

A carbon chain scission occurs also during the boiling of dibromofluoro-
acetic acid with water, which yields dibromofluoromethane (*984*). On the other
hand, trifluoroacetic acid does not give the fluoroform reaction (*436*).

Fluoroform is also formed from derivatives carrying the trifluoromethyl
group on sulphur (*378*) or phosphorus (*72, 825*).

455 $\quad CF_3SCl \xrightarrow[95°]{NaOH} CHF_3 \xleftarrow{\ 15\%\ NaOH\ }_{95°} CF_3-SO_2Na$ [*378*]
$\qquad\qquad\qquad\qquad\quad 99\cdot8\%$

456 $\quad (CF_3)_2P-OH \xrightarrow{H_2O} CHF_3 + CF_3-P(OH)_2 \xrightarrow{H_2O} CHF_3 + H_3PO_3$ [*72*]

$$\xleftarrow[20°]{H_2O} (CF_3)_3PO \xrightarrow[20°]{10\%\ NaOH}$$

[*825*]

457 $\quad CHF_3 + (CF_3)_2P\begin{smallmatrix}O\\\\OH\end{smallmatrix} \qquad 2\ CHF_3 + CF_3P\begin{smallmatrix}OH\\O\\OH\end{smallmatrix}$
$\qquad\qquad\qquad\qquad\qquad\qquad\qquad\qquad\text{quant.}$

During the treatment of pentachlorofluoroacetone and tetrachloro-$\alpha,\alpha$-di-
fluoroacetone with sodium hydroxide, fission takes place between the carbonyl
and the trichloromethyl group thus affording chloroform and dichlorofluoro-
acetic acid or chlorodifluoroacetic acid, respectively (*714*).

458 $\quad CCl_2F-CO-CCl_3 \xrightarrow[40°]{20\%\ NaOH} CCl_2F-COOH + CHCl_3$ [[*714*]
$\qquad\qquad\qquad\qquad\qquad\qquad\quad 95\% \qquad\quad 80\%$

## THE FRIEDEL–CRAFTS SYNTHESIS

Fluorinated derivatives can participat ein the Friedel–Crafts reaction either
as passive components (they can be alkylated or acylated), or as active components (they can alkylate or acylate). Fluorine atoms in alkylating or acylating components can either be split out during the reaction in the form of
hydrogen fluoride, or else they can stay in the molecule.

14

## Fluorinated Compounds as Passive Components in the Friedel–Crafts Reaction

Aromatic fluorinated compounds are alkylated or acylated without difficulties. An example is the synthesis of bis(p-fluorophenyl) dichloromethane from fluorobenzene and carbon tetrachloride (542).

459   2 F—⟨⟩ + CCl₄  $\xrightarrow[15°]{\text{AlCl}_3,\ \text{CS}_2}$  F—⟨⟩—CCl₂—⟨⟩—F   59%    [542]

Chloral reacts with two moles of fluorobenzene and yields 1,1-bis(p-fluorophenyl)-2,2,2-trichloroethane (344).

460   CCl₃—CHO + 2 ⟨⟩—F  $\xrightarrow{\text{H}_2\text{SO}_4}$  CCl₃—CH⟨⟨⟩—F, ⟨⟩—F⟩    33% [344]

An example for acylation is the reaction between fluorobenzene and acetyl chloride which affords p-fluoroacetophenone (149).

461   F—⟨⟩ + ClCO—CH₃  $\xrightarrow{\text{AlCl}_3,\ \text{CS}_2}$  F—⟨⟩—CO—CH₃   76%    [149]

## Fluorinated Compounds as Active Components in the Friedel–Crafts Reaction

More interesting than the previous type are Friedel–Crafts syntheses in which fluorinated compounds act as active components.

Reactions will first be discussed which are accompanied by loss of hydrogen fluoride. In this occasion, a very curious reaction should be mentioned which has been discovered only recently. Perchloryl fluoride, which can be considered as a fluoride of perchloric acid, converts aromatic hydrocarbons to an entirely new type of compound that will cause a serious problem in chemical nomenclature (528):

462          ⟨⟩ + FClO₃  $\xrightarrow{\text{AlCl}_3}$  ⟨⟩—ClO₃          [528]

Syntheses using alkyl fluorides and acyl fluorides are not very advantageous, since the fluorinated derivatives are, as a rule, less accessible than the other halogen derivatives; the yields are very often unsatisfactory, and the escaping hydrogen fluoride requires special precautions. However, some interesting reactions have been accomplished in this field.

Alkyl fluorides and acyl fluorides form, with an equivalent amount of boron trifluoride, alkyl or acyl fluoroborates, respectively, which are stable at low temperatures and which readily alkylate or acylate aromatic hydrocarbons (*782, 787*).

**463** [*782, 787*]

$$C_2H_5F + BF_3 \xrightarrow{-50° \text{ to } -110°} \overset{\oplus}{C_2H_5} \overset{\ominus}{BF_4} ; \quad CH_3{-}COF + BF_3 \longrightarrow \overset{\oplus}{CH_3{-}CO} \overset{\ominus}{BF_4}$$
m.p. −105°                                    decompn. 20°

**464** $C_2H_5F +$ ⟨⟩$-CH_3 \xrightarrow[-20° \text{ to } -80°]{BF_3}$ $C_2H_5-$⟨⟩$-CH_3$ 81% [*787*]

Fluorine seems to react preferentially to other halogens in the Friedel–Crafts alkylations (*782*):

**465** ⟨⟩ $+ FCH_2{-}CH_2Cl \xrightarrow{BCl_3}$ ⟨⟩$-CH_2{-}CH_2Cl$ [*782*]

Whereas the sequence of halogen reactivity in alkylations is $F > Cl > Br > I$, the order is reversed in acylations (*157*).

A modification of the Friedel–Crafts reaction is a fluorine analogue of the Gattermann-Koch synthesis of aromatic aldehydes. In contrast to the unstable formyl chloride which exists only as a complex with aluminium chloride, or with cuprous chloride, formyl fluoride is a relatively stable compound (*767*) (Equation 126, p. 105). The reaction requires catalysis by boron trifluoride but even so the results are unsatisfactory.

**466** $HCOF +$ ⟨⟩$-CH_3 \xrightarrow{BF_3}_{-80°}$ $CHO-$⟨⟩$-CH_3$ 30% [*784*]

Examples of alkylations with fluorine remaining in the alkylating group are not very common but they do not lack in interest. Fluoromethanol reacts with benzene to give benzyl fluoride (*791*). The same compound seems to originate from the action of paraformaldehyde and anhydrous hydrogen fluoride on benzene (*791*).

**467** $FCH_2OH +$ ⟨⟩ $\xrightarrow[\text{reflux}]{ZnCl_2}$ ⟨⟩$-CH_2F$ 58·5% [*791*]

An audacious experiment is the Friedel–Crafts reaction between benzene, chlorotrifluoroethylene and boron trifluoride which is run at high temperature and affords low yields of α,β,β-trifluorostyrene (*856*).

**468** ⟨⟩ $+ ClCF{=}CF_2 \xrightarrow{BF_3}_{550°}$ ⟨⟩$-CF{=}CF_2$ 5·5% [*856*]

More common are acylations involving fluorinated acyl halides. The Friedel–Crafts reaction allows the synthesis of all three fluorinated acetophenones

14*

with fluorine atoms bound in the methyl group: monofluoro- (*90, 337*), di-fluoro- (*856*), and trifluoroacetophenones (*183, 958*).

**469**

$$\text{CH}_3\text{—}\langle = \rangle\text{—CO—CH}_2\text{F} \quad 81\%$$

$$\xleftarrow[\text{AlCl}_3, \text{CH}_2\text{Cl}_2, 0°]{\text{ClCOCH}_2\text{F}} \langle = \rangle \xrightarrow[\text{AlCl}_3, 10—20°]{\text{ClCOCF}_3} \quad [90, 183, 856]$$

$$\text{ClCOCHF}_2 \,\big|\, \text{AlCl}_3,$$

$$\langle = \rangle\text{—CO–CHF}_2 \quad 69\%$$

$$\langle = \rangle\text{—CO—CF}_3 \quad 64\%$$

Acylations using perfluorocarboxylic acid chlorides are carried out at −10°C to 10°C with low-boiling acid chlorides, and at 50°C with higher-boiling acid chlorides, and give yields ranging from 21 to 65% (*952*).

**470**                                                     [*952*]

$$\text{CH}_3\text{—}\langle = \rangle + \text{C}_5\text{F}_{11}\text{—COCl} \xrightarrow[50°]{\text{AlCl}_3} \text{CH}_3\text{—}\langle = \rangle\text{—CO—C}_5\text{F}_{11} \quad 65·5\%$$

This reaction is not limited to monocarboxylic acid derivatives. Perfluorodi-carboxylic acid dichlorides give rise to diketo derivatives (*270*):

**471**   
$$\text{CF}_2\langle\begin{array}{l}\text{COCl}\\\text{COCl}\end{array} \;+\; \begin{array}{l}\langle = \rangle\\\langle = \rangle\end{array} \xrightarrow[10—20°]{\text{AlCl}_3} \text{CF}_2\langle\begin{array}{l}\text{CO—}\langle = \rangle\\\text{CO—}\langle = \rangle\end{array} \quad 68\% \quad [270]$$

The trifluoroacetyl group has been introduced into aromatic compounds by means of trifluoroacetyl chloride (*183, 958*), trifluoroacetic anhydride (*187*) or trifluoroacetonitrile (*1080*), this last by the Houben-Hoesch procedure.

**472**   
$$\begin{array}{c}\text{OH}\\\langle = \rangle\\\text{OH}\end{array} + \text{CF}_3\text{—CN} \xrightarrow[0° \to 100°]{\text{ZnCl}_2, \text{HCl (g)}, (\text{C}_2\text{H}_5)_2\text{O}} \begin{array}{c}\text{CF}_3\\|\\\text{CO}\quad\text{OH}\\\langle = \rangle\\\text{OH}\end{array} \quad 63\% \quad [1080]$$

*Procedure 27*

    *The Friedel-Crafts Acylation*
    *Preparation of ω,ω,ω-Trifluoroacetophenone from Benzene and Trifluoroacetyl Chloride (183)*
    (a) *Preparation of Trifluoroacetyl Chloride.* A mixture of 170 g (1·25 moles) of sodium trifluoroacetate and 280 g (1·83 moles) of phosphorus oxychloride is heated for 21 hours at 100°C under a reflux condenser connected to a dry-ice trap. The yield of trifluoroacetyl chloride is 147 g (1·1 moles) (89%).
    (b) *Acylation of Benzene.* Trifluoroacetyl chloride (180 g, 1·35 moles) is passed into

a stirred mixture of 214 g (1·6 moles) of aluminium chloride, 125 g (1·6 moles) of benzene, and 490 ml of carbon disulphide over a period of 4 hours. The mixture is cooled to 2–3°C in the beginning and to 10°C toward the end of the reaction. It is then decomposed with 1900 g of ice and 335 ml. of concentrated hydrochloric acid, the organic layer is separated, the aqueous layer is extracted with ether, the combined solutions of the product in benzene and ether are filtered, dried and distilled. $\omega,\omega,\omega$-Trifluoroacetophenone boiling over a range of 66–67°C at 37 mm is obtained in a yield of 150 g (0·86 mole) (64%).

## REACTIONS OF ORGANOMETALLIC COMPOUNDS

In the chapter dealing with fluorinated organometallics, distinction will be made between two types of reactions. Those in which a fluorinated compound is the passive component which reacts with an ordinary non-fluorinated organometallic compound, and those in which fluorine atom is present in the organometallic compound itself. The latter ones are most interesting when formed from the perfluoro-derivatives. Such reactions have been discovered only recently and are of great help for introducing perfluoroalkyl groups into organic compounds. As in the chemistry of non-fluorinated derivatives, the organomagnesium compounds are the best known. Fluorinated lithium compounds are still rare but are attracting increased attention (p. 222). As to other organometallics, compounds of zinc, cadmium, mercury, and metalloids will be mentioned only briefly.

### Organomagnesium Compounds

Reactions of the Grignard reagents with fluorinated compounds will be followed by the description of formation and reactions of the Grignard reagents prepared from fluorinated compounds.

*Reactions of fluorinated compounds with the Grignard reagents.* The Grignard synthesis is a convenient method for introducing a fluorinated group into an organic molecule. Starting material for such preparations are fluorinated aldehydes (*585, 657, 662*), fluorinated ketones (*90*), fluorinated acids (*1011*), fluorinated esters (*332, 479, 657, 844*) and fluorinated nitriles (*543, 663*). The course of the reaction is normal. However, reduction occurs to a considerable extent with perfluoroaldehydes and ketones (p. 166). It can be partly suppressed by the addition of magnesium bromide (*662*) (Table 30).

TABLE 30

*Results of the Reaction of Perfluoroaldehydes with Ethylmagnesium Iodide at 0°C* (*662*)

| Reaction Product | $C_2F_5CH(OH)C_2H_5$, % | $C_2F_5CH_2OH$, % |
|---|---|---|
| $C_2F_5CHO$ Alone | 34 | 55 |
| $C_2F_5CHO$ With $MgBr_2$ Added | 51 | 16 |

Methyl and ethyl trifluoroacetate react with methylmagnesium bromide and afford dimethyltrifluoromethylcarbinol in high yields (*332, 479*).

$$\textbf{473} \quad CF_3—COOCH_3 + 2\,CH_3MgBr \xrightarrow{(C_2H_5)_2O} \quad CF_3—\overset{\overset{\displaystyle CH_3}{|}}{\underset{\underset{\displaystyle CH_3}{|}}{C}}—OH \quad 80\% \qquad [\textit{332, 479}]$$

Free trifluoroacetic acid yields trifluoromethyl ketones in fair yields, even if as much as three equivalents of the Grignard reagent are applied (*215, 1011, 1012*) (Table 31).

TABLE 31

*Yields of Trifluoromethyl Alkyl Ketones in the Reaction of Trifluoroacetic Acid with Alkylmagnesium Halides* (*1011*)

| Alkyl ...................... | $CH_3$ | $C_2H_5$ | $(CH_3)_2CH$ | $C_4H_9$ | $(CH_3)_3C$ | $C_6H_5$ |
|---|---|---|---|---|---|---|
| Trifluoromethyl alkyl ketone, % | 56 | 52 | 42 | 63 | 25 | 60 |

In addition to the ketones, secondary alcohols are formed by the action of the aliphatic Grignard reagents (reduction product), and tertiary alcohols, when aromatic organomagnesium compounds have been used (*215*).

**474**                                                                   [*215*]

$$CF_3—COOH + 3\,C_3H_7MgBr \xrightarrow[35°]{(C_4H_9)_2O} CF_3—CO—C_3H_7 + CF_3—\underset{\underset{\displaystyle OH}{|}}{CH}—C_3H_7$$
$$51\cdot2\% \qquad\qquad\qquad 11\cdot8\%$$

$$CF_3—COOH + 3\,C_6H_5MgBr \xrightarrow{(C_2H_5)_2O} CF_3—CO—C_6H_5 + CF_3—\overset{C_6H_5}{\underset{\underset{\displaystyle OH}{|}}{C}}{<}_{C_6H_5}$$
$$\textbf{475} \qquad\qquad\qquad\qquad 58\% \qquad\qquad\qquad [\textit{215}]$$
$$6\cdot7\%$$

Perfluoroalkyl ketones are also formed in the reaction of the Grignard reagents with perfluorocarboxylic acid nitriles (*543, 663*) (Equation 312, p. 167).

**476**                                                       [*543*]

$$CF_3—CN + \langle \rangle—CH_2MgCl \xrightarrow{(C_2H_5)_2O} CF_3—CO—CH_2—\langle \rangle \quad 38\cdot4\%$$

The addition of the Grignard reagents occurs not only across a carbonyl bond but also to the *beta*-position of a $\alpha,\beta$-double bond. In the reaction of ethylmagnesium bromide with 1-nitro-1H,2H-heptafluoro-1-pentene, the alkyl is fixed to the carbon atom *beta* to the nitro group (*186*).

477    $C_3F_7$—CH=CH—$NO_2$ + $C_2H_5MgBr$  →  $C_3F_7$—CH—$CH_2$—$NO_2$    53%    [*186*]
$\phantom{477 \quad C_3F_7—CH=CH—NO_2 + C_2H_5MgBr  →  C_3F_7—CH}$|
$\phantom{477 \quad C_3F_7—CH=CH—NO_2 + C_2H_5MgBr  →  C_3F_7—CH—}C_2H_5$

The Grignard reagents react avidly with fluorinated olefins. The primary addition across the double bond is usually followed by splitting off a halide anion, so that an unsaturated compound results (*1027*).

478                                                                 [*1027*]

$CF_2$=$CCl_2$ + ⟨⟩—MgBr → [⟨⟩—$CF_2$—$\overline{C}Cl_2$]MgBr  $\xrightarrow{\text{—MgBrF}}$  ⟨⟩—CF=$CCl_2$
$\phantom{xxxxxxxxxxxxxxxxxxxxxxxxxxxxxxxxxxxxxxxxxxxxxxxxxxxxxxxxxxxxxxxxxxxx}$63%

Fluorinated propenes behave similarly. While 1-iodoperfluoro-2-propene gives with methylmagnesium bromide, 4H,4H,4H-pentafluoro-1-butene as the only product (*1027*),

479            $CF_2$=CF—$CF_2I$ + $CH_3MgBr$  →  $CF_2$=CF—$CF_2$—$CH_3$        [*1027*]

the analogous chloro-derivative of propene affords in the reaction with phenyl-magnesium bromide a mixture, in which the product with a conjugated double bond prevails (*1027*).

480                                        $CF_2$=CF—$CF_2$—⟨⟩  1 part    [*1027*]

$CF_2$=CF—$CF_2Cl$ + ⟨⟩—MgBr  $\xrightarrow{20°}$
$\phantom{xxxxxxxxxxxxxxxxxxxxxxxxxxxxxxxx}$⟨⟩—CF=CF—$CF_2Cl$  2 parts

The last example shows that the reaction proceeds via the addition of the Grignard reagent across the double bond, and not, as could also be explained, by a metathetical reaction of the Grignard compound with the allylic halogen.

Less common than the addition reactions of the Grignard reagents are the reactions of organomagnesium compounds with halogen derivatives. They require sufficiently reactive halogens. Fluorine is replaced by the alkyl group of a Grignard reagent, only if it is sufficiently activated, and even so only at higher temperatures (*90*).

$\phantom{xxxxxx}$ $\xrightarrow[0°]{C_6H_5MgBr}$ $\phantom{xx}$ ⟨⟩—CO—$CH_2F$ $\xrightarrow[\text{heating}]{C_6H_5MgBr}$

481   ↓                                                                    ↓   [*90*]

$\phantom{xx}$OH
$\phantom{xxx}$|
⟨⟩—C—⟨⟩  65%                          ⟨⟩—CO—$CH_2$—⟨⟩
$\phantom{xxx}$|
$\phantom{xx}CH_2F$

Aromatic fluorine in o-fluorobenzaldehyde is, according to expectation, perfectly stable, and the reaction with methyl magnesium bromide gives 78–83% of o-fluorophenylmethylcarbinol (585).

*Preparation and reactions of fluorinated Grignard reagents.* Success in the preparation of the Grignard reagents from fluorinated compounds depends largely on the reactivity of the halogen and on its location in the fluorinated molecule, especially on its distance from the fluorine atoms. The reactivity drops from iodine to fluorine. As a matter of fact, organically bound fluorine does not seem to react with magnesium. Benzyl fluoride, for instance, does not give benzylmagnesium fluoride (93). However, an interconversion between a fluoride and an alkylmagnesium halide probably takes place in the sense of the following equation:

482          $R—F + R'MgX \longrightarrow RMgF + R'—X$

Such reaction seems to take place when treating 6-fluorohexylmagnesium chloride with ethyl chloroformate for beside the main product, ethyl 7-fluoro-heptanoate, a minor amount of diethyl suberate is isolated (509).

483                                                                      [509]

$F(CH_2)_6MgCl + ClCOOC_2H_5 \longrightarrow F(CH_2)_6COOC_2H_5 + C_2H_5OCO(CH_2)_6COOC_2H_5$
                                                       48%                    6%

Another example is the reaction of 6-fluorohexylhalides with magnesium. In addition to the expected 6-fluorohexylmagnesium halides, varying amounts of 6-halogenohexylmagnesium fluoride and 6-fluoromagnesylhexylmagnesium halide seem to arise in the reaction mixture, since not only 7-fluoroheptanoic acid, but also 7-halogenoheptanoic acid and suberic acid have been produced on carbonation (816).

484                                                                      [816]

$F(CH_2)_6X \xrightarrow[2.CO_2]{1.\,Mg,\,(C_2H_5)_2O} F(CH_2)_6COOH + X(CH_2)_6COOH + COOH(CH_2)_6COOH$

|        | $F(CH_2)_6COOH$ | $X(CH_2)_6COOH$ | $COOH(CH_2)_6COOH$ |
|--------|------|-------|------|
| Cl     | 64%  | 0%    | 18%  |
| X = Br | 31·5% | 22·5% | 12%  |
| I      | 0%   | 23·4% | 0%   |

The above-mentioned reaction requires very pure fluorohalogen derivatives, and a 40–50% excess of magnesium activated by the reaction with butyl chloride; it gives best results when carried out in a nine-fold volume of ether, based on the volume of the fluorohalide. Fluoroalkyl chlorides give the highest yields (63–73%) of the normal product, whereas fluorobromides and especially fluoroiodides tend to give side reactions (816).

In the example quoted, the fluorine and halogen atoms were so far apart that they reacted independently of each other. The distance between the halogen atom and fluorine is critical for the preparation of the Grignard reagents. The closer they are, the more complications are to be expected. Whereas 1-bromo-2-perfluoropropylethane react normally and yields the corresponding fluorinated organomagnesium compound *(842)*,

485    $CF_3$—$CF_2$—$CF_2$—$CH_2$—$CH_2Br$ $\xrightarrow{\text{Mg}}$ $CF_3$—$CF_2$—$CF_2$—$CH_2$—$CH_2MgBr$    [*842*]

1-chloro-3,3,3-trifluoropropane gives the corresponding Grignard reagent only when the reaction with magnesium is carried out under cooling to 20°C *(669)*. Without cooling 1,1-difluorocyclopropane is formed *(669)*.

486
$\xleftarrow[\quad 20° \quad]{\text{Mg}}$ $CF_3$—$CH_2$—$CH_2Cl$ $\xrightarrow[\text{spont.}]{\text{Mg}}$    [*669*]

$CF_3$—$CH_2$—$CH_2MgCl$

$CF_2$————$CH_2$
$\diagdown CH_2 \diagup$

Though similar cyclopropane derivatives were not isolated in the reaction of magnesium with 1-bromo- and 1-iodo-3,3,3-trifluoropropane, low yields of the reaction of the formed Grignard reagents with esters suggest that ring closure may compete *(649)*.

If the halogen atom is bound to a carbon adjacent to a carbon atom carrying fluorine, the Grignard reagent is not formed. Either the compound does not react at all *(431)*, or else both the halogen and fluorine atoms are split off to form an olefin *(310, 431)*.

487|    $CF_3$—$CH_2I$ $\xrightarrow{\text{Mg, (C}_2\text{H}_5)_2\text{O}}$ $CF_2$=$CH_2$  90%    [*310*]

The most important discovery in the field of fluorinated organometallic reagents is the preparation and reactions of the *Grignard reagents from perfluoroalkyl halides (447, 449)*. The older attempts aiming at the perfluoroalkylmagnesium halides were unsuccesful especially because they were carried out with compounds of insufficient reactivity, such as trifluoromethyl chloride *(953)*, and conditions used were not favourable for the reaction. The preparation of perfluoroalkylmagnesium halides requires especially pure chemicals and definite procedures. Another reason because of which the previous experiments did not succeed, lies not in the lack of reactivity of the perfluoroalkyl halide but in the limited stability of the Grignard reagent formed. Elimination of magnesium fluorohalide takes place already at room temperature, so that the reaction of perfluoroalkyl halide with magnesium has to be carried out at as low a temperature as possible *(447)*. Much experience has been gained with the preparation and reactions of perfluoropropylmagnesium iodide *(362, 367,*

*447, 449, 654, 841*). The following directions have been outlined (*449*): The reaction between magnesium and perfluoroalkyl iodide diluted with 1300 ml. (per mole) of diethyl ether or a still larger proportion of tetrahydrofuran, is initiated at room temperature. As soon as it takes off, the temperature is lowered to −80°C and kept there for at least 24 hours. The consumption of magnesium is completed by raising the temperature progressively to −60°C to −30°C. Acyl halides, aliphatic aldehydes and ketones are best added before the temperature is raised to −50°C. For such compounds which do not undergo the aldol-type condensation in the presence of the Grignard reagents, the technique used for allylic compounds is more suitable: simultaneous addition of the perfluoroalkyl iodide and an equivalent amount of the carbonyl compounds in 1300 ml. of ether (per mole), once the reaction with magnesium has been initiated (*449*).

Magnesium which is to be used in the preparation of perfluoroalkylmagnesium should be spectroscopically pure. In such case the reaction with perfluoroalkyl iodide starts spontaneously at room temperature. If activation of magnesium is necessary, the best way seems to be heating the metal with 10% of iodine at 150°C, or starting the reaction with 10–15% excess of magnesium and 10–15% of ethyl bromide or bromobenzene in ether, pouring off the solution after the reaction has proceeded, and adding an ethereal solution of perfluoropropyl iodide to the metal etched in this way (*362*). The reaction should be carried out with exclusion of oxygen, carbon dioxide and water vapour (*362, 367*).

Careful control of the reaction temperature is essential, since the perfluoropropylmagnesium iodide formed suffers decomposition already at room temperature (*449*). The effect of temperature upon the yield of perfluoropropylmagnesium iodide is evident from Table 32 (*362*).

TABLE 32

*Effect of Temperature on the Yield of the Grignard Reagent Prepared from Perfluoropropyl Iodide (362)*

| Temperature, °C | −80 | −60 | −40 | −20 | 0 | 10 | 20 | 40 |
|---|---|---|---|---|---|---|---|---|
| Yield of $C_3F_7MgI$, % ...... | 37 | 55 | 59 | 64 | 58 | 44 | 41 | 27 |
| Recovered $C_3F_7I$, % ....... | 23 | 9 | 10 | 12 | 7 | 9 | 3 | 0 |

Solvents used for the preparation of perfluoroalkylmagnesium iodide are diethyl ether, dibutyl ether, tetrahydrofuran, tetrahydropyran, trimethylamine and triethylamine. Less basic solvents such as perfluorodibutyl ether or perfluorotriethylamine seem to inhibit the reaction (*449*). Table 33 compares the results of the formation of perfluoropropylmagnesium iodide in different ethers (*362*).

TABLE 33

*Effect of Solvent on the Yield of Perfluoropropylmagnesium Iodide at Various Temperatures* (*362*)

| Solvent | Yield of $C_3F_7MgI$ at Temperature, °C | | | | |
|---|---|---|---|---|---|
| | −60 | −30 | 0 | 20 | 40 |
| $(C_2H_5)_2O$ | 58 | 68 | 59 | 50 | 33 |
| $(C_4H_9)_2O$ | 62 | 66 | 61 | 49 | 27 |
| ⬡O | 64 | 72 | 65 | 57 | 48 |

In addition to the temperature and solvents in the preparation of perfluoropropylmagnesium iodide, the yield of the Grignard reagent is influenced by the concentration of the iodide in the solvent (Table 34) (*362*).

TABLE 34

*Effect of Concentration of Perfluoroalkyl Iodide on the Yield of the Grignard Reagent* (*362*) (2·5 g of Magnesium, 5 g of Perfluoropropyl Iodide)

| Amount of Ether, ml. | 5 | 20 | 50 | 100 | 150 |
|---|---|---|---|---|---|
| Yield of Perfluoropropyl-magnesium Iodide, % | 6 | 15 | 32 | 43 | 59 |

The perfluorinated Grignard reagents react with a variety of compounds, such as carbon dioxide and organic carbonyl compounds, as can be seen from the following examples (*367, 449, 841*):

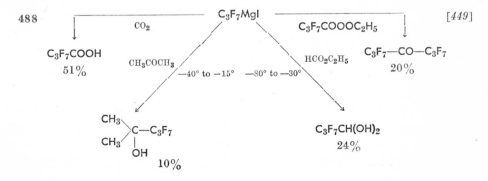

Even for the reaction of the perfluoropropyl iodide with carbonyl compounds, the maintenance of low temperature is necessary to ensure good yields (Table 35) (*362*).

Table 35

*Effect of Temperature on the Reaction of Perfluoropropylmagnesium Iodide with Carbon Dioxide (362)*

| Temperature, °C | −40 | −20 | 0 | 20 |
|---|---|---|---|---|
| Yield of Perfluorobutyric acid, % | 57 | 65 | 56 | 43 |

*Procedure 28.*

*Grignard Reaction with Perfluoroalkyl Iodides*
*Preparation of 1,1-Dihydroperfluoro-1-butanol (362)*

(*a*) *Preparation of Perfluoropropylmagnesium Iodide.* A mixture of 3·5 g (0·145 mole) of spectroscopically pure magnesium and 3·5 g (0·014 mole) of iodine is heated to 150°C. After cooling, 150 ml. of dibutyl ether is added, the mixture is cooled to −25°C, and 10 g (0·034 mole) of perfluoropropyl iodide is added with stirring over a period of 8 hours.

(*b*) *Reaction of Perfluoropropylmagnesium Iodide with Formaldehyde.* Gaseous formaldehyde (2·0 g 0·067 mole) is passed in the course of 2 hours into the solution of perfluoropropylmagnesium iodide in dibutyl ether at −25°C. The mixture is stirred for 8 hours during which time the temperature is allowed to rise to 20°C. After heating the mixture shortly at 60°C and distilling off volatile components, the mixture is decomposed by pouring into an excess of acid. The ether layer is separated, dried with phosphorus pentoxide, and distilled. Redistillation of the fraction boiling at 95–105°C with phosphorus pentoxide yields 42% of 1H,1H-perfluoro-1-butanol, 17% of heptafluoropropane, 4% of hexafluoropropene, and 24% of the recovered perfluoropropyl iodide.

The yields of the products are lowered by incomplete reaction, by reduction *(367)*, and by olefin formation (*362*) (Table 36).

Table 36

*Composition of the Product of the Reaction Between Perfluoropropylmagnesium Iodide and Formaldehyde or Acetaldehyde (362)*

| Aldehyde | Recovered Perfluoropropyl Iodide, % | Reaction Product, % | 1H-heptafluoropropane, % | Hexafluoropropene, % |
|---|---|---|---|---|
| $CH_2O$ | 24 | 42 | 17 | 4 |
| $CH_3CHO$ | 8 | 37 | 21 | 2 |

Another example of a perfluorinated Grignard reagent is the preparation of trifluoromethylmagnesium iodide. After many difficulties it was carried out in tetrahydrofuran at $-30°C$ to $-10°C$ in a maximum yield of 45%. The reactivity of trifluoromethylmagnesium iodide is approximately one thousandth of that of perfluoropropylmagnesium iodide (*372*). Yields of some of its reactions are listed in Table 37 (*372*).

TABLE 37

*Reactions of Trifluoromethylmagnesium Iodide with Organic Compounds* (*372*)

| Compound | CH₂—CH₂ \\ / O | $CH_3COCl$ | $CF_3COCl$ | $CH_3CN$ |
|---|---|---|---|---|
| Product | $CF_3\text{-}CH_2\text{-}CH_2OH$ | $CH_3COCF_3$ | $CF_3COCF_3$ | $CH_3COCF_3$ |
| Yield, % | 57 | 59 | 49 | 38 |
| Compound | $CF_3CN$ | $C_2H_5COCl$ | $C_3F_7COCl$ | |
| Product | $CF_3COCF_3$ | $C_2H_5COCF_3$ | $C_3F_7COCF_3$ | |
| Yield, % | 33 | 47 | 43 | |

Another route to perfluorinated Grignard reagents is an interconversion between perfluoroalkyl iodides and the Grignard reagents. Thus perfluoropropyl iodide reacts with phenylmagnesium bromide in ether and gives perfluoropropylmagnesium bromide, stable only at low temperatures (*654*) (Table 38).

TABLE 38

*Spontaneous Decomposition of Perfluoropropylmagnesium Bromide* (*654*)

| Temperature, °C | Time, Hrs. | Decomposition, % |
|---|---|---|
| 2 | 6 | 92 |
| $-20$ | 12 | 49 |
| $-73$ | 24 | 10 |

By the same method, perfluoroethylmagnesium bromide and dimethylperfluoroethylcarbinol therefrom was prepared by the simultaneous reaction with acetone (*671*).

489                          [*671*]

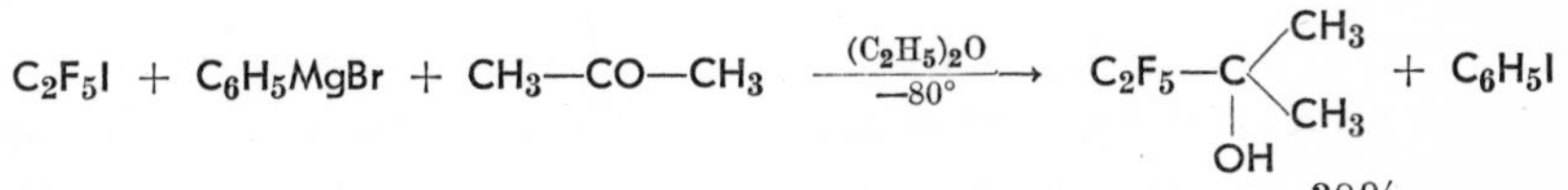

$$C_2F_5I + C_6H_5MgBr + CH_3\text{---}CO\text{---}CH_3 \xrightarrow[-80°]{(C_2H_5)_2O} C_2F_5\text{---}\underset{\underset{OH}{|}}{C}\big(\overset{CH_3}{\underset{CH_3}{}}\big) + C_6H_5I$$

38%

More recently, successful preparation of perfluorovinyl magnesium bromide (544, 578) and perfluorovinylmagnesium iodide (578, 805) has been reported:

**490**                                                                            [578]

$$CF_2{=}CFI \xrightarrow[-20°]{Mg,\,(C_2H_5)_2O} [CF_2{=}CFMgI]$$

$$\xrightarrow{H_2SO_4} CF_2{=}CHF \qquad 69\%$$

$$\xrightarrow{CO_2} CF_2{=}CF{-}COOH \quad 37{\cdot}7\%$$

Fluorinated compounds give acetylenic Grignard reagents. Thus trifluoropropyne and ethylmagnesium bromide yield trifluoropropynylmagnesium bromide, which reacts with acetone to give dimethyltrifluoropropynylcarbinol (464).

**491**                                                                            [464]

$$CF_3{-}C{\equiv}CH \xrightarrow[(C_2H_5)_2O]{C_2H_5MgBr} CF_3{-}C{\equiv}CMgBr \xrightarrow{CH_3COCH_3} CF_3{-}C{\equiv}C{-}\underset{\underset{CH_3}{|}}{\overset{\overset{CH_3}{|}}{C}}{-}OH \quad 75\%$$

In the aromatic series, the Grignard synthesis with $m$-trifluorophenylmagnesium bromide follows a regular course (958, 1013).

**492**                                                                            [1013]

A surprising achievement in the field of Grignard reagents has been accomplished recently, viz. the conversion of bromopentafluorobenzene and iodopentafluorobenzene to pentafluorophenylmagnesium bromide or iodide, respectively. The preparation does not seem to require any special precautions (774).

**493**                                                                            [774]

## Organolithium Compounds

Organolithium compounds differ from the Grignard reagents by smaller space occupancy and by higher reactivity. Some differences result, partly favourable, partly undesirable.

The relatively small volume occupied by organolithium compounds allows smooth addition reactions to carbonyl groups. In contrast to the Grignard reagents, addition prevails over reduction (*662*) (Table 39).

TABLE 39

*Results of the Reaction between Perfluorobutyraldehyde and Ethylmagnesium Bromide or Ethyllithium (662)*

| Organometallic | Addition Product, % $C_3F_7CH(OH)C_2H_5$ | Reduction Product, % $C_3F_7CH_2OH$ |
|---|---|---|
| $C_2H_5MgBr$ ............. | 19 | 61 |
| $C_2H_5Li$ ............... | 68 | 0 |

High reactivity of the organolithium compound accounts for their reactions even with firmly bound aromatic fluorine. Fluorobenzene and phenyl lithium give diphenyl (*541*).

494     ⟨benzene⟩—F + Li—⟨benzene⟩ → ⟨benzene⟩—⟨benzene⟩   52%     [*541*]

The reaction of phenyl lithium with trifluoroacetic acid affords $\omega,\omega,\omega$-trifluoroacetophenone only when carried out at low temperatures (*681*). Otherwise a complex mixture of benzoic acid, benzophenone, triphenylmethane and tetraphenylethylene arises (*680*). Also, the reaction of phenyl lithium with perfluoropropionic acid does not yield the expected products. Among other products 1,2-difluoro-1,2-diphenylethylene has been isolated from the reaction mixture. The same product was obtained from phenyllithium and tetrafluoroethylene (*682*).

495                                                                                 [*682*]

$CF_3$—$CF_2$—COOH $\xrightarrow[\text{reflux}]{\substack{C_6H_5Li \\ (C_2H_5)_2O}}$ ⟨benzene⟩—CF=CF—⟨benzene⟩ $\xleftarrow[\text{reflux}]{\substack{C_6H_5Li \\ (C_2H_5)_2O}}$ $CF_2$=$CF_2$

                                25%                         18%

On the other hand, methyl perfluorobutyrate and phenyllithium afforded quite regularly a 84% yield of diphenylperfluoropropylcarbinol (*681*).

In the preparation of lithium derivatives from fluorinated compounds, less irregularities will be encountered.

The direct replacement of an aromatic halogen by lithium by the action of lithium metal upon a halogenated benzotrifluoride has not been successful (*313*). However, fluorinated aromatic lithium compounds are smoothly prepared by the interconversion of the halogen derivatives with butyl lithium (*308,313*).

The reaction of o-bromofluorobenzene with butyl lithium has to be run at low temperatures (308).

Trifluoromethylphenyllithium is easily obtained by treatment of m-bromo-benzotrifluoride with butyl lithium (313).

In this way even 2,4,6-tris(trifluoromethyl)chlorobenzene affords the corresponding lithium derivative (674).

In addition to the interconversion of butyl lithium with halogen derivatives, the replacement of hydrogen by means of butyl lithium is also possible in the aromatic series. Fluorobenzene and butyl lithium react in tetrahydrofuran at low temperatures to give o-fluorophenyllithium, as evidenced by the formation of o-fluorobenzoic acid on carbonation (312).

In benzotrifluoride the replacement of a hydrogen by lithium is easier than in benzene itself. The lithium seems to be localized preponderantly ortho-to the perfluoromethyl group (880).

Very important is the preparation of lithium derivatives of fluorocarbons. Methyllithium and perfluoropropyl iodide give perfluoropropyllithium at low

temperatures (*843*). At room temperature the compound decomposes to perfluoropropene. Its reactions with sulphuric acid and organic carbonyl compounds take place somewhat more easily than those with perfluoropropylmagnesium iodide (*843*).

**501**   $C_3F_7I$  $\xrightarrow[-74°]{CH_3Li,(C_2H_5)_2O}$  $C_3F_7Li$  $\xrightarrow{20°}$  $C_3F_6$   93%   [*843*]

$\downarrow$ $H_2SO_4.H_2O$ reflux    $\downarrow$ $C_2H_5CHO$ —42°

$C_3F_7H$   63%    $C_3F_7$—CH—$C_2H_5$   50%
                                                    |
                                                   OH

An interesting reaction was observed between perfluoropropyllithium and
methyl perfluorobutyrate. Only a small proportion of the normal addition
product was obtained, whereas the main product was di(perfluoropropyl)car-
binol. The reduction of the ketone formed primarily evidently took place at
the expense of the solvent, for some ethylvinyl ether was found among the
products (*668*).

**502**                                                                [*668*]

$C_3F_7Li + C_3F_7COOCH_3$  $\xrightarrow[-35° \text{ to } -50°]{(C_2H_5)_2O}$  $(C_3F_7)_2CO + (C_3F_7)_2CHOH + C_3F_6$
                                                                      8·7%        42%        25%

## Organozinc Compounds

The preparation of perfluoroalkylzinc iodides is somewhat tedious. In some
solvents the reaction between perfluoropropyl iodide does not take place at
all, or it produces perfluorohexane (*400, 439*). Only ethers such as diethylether,
dibutyl ether, tetrahydrofuran and dioxane have been found to be suitable
for the preparation of perfluoropropylzinc iodide which was isolated in 50–75%
yields when the reaction was carried out at 0°C (*400, 730*). The compound
reacts like its organomagnesium or organolithium analogue, but seems less
reactive, since it is not attacked by oxygen (*730*) or by carbon dioxide (*400,
730*).

**503**  $C_3F_7I$  $\xrightarrow[0°]{Zn, dioxan}$  50—60%  $C_3F_7ZnI$  $\xrightarrow{180—200°}$  $C_3F_6$   58%   [*400*]

$\downarrow$ $Cl_2$      HCl 150°  $H_2O$ 100°  $C_2H_5OH$ 120°      $\downarrow$ $CH_3COCl$

$C_3F_7Cl$   89%       93%  96·2%  71%       $C_3F_7$—CO—$CH_3$   18%
                         $C_3F_7H$

An important application of zinc is the Reformatsky reaction, which has been carried out not only with bromoesters and fluorinated aldehydes (*664*) and ketones (*656*),

504                                                                                    [*656*]

$$\text{C}_3\text{F}_7\text{—CO—}\langle\!=\!\rangle + \text{BrCH}_2\text{—COOC}_2\text{H}_5 \xrightarrow{\text{Zn, C}_6\text{H}_6} \text{C}_3\text{F}_7\text{—}\overset{\displaystyle \text{CH}_2\text{—COOC}_2\text{H}_5}{\underset{\displaystyle \text{OH}}{\text{C}}}\langle\!=\!\rangle \quad 49\%$$

but also with bromofluoroacetate as the active component (*656*):

505                                                                                    [*656*]

$$\text{CH}_3\text{—CH}=\text{CH—CHO} + \text{BrCHF—COOC}_2\text{H}_5 \xrightarrow{\text{Zn, C}_6\text{H}_5} \text{CH}_3\text{—CH}=\text{CH—}\overset{\displaystyle \text{OH}}{\underset{}{\text{CH}}}\text{—CHF—COOC}_2\text{H}_5$$
$$61\%$$

## Organocadmium Compounds

Dialkylcadmium was used for the preparation of fluorinated ketones from the chlorides of fluorinated acids (*159*).

506        $\text{CF}_3\text{—COCl} + (\text{C}_6\text{H}_{13})_2\text{Cd} \xrightarrow[\text{reflux}]{\text{C}_6\text{H}_6} \text{CF}_3\text{—CO—C}_6\text{H}_{13} \quad 22\%$        [*159*]

## Organomercuric Compounds

Organomercuric compounds show some interesting reactions. Aromatic mercurials are obtained by treating aryl fluorides with mercuric acetate (*224*),

$$\langle\!=\!\rangle\text{—F} \xrightarrow[\text{reflux}]{\text{Hg(OCOCH}_3)_2,\ \text{CH}_3\text{COOH}}$$

507                                                                                    [*224*]

$$\left[\text{F—}\langle\!=\!\rangle\text{—Hg—OCO—CH}_3\right] \xrightarrow{\text{HCl}} \text{F—}\langle\!=\!\rangle\text{—HgCl} \quad 11\%$$

aliphatic perfluoromercurihalides are easily formed by the action of perfluoroalkyl iodides on mercury, especially under ultra-violet irradiation (*238*).

508        $\text{CF}_3\text{I}$
$\xrightarrow{\text{Hg, 260—290°}} 22\%$
$\xrightarrow{\text{Hg, }h\nu} 50\text{—}80\% \quad \text{CF}_3\text{HgI}$        [*238*]
$\xrightarrow[110°]{\text{Hg, C}_6\text{F}_{11}\text{CF}_3} 80\%$

Perfluoroalkylmercuriiodides are converted to di(perfluoroalkyl)mercury by metallic amalgams (*239*).

509                $\text{CF}_3\text{HgI} \xrightarrow{\text{AgHg}_x} (\text{CF}_3)_2\text{Hg} \quad 80\%$                [*239*]

## Compounds of Phosphorus and Arsenic

Still easier than the reaction of trifluoromethyl iodide with mercury is its reaction with phosphorus or arsenic. Mixtures of mono-, bis- and tris (trifluoromethyl)phosphines or arsines are formed which readily split off fluoroform on alkaline hydrolysis (*70*).

510        X + n CF$_3$I $\xrightarrow{220°}$ CF$_3$XI + (CF$_3$)$_2$XI + (CF$_3$)$_3$X        [*70*]

X = P, As        NaOH

CF$_3$H

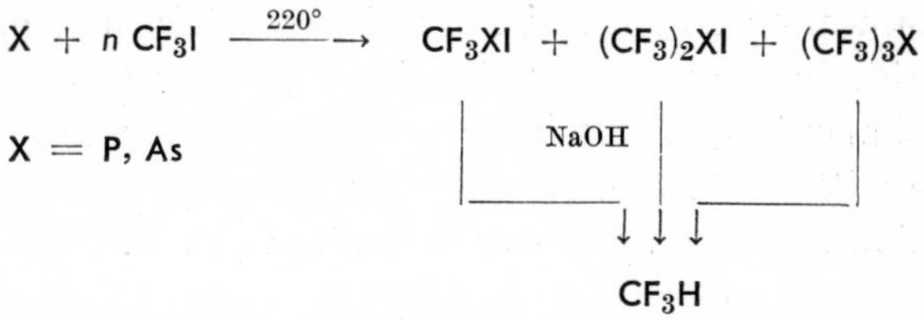

### BASIC CONDENSATIONS

The paragraph on basic condensations will include the aldol condensation of aldehydes and ketones, the Perkin and Knoevenagel syntheses, the cyanohydrin synthesis, the Mannich reaction, the Michael addition and the Claisen condensation of esters with ketones, esters or nitriles. Fluorine atoms may be contained in either of the reagents, in that having a carbonyl function, in that which condenses with the carbonyl groups, or in both.

## Aldol Condensation of Aldehydes and Ketones

Perfluorobutyraldehyde condenses (in form of its hydrate) with nitromethane to 1-nitro-3,3,4,4,5,5,5-heptafluoro-2-pentanol, which loses water only in an energetic dehydration (*186*).

511   C$_3$F$_7$—CH(OH)$_2$ + CH$_3$—NO$_2$ $\xrightarrow[50-60°]{K_2CO_3}$ C$_3$F$_7$—CH(OH)—CH$_2$—NO$_2$  75%   [*186*]

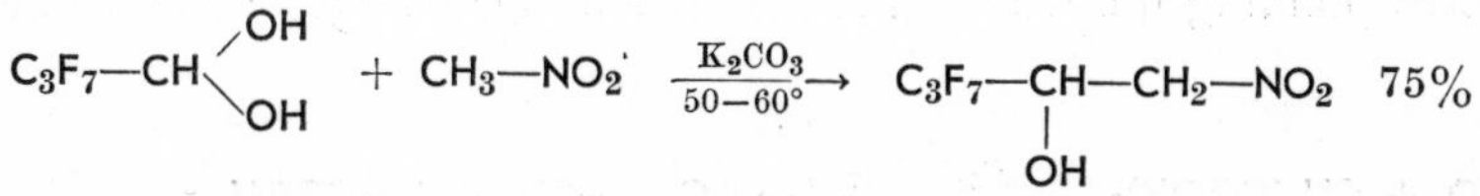

Trifluoroacetone suffers autocondensation which leads to 1,1,1,5,5,5-hexafluoro-2-methyl-2-pentanol-4-one. Sodium ethoxide in ether or sodamide in chloroform are used as the base needed. Aqueous alkali decreases the yield by cleaving the product to fluoroform (*632*).

512        [*632*]

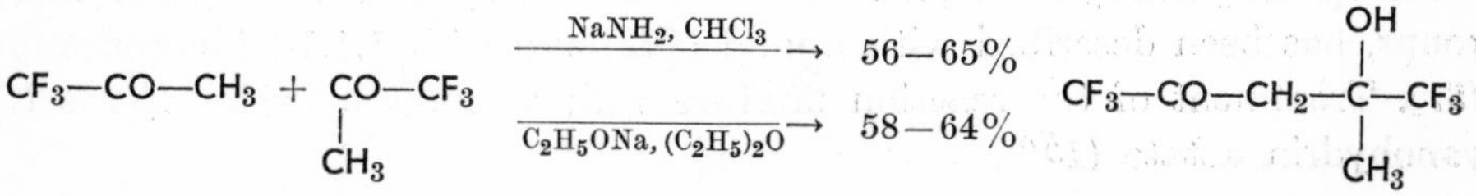

CF$_3$—CO—CH$_3$ + CO(CH$_3$)—CF$_3$ $\xrightarrow{NaNH_2, CHCl_3}$ 56—65%  $\xrightarrow{C_2H_5ONa, (C_2H_5)_2O}$ 58—64%   CF$_3$—CO—CH$_2$—C(OH)(CH$_3$)—CF$_3$

Aqueous alkaline solution brings about the condensation of benzaldehyde with ω-fluoroacetophenone which affords α-fluorochalcone (*92*).

**513**                                                                                [*92*]

$$\langle\rangle\text{—CHO} + \text{FCH}_2\text{—CO—}\langle\rangle \xrightarrow[<23°]{\text{NaOH, C}_2\text{H}_5\text{OH}} \langle\rangle\text{—CH=CF—CO—}\langle\rangle \quad 40\%$$

Condensations of aldehydes with acids or their derivatives represent a series of important preparations. *The Perkin synthesis* of benzaldehyde and fluoroacetanhydride in the presence of sodium fluoroacetate gave only a 2% yield of α-fluorocinnamic acid (*86*). Much more successful was the condensation of benzaldehyde with ethyl fluoroacetate by means of sodium hydride (*86*), or with methyl fluoroacetate by means of sodium (*1008*).

**514**                                                                              [*1008*]

$$\langle\rangle\text{—CHO} + \text{CH}_2\text{F—COOCH}_3 \xrightarrow[20°]{\text{Na, (C}_2\text{H}_5)_2\text{O}} \langle\rangle\text{—CH=CF—COOCH}_3 \quad 47\%$$

*The Knoevenagel reaction* of *p*-fluorobenzaldehyde with malonic acid in the presence of ammonium acetate gave only a low yield of *p*-fluorocinnamic acid and the main product was found to be β-amino-β-(*p*-phenyl)propionic acid (*689*).

**515**                                                                                [*689*]

$$\text{F—}\langle\rangle\text{—CHO}$$

$$\xrightarrow[\text{CH}_3\text{COONH}_4]{\text{CH}_2(\text{COOH})_2} \text{F—}\langle\rangle\text{—CH=CH—COOH} + \text{F—}\langle\rangle\text{—CH—CH}_2\text{—COOH}$$
$$15\% \qquad\qquad\qquad\qquad\qquad\qquad\qquad\qquad \underset{\text{NH}_2}{|} \quad 63\%$$

Similar condensations of perfluorobutyraldehyde or perfluorocaprylaldehyde with malonic acid gave very stable β-hydroxyacids, which are dehydrated to the corresponding unsaturated acids only under energetic conditions (*673*).

**516**                                                                                [*673*]

$$\text{C}_3\text{F}_7\text{—CHO} + \text{CH}_2(\text{COOH})_2 \xrightarrow[\text{reflux}]{\text{C}_5\text{H}_5\text{N,C}_7\text{H}_8} \text{C}_3\text{F}_7\text{—CH—CH}_2\text{—COOH}$$
$$\underset{\text{OH}}{|} \quad 88\%$$

$$\xrightarrow{\text{P}_2\text{O}_5} \text{C}_3\text{F}_7\text{—CH=CH—COOH}$$
$$84\%$$

*The cyanohydrin synthesis*, i.e. the addition of hydrogen cyanide to carbonyl groups, has been described with fluoral (*642*) and with 1,1,1-trifluoroacetone (*197*). Treatment of the reaction mixture with acetic anhydride affords the cyanohydrin actate (*150*).

517      $CF_3-CO-CH_3 + HCN$   $\xrightarrow[\text{2. H}_2\text{SO}_4,\ 0°]{\text{1. NaCN, 0°}}$   $CF_3-\underset{\underset{OH}{|}}{\overset{\overset{CN}{|}}{C}}-CH_3$   62·5%     [*197*]

On the other hand, $\omega,\omega,\omega$-trifluoroacetophenone does not accept hydrocyanic acid (*958*).

*The Mannich reaction* of $\alpha,\alpha,\alpha$-trifluoroacetone with formaldehyde and a dialkylamine gives interesting products, which are formulated as follows (*325*).

518      [*325*]

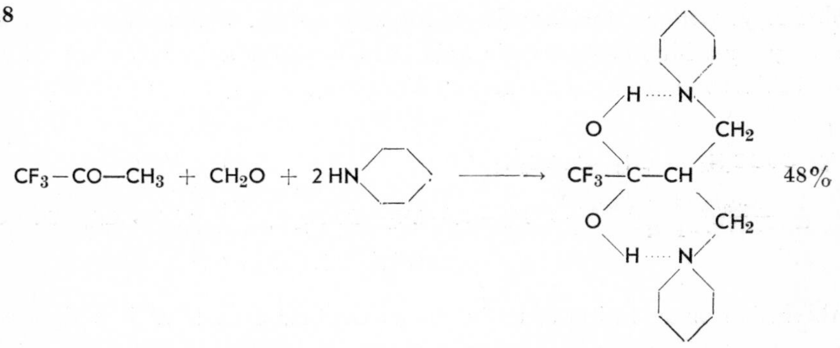

$CF_3-CO-CH_3 + CH_2O + 2\ HN\langle\ \rangle$   $\longrightarrow$     48%

*The Michael addition* was described with diethyl malonate and ethyl perfluoropropylacrylate (*673*).

519      [*673*]

$C_3F_7-CH{=}CH-COOC_2H_5 + CH_2(COOC_2H_5)_2$   $\longrightarrow$   $C_3F_7-\underset{\underset{CH(COOC_2H_5)_2}{|}}{CH}-CH_2-COOC_2H_5$   90%

## The Claisen Condensation of Fluorinated Esters

The Claisen condensation of trifluoroacetates with ketones is a very important method for preparing trifluoromethyl ketones (*873*). It is carried out in ether with sodium methoxide or ethoxide or with sodium hydride as the base, and gives very good yields. It can be applied to aliphatic (*466, 873*), aliphatic-aromatic (*39, 873*), and aliphatic-heterocyclic ketones (*38, 780*).

520      [*466*]

$CF_3-COOC_2H_5 + CH_3-CO-CH_3$   $\xrightarrow[\text{reflux 50 hrs.}]{\text{C}_2\text{H}_5\text{ONa, (C}_2\text{H}_5)_2\text{O}}$   $CF_3-CO-CH_2-CO-CH_3$   70%

521      [*39*]

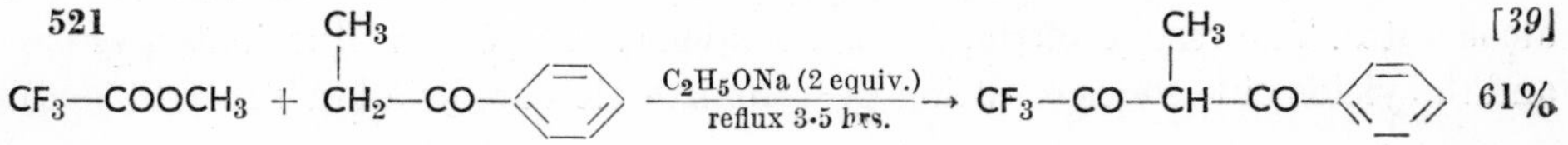

$CF_3-COOCH_3 + \underset{\underset{}{}}{CH_2}\overset{\overset{CH_3}{|}}{{-}}CO-\langle\!=\!\rangle$   $\xrightarrow[\text{reflux 3·5 hrs.}]{\text{C}_2\text{H}_5\text{ONa (2 equiv.)}}$   $CF_3-CO-\overset{\overset{CH_3}{|}}{CH}-CO-\langle\!=\!\rangle$   61%

**522**                                                                          [*780*]

$$\text{[thienyl]}-CO-CH_3 + CF_3-COOC_2H_5 \xrightarrow[\text{reflux}]{CH_3ONa, C_6H_6} \text{[thienyl]}-CO-CH_2-CO-CF_3 \ 91.5\%$$

A trifluoroacetate and trifluoroacetone afford 1,1,1,5,5,5-hexafluoroacetyl-acetone (*466*).

**523**                                                                          [*466*]

$$CF_3-COOC_2H_5 + CH_3-CO-CF_3 \xrightarrow[(C_2H_5)_2O]{C_2H_5ONa} CF_3-CO-CH_2-CO-CF_3 \ 72\%$$

Similarly, trifluoroacetates condense with esters. Ethyl acetate gives the ester of γ,γ,γ-trifluoroacetoacetic acid, which forms a hydrate and is much more stable than acetoacetic acid (*543, 1001*).

**524**                                                                          [*1001*]

$$CF_3-COOC_2H_5 + CH_3-COOC_2H_5$$

$$\xrightarrow[2.30\%\ H_2SO_4]{2.C_2H_5ONa,\ (C_2H_5)_2O} CF_3-CO-CH_2-COOC_2H_5 \xrightarrow[70°]{10\%\ H_2SO_4} CF_3-CO-CH_3$$

$$70\%$$

Alkylation or acylation of trifluoroacetoacetate affords various fluorinated acetoacetates, whose fission yields the expected acids or ketones, as in the case of the non-fluorinated compounds.

Whereas trifluoroacetates act only as acceptors in the Claisen condensation, monofluoroacetates act also as active components adding to the carbonyl group. Thus ethyl fluoroacetate and diethyl oxalate give ethyl fluorooxalacetate (*101, 879*).

**525**                                                                          [*879*]

$$C_2H_5OCO-COOC_2H_5 + CH_2F-COOC_2H_5 \xrightarrow[(C_2H_5)_2O]{C_2H_5ONa} C_2H_5OCO-CO-CHF-COOC_2H_5$$

$$55\%$$

In the autocondensation of ethyl fluoroacetate, the compound acts both as addend and acceptor giving rise to a α,γ-difluoroacetoacetate (*912*).

**526**                                                                          [*912*]

$$CH_2F-COOCH_3 + CH_2F-COOCH_3 \xrightarrow{Na,\ CH_3OH} CH_2F-CO-CHF-COOCH_3 \ 35\%$$

The Claisen condensations give very good results, especially when sodium hydride is used as a catalyst (*660*), because its reaction with the acidic hydrogen of the esters is irreversible, in contrast to the reversible reaction with sodium alcoholates. The results of the Claisen condensation of some fluorinated esters, and the yields of ketones resulting by ketonic fission are listed in Table 40 (*660*).

TABLE 40

*Results of the Claisen Condensation of Fluorinated Acetates and of the Ketonic Fission of the Fluorinated Acetoacetates (662)*

| Fluorinated Acetoacetate | Reaction Temp., °C | Yield % | Fluorinated Acetone | Yield % |
|---|---|---|---|---|
| $CF_3COCF_2COOC_2H_5$ | 75 | 20 | $CF_3COCHF_2$ | 35 |
| $CHF_2COCF_2COOC_2H_5$ | 70 | 81 | $CHF_2COCHF_2$ | 62 |
| $CF_3COCHFCOOC_2H_5$ | 50 | 86 | $CF_3COCH_2F$ | 67 |
| $CF_3COCH_2COOC_2H_5$ | 50 | 84 | | |
| $CHF_2COCHFCOOC_2H_5$ | 30 | 68 | $CHF_2COCH_2F$ | 65 |
| $CHF_2COCH_2COOC_2H_5$ | 40 | 83 | | |
| $CH_2FCOCHFCOOC_2H_5$ | 40 | 69 | $CH_2FCOCH_2F$ | 74 |

*Procedure 29*

*Claisen Condensation of Fluorinated Esters*

*Preparation of Ethyl α,γ,γ,γ-Tetrafluoroacetoacetate from Ethyl Fluoroacetate and Trifluoroacetate (660)\**

A 100 ml. three-necked flask fitted with a gas-tight stirrer, a dropping funnel and a reflux condenser is charged with 28·4 g (0·2 mole) of ethyl trifluoroacetate and 2·4 g (0·1 mole) of sodium hydride, and 10·6 g (0·1 mole) of ethyl fluoroacetate is added dropwise at 50°C in the course of 6 hours while stirring the mixture vigorously under nitrogen. Stirring at room temperature is continued for two more hours after the addition of the fluoroacetate has been completed, and the mixture is allowed to stand overnight. The resulting dark-brown viscous liquid is poured over a mixture of 50 g of ice and 10 g of concentrated sulphuric acid. The ester layer is extracted with three 50 ml. portions of ether, the ether solution of the product is dried with anhydrous calcium sulphate and distilled to give 10 g (0·05 mole) (50%) of ethyl α,γ,γ,γ-tetrafluoroacetoacetate which distils at 137–140°C. Some small amount of ethyl trifluoroacetate is recovered from the lower boiling fraction.

Ethyl trifluoroacetate has been condensed not only with ketones and esters but also with nitriles possessing α-hydrogen atoms. Benzylcyanide afforded thus α-(trifluoroacetyl)benzylcyanide and on further treatment trifluoromethyl benzyl ketone (766).

\* Carried out by J. Salák.

Surprisingly, attempted acyloin condensation of ethyl trifluoroacetate yielded ethyl $\gamma,\gamma,\gamma$-trifluoroacetoacetate which was probably formed by reduction of the primary ethyl pentafluoroacetoacetate. It is known that ethyl pentafluoroacetoacetate is partly reduced to ethyl $\gamma,\gamma,\gamma$-trifluoroacetoacetate under comparable conditions (659) (Equation 301, p. 164).

**528**    $2 \text{ CF}_3\text{—COOC}_2\text{H}_5 \xrightarrow{\text{Na, (C}_2\text{H}_5\text{)}_2\text{O}} \text{CF}_3\text{—CO—CH}_2\text{—COOC}_2\text{H}_5$    25—30%    [659]

Finally a curious reaction should be quoted here, viz. the conversion of ethyl perfluorobutyrate to bis(perfluoropropyl)ketone by means of sodium (403).

**529**    $\text{C}_3\text{F}_7\text{—COOC}_2\text{H}_5 \xrightarrow[\text{2. H}_2\text{SO}_4]{\text{1. Na, (C}_2\text{H}_5\text{)}_2\text{O}} \text{C}_3\text{F}_7\text{—CO—C}_3\text{F}_7$    60%    [403]

## MOLECULAR REARRANGEMENTS

The paragraph dealing with molecular rearrangements will include double bond shifts, transfer of halogens from one carbon atom to another, changes in the sequence of individual carbon atoms in the carbon chain, and shifts of elements or whole groups from nitrogen or oxygen atom to carbon.

### Double Bond Shifts

Double bond shifts were observed during the thermal decomposition of alkaline salts of perfluorocarboxylic acids. While sodium salts yield only perfluoroolefins with a terminal double bond, the decomposition of potassium salts affords a mixture of olefins with a terminal double bond and with a double bond shifted inward (386).

**530**    $\text{CF}_3\text{—CF}_2\text{—CF}_2\text{—CF}_2\text{—COOK} \xrightarrow[\text{10 mm}]{270°}$    $\begin{array}{ll} \text{CF}_3\text{—CF}_2\text{—CF}=\text{CF}_2 & 43\% \\ \text{CF}_3\text{—CF}=\text{CF—CF}_3 & 48\% \end{array}$    [386]

During the dehalogenation of 2,3-dichlorohexafluoro-2-butene with zinc, formation of a small amount of a compound which is probably perfluoro-1,2-butadiene was noticed (441). Spontaneous shift of a double bond takes place in methyl perfluoroallyl ether, which changes slowly to methyl perfluoropropenyl ether (229).

**531**    $\text{CF}_2=\text{CF—CF}_2\text{—OCH}_3 \xrightarrow[\text{few days}]{20°} \text{CF}_3\text{—CF}=\text{CF—OCH}_3$    [229]

An allylic shift of double bond occurs in the reaction of 1-chloropentafluoro-2-propene with antimony chlorofluoride (465).

**532**    $\text{CF}_2\text{Cl—CF}=\text{CF}_2 \xrightarrow[125°]{\text{SbF}_3, \text{Cl}_2} \text{CFCl}=\text{CF—CF}_3$    82%    [465]

A similar isomerization takes place in the treatment of perfluoroallyl chloride with sodium iodide. This reaction most probably follows the course outlined below *(735)*.

**533** [735]

$$CF_2{=}CF{-}CF_2Cl \xrightarrow[CH_3COCH_3]{NaI} [CF_2{-}CF{-}CF_2Cl] \dashrightarrow CF_2{-}CF{=}CF_2$$

An interesting bond shift was observed during the conversion of 3,5-cyclo-cholestan-6β-ol to 3β-fluoro-5-cholestene *(940)*.

**534**

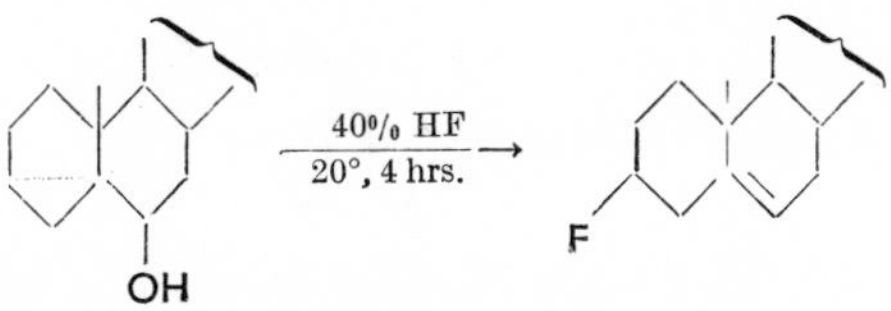

$$\xrightarrow[20°, 4 \text{ hrs.}]{40\% \text{ HF}}$$

75%  [940]

## Transfer of Halogens

Heating of polyfluoropolyhaloparaffins or olefins in the presence of strong acidic catalysis such as aluminium chloride, aluminium bromide, ferric chloride, titanium tetrachloride, antimony pentachloride etc. results in mutual interchange of halogen atoms in which fluorine atoms tend to concentrate at one spot and form trifluoromethyl clusters *(472, 711, 715, 734)*.

**535** [734]

$$CClF_2{-}CCl_2F \xrightarrow[55{-}60°]{AlCl_3} CF_3{-}CCl_3 + CClF_2{-}CCl_3 + CCl_2F{-}CCl_3 + CClF_2{-}CCl_2$$

50%    40%    5%    5%

conversion 50%

## Changes in the Carbon Chain

In the fluorination of heptane with cobalt trifluoride at 275–300°C, small amounts of perfluorohexane, perfluoroethylcyclopentane and perfluorodimethylcyclopentane were isolated in addition to the main product, perfluoroheptane *(287)*.

**536** [287]

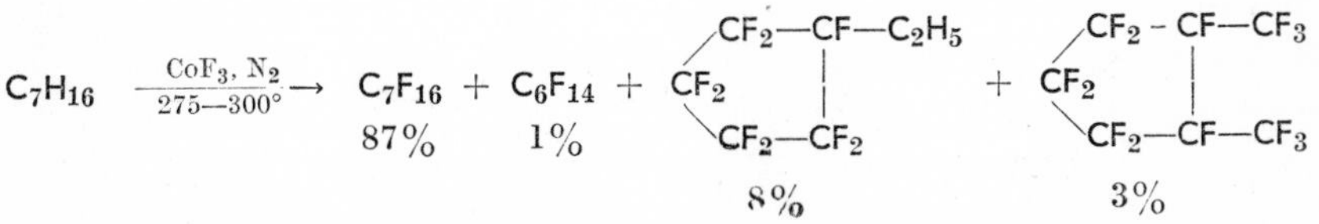

$$C_7H_{16} \xrightarrow[275{-}300°]{CoF_3, N_2} C_7F_{16} + C_6F_{14} + \begin{matrix} CF_2{-}CF{-}C_2H_5 \\ | \quad\quad | \\ CF_2{-}CF_2 \end{matrix} + \begin{matrix} CF_2{-}CF{-}CF_3 \\ | \quad\quad | \\ CF_2{-}CF{-}CF_3 \end{matrix}$$

87%    1%    8%    3%

The addition product of cyclopentadiene and tetrafluoroethylene, 1,1,2,2-tetrafluoro-(3,4 : 3′,4′)-cyclopentenocyclobutane, rearranges at high temperatures to a mixture of isomeric tetrafluorocycloheptadienes (223).

**537**                                                                  [223]

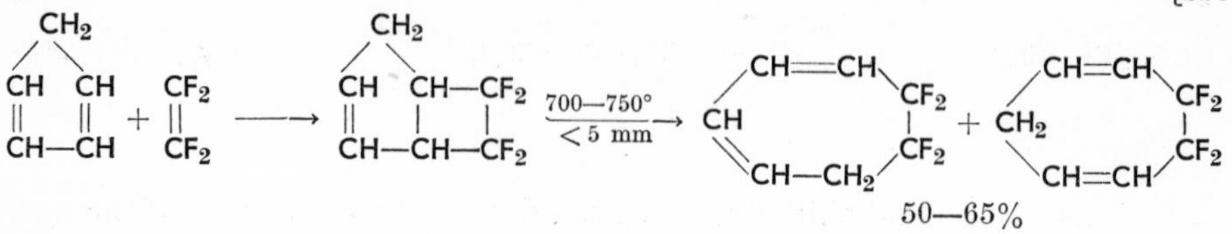

50—65%

A similar reaction takes place during the pyrolysis of 2,2,3,3-tetrafluoro-1-vinylcyclobutane which affords 3,3,4,4-tetrafluoro-1-cyclohexene (222).

**538**

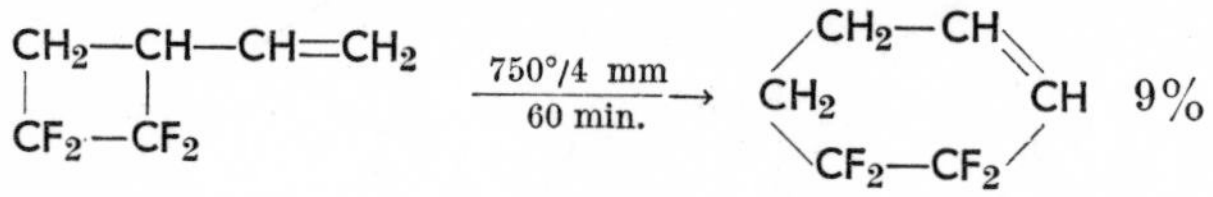

9%                          [222]

The reaction between 1,1-diphenylethylene, anhydrous hydrogen fluoride and lead tetraacetate is accompanied by a rearrangement so that the product is not the expected 1,2-difluoro-1,1-diphenylethane (214) but 1,1-difluoro-1,2-diphenylethane (115).

**539**

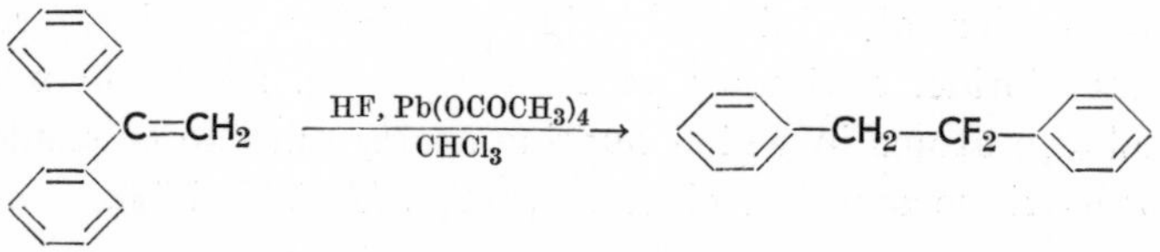

[115]

A similar rearrangement occurs during the reaction of 1,1-bis(p-chlorophenyl)-2,2,2-trichloroethane with anhydrous hydrogen fluoride and mercuric oxide; it affords 1,1,2-trifluoro-1,2-bis(p-chlorophenyl)ethane (181).

**540**                                                                  [181]

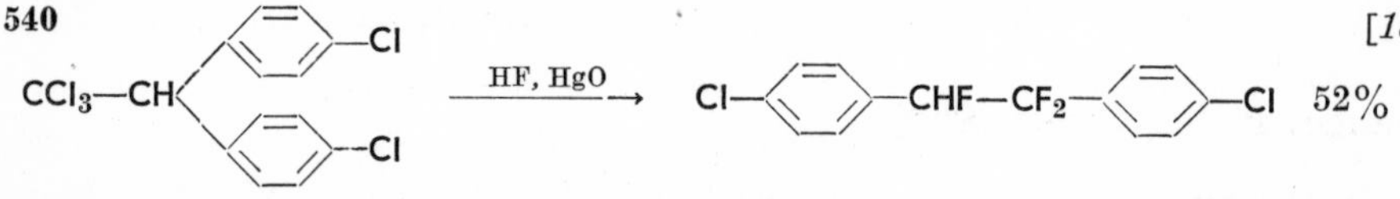

52%

The same kind of a rearrangement seems to cause the formation of di-p-fluorobenzil when 1,1-bis(p-fluorophenyl)-1-acetoxy-2,2,2-trichloroethane is treated with sulphuric acid (545).

**541**                                                                  [545]

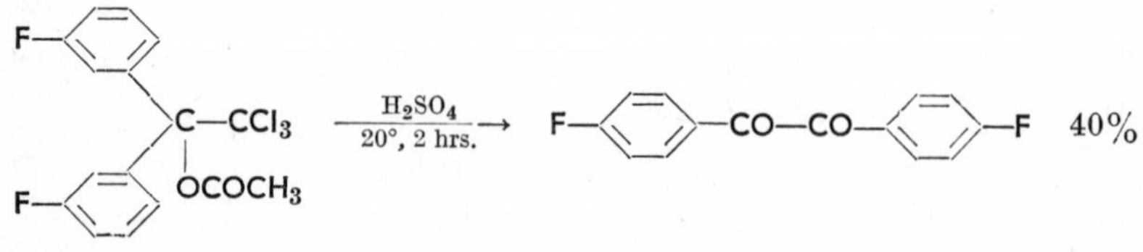

40%

The *Favorsky rearrangement* of α-fluoroketones leads to non-fluorinated acid esters (*553*).

542      (structure) $\xrightarrow[\text{(C}_2\text{H}_5)_2\text{O}]{\text{CH}_3\text{ONa}}$ (structure, COOCH$_3$)    40%      [*553*]

The *Wolff rearrangement* is used to extend the chain of fluorinated carboxylic acids by the Arndt-Eistert procedure. While trifluoroacetyl chloride and diazomethane gives a satisfactory yield of ethyl $\beta,\beta,\beta$-trifluoropropionate (*135*),

543                                                             [*135*]

$$CF_3\text{—COCl} \xrightarrow[0 \to 20°]{CH_2N_2} CF_3\text{—CO—CHN}_2 \xrightarrow{Ag_2O,\ C_2H_5OH} CF_3\text{—CH}_2\text{—COOC}_2H_5$$
                                       62·5%                            40·4%

perfluoropropionyl chloride, perfluorobutyryl chloride, and perfluorovaleroyl chloride have proved failures (*802*). Perfluoropropylacetyl chloride gave only a negligible amount of methyl $\beta$-perfluoropropylpropionate, but the latter was extended to its next higher homologue in 33% yield (*802*).

The *Curtius degradation* of acid azides yields either an isocyanate, or an amine. $\gamma,\gamma,\gamma$-Trifluorobutyric acid yields 3,3,3-trifluoropropylamine (*482*),

$$CF_3\text{—CH}_2\text{—CH}_2\text{—COCl} \xrightarrow[\text{2. H}_2\text{SO}_4,\ 55-65°]{\text{1. NaN}_3,\ C_6H_6,\ 55-65°} \searrow 81\%$$

544                                          $CF_3\text{—CH}_2\text{—CH}_2\text{—NH}_2$  [*482*]

$$CF_3\text{—CH}_2\text{—CH}_2\text{—CONH}_2 \xrightarrow{\text{NaOBr}} \nearrow 35\%$$

while the azide of perfluorobutyric acid rearranges to a perfluoropropylisocyanate (*46, 482, 524*). Trifluoroacetic azide and perfluoropropionic azide are described as explosive (*524*). Nevertheless trifluoroacetic azide was successfully rearranged to trifluoromethylisocyanate (*46, 1096*).

545             $CF_3\text{—CON}_3 \xrightarrow{165°,\ 2\ \text{hrs}} CF_3\text{—N=C=O}$             [*46*]
                                                80%

*Procedure 30.*

*Curtius Degradation*
*Preparation of Perfluoropropylisocyanate from Perfluorobutyryl Chloride (46)*

Perfluorobutyryl chloride (1·65 g, 0·0071 mole), 1·2 g (0·018 mole) of dry sodium azide and 7 ml. of anhydrous toluene are sealed in the absence of air in a 200 ml. thick-walled glass tube; the tube is cooled for 10 minutes to −22°C, heated to 20°C in the course of two hours, shaken for 12 hours at room temperature, and finally heated for 2 hours at 110°C. Vacuum distillation of the content of the tube into dry-ice traps affords 1·22 g (0·0058 mole) (82%) of perfluoropropylisocyanate, b. p. 24·5°C.

Fluorinated acid amides are degraded by alkaline hypohalites in ways which vary with the reaction conditions and the distance of the fluorine atoms from the amide group. $\beta,\beta,\beta$-Trifluoropropionamide yields only 3% of 2,2,2-trifluoroethylamine, while $\gamma,\gamma,\gamma$-trifluorobutyramide gives 35% of 3,3,3-trifluoropropylamine (482).

The *Hofmann degradation* of perfluorocarboxylic acid amides deviates sometimes from the normal course. Trifluoroacetamide does not give, on treatment with a hypobromite, trifluoromethylamine, but hexafluoroethane (337, 524, 525), or more probably bromotrifluoromethane (47). Perfluoropropionamide similarly affords perfluoroethyl bromide (524). The reaction of perfluorobutyramide depends on the hypohalite used. Sodium hypochlorite gives a mixture of perfluoropropyl chloride with perfluoropropene, pentafluoroethane, and 1H-heptafluoropropane. Sodium hypobromite produces perfluoropropyl bromide in a fair yield, and sodium hypoiodide converts perfluorobutyramide to 1H-heptafluoropropane (525).

546

$$\xrightarrow[105°]{\text{NaOCl}} \quad CF_3-CF_2-CF_2-CONH_2 \quad \xrightarrow[\text{distillation}]{\text{NaOI}}$$

$$\text{NaOBr} \Big| 95°$$

[525]

$$CF_3-CF_2-CF_2Cl \qquad CF_3-CF_2-CF_2Br \qquad CF_3-CF_2-CF_2H$$
$$+\ C_3F_6 + C_3F_7H + C_2F_5H \qquad 65\text{—}70\% \qquad 25\text{—}40\%$$

When the crystalline sodium salt of N-bromoperfluorobutyramide is decomposed in dry state, it yields perfluoropropyl isocyanate (47), while the decomposition with water gives perfluoropropyl bromide (47).

547                                                                                      [47]

$$C_2F_5-CF_2-CONH_2 \xrightarrow[\text{reflux}]{Ag_2O,\ (C_2H_5)_2O} C_2F_5-CF_2-CONHAg \xrightarrow[CF_3COOH]{Br_2} C_2F_5-CF_2-CONHBr$$
$$98\% \qquad\qquad\qquad 75\%$$

$$\text{NaOH} \Big| 5\text{—}10°$$

$$C_2F_5-CF_2-N=C=O \qquad \begin{array}{c} C_2F_5-CF_2 \\ | \\ Br \end{array} \xleftarrow[30\ \text{min.}]{H_2O,\ 100°} \begin{bmatrix} C_2F_5-CF_2\!\!+\!\!C=O \\ | \quad | \\ Br\!+\!N| \end{bmatrix} Na$$
$$83\% \qquad\qquad 91\% \qquad\qquad\qquad\qquad 99\%$$

$$\uparrow \qquad\qquad\qquad 165-170°$$

In the reaction of trifluoroacetyl chloride with benzylzinc chloride $\omega,\omega,\omega$-trifluoro-*o*-methylacetophenone was obtained as a main product. Such rearrangements are common in Grignard condensations with benzyl halides (543).

**548** *543]*

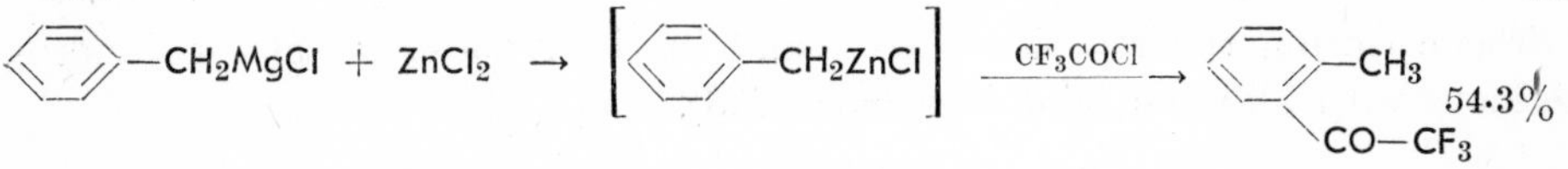

$$\text{C}_6\text{H}_5\text{—CH}_2\text{MgCl} + \text{ZnCl}_2 \rightarrow \left[\text{C}_6\text{H}_5\text{—CH}_2\text{ZnCl}\right] \xrightarrow{\text{CF}_3\text{COCl}} \text{C}_6\text{H}_4(\text{CH}_3)(\text{CO—CF}_3) \quad 54.3\%$$

## Shifts from Nitrogen or Oxygen to Carbon

In the aromatic series, rearrangements occur when a reaction affects a nitrogen or an oxygen atom directly attached to the ring. Thus phenylhydroxylamine rearranges in hydrogen fluoride to *p*-fluoroaniline (Equation 171, p. 118). Of the same nature is the formation of fluorinated aromatic hydroxyacids by the Kolbe process (*158, 407, 408*).

**549** *[158,408]*

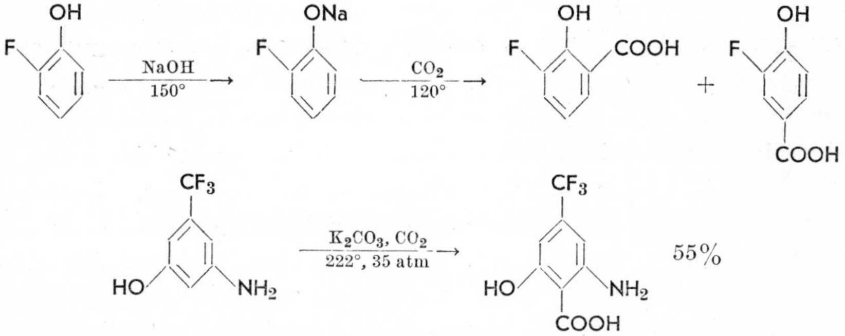

$$\text{(}F\text{)(}OH\text{)C}_6\text{H}_4 \xrightarrow[150°]{\text{NaOH}} \text{(}F\text{)(}ONa\text{)C}_6\text{H}_4 \xrightarrow[120°]{\text{CO}_2} \text{(}F\text{)(}OH\text{)(COOH)C}_6\text{H}_3 + \text{(}F\text{)(}OH\text{)(COOH)C}_6\text{H}_3$$

$$\text{CF}_3\text{-}C_6\text{H}_3(OH)(NH_2) \xrightarrow[222°, \ 35\ \text{atm}]{\text{K}_2\text{CO}_3,\ \text{CO}_2} \text{CF}_3\text{-}C_6\text{H}_2(HO)(NH_2)(COOH) \quad 55\%$$

### ALKYLATION, ARYLATION AND ACYLATION

In the following paragraph, such reactions will be discussed in which fluorinated derivatives alkylate, arylate or acylate organic compounds at oxygen, sulphur, nitrogen or carbon atoms.

## Alkylation

Fluorohalo-derivatives combine with *alcohols* and *phenols* and form fluorinated ethers. An example is the formation of ethyl 2,4-dichlorophenoxyfluoroacetate from sodium 2,4-dichlorophenoxide and ethyl chlorofluoroacetate (*686*). Evidently chlorine is more reactive than fluorine and reacts preferentially.

**550** *[686]*

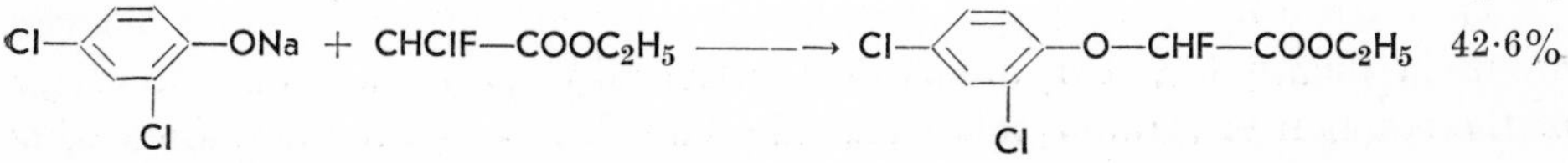

$$\text{Cl}_2\text{C}_6\text{H}_3\text{—ONa} + \text{CHClF—COOC}_2\text{H}_5 \rightarrow \text{Cl}_2\text{C}_6\text{H}_3\text{—O—CHF—COOC}_2\text{H}_5 \quad 42.6\%$$

The same reactivity of chlorine and bromine was observed in the reaction of dibromodifluoromethane with potassium phenoxide, which yielded a small amount of phenyl difluoromethyl ether (one bromine atom was replaced by hydrogen during the reaction) (173), and in the reaction of chlorodifluoromethane with alkaline alkoxides (497, 498).

551          CHClF$_2$ $\xrightarrow{\text{Na, CH}_3\text{OH}}$ CH$_3$—O—CHF$_2$ + HC(OCH$_3$)$_3$          [497]

In the last reaction the substitution is not limited to chlorine but affects fluorine to a certain extent (498). Fluorine also reacts during the reaction of 1,1,1-trifluoropropene with sodium alkoxide, which produces ethyl α,α-difluoroallyl ether in a good yield (481). The true mechanism of this particular alkylation is unknown.

552          [481]

CH$_2$=CH—CF$_3$ $\xrightarrow[100°, 21 \rightarrow 8 \text{ atm}]{\text{C}_2\text{H}_5\text{ONa, C}_2\text{H}_5\text{OH}}$ CH$_2$=CH—CF$_2$—OC$_2$H$_5$ + C$_2$H$_5$O—CH$_2$—CH$_2$—CF$_3$

74%                    7·6%

Alkylations of alkoxides and phenoxides with aliphatic chlorofluoro-derivatives represent in some cases a sequence of two reactions, elimination of a hydrogen halide followed by nucleophilic addition of the alkoxide or phenoxide ion to the olefin thus formed. This assumption is substantiated by the fact that sometimes the same ether results from the olefin obtained in a separate dehydrohalogenation. This seems to be true in the following examples (629, 1020, 1101).

553          [629, 1020]

CHCl$_2$—CF$_2$Cl $\xrightarrow[15-20°, \text{reflux}]{\text{C}_6\text{H}_5\text{ONa, CH}_3\text{COCH}_3}$ CHCl$_2$—CF$_2$—O—⟨  ⟩—$\xrightarrow[\text{reflux}]{\text{C}_6\text{H}_5\text{ONa,CH}_3\text{COC}_2\text{H}_5}$ CCl$_3$—CF$_2$Cl

79%                              30%

554          CHF$_2$—CF$_2$Cl $\xrightarrow[120°]{\text{C}_2\text{H}_5\text{ONa}}$ CHF$_2$—CF$_2$—OC$_2$H$_5$ $\xleftarrow{\text{C}_2\text{H}_5\text{OH}}$ CF$_2$=CF$_2$          [1101]

66−70%                    90%

*Alkylation on sulphur* was observed in the reaction of chlorodifluoromethane with sodium thiophenoxide (497).

555          CHClF$_2$ + C$_6$H$_5$SH $\xrightarrow{\text{Na, CH}_3\text{OH}}$ C$_6$H$_5$—S—CHF$_2$          [497]

An interesting contribution to the knowledge of the reactivity of halogen atoms in fluorinated compounds during the alkylation reactions is a study of the reaction of sodium thiophenoxide with fluoroethyl iodides. Increasing electron traction, due to the increasing number of fluorine atoms in the methyl group of ethyl iodide, is reflected in the decreased reaction rate (*496*) (Table 41).

TABLE 41

*Effect of Substitution by Fluorine upon the Rate of the Reaction between Sodium Phenoxide with Fluorinated Ethyl Iodides at 20°C (496)*

| Alkyl Iodide | $CH_3CH_2I$ | $CH_2FCH_2I$ | $CHF_2CH_2I$ | $CF_3CH_2I$ |
|---|---|---|---|---|
| $K \times 10^5$ (l./mole. sec) | 2600 | 166 | 7·34 | 0·149 |

Another example for S-alkylation is the reaction of $\alpha,\omega$-fluorohaloparaffins with alkaline thiocyanates, which yields $\omega$-fluoroalkylisothiocyanates (*510*).

556    $F(CH_2)_6Br \xrightarrow[\text{reflux}]{\text{KSCN, } C_2H_5OH} F(CH_2)_6SCN$   95%    [*510*]

The reaction of perfluoroalkyl iodides with sulphur which yields bis(perfluoroalkyl)disulphides and trisulphides can also be regarded as sulphur-alkylation (*128, 404*).

557    $C_3F_7I + S \xrightarrow[\text{autoclave}]{250°} C_3F_7-S-S-C_3F_7 + C_3F_7-S-S-S-C_3F_7$    [*404*]

47%          18·7%

*Nitrogen alkylations* occur in nitrites (*815*) (Equation 404, p. 192), in ammonia (*989*),

558    $CHF_2-CH_2Br \xrightarrow[\text{125—145°}]{NH_3, C_2H_5OH} CHF_2-CH_2-NH_2$    [*989*]

and in amines (*212, 598*).

559    [*212*]

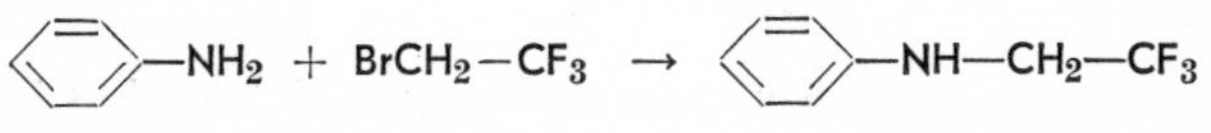

560    $CH_2F-CH_2Br + (CH_3)_3N \xrightarrow[60°]{} CH_2F-CH_2-\overset{\oplus}{N}(CH_3)_3 \ \overset{\ominus}{Br}$    [*598*]

In these reactions, too, other halogens are more effective than fluorine. Surprisingly the fluoromethyl group in 2-trifluoromethylbenzoimidazole readily combines with *o*-phenylene diamine (*604*).

561    [*604*]

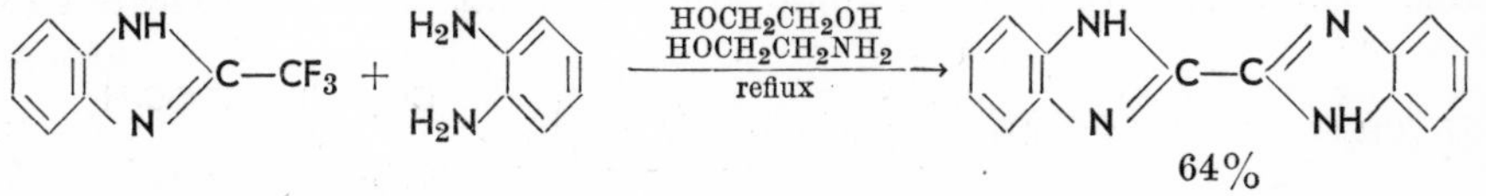

64%

The reaction of ethyl α-bromo-γ,γ,γ-trifluorobutyrate with sodium azide produces ethyl α-azido-γ,γ,γ-trifluorobutyrate in fair yield (*1062*).

**562**                                                                                                    [*1062*]

$$CF_3—CH_2—CHBr—COOC_2H_5 \xrightarrow[\text{reflux}]{NaN_3,\ C_2H_5OH,\ H_2O} CF_3—CH_2—CH—COOC_2H_5 \quad 63\%$$

$$\underset{N{=}N{=}N}{|}$$

*Carbon alkylations* occur with cyanides (*815*)

**563**              $F(CH_2)_6X \xrightarrow[\text{reflux 7·5 hrs.}]{NaCN,\ 80\%\ C_2H_5OH} F(CH_2)_6CN \quad 90\%$            [*815*]

and in the acetoacetic and malonic ester series. In α,ω-fluorohalogenoparaffins chlorine or bromine react preferentially and exclusively (*139, 502, 815*).

**564** 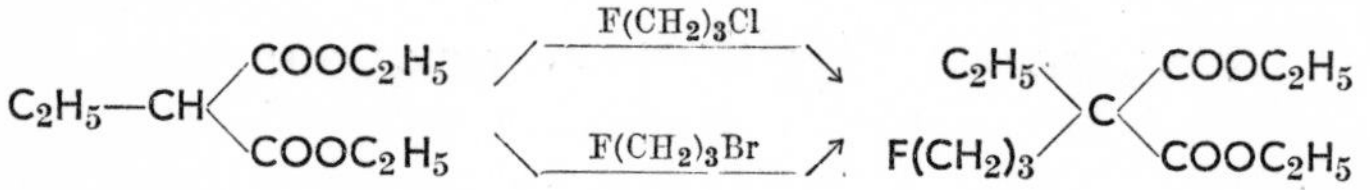
$$C_2H_5—CH{\Big\langle}{COOC_2H_5 \atop COOC_2H_5} \quad {\overset{F(CH_2)_3Cl}{\underset{F(CH_2)_3Br}{\rightleftharpoons}}} \quad {C_2H_5 \atop F(CH_2)_3}{\Big\rangle}C{\Big\langle}{COOC_2H_5 \atop COOC_2H_5}$$
[*139*]

**565**        $F(CH_2)_5X + CH_2(COOC_2H_5)_2 \longrightarrow F(CH_2)_5CH(COOC_2H_5)_2$        [*502*]

*Procedure 31.*

*Alkylation at a Carbon Atom*
*Preparation of 7-Fluoroenanthonitrile from 1-Bromo-6-fluorohexane (815)*

A mixture of 12 g (0·245 mole) of sodium cyanide and 30 g (0·164 mole) of 1-bromo-6-fluorohexane is refluxed with 35 ml. of 80% ethanol for 7·5 hours. After cooling, the sodium bromide is removed by filtration and washed with ethanol. The filtrate is evaporated on the steam-bath, the residue is cooled and diluted with an equal volume of water, the organic layer is separated and the aqueous layer is extracted with ether. The ethereal solution of the product is dried over anhydrous calcium chloride and distilled to give 19·1 g (0·148 mole) of 7-fluoroenanthonitrile.

Ethyl fluoroacetate can be alkylated by benzyl bromide in the presence of sodium ethoxide (*89*).

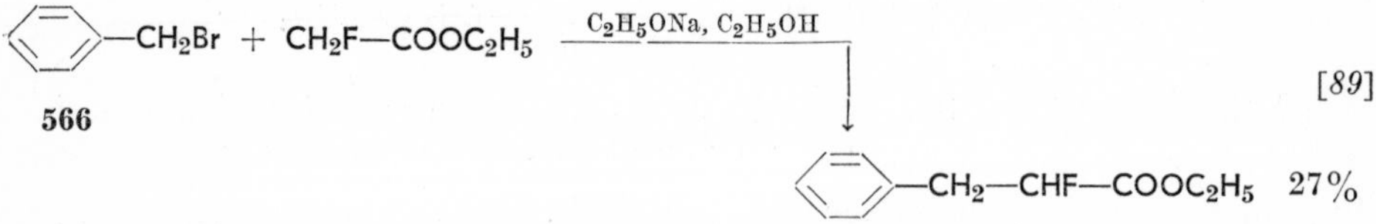

$$\langle\!\!\!\bigcirc\!\!\!\rangle—CH_2Br + CH_2F—COOC_2H_5 \xrightarrow{C_2H_5ONa,\ C_2H_5OH}$$

**566**                                                                                                    [*89*]

$$\downarrow$$

$$\langle\!\!\!\bigcirc\!\!\!\rangle—CH_2—CHF—COOC_2H_5 \quad 27\%$$

Similar reactions with allyl bromide are somewhat more complicated (*89*):

**567**  $CH_2F—COOC_2H_5 + CH_2{=}CH—CH_2Br \xrightarrow[C_2H_5OH]{C_2H_5ONa}$  $\begin{array}{l} CH_2—CF—COOC_2H_5 \\ \ \ |\qquad | \\ CH_2—CH_2 \quad 45\% \end{array}$  [*89*]

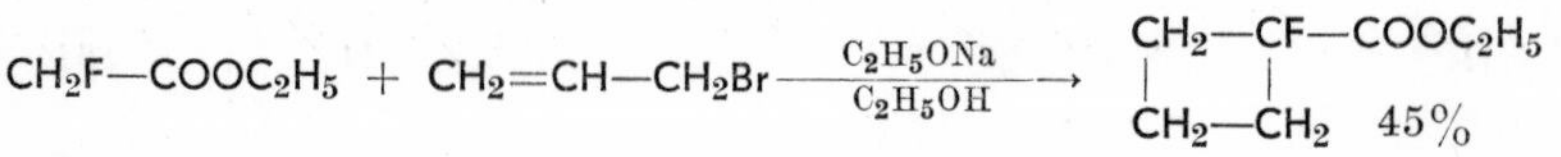

In the paragraph on alkylation reactions, cleavage of an epoxide ring (oxirane ring) will be included. When one of the epoxide ring carbon atoms carries a trifluoromethyl group, the bond between oxygen and the more distant carbon atoms is cleaved (*643, 1061*).

568 $$CF_3-\underset{\underset{O}{\times}}{\overset{\overset{CH_3}{|}}{C}}-CH-CH_3 \quad \xrightarrow[C_2H_5ONa]{C_2H_5OH,\ H_2SO_4} \quad \begin{array}{l} 72\% \\ 63\cdot5\% \end{array} \quad CF_3-\underset{\underset{OH}{|}}{\overset{\overset{CH_3}{|}}{C}}-\underset{\underset{OC_2H_5}{|}}{CH}-CH_3 \qquad [643]$$

This rule holds true even when the far carbon atom carries a carbethoxy group (*1061*).

569 $$CF_3-\underset{\overset{}{\diagdown O\diagup}}{CH-CH}-COOC_2H_5 \quad \xrightarrow{NH_3} \quad CF_3-\underset{\underset{OH}{|}}{CH}-\underset{\underset{NH_2}{|}}{CH}-COOC_2H_5 \quad 13\% \qquad [1061]$$

## Arylation

Aromatic fluorinated compounds act as arylating reagents only when they contain sufficiently reactive halogens. The presence of a trifluoromethyl group on the benzene ring increases somewhat the reactivity of a halogen atom bound in *ortho-* or *meta*-positions. Chlorinated or brominated benzotrifluorides treated with sodium methyl mercaptide give the corresponding sulphides readily (*635*).

570 $\xrightarrow[210°,\ 40\ atm]{CH_3SNa}$ 82% [635]

Similarly, both *o-* and *m*-chlorobenzotrifluoride give the same product, *m*-aminobenzotrifluoride, on treatment with sodamide (*67*).

571 $\xrightarrow[-33°]{NaNH_2,\ NH_3(l)}$ 52% 35% $\xrightarrow[-33°]{NaNH_2,\ NH_3(l)}$ [67]

Very important are arylations by means of 2,4-dinitrofluorobenzene. Alcohols afford 2,4-dinitrophenyl ethers (*95*) which can be used for characterization (*1078*). The reactivity of 2,4-dinitrofluorobenzene with sodium methoxide is approximately six hundred times greater than that of 2,4-dinitrochlorobenzene (*503*).

The higher degree of reactivity of the 2,4-dinitrofluoro-derivative (in comparison to the chloro-compound) has made 2,4-dinitrofluorobenzene a favorite reagent for temporary protection of amino groups. The reaction with primary amines at 30°C is 100–200 times and at 50°C 20–70 times faster than the similar reaction with 2,4-dinitrochlorobenzene (167, 168):

**572**

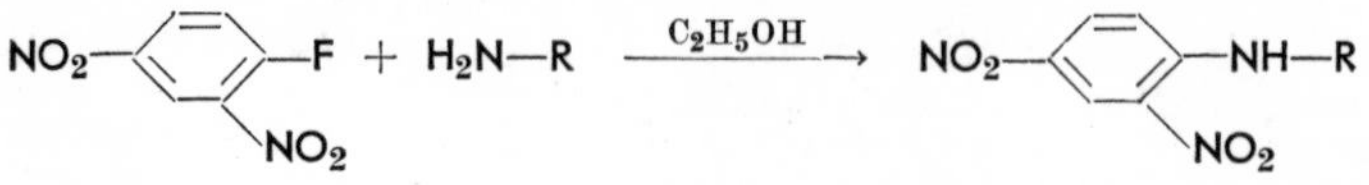

The reaction is practically quantitative and it is sufficiently fast at normal temperature to allow satisfactory estimation of amino groups in peptides.

Recently perfluorobenzene was found to react readily with nucleophilic reagents and yields the corresponding perfluorophenyl derivatives. The reaction with sodium methoxide affords pentafluoroanisol (283, 859), with ethanolic potassium hydroxide perfluorophenetol (97, 859).

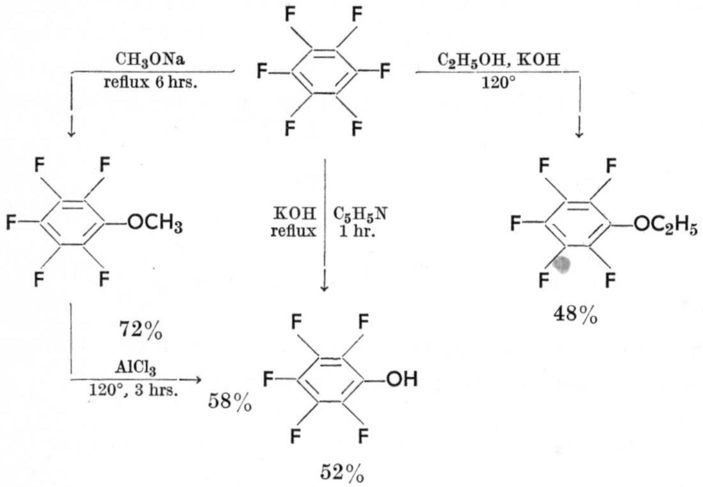

Treatment of hexafluorobenzene with sodamide in liquid ammonia affords pentafluoroaniline in high yield (284).

**574**

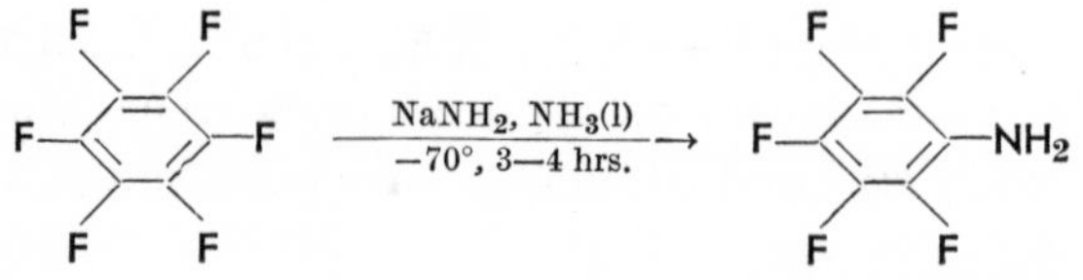

## Acylation

The most frequently used acylating reagent is trifluoroacetic anhydride. This compound readily acylates hydroxy-derivatives. The trifluoroacetates thus formed undergo hydrolysis and alcoholysis easily (*121*). For this reason they are often used in the carbohydrate chemistry for temporary blocking of the hydroxyl groups (*1075*) (p. 280).

575

[*121*]

Still more important is the trifluoroacetylation of amino-derivatives. The amino-group in amino acids can be temporarily protected; this is very useful in the syntheses of polypeptides. Acylations are carried out in anhydrous trifluoroacetic acid or by trifluoroacetic anhydride (*116, 1072, 1074*); phenyl trifluoroacetate (*1076*) and ethyl S-trifluorothioacetate can also be employed (*915*). In alkaline medium this ester acylates the *omega*-amino group of diamino acids, while trifluoroacetic anhydride causes preferential acylation in the *alpha*-amino group (*915, 1072*).

576
$$CF_3-COOC_6H_5$$
$$(CF_3-CO)_2O \qquad \xrightarrow{\quad NH_2-CH_2-COOH \quad} \qquad CF_3-CONH-CH_2-COOH$$
$$CF_3-COSC_2H_5$$

The trifluoroacetyl-derivatives are very useful not only in the synthesis of polypeptides where they "activate" the carboxyl group but also in the isolation because their vapour pressure is relatively high and they can be vacuum distilled (*1073*). Moreover selective hydrolysis allows sometimes part-degradation and decarbobenzylation (*1077*).

Of importance is the preparation of fluorinated amides from fluorinated esters. Fluoroacetamide is prepared for use as an analytical standard (*142*); trifluoroacetamide is formed by treatment of ethyl trifluoroacetate with an ether (*655*) or methanol (*309*) solution of ammonia.

577 $\qquad CH_2F-COOCH_3 \xrightarrow{\quad NH_3 \text{ aq.} \quad} CH_2F-CONH_2 \quad 100\%$ [*142*]

Fluorinated acid chlorides and sodium azide afford fluorinated acid azides (*482*),

578 $\qquad CF_3-CH_2-CH_2-COCl \xrightarrow[55-65°]{\quad NaN_3, C_6H_6 \quad} CF_3-CH_2-CH_2-CON_3$ [*482*]

fluorinated primary amines and phosgene give fluorinated isocyanates (*739,874*).

579 $\qquad CF_3-CH_2-NH_2 \xrightarrow[5° \to 55°]{\quad COCl_2, (CH_2OC_2H_5)_2 \quad} CF_3-CH_2-N=C=O \quad 19\%$ [*874*]

16*

Formyl fluoride reacts with secondary amines and converts them to the hydrofluorides of the corresponding disubstituted formamides in yields ranging from 80 to 90% (*785*).

580             $NH(C_2H_5)_2$ + HCOF $\xrightarrow[< 0°]{(C_2H_5)_2O}$ $HCON(C_2H_5)_2 \cdot HF$  84%          [*785*]

Acylations at carbon atoms occur in the reaction of fluorinated acid chlorides with diazomethane which give diazoketones (*135*) (Equation 543, p. 235). Similarly, formyl fluoride and diazomethane afford an unstable diazoacetaldehyde, which gives fluoroacetaldehyde by treatment with anhydrous hydrogen fluoride (*783*).

581                                                                              [*783*]

HCOF + $CH_2N_2$ $\xrightarrow{(C_2H_5)_2O}$ $[HCOCHN_2]$ $\xrightarrow{HF}$ $CH_2F$—CHO  24·5%

Fluorinated acid halides react with copper or silver cyanide to yield nitriles of fluorinated α-ketoacids (*101*, *824*).

582             $CH_2F$—COBr $\xrightarrow{CuCN}$ $CH_2F$—CO—CN  25%          [*101*]

583             $CF_3$—COCl $\xrightarrow[80—95°, \text{ sealed}]{AgCN}$ $[CF_3$—CO—CN$]_2$  80%          [*824*]

An analogous reaction with silver thiocyanate gives fluorinated ketoacids thiocyanates (*824*).

584             $CF_3$—COCl $\xrightarrow[60-80°]{AgSCN}$ $CF_3$—CO—SCN  60%          [*824*]

## ADDITIONS

Compounds containing double and triple bonds accept many inorganic and organic compounds to form saturated derivatives. Some examples have already been described under various labels: addition of hydrogen in the chapter on reduction, additions of halogens, hydrogen halides and hypohalides in the chapter on halogenation, addition of nitrosyl chloride in nitrosation, addition of bisulphite in sulphonation, and additions of the Grignard reagents in the chapter dealing with organometallics. However, some of inorganic and organic compounds which also add to unsaturated derivatives have escaped this classification, and their addition reactions will be discussed in the following paragraphs. Some of the additions occur spontaneously but the majority of them require catalysis by acids or bases, or else initiation by radicals. Additions to olefins, to acetylenes and to other compound will be described separately. A further division will be made between products with or without a new carbon–carbon bond.

## Additions to Double Bonds

Fluorinated olefins accept a variety of inorganic and organic compounds. The former follow usually a polar mechanism, the latter a polar or a radical mechanism. As a consequence of the negativity of fluorine, the electron density at the double bond in fluorinated olefins is decreased, so that electrophilic additions (hydrogen halides etc.) become more difficult while nucleophilic additions (alkoxides, amines) become easier.

*Additions forming no new carbon–carbon bonds.* Additions of many inorganic compounds have already been mentioned elsewhere. In this paragraph only the addition of ammonia, hydrazine, phosphine, sulphur trioxide, sulphur monochloride, trichlorosilane, and of some organic compounds will be briefly described.

In the addition of *ammonia* to ethyl $\gamma,\gamma,\gamma$-trifluorocrotonate the $NH_2$ anion links to the *beta*-carbon. The trifluoromethyl induction, which in itself should cause "reversal" in the direction of addition, is evidently weaker than the mesomeric action of the carbethoxy group (*1064*).

585 $\quad CF_3-CH{=}CH-COOC_2H_5 \quad \xrightarrow[100°]{NH_3} \quad CF_3-\underset{\underset{NH_2}{|}}{CH}-CH_2-CONH_2 \quad 95\% \qquad$ [*1064*]

However when two trifluoromethyl groups are in competition with one carboxyl group, the addition of ammonia occurs to the *alpha*-position to the carboxyl (*563*).

586 $\quad \underset{CF_3}{\overset{CF_3}{>}}C{=}CH-COOH \quad \xrightarrow{NH_3} \quad \underset{CF_3}{\overset{CF_3}{>}}CH-\underset{\underset{NH_2}{|}}{CH}-COOH \qquad$ [*563*]

The addition of ammonia to perfluorinated olefins is accompanied by simultaneous loss of hydrogen fluoride, so that an amide or a nitrile are the final reaction product (*570, 608*).

587 $\quad CF_3-CF_2-CF{=}CF_2 \quad \xrightarrow{NH_3} \quad [CF_3-CF_2-CHF-CF_2-NH_2] \qquad$ [*608*]

$\qquad\qquad\qquad\qquad\qquad\qquad \overset{\displaystyle |}{\underset{\rule{2cm}{0.4pt}}{\;\;-2HF}} \quad CF_3-CF_2-CHF-CN \quad 20\%$

588 $\quad \underset{CF_3}{\overset{CF_3}{>}}C{=}CF_2 \quad \xrightarrow[-60°\ -0°]{NH_3,\ (C_2H_5)_2O} \quad \underset{CF_3}{\overset{CF_3}{>}}CH-CONH_2 \quad + \quad \underset{CF_3}{\overset{CF_3}{>}}CH-CN \qquad$ [*570*]

$\qquad\qquad\qquad\qquad\qquad\qquad\qquad\qquad\qquad\qquad 13\% \qquad\qquad\qquad 21\%$

In the case of tetrafluoroethylene the nitrile thus formed immediately trimerizes to tris(difluoromethyl)-*sym*-triazine, the hydrolysis of which represents a convenient method for the preparation of difluoroacetic acid (*180, 467*).

**589**          [*467*]

$$CF_2{=}CF_2 \xrightarrow[\text{20°, 35 atm}]{\text{NH}_3/\text{Cu}^{++}} [CHF_2{-}CF_2{-}NH_2 \xrightarrow{-2\ \text{HF}} CHF_2{-}CN] \rightarrow$$

100%

The addition and trimerization are extremely exothermic and careful control of temperature (by applying dry-ice cooling) is necessary to prevent explosion (*434*).

Dehydrofluorination occurs also in the addition of *hydrazine* to perfluoroolefins (*160*).

$$CF_3{-}CF{=}CF_2 \xrightarrow[\text{60°, 8 hrs.}]{\text{N}_2\text{H}_4 \cdot \text{H}_2\text{O, (C}_2\text{H}_5)_2\text{O}} CF_3{-}CHF{-}CF{=}NNH_2 \ \ 66\%$$

**590**          [*160*]

$$\xrightarrow[\text{70\%}]{\text{CH}_3\text{COOH}} $$

In contrast to the addition of ammonia, the addition of *phosphine* to perfluoroolefins seems to be of a radical nature and telomerization occurs concurrently (*813*).

[*813*]

$$CF_2{=}CF_2 \xrightarrow[\text{150°, 8 hrs.}]{\text{PH}_3} CHF_2{-}CF_2{-}PH_2 + CHF_2{-}CF_2{-}PH{-}CF_2{-}CHF_2 +$$

**591**      53%               7%    + $PH_2{-}CF_2{-}CF_2{-}PH_2$ 9%

*Sulphur monochloride* and tetrafluoroethylene yield a complex mixture of sulphur derivatives (*568*).

                                 $ClCF_2{-}CF_2SCl$ + $(ClCF_2{-}CF_2)_2S$

                                     12·3%           11·5%

**592**    $CF_2{=}CF_2 + S_2Cl_2 \xrightarrow[\text{autoclave}]{100{-}120°}$                             [*568*]

                                   $(ClCF_2{-}CF_2S)_2$ + $(ClCF_2{-}CF_2S)_2S$

                                   28·8%           6·9%

The direction of addition of *sulphur trioxide* to chlorotrifluoroethylene is random (*511*).

**593**                             $CClF{-}CF_2$      $CClF{-}CF_2$     [*511*]

$$CClF{=}CF_2 + SO_3 \xrightarrow[\text{reflux 3 hrs.}]{-28°\ \text{to}\ -21°}$$

                                      53·7%           25·2%

In contrast to the preceding reaction, the course of which is determined by the polarity of the double bond, the addition of *trichlorosilane* is of radical character (*305, 672*).

**594**                                                                 [*305, 672*]

$$CF_3—CH=CH_2 + SiHCl_3 \quad \overset{\displaystyle \overset{hv}{\underset{20°,\ 96\ hrs.}{\diagup}} \searrow\ 91\%}{\underset{\displaystyle \underset{125—130°}{[(CH_3)_3CO]_2} \nearrow\ 75·3\%}{}} \quad CF_3—CH_2—CH_2—SiCl_3$$

*Organic silicon compounds* behave similarly when added to fluorinated olefins (*304*).

**595**   $CF_2=CF_2 + H_2Si(CH_3)_2 \xrightarrow[\ 19\ hrs.\ ]{hv} CHF_2—CF_2—SiH(CH_3)_2$   54%   [*304*]

*Alcohols* and *phenols* again follow a polar mechanism; their nucleophilic addition to fluorinated olefins requires basic conditions and leads to fluorinated ethers (*1101*) (Equation 554, p. 238).

*Procedure 32.*

*Polar Addition to Fluoroolefins*
*Preparation of α,α,β,β,β′,β′,β′,–Heptafluorodiethyl Ether by the Addition of 2,2,2–Tri-fluoroethanol to Tetrafluoroethylene (480)*

A mixture of 50 g (0·5 mole) of 2,2,2-trifluoroethanol, 1·5 g (0·06 mole) of sodium, and 75 g (0·75 mole) of tetrafluoroethylene is stirred in an autoclave while being heated for 16 hours at 180°C and 40 atm. After cooling, the autoclave is vented through a dry-ice trap (3·5 g of a condensate being thus collected), and the liquid content of the auto-clave is distilled to give 75 g of the crude ether boiling over a range of 51—61°C. Total yield combined with the dry-ice trap condensate represents 78·5 g (0·78 mole) (78·5%) of α,α,β,β,β′,β′,β′-heptafluorodiethyl ether. Decomposition of the distillation residue with water gives 4 g (0·04 mole) of recovered 2,2,2-trifluoroethanol.

With unsymmetrical olefins the direction of addition is determined by the polarity of the double bond, which again is the combination of the inductive and electromeric effects of the fluorine atoms. The negative alkoxide ion adds to the carbon atom with lower electron density.

In 1,1,1-trifluoro-2-propene, the inductive effect of the trifluoromethyl group directs the ethoxide anion to the end carbon of the double bond (*481*) (Equation 552, p. 238). In perfluoropropene both the inductive effect of the trifluoromethyl group and the electromeric effect of vinylic fluorine shift the π-electrons to the central carbon atom and the alkoxide joins the terminal —$CF_2$— group (*570, 576*):

**596**   $CF_3—CF=CF_2 \xrightarrow[\ 60°\ ]{CH_3OH,\ KOH} CF_3—CHF—CF_2—OCH_3$   83%   [*576*]

In 1,1-dichlorodifluoroethylene and in chlorotrifluoroethylene the electromeric effect of fluorine overshadows that of chlorine, and the alkoxide adds to the $=CF_2$ side (258, 810, 1019).

**597**    $CCIF{=}CF_2 \xrightarrow[\text{2·5 hrs.}]{C_2H_5OH,\ C_2H_5ONa} CHCIF{-}CF_2{-}OC_2H_5$    88—92%    [258]

Fluorinated ethers formed by the addition of alcohols to fluorinated olefins tend to split off hydrogen fluoride or hydrogen halide and yield enol ethers (570, 588, 809, 1019). The ratio between the primary addition product and the secondarily formed enol ether depends on the nature of the fluorinated olefin, on the nature of the alcohol, and on the reaction conditions. The results of adding various alcohols to perfluoroisobutylene are shown in the following equation and Table 42.

**598**    $\begin{array}{c} CF_3 \\ \phantom{x} \\ CF_3 \end{array}\!\!\!\!\!\!\!\!\searrow\!\!C{=}CF_2 \xrightarrow[20-40°]{ROH} \begin{array}{c} CF_3 \\ \phantom{x} \\ CF_3 \end{array}\!\!\!\!\!\!\!\!\searrow\!\!CH{-}CF_2{-}OR + \begin{array}{c} CF_3 \\ \phantom{x} \\ CF_3 \end{array}\!\!\!\!\!\!\!\!\searrow\!\!C{=}CF{-}OR$    [588]

TABLE 42

*Results of Addition of Alcohols to Perfluoroisobutylene (588)*

| Alcohol | Conversion, % | Yield of $(CF_3)_2CHCF_2OR$,% | Yield of $(CF_3)_2C{=}CFOR$,% |
|---|---|---|---|
| $CH_3OH$ | 95 | 70 | 0 |
| $C_2H_5OH$ | 95 | 59 | 6 |
| $CH_3CH_2CH_2OH$ | 77 | 43 | 14 |
| $(CH_3)_2CHOH$ | 80 | 35 | 26 |
| $CH_3CH_2CH_2CH_2OH$ | 70 | 52 | 27 |
| $CH_2{:}CHCH_2OH$ | 60 | 56 | 0 |
| $ClCH_2CH_2OH$ | 70 | 51 | 0 |
| $FCH_2CH_2OH$ | 65 | 60 | |

The addition of phenol to 1,1-dichlorodifluoroethylene gives the normal ether at low temperature and the enol ether at high temperature (1020).

$$CCl_2{=}CF_2 + OH{-}\!\!\!\bigcirc$$

**599**    $\xleftarrow{\quad 10°\quad}\xrightarrow[]{\text{KOH}\ |\ CH_3COCH_3}\xrightarrow{\quad 70°\quad}$    [1020]

$CHCl_2{-}CF_2{-}O{-}\!\!\!\bigcirc$    60%          $CCl_2{=}CF{-}O{-}\!\!\!\bigcirc$    62%

Loss of a hydrogen halide occurs easily with ethers derived from fluorinated cyclobutenes. Not only are the primary addition products difficult to isolate but the enol ethers formed by the loss of hydrogen fluoride accept another molecule of the alcohol and lose a second hydrogen fluoride, so that enediol diethers are found as the final products (*50, 806*).

**600** [*50*]

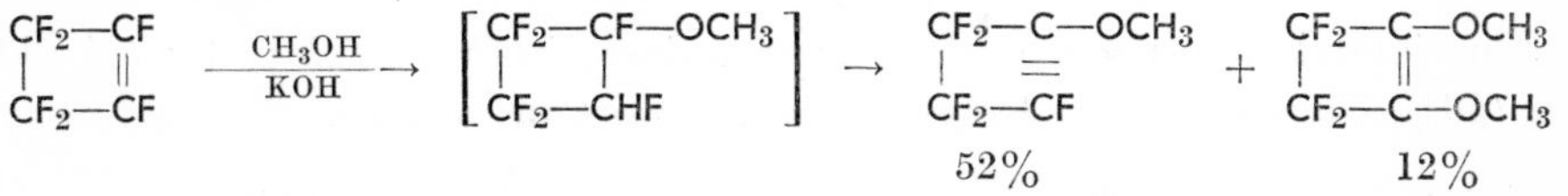

$$\begin{array}{c}CF_2\!-\!CF\\|\quad\;\;\|\\CF_2\!-\!CF\end{array}\;\xrightarrow[\text{KOH}]{\text{CH}_3\text{OH}}\;\left[\begin{array}{c}CF_2\!-\!CF\!-\!OCH_3\\|\quad\;\;|\\CF_2\!-\!CHF\end{array}\right]\;\rightarrow\;\begin{array}{c}CF_2\!-\!C\!-\!OCH_3\\|\qquad\;=\\CF_2\!-\!CF\end{array}\;+\;\begin{array}{c}CF_2\!-\!C\!-\!OCH_3\\|\qquad\;\|\\CF_2\!-\!C\!-\!OCH_3\end{array}$$

$$\qquad\qquad\qquad\qquad\qquad\qquad\qquad\qquad\qquad\qquad\qquad 52\%\qquad\qquad\qquad 12\%$$

In the reaction of fluorinated olefins with alcohols liberation of hydrogen fluoride or halide from the addition products is not the only side-reaction. $\alpha,\alpha$-Difluoroethers undergo thermal fission to alkyl fluoride and acyl fluoride according to the following equation (*1019*):

**601**

$$CHCl_2\!-\!CF_2\!-\!OCH\!\!\begin{array}{c}\diagup CH_3\\\diagdown CH_3\end{array}\;\xrightarrow[\text{heating}]{}\;CHCl_2\!-\!COF\;+\;FCH\!\!\begin{array}{c}\diagup CH_3\\\diagdown CH_3\end{array}\qquad[\textit{1019}]$$

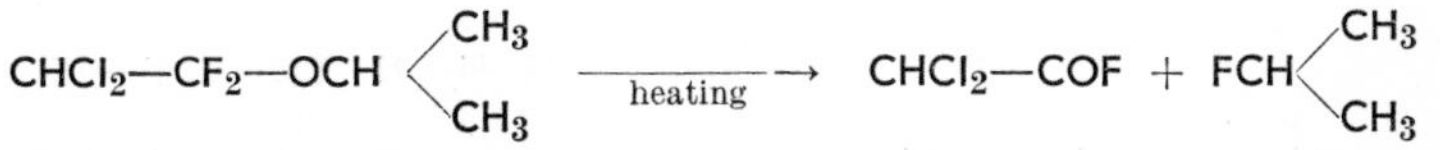

Hydrolysis of such ethers with mineral acids gives carboxylic acids or their esters (Equation 439, p. 204).

Like alcohols or phenols, *mercaptans* (*865*) or *amines* (*174, 565, 858*) add to fluorinated olefins. Here, too, loss of hydrogen halides leads to unsaturated derivatives, especially in the case of amines.

**602**

$$\xleftarrow[20°]{R\!-\!SH,\ NaSH}\quad\begin{array}{c}CF\!=\!CF\\|\quad\;\;|\\CF_2\!-\!CF_2\end{array}\quad\xrightarrow{R\!-\!NH\!-\!R}$$

$$\begin{array}{c}CF\!=\!C\!-\!SR\\|\quad\;\;|\\CF_2\!-\!CF_2\end{array}\qquad\qquad\begin{array}{c}CF\!=\!C\!-\!N\!\!\begin{array}{c}\diagup R\\\diagdown R\end{array}\\|\quad\;\;|\\CF_2\!-\!CF_2\end{array}\qquad[\textit{858, 865}]$$

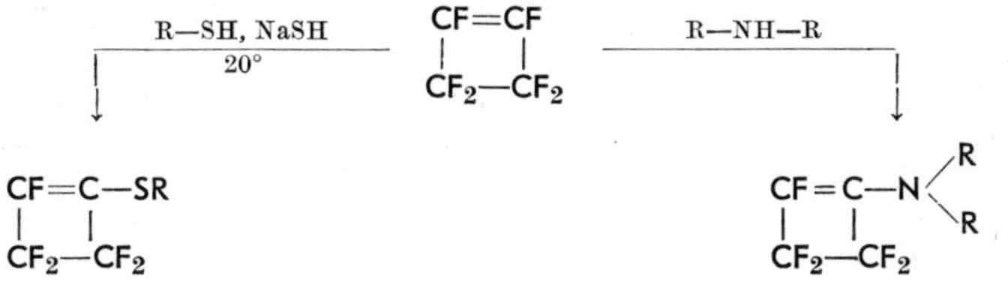

As with the alcohols, the ratio between the primary addition product and the final unsaturated amine (enamine) depends largely on the structure of the fluorinated olefin (*563*) (Table 43).

TABLE 43

*Results of Addition of Secondary Amines to Fluorinated Olefins (563)*

| Fluorinated Olefin RR′C=CF₂ | Product | |
|---|---|---|
| | RR′CHCF₂NR₂″ | RR′C=CFNR₂″ |
| R = F,    R′ = F | 100% | 0% |
| R = F,    R′ = CF₃ | 70% | 30% |
| R = CF₃,  R′ = CF₃ | 0% | 100% |

Both the primary addition products and the enamines are readily converted to substituted amides (*180, 565, 570*).

$$CF_2{=}CF_2 + C_6H_5{-}NH_2 \qquad\qquad\qquad\qquad\qquad [180]$$
**603**

$$\xrightarrow[130°, 8 \text{ hrs.}]{H_2O,Na_2B_4O_7} \quad [CHF_2{-}CF_2{-}NH{-}C_6H_5] \xrightarrow{K_2CO_3} CHF_2{-}CONH{-}C_6H_5 \quad 71\%$$

Fluorinated dienes accept alcohols or amines either by *1,2*- or by *1,4*-addition (*564*). Subsequent hydrolysis affords esters or amides.

$$\text{604} \quad \xleftarrow[90-100°, 4 \text{ hrs.}]{C_2H_5OH/N(C_2H_5)_3} \quad CF_2{=}CF{-}CF{=}CF_2 \quad \xrightarrow{(C_2H_5)_2NH, (C_2H_5)_2O} \qquad [564]$$

$$C_2H_5O{-}CF_2{-}CF{=}CF{-}CHF_2 \quad 53{\cdot}5\% \qquad\qquad (C_2H_5)_2N{-}CF{=}CF{-}CF{=}CF_2 \quad 81\%$$

$$\downarrow \begin{smallmatrix} H_2O & (C_2H_5)_2O \\ \text{few} & \text{min.} \end{smallmatrix}$$

$$(C_2H_5)_2N{-}CO{-}CHF{-}CF{=}CF_2 \quad 86\%$$

An interesting reaction taking place between a fluorinated olefin and *trialkylphosphite* leads to alkyl fluoride and dialkyl fluoroalkene phosphonate (*574*).

**605**

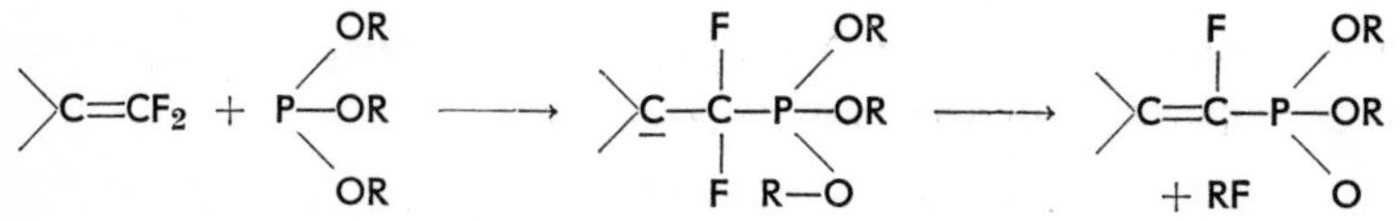

*Additions accompanied by the formation of carbon–carbon bonds.* The main interest in additions with formation of a carbon–carbon bond is focused on the additions of halogenated paraffins to olefins or halogenated olefins, such as:

**606**

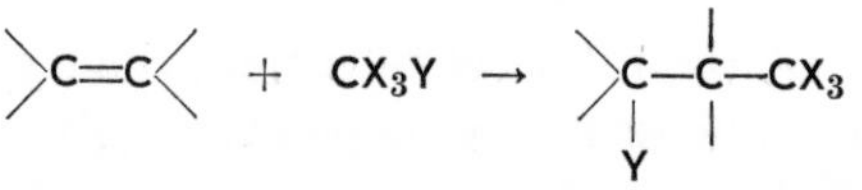

They must be preceded by the cleavage of the halogenated paraffin into a halogen and a carbon-containing residue. This cleavage may be accomplished by using polar catalysts, such as aluminium chloride which effects polar dissociation. The reaction is usually complicated by subsequent halogen exchange (*179*). Much more universal and reliable is homolytic fission, which is carried out by ultra-violet irradiation or with an organic peroxide such as dibenzoyl peroxide. In this manner halogenated paraffins are split into halogen atoms and very reactive halogenoalkyl radicals (*393*), which attack the olefin at that carbon atom of the double bond which is sterically more accessible.

Thus the addition direction of the halogenoalkyl radical parallels that of bromine atom in the radical addition of hydrogen bromide to olefins (*463*).

607 $\xrightarrow[hv,\,20°]{\text{HBr}}$ CF$_3$—CH=CH$_2$ $\xrightarrow[hv]{\text{CCl}_3\text{Br}}$ [*463*]

CF$_3$—CH$_2$—CH$_2$Br  90%    CF$_3$—CHBr—CH$_2$—CCl$_3$  76%

Sometimes the direction of addition is specific and unambiguous (*398*),

608 $\xrightarrow[]{\text{HBr}}$ CF$_2$=CHCl $\xrightarrow[hv]{\text{CF}_3\text{I}}$ [*398*]

CBrF$_2$—CH$_2$Cl  99%    CF$_3$—CF$_2$—CHClI  94%

in other cases it splits between the two possible paths, with prevalence of one (*399*).

609 $\xrightarrow[hv]{\text{HBr}}$ CF$_2$=CHF $\xrightarrow[hv]{\text{CF}_3\text{I}}$ [*399*]

CF$_2$Br—CH$_2$F  42%    CF$_3$—CF$_2$—CHFI  20%
CF$_2$H—CHFBr  58%    CF$_3$—CHF—CF$_2$I  80%

A brief survey of data in this field appears in Table 44 where the bold-face halogens are those which split off from the halogenoparaffins, the bold-face carbon atoms in the olefins are those to which a halogenoalkyl radical becomes attached, and the numbers in the columns are references listed on p. 358.

TABLE 44

*Addition of Haloparaffins to Olefins and Haloolefins*

| Haloparafin*) | CHCl$_3$ | CCl$_4$ | CCl$_3$Br | CCl$_3$I | CF$_3$I | CF$_2$Br$_2$ | CHFCl$_2$ | CF$_2$ClBr | C$_2$F$_5$I | C$_3$F$_7$I | CF$_2$BrCFClBr |
|---|---|---|---|---|---|---|---|---|---|---|---|
| Olefin**) | | | | | Reference Number on p. 358 | | | | | | |
| CH$_2$=CH$_2$ | | | | | 361 379 | 1024 | | | | | |
| CH$_2$=CHF | | | | | | 1026 | | | | | |
| CH$_2$=CF$_2$ | | | | | 395 | | | 1025 | 397 | | 1022 |
| CHF=CF$_2$ | | | | | 399 | | | | | | |

*continued*

Table 44 — *continued*

| Haloparafin* | CHCl₃ | CCl₄ | CCl₃Br | CCl₃I | CF₃I | CF₂Br₂ | CHFCl₂ | CF₂ClBr | C₂F₅I | C₃F₇I | CF₂BrCFClBr |
|---|---|---|---|---|---|---|---|---|---|---|---|
| Olefin** | | | | | Reference Number on p. 358 | | | | | | |
| $CF_2=CHCl$ | | | | | 398 | | | | | | |
| $CF_2=CFCl$ | 179 | 179 | 456 457 | | 394 456 | | 179 | | | | |
| $CF_2=CF_2$ | 179 | 179 | | | 354 371 | | 179 | | 371 | | |
| $CH_2=CH-CH_3$ | | | | | | 1024 | | 1025 | | | |
| $CHCl=CH-CH_3$ | | | 1018 | | | | | | | | |
| $CH_2=C=CH_2$ | | | | | 382 | | | | | | |
| $CH_2=CH-CH_2Cl$ | | | | | 382 | | | | | | 1021 |
| $CF_2=CH-CH_3$ | | | | | 370 | | | | | | |
| $CH_2=CH-CF_3$ | | | 463 | 366 | 463 | | | | | | |
| $CF_2=CH-CF_3$ | | | | | 397 | | | | | 397 | |
| $CF_2=CF-CF_3$ | | | | | 369 | | | | | | |
| $CH_2=C\begin{smallmatrix}CH_3\\CH_3\end{smallmatrix}$ | | | | | | | | 1025 | | | |

\* Bold-face typed halogen splits off during the dissociation.
\*\* Bold-face typed carbon combines with the paraffin radical carbon.

Halogenated paraffins, containing different halogen atoms, split their weakest bonds. Thus iodine and bromine dissociate more easily than chlorine or fluorine, and chlorine more easily than fluorine.

The ease of addition of halogenated paraffins to various fluorinated olefins depends on the nature of the olefin. For the radical addition of trifluoroiodomethane to fluorinated olefins the following sequence of decreasing reactivity was obtained (*386*).

$$CF_2{=}CH_2 > CF_2{=}CF_2 > CF_2{=}CHCl > CF_2{=}CFCl >$$

610

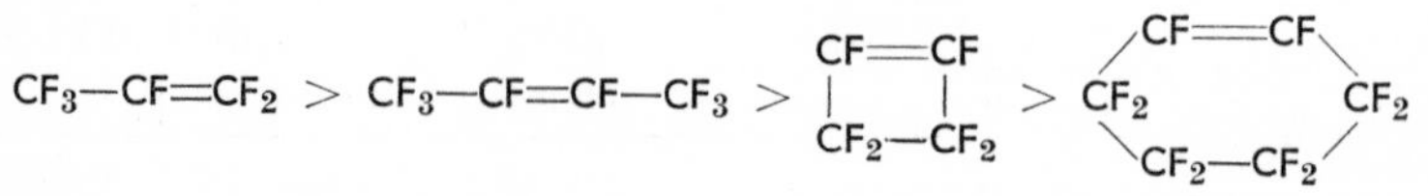

$$CF_3{-}CF{=}CF_2 > CF_3{-}CF{=}CF{-}CF_3 > \begin{smallmatrix}CF{=}CF\\ |\quad\ |\\ CF_2{-}CF_2\end{smallmatrix} > \begin{smallmatrix}/CF{=}CF\backslash\\ CF_2\qquad CF_2\\ \backslash CF_2{-}CF_2/\end{smallmatrix} \qquad [386]$$

The direction of addition to the double bond of unsymmetrical olefins is difficult to predict. From the experimental data listed in Table 44 it is evident that the halogenoalkyl radical becomes attached to the end of the double bond which is sterically less hindered. The steric hindrance of elements or groups increases in the series:

**611**  $H < F < Cl < CH_3 < CF_3$                   *[398,399,805]*

$$CH_2 < CHF < CHCl < CF_2 < \begin{matrix} CClF \\ CCl_2 \end{matrix}$$

The reaction of olefins with halogenated paraffins under radical conditions shows the typical features of a chain reaction. Irradiation, heat and especially addition of a peroxide originates a halogenoalkyl radical which attacks the olefin and forms another radical. This one is stabilized by either extracting a weakly-bonded halogen atom from the halogenated paraffin, or else adding another molecule of the olefin forming thus a dimeric radical which again can react with the original halogenated paraffin, or with another molecule of the olefin.

**612**

$$CF_3I \xrightarrow{h\nu} CF_3\cdot + I\cdot$$

$$CF_3\cdot + CF_2{=}CF_2 \longrightarrow CF_3{-}CF_2{-}CF_2\cdot$$

| | | | $CF_3I$ |
|---|---|---|---|
| $CF_2{=}CF_2$ | | | |

$$CF_3{-}CF_2{-}CF_2{-}CF_2{-}CF_2\cdot \qquad\qquad CF_3{-}CF_2{-}CF_2I + CF_3\cdot$$

$CF_3I$ | | $CF_2{=}CF_2$

$$CF_3{-}CF_2{-}CF_2{-}CF_2{-}CF_2I + CF_3\cdot \qquad CF_3{-}CF_2{-}CF_2{-}CF_2{-}CF_2{-}CF_2{-}CF_2\cdot$$

The product of the addition of a halogenated paraffin to olefins is always accompanied by the so called telomers, i.e. compounds containing multiple basic units of the starting olefin. Their proportion largely depends on the nature of the reactants and on the conditions used, especially on the ratio between the halogenated paraffin and olefin. Formation of the telomers may be suppressed by using a large excess of the haloparaffin (354, *371*).

**613**                                         *[354]*

$$CF_3I + CH_2{=}CH_2 \begin{array}{c} \nearrow^{h\nu} \; 82\% \\ \xrightarrow{Hg,\, 20-30°} 78\% \\ \searrow_{>200°} \; 75\% \end{array} CF_3{-}CH_2{-}CH_2I + CF_3{-}CH_2{-}CH_2{-}CH_2{-}CH_2I$$

Only a few examples will be quoted to show experimental conditions and results and yields of the products:

**614**      $CF_3I + CF_3-CH=CF_2 \xrightarrow[20°, 4 \text{ days}]{h\nu} CF_3-CH-CF_2I$   80%          [397]

$\qquad\qquad\qquad\qquad\qquad\qquad\qquad\qquad\quad$ $|$
$\qquad\qquad\qquad\qquad\qquad\qquad\qquad\qquad\quad CF_3$

**615**      $CF_2ClBr + CH_2=CF_2 \xrightarrow[100°, 4 \text{ hrs.}]{(C_6H_5COO)_2} CF_2Cl-CH_2-CF_2Br$   35%          [1025]

**616**      $CCl_3Br \longleftarrow CF_2=CFCl \xrightarrow[h\nu]{5 CF_3I}$                    [394, 457]

$CCl_3-CF_2-CFClBr$                                $CF_3-CF_2-CFClI$   85%

**617**      $CF_2=CF_2 + C_2F_5I \xrightarrow[20°]{h\nu} C_2F_5-CF_2-CF_2I + C_2F_5(CF_2-CF_2)_2I$          [371]

$\quad$ 0·02 mole    0·1 mole          91%          4%

*Procedure 33.*

*Radical Addition of Perfluoroalkyl Halides to Olefins*
*Preparation of 1-Chloro-1-iodohexafluoropropane from Trifluoroiodomethane and Chlorotrifluoroethylene (394)*

A quartz tube is charged with 15·4 g (0·079 mole) of trifluoroiodomethane and 2 g (0·017 mole) of chlorotrifluoroethylene, evacuated, sealed, and irradiated with ultraviolet light in such a manner that the rays do not pass through the liquid phase. After 4 days, vacuum distillation recovers 12·4 g (0·063 mole) (80%) of trifluoroiodomethane and affords 4 g (0·013 mole) of 1-chloro-1-iodohexafluoropropane, b.p. 58°C at 330 mm, and 0·6 g (0·0014 mole) (9%) of 1,3-dichloro-1-iodononafluoropentane, b.p. 68°C at 32 mm. The yield of 1-chloro-1-iodohexafluoropropane is 85% based on the reacted trifluoroiodomethane, and 76·5% based on the chlorotrifluoroethylene.

*Trifluoromethanesulphenyl pentafluoride* ($CF_3SF_5$) reacts at high temperatures with perfluoroolefins (*219*) and perfluoroimines (*220*), so that sulphur tetrafluoride is split out and one end of the double bond is combined with fluorine, the other with trifluoromethyl group:

**618**      $CF_3-N=CF_2 + CF_3-SF_5 \xrightarrow[1 \text{ atm}]{519°} CF_3-N-CF_3 + SF_4 + C_2F_6$          [220]

$\qquad\qquad\qquad\qquad\qquad\qquad\qquad\qquad\quad$ $|$
$\qquad\qquad\qquad\qquad\qquad\qquad\qquad\qquad\quad CF_3$   92%
$\qquad\qquad\qquad\qquad\qquad\qquad\qquad\qquad\quad$ 50%

The radical addition to olefins is not limited to halogenated paraffins. Under similar conditions, in the presence of dibenzoyl peroxide, *alcohols* add in a way different from their polar addition and form a new carbon–carbon bond (*610*).

**619**  [*610*]

$$CF_3-CF_2-CF=CF_2 + CH_3OH \xrightarrow[110-120°]{(C_6H_5COO)_2} CF_3-CF_2-CHF-CF_2-CH_2OH \quad 90\%$$

A similar addition of *aldehydes* affords ketones (*610*).

**620**  [*610*]

$$CF_3-CF=CF_2 + CHO-C_3H_7 \xrightarrow[80°]{(C_6H_5COO)_2} CF_3-CHF-CF_2-CO-C_3H_7 \quad 70\%$$

In the *Prins reaction* between tetrafluoroethylene and formaldehyde, the primary addition product is immediately hydrolyzed to $\alpha,\alpha$-difluoro-$\beta$-hydroxypropionic acid (*653*).

**621**  [*653*]

$$CF_2=CF_2 + CH_2O \xrightarrow[85°]{95\% H_2SO_4} \left[\begin{matrix} CH_2-CF_2-CF_2-OH \\ | \\ OH \end{matrix}\right] \to \begin{matrix} CH_2-CF_2-COOH \\ | \\ OH \end{matrix} \quad 20\%$$

*Acyl chlorides* add under polar conditions (in the presence of aluminium chloride) to fluorinated olefins and give $\beta$-chloroketones which are easily dehydrohalogenated to $\alpha,\beta$-unsaturated ketones (*577*).

**622**  [*577*]

$$CH_3-COCl + CHF=CF_2 \xrightarrow{AlCl_3} \underset{34\%}{CH_3-CO-CHF-CF_2Cl} \xrightarrow{(C_2H_5)_3N} \underset{21\%}{CH_3-CO-CF=CF_2}$$

*Sodium cyanide* in aqueous acetonitrile adds to fluorinated olefins and subsequent hydrolysis of the products with dilute sulphuric acid gives mixtures of fluorinated acids and amides (*257*).

**623**  [*257*]

$$CClF=CF_2 \xrightarrow[70-80°]{\underset{CH_3CN}{NaCN}} \xrightarrow{50\% H_2SO_4} \underset{19\%}{CHClF-CF_2-CONH_2} + \underset{62·5\%}{CHClF-CF_2-COOH}$$

An interesting compound, *trifluoromethylhypofluorite*, adds fluorine to one end of a double bond, and the trifluoromethoxy group to the other end (*849*).

**624**

$$\begin{matrix} & CF=CF & \\ & \diagup \quad \diagdown & \\ CF_2 & & CF_2 \\ & \diagdown \quad \diagup & \\ & CF_2 & \end{matrix} \xrightarrow{CF_3OF} \begin{matrix} & & OCF_3 \\ & & \diagup \\ & CF_2-CF & \\ & \diagup \quad \diagdown & \\ CF_2 & & CF_2 \\ & \diagdown \quad \diagup & \\ & CF_2 & \end{matrix} \quad 29\%$$   [*849*]

Addition of *diazomethane* to fluorinated olefins gives the expected pyrazoline derivatives, the heating of which forms cyclopropane rings (*742*).

625

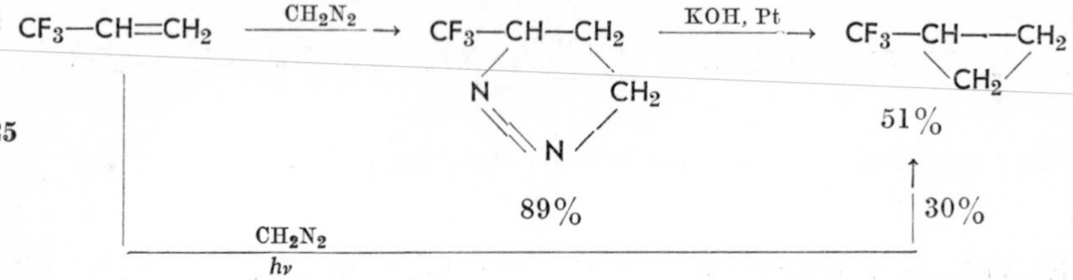

The reaction between tetrafluoroethylene and *trifluoronitrosomethane* or *heptafluoronitrosopropane* brings interesting results. In addition to a linear copolymer, the main product is perfluoro(2-alkyl-1,2-oxazetidine), a heterocyclic compound with a four-membered ring (*43, 45*).

626

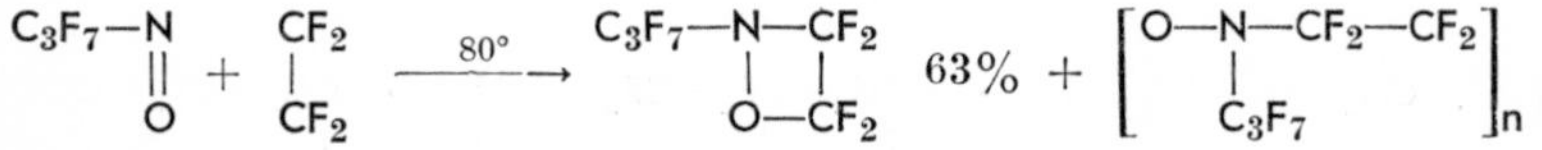

With unsymmetrical fluoroolefins, the addition occurs often in both possible directions, in varied extent (*48*).

627

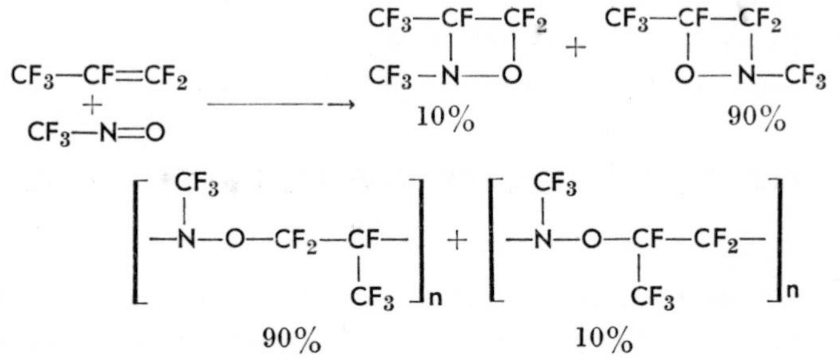

## Additions of Olefins to Olefins and Dienes

Perfluorovinyl iodide adds to some olefins as iodine at one end of the double bond and the trifluorovinyl group at the other end (*805*).

628

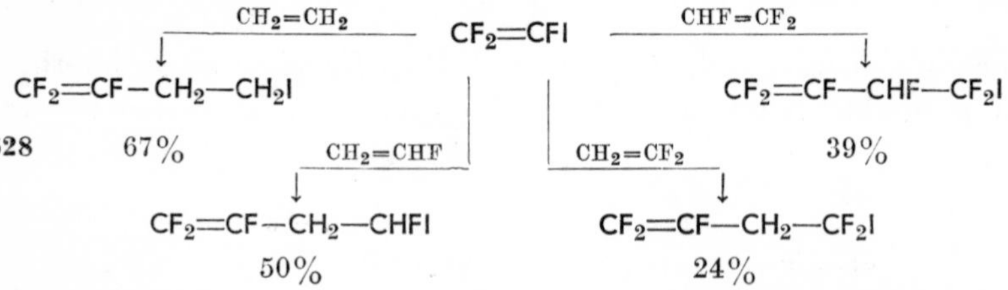

A remarkable property of fluorinated olefins is their dimerization to cyclobutane derivatives. In this respect they differ sharply from the non-fluorinated olefins which form linear dimers. Whereas trichloroethylene affords chlorinated butenes (*477*), 1,1-dichlorodifluoroethylene (*477*), chlorotrifluoroethylene (*477*), tetrafluoroethylene (*721*), and perfluoropropene (*136*) form fluorinated cyclo-butanes, in which the components are combined "head to head" (*477*).

$$
\textbf{629} \quad
\begin{array}{c}
CF_2{=}CFCl \\
\\
CF_2{=}CFCl
\end{array}
\xrightarrow[\text{autoclave}]{200°}
\begin{array}{c}
CF_2{-}CFCl \\
|\qquad | \\
CF_2{-}CFCl
\end{array}
\; 80{-}85\% \qquad [477]
$$

$$
\textbf{630} \quad
\begin{array}{c}
CF_2{=}CF_2 \\
\\
CF_2{=}CF_2
\end{array}
\xrightarrow{550°}
\begin{array}{c}
CF_2{-}CF_2 \\
|\qquad | \\
CF_2{-}CF_2
\end{array}
\; 71\% \qquad [721]
$$

In the case of perfluoropropene, head to tail dimerization was also observed.

$$
\textbf{631} \quad
\begin{array}{c}
CF_3{-}CF{=}CF_2 \\
\\
CF_3{-}CF{=}CF_2
\end{array}
\xrightarrow[\text{18 hrs.}]{400°}
\begin{array}{c}
CF_3{-}CF{-}CF_2 \\
|\qquad\quad | \\
CF_3{-}CF{-}CF_2
\end{array}
+
\begin{array}{c}
CF_3{-}CF{-}CF_2 \\
|\qquad\quad | \\
CF_2{-}CF{-}CF_3
\end{array}
\qquad [136]
$$
$$
48\%
$$

The same cyclo-dimerization tendency converts perfluoro-1,5-hexadiene to perfluorobicyclo-(0,2,2)-hexane (*266*)

$$
\textbf{632} \quad
\begin{array}{c}
CF_2{=}CF{-}CF_2 \\
|\\
CF_2{=}CF{-}CF_2
\end{array}
\xrightarrow[\text{6 min.}]{450°}
\begin{array}{c}
CF_2{-}CF{-}CF_2 \\
|\qquad|\qquad| \\
CF_2{-}CF{-}CF_2
\end{array}
\begin{array}{l}
\text{yield } 70\% \\
\\
\text{recovered } 27\%
\end{array}
\qquad [266]
$$

and perfluoro-1,3-butadiene to a still more astonishing compound, perfluoro-tricyclo-(4,2,0,0^{2,5})-octane (m.p. 40°C, b.p. 80°C), whose structure has been proposed on the basis of its inertness toward chlorine or potassium permanganate and its X-ray diffraction pattern (*857*).

$$
\textbf{633} \qquad\qquad\qquad\qquad\qquad\qquad\qquad\qquad\qquad\qquad\qquad\qquad [857]
$$
$$
\begin{array}{c}
CF_2{=}CF{-}CF{=}CF_2 \\
\\
CF_2{=}CF{-}CF{=}CF_2
\end{array}
\xrightarrow[\substack{\text{164 hrs.}\\ \text{sealed}}]{200°}
\begin{array}{c}
CF_2{-}CF{-}CF{=}CF_2 \\
|\qquad|\qquad|\qquad| \\
CF_2{-}CF{-}CF{=}CF_2
\end{array}
\rightarrow
\begin{array}{c}
CF_2{-}CF{-}CF{-}CF_2 \\
|\qquad|\qquad|\qquad| \\
CF_2{-}CF{-}CF{-}CF_2
\end{array}
\; 65\%
$$

On the other hand, a tetramer of hexafluoro-2-butyne, to which a similar skeleton had been ascribed, seems to have a cyclooctatetraene ring according to new measurements (*136, 233*).

$$
\textbf{634} \qquad 4\; CF_3{-}C{\equiv}C{-}CF_3 \xrightarrow[\text{31 hrs.}]{320°} \qquad\qquad\qquad\qquad\qquad [136, 233]
$$

The tendency of forming a four-membered ring is very pronounced, even in the reaction of fluorinated olefins with the non-fluorinated ones (*41, 178*). The yields of such a reaction between tetrafluoroethylene and various unsaturated compounds in the presence of quinol as a polymerization inhibitor are listed in Table 45.

TABLE 45

*Reaction of Tetrafluoroethylene with Unsaturated Compounds at 150° under Autogenous Pressure (around 45 atm.) (178)*

| Unsaturated Compound | Yield of the Cyclobutane Derivative, % | Unsaturated Compound | Yield of the Cyclobutane Derivative, % |
|---|---|---|---|
| $CH_2=CH_2$ | 40 | $CH_2=CH-CH=CH_2$ | 90 |
| $CH_3CH=CH_2$ | 72 | $CH_2=CF-CH=CH_2$ | 35 |
| $(CH_3)_2C=CH_2$ | 30 | $CH_3COCH=CH_2$ | 18 |
| $C_6H_5CH=CH_2$ | 85 | $CH_2=CHCN$ | 84 |
| $CH_2=CHCl$ | 23 | | |
| $CH_2=CCl_2$ | 46 | | |
| $CHCl=CCl_2$ | 18 | | |

## The Diels–Alder Reaction

In the Diels–Alder reaction, fluorinated compounds act both as dienes and as dienophilic components. The formation of cyclobutane derivatives which has just been mentioned is a dangerous competitive side-reaction in the Diels-Alder synthesis with fluorinated olefins. Therefore only such fluorinated components that do not dimerize to cyclobutanes can be used successfully. Fluoroprene reacts with acrolein or methyl vinyl ketone and gives 4-fluoro-$\Delta^3$-tetrahydrobenzaldehyde or 4-fluoro-$\Delta^3$-tetrahydroacetophenone, respectively (*835*).

The reactions of tetrachloro-1,1-difluoro-2,4-cyclopentadiene (Equation 636) with some unsaturated compounds are shown in Table 46 (*676*).

TABLE 46

*Results of the Diels–Alder Reaction of 1,1-Difluorotetrachloro-2,4-cyclopentadiene with Some Unsaturated Compounds (676)*

| Unsaturated Compound | Yield of th Adduct % | Unsaturated Compound | Yield of the Adduct % |
|---|---|---|---|
| 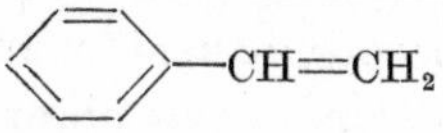 ⬡—CH=CH$_2$ | 95 | CH$_2$=CH·COOH | 72 |
| 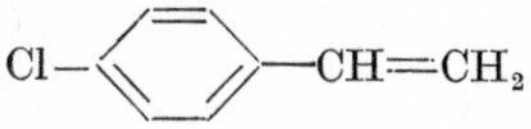 Cl—⬡—CH=CH$_2$ | 75 | CH—CO<br>‖    O<br>CH—CO | 88 |
| CH=CH<br>\|    CH$_2$<br>CH=CH | 74 | 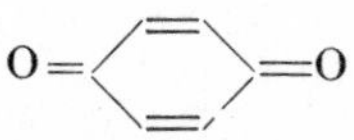 O=⬡=O | 56 |

The addition product of tetrachloro-1,1-difluoro-2,4-cyclopentadiene and methyl acetylenedicarboxylate suffers pyrolytic split of the difluoromethylene bridge and an aromatic derivative is formed as final product (*676*).

**636**                                                                   [*676*]

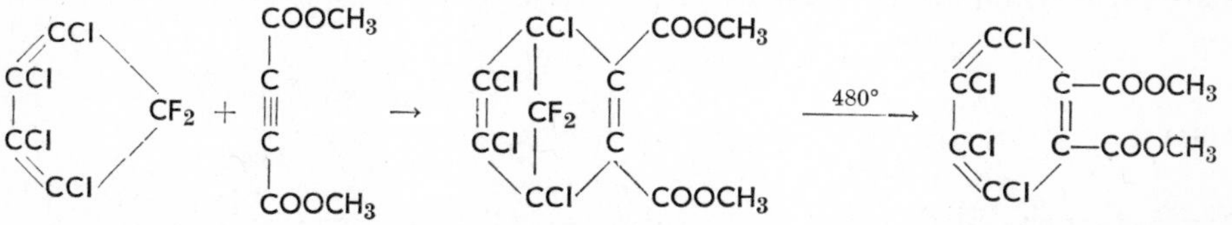

As dienophilic components, the only fluorinated olefins which can be used are those which react more rapidly with the dienes than with themselves and which do not possess too strong a tendency to form four-membered rings, as in the case of perfluoroacrylonitrile and butadiene (*613*).

**637**                                                                   [*613*]

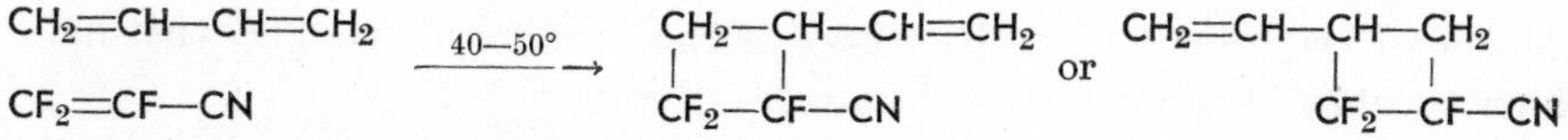

Successful Diels–Alder syntheses have been carried out between perfluoropropene and butadiene, cyclopentadiene or anthracene in yields of 69, 64 and 76%, respectively (*647*).

**638** 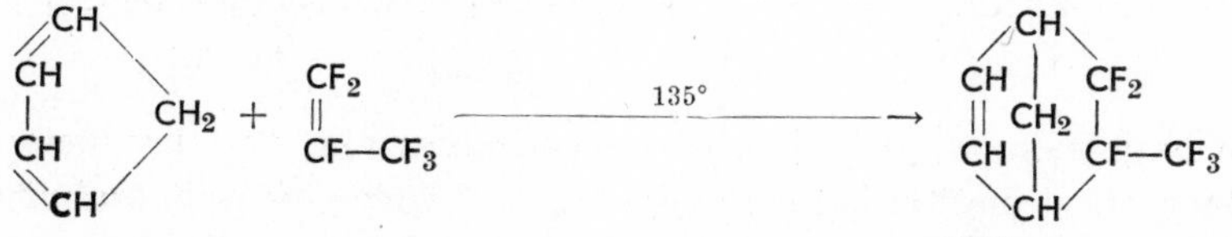                                                                   [*647*]

69%

The reactivity of the fluorinated dienophiles toward the mentioned dienes drops along the series (647):

**639**        $CF_3-CF=CF_2 > C_2F_5-CH=CH_2 > CF_3-CH=CH_2 >$        [647]

$$CH_2F-CH=CH_2 > \underset{CH_3}{\overset{CF_3}{>}}C=CH_2$$

In the reaction of cyclopentadiene with $\gamma,\gamma,\gamma$-trifluorocrotonic acid, a somewhat unusual ratio between the resulting *exo*- and *endo*-isomers was observed (648).

**640**

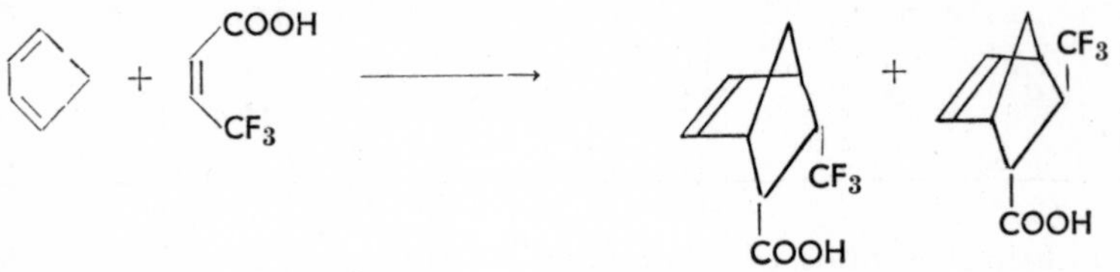

66%   exo        33%   endo        [648]

The reaction of butadiene with trifluoroacetonitrile leads by simultaneous dehydrogenation to an aromatic derivative, $\alpha$-trifluoromethylpyridine (540).

**641**

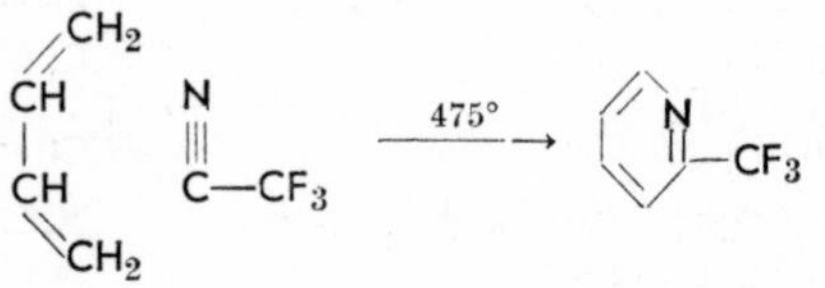

conversion   13%
yield        88%        [540]

## Additions to Triple Bonds

As for the olefins, a distinction will be made between reactions with or without formation of a new carbon–carbon bond.

*Additions forming no carbon–carbon bonds.* Additions of halogens and hydrogen halides to acetylenes were discussed in the paragraph dealing with halogenations (p. 176, 188).

*Hydration* of fluorinated acetylenes affords aldehydes or ketones. Addition of water to trifluoropropyne in the presence of mercuric sulphate gives a mixture of $\beta,\beta,\beta$-trifluoropropionaldehyde and $\alpha,\alpha,\alpha$-trifluoroacetone (379, 464).

**642**        $CF_3-C\equiv CH \xrightarrow[20°]{H_2SO_4/HgSO_4} CF_3-CH_2-CHO + CF_3-CO-CH_3$        [379]

28%        61%

The hydration of pentafluoro-1-butyne gives also two products: 16% of pentafluoroethyl methyl ketone and 17% of $\beta,\beta,\gamma,\gamma,\gamma$-pentafluorobutyric acid (379).

The course of the reactions of organic compounds with trifluoropropyne is consistently simpler. *Addition of methanol* or *ethanol* in the presence of sodium gives exclusively enol ethers of $\beta,\beta,\beta$-trifluoropropionaldehyde, as could be expected from analogy with 1,1,1-trifluoropropene (*363*) (p. 247).

643  $\quad\quad$ CF$_3$—C≡CH  $\xrightarrow[20°]{\text{CH}_3\text{OH, Na}}$  CF$_3$—CH=CH—OCH$_3$  92%  $\quad\quad$ [*363*]

*Amines (363)* and *acids (478)* react in a similar manner.

644  $\quad\quad$ CF$_3$—C≡CH  $\xrightarrow{(\text{C}_2\text{H}_5)_2\text{NH}}$  CF$_3$—CH=CH—N(C$_2$H$_5$)$_2$  28%  $\quad\quad$ [*363*]

645 CF$_3$—C≡C—CF$_3$  $\xrightarrow[55-60°]{\text{CH}_3\text{COOH, CH}_3\text{COONa}}$  CF$_3$—CH=C—CF$_3$  $\quad$ [*478*]
$\quad\quad\quad\quad\quad\quad\quad\quad\quad\quad\quad\quad\quad\quad\quad\quad\quad\quad\quad\quad\quad$ | $\quad$ 40—55%
$\quad\quad\quad\quad\quad\quad\quad\quad\quad\quad\quad\quad\quad\quad\quad\quad\quad\quad\quad\quad\quad$ OCO—CH$_3$

When the addition of acetic acid is carried out at elevated temperatures, acetyl fluoride and $\alpha,\alpha,\alpha$-trifluoroacetone are formed (*478*).

*Additions accompanied by the formation of carbon-carbon bonds.* Under free radical conditions, acetylenes accept *halogenated paraffins* in the same way as olefins, but instances are not as numerous. Acetylene, propyne and 1,1,1-trifluoropropyne were allowed to react with trifluoroiodomethane, or less frequently with trichloroiodomethane, perfluoroethyl iodide, or perfluoropropyl iodide. Table 47 gives references to the papers in which these combinations have been described.

TABLE 47

*Addition of Haloparaffins to Acetylenes*

| Haloparaffin<br>Acetylene*) | CCl$_3$I | CF$_3$I | C$_2$F$_5$I | C$_3$F$_7$I |
|---|---|---|---|---|
| | | Reference Number on p. 358. | | |
| CH≡CH | | 358 | 379<br>380 | 357 |
| CH≡C—CH$_3$ | | 381<br>615 | | |
| CH≡C—CF$_3$ | 366 | 363<br>615 | | 615 |

*) Bold-face typed carbon combines with the paraffin radical carbon.

The direction of halogenoparaffin addition to terminal acetylenes has so far been uniform. Both propyne and trifluoropropyne accept the trihalogeno-alkyl group at the open end of their carbon chain (*363, 366, 381, 615*).

The reactions with acetylenes are carried out at temperature around 200°C and better still under ultra-violet irradiation. They do not take place at room temperature. As with olefins, the normal addition products of acetylenes are accompanied by telomers, products of the reaction of the primarily formed radical with two or more acetylene molecules (*358*) (p. 253).

646

$$\underset{h\nu}{\overset{CF_3I}{\longleftarrow}} \quad CH\equiv CH \quad \underset{220°}{\overset{C_3F_7I}{\longrightarrow}} \qquad [357,358,380]$$

$$C_2F_5I \,\big|\, 240°$$

$$CF_3-CH{=}CHI \;\; 79\% \qquad C_2F_5-CH{=}CHI \;\; 72\% \qquad C_3F_7-CH{=}CHI \;\; 89\%$$

647

$$CH_3-C{\equiv}CH \;\; \underset{h\nu}{\overset{CF_3I}{\longrightarrow}} \;\; CH_3-CI{=}CH-CF_3 \;\; 91\% \qquad\qquad [381]$$

648

$$\underset{h\nu}{\overset{CCl_3I}{\longleftarrow}} \quad CF_3-C{\equiv}CH \quad \overset{CF_3I}{\longrightarrow} \qquad [366,615]$$

$$CF_3-C\,I{=}CH-CCl_3 \;\; 74\% \qquad\qquad CF_3-CI{=}CH-CF_3 \;\; 38\%$$

The addition reactions of acetylene with perfluoroalkyl iodides have considerable preparative importance. They allow the preparation in a few operations of various substituted fluorinated olefins or paraffins, whose synthesis by other methods would be tedious and laborious (p. 131,132).

## Other Addition Reactions of Fluorinated Derivatives

A very significant reaction of polyfluoro and especially perfluoroaldehydes and ketones is the *addition of water* to the carbonyl group and the formation of hydrates (*403*). Such hydrates are much more stable than, for example, the well known chloral hydrate. Often the hydrates of fluorinated aldehydes and ketones are obtained as primary products in various reactions (*523, 1001*), and their dehydration presents in some cases, such as that of hexafluoroace-tone, a very difficult problem and requires the most energetic dehydrating agents (phosphorus pentoxide) (*479*).

649

$$\underset{C_3F_7}{\overset{C_3F_7}{>}}C{=}O \;+\; H_2O \;\;\rightarrow\;\; \underset{C_3F_7}{\overset{C_3F_7}{>}}C\underset{OH}{\overset{OH}{<}} \qquad\qquad [403]$$

*Addition of diazomethane* to fluorinated alcohols gives fluorinated ethers (*480*), if the hydrogen in the hydroxyl group is sufficiently acidic. This requirement is met in the alcohols with an adjacent perfluoroalkyl group. The yields can be increased by the addition of aluminium isopropoxide (*480*).

650   $CF_3-CH_2OH + CH_2N_2$ ⟨ 27% / Al[OCH(CH₃)₂] → 75% ⟩ $CF_3-CH_2-OCH_3$   [*480*]

Finally the addition of *trifluoromethyl iodide to nitric oxide* should be mentioned. In the presence of mercury it gives trifluoronitrosomethane at room temperature, and trifluoronitromethane at 100°C (*43, 536*). This has already been discussed in the paragraph on nitration and nitrosation (p. 192).

A non-stoichiometric addition of *trifluoromethyliodide to phosphorus and arsenic* yields trifluoromethyl iodophosphines (*147*) and trifluoromethyl iodoarsines (*240*).

651   $[CF_3-COOAg + I_2]$ → $[CF_3I]$ $\xrightarrow[red]{P}$ $CF_3PI_2$ + $(CF_3)_2PI$ + $(CF_3)_3P$   [*147*]
                                                     6.4%        11%          9.5%

652                $CF_3I + As$ → $CF_3AsI_2$                                    [*240*]

## ELIMINATIONS

Eliminations are characterized by loss of two elements, either from the same carbon atom (*1,1*-elimination), from two carbon atoms of the same molecule (*1,2*- or *1,3*-elimination), or else from two or more various molecules. The elimination reactions will include dehalogenation, dehydrohalogenation, dehydration, decarboxylation and decarbonylation.

### Dehalogenation

Dehalogenation of fluorinated halogen derivatives occurs most frequently at two adjacent carbon atoms and results in the formation of a double or triple bond. The splitting off of three pairs of halogens from a fluorinated cyclane leads to fluorinated aromatics. If a halogen molecule is split off from two more distant carbon atoms, a ring is formed. Finally dehalogenation may occur intermolecularly, i.e. halogens are eliminated from two or more molecules and these combine in a new carbon–carbon bond. This type of dehalogenation is represented for example by the Wurtz-Fittig reaction. All types of dehalogenations can be effected by the action of metals, most frequently zinc or magnesium, less often by copper, nickel, by organometallic compounds, or catalytic hydrogenation.

*Intramolecular dehalogenation.* Splitting off of two halogens, chlorine, bromine or iodine atoms, or their combinations, from two adjacent carbon atoms results in the formation of an olefin. Owing to the firmer carbon–fluorine bond, the other halogens are eliminated preferentially. The reactions are carried out by treating the halogenated compounds with granulated or mossy zinc or better still zinc dust in alcohol, ether, dioxane or other solvents. Sometimes the reaction requires activation of the zinc by small amounts of hydrogen chloride, zinc chloride, stannous chloride or sodium iodide. Once started, the dehalogenation proceeds smoothly, sometimes vigorously, and almost quantitatively. It is therefore used very often not only for the preparation, but also for the isolation and purification of fluorinated olefins and other unsaturated compounds.

Dehalogenation of 1,2-dibromo-1-fluoroethane gives vinyl fluoride (*986*), that of 1,2-dichloro-1,1-difluoroethane vinylidene fluoride (*644*).

**653** $CF_2Cl—CH_2Cl$ $\xrightarrow[\substack{CH_3CONH_2,C_4H_9CHEtCH_2OH \\ 145°}]{Zn/NaI}$ $CF_2{=}CH_2$ conversion 76% yield 97% [*644*]

Treatment of 1,2-dichloro-1,1,2-trifluoroethane with zinc affords trifluoroethylene (*202*).

**654** $CF_2Cl—CHFCl$ $\xrightarrow[\substack{polyglycol\ ether\ stearate,\ H_2O \\ 90°,\ 16\ hrs.}]{Zn/SnCl_2}$ $CF_2{=}CHF$ conversion 42·9% [*202*] yield 99%

Dehalogenation of 1,1,2-trichlorotrifluoroethane to chlorotrifluoroethylene (*65, 202, 621*) will be described in detail in Chapter X which deals with plastics (Equation 732, p. 343). Dehalogenation of 1,2-dichlorotetrafluoroethane (*73, 202, 621*) is more difficult and requires special conditions, such as the use of acetic anhydride as solvent (*434*).

The excellent results given by zinc dehalogenations make other dehalogenating reagents superfluous in the laboratory. In industrial production, however, dehalogenation with zinc is not ideal. Handling of large amounts of zinc is inconvenient, and zinc is not cheap. Dehalogenation with iron, although patented (*917*), does not give satisfactory results (*877*). On the other hand, there is much to be expected from catalytic dehalogenation with hydrogen over various metals (*114, 171, 690, 891*) (Table 48).

TABLE 48

*Dehalogenation of 1,1,2-Trichlorotrifluoroethane to Chlorotrifluoroethylene by Means of Hydrogen and Various Catalysts (690)*

| Catalyst | Substrate | Temp., °C | Yield, % |
|---|---|---|---|
| Cu (gauze) | — | 580 | 78 |
| Cu | MgO | 550 | 76·6 |
| Cu—Co | MgO | 480 | 93·2 |
| Co—Cu | $MgF_2$ | 470 | 96 |
| Co | MgO | 540 | 86·4 |
| Pt | $Al_2O_3$ | 540 | 77·4 |
| Ag (gauze) | — | 560 | 80·1 |

Though practical experience with this method is still scanty, the results look promising.

Dehalogenation of fluorinated vicinal dihalogenoolefins affords fluorinated acetylenes in good yields (*441, 462*) (Equation 223, p. 135).

**655** $\qquad CF_3—CBr{=}CHBr \xrightarrow[\text{reflux}]{\text{Zn, } C_2H_5OH} CF_3—C{\equiv}CH \ 97\%$ $\qquad$ [*462*]

Elimination of fluorine is much rarer than that of chlorine, bromine or iodine, but it has been effected in a few instances. Although other halogens are eliminated preferentially, fluorine, either alone or together with an atom of another halogen, is split off when there is no other possibility. Thus fluorine and chlorine were eliminated by means of zinc in alcohol (*629, 651*),

**656** 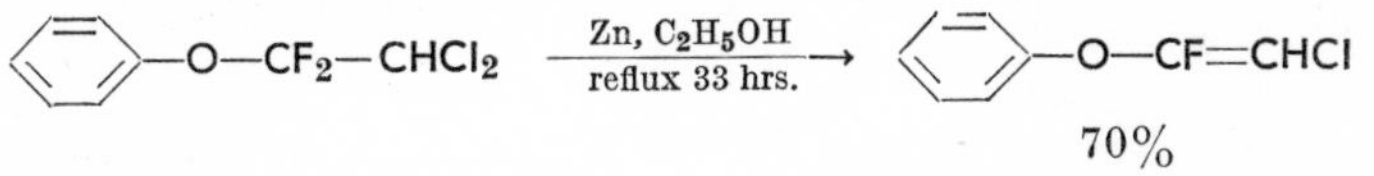 [*629*]

70%

**657** 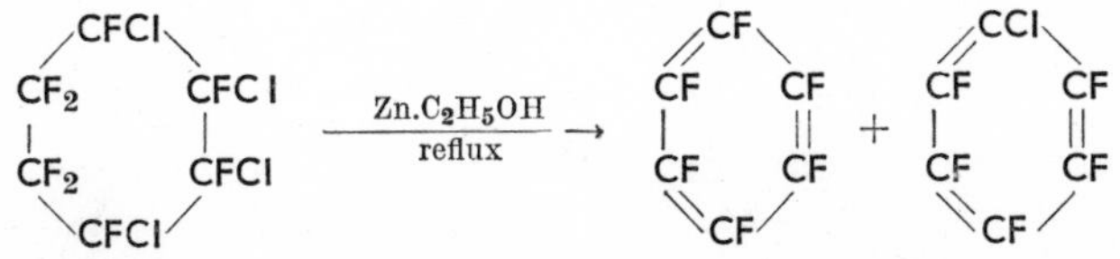 [*651*]

fluorine and iodine by treatment with zinc in acetic acid (*633*), magnesium (*310*), or with phenylmagnesium bromide (*1007*).

**658** $\quad CF_3—CF_2—CF_2—CH_2I \xrightarrow[\text{reflux}]{\text{Zn, } CH_3COOH} CF_3—CF_2—CF{=}CH_2 \ 93\%$ $\quad$ [*633*]

**659** $\qquad CF_3—CH_2I \xrightarrow{\text{Mg,}(C_2H_5)_2O} CF_2{=}CH_2 \ 90\%$ $\qquad$ [*310*]

**660** $\quad CHF_2—CH_2I \xrightarrow[\text{reflux}]{C_6H_5MgBr, (C_2H_5)_2O} CHF{=}CH_2 \ 98\%$ $\quad$ [*1007*]

Surprising results were obtained by passing fluorinated hydroaromatic compounds over a nickel or iron gauze at temperatures ranging from 250 to 660°C. Under these conditions fluorine was readily liberated and an aromatic perfluoro-derivative was formed. The metal gauze is regenerated by passing a stream of hydrogen at temperatures at 300–600° (29, 145, 303, 1028).

661

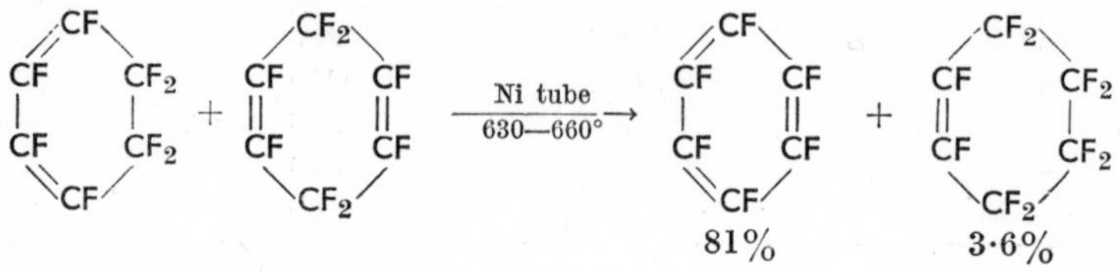

$$\text{(structure)} \quad + \quad \text{(structure)} \quad \xrightarrow[\text{630—660°}]{\text{Ni tube}} \quad \text{(structure)} \quad + \quad \text{(structure)} \qquad [29]$$

81%    3·6%

The reaction is general for the aromatization of saturated perfluorohydro-aromatic derivatives (303). In this respect it parallels the classical dehydrogenation of hydroaromatic systems to aromatic compounds. Perfluorocyclo-hexane at 550°C over iron gauze affords hexafluorobenzene (1028), perfluoro-decalin at 500°C perfluoronaphthalene in a yield of 51% (303), perfluoro-piperidine perfluoropyridine (145).

662

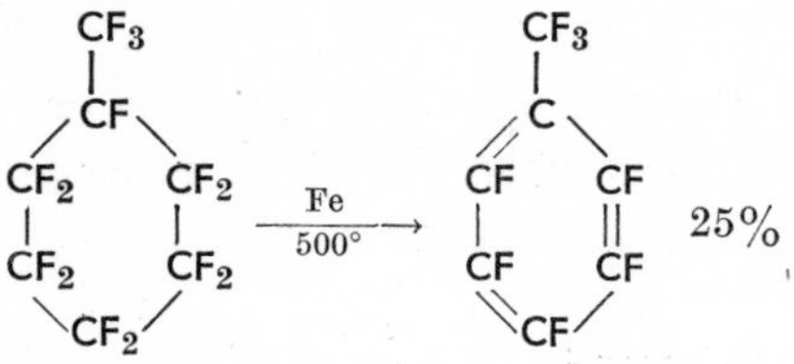

$$\text{(structure)} \quad \xrightarrow[\text{500°}]{\text{Fe}} \quad \text{(structure)} \quad 25\% \qquad [303]$$

663

$$\text{(structure)} \quad \xrightarrow[\text{560°}]{\text{Ni}} \quad \text{(structure)} \qquad [145]$$

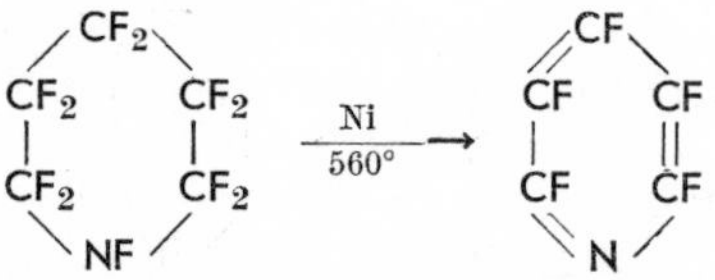

Dehalogenation, accompanied by the formation of a three-membered ring, has been mentioned elsewhere (669) (Equation 486, p. 217).

An interesting example of dehalogenation occuring at one carbon atom is the pyrolysis of tribromofluoromethane which gives hexafluorobenzene (97, 208, 209, 423) (Equation 232, p. 139).

*Intermolecular dehalogenation.* Synthetic dehalogenations, reactions in which halogen elimination results in combining two or more molecules, are, with a few exceptions, of much less practical value than the dehalogenations leading to olefins, acetylenes or aromatic fluoroderivatives.

Combination occurs in the treatment of perfluoropropyl iodide with zinc in acetic anhydride (*439*), and of perfluoroallyl iodide with zinc in dioxane (*266*).

**664**  $2 \; CF_2\!\!=\!\!CF\!\!-\!\!CF_2I \; \xrightarrow[\text{reflux}]{\text{Zn, dioxane}} \; CF_2\!\!=\!\!CF\!\!-\!\!CF_2\!\!-\!\!CF_2\!\!-\!\!CF\!\!=\!\!CF_2 \; 83\% \qquad [\textit{266}]$

Under similar conditions, zinc doubles the molecule of 1,2-dichloro- 1,2,2-trifluoro-1-iodoethane to 1,2,3,4-tetrachlorohexafluorobutane (*471*). The same product is obtained from the same starting material by means of mercury or ultra-violet irradiation (*365*). When the reaction is carried out in dioxane at 100°C, intramolecular dehalogenation follows and the final product is perfluoro-1,3-butadiene (*365*).

**665**  $\qquad\qquad$ Zn, (CH$_3$CO)$_2$O, CH$_2$Cl$_2$ $\qquad\qquad\qquad$ [*365, 471*]

$\nearrow \qquad\qquad\qquad\qquad \searrow 51\%$

$CF_2Cl\!\!-\!\!CFClI \; \xrightarrow[\text{hv 20°}]{\text{Hg}} \; 82\% \; CF_2Cl\!\!-\!\!CFCl\!\!-\!\!CFCl\!\!-\!\!CF_2Cl \; \xrightarrow{\text{Zn, C}_2\text{H}_5\text{OH}} \; CF_2\!\!=\!\!CF\!\!-\!\!CF\!\!=\!\!CF_2 \; 98\%$

$\Big| \quad \searrow \underset{25\text{---}50°}{\text{Zn, dioxane}} \nearrow \qquad\qquad \searrow \qquad\qquad \nearrow \qquad \Big\uparrow \quad 60\%$

$\qquad\qquad\qquad\qquad\qquad \underset{100°}{\text{Zn, dioxane}}$

Practical reasons led to attempting the dehalogenation of dichlorodifluoromethane to obtain fluorinated ethanes or ethylenes. High temperature reactions of this compound with hydrogen (*80*) (Equation 283, p. 160) or with tellurium (*24*), give but low yields of desired products.

**666**  $CCl_2F_2 \; \xrightarrow[500°]{\text{Te}} \; CClF_2\!\!-\!\!CClF_2 \; + \; CF_2Cl\!\!-\!\!CF_2\!\!-\!\!CF_2Cl \; + \; CF_2\!\!=\!\!CF_2 \qquad [\textit{24}]$
$\qquad\qquad\qquad\qquad\quad 61\cdot2\% \qquad\qquad\quad 22\cdot4\% \qquad\qquad\quad 16\cdot4\%$

Aromatic fluorohalogenoderivatives are able to combine on treatment with metals. Thus boiling of bis(*p*-fluorophenyl)dichloromethane with copper dust in pyridine affords 60% of tetrakis(*p*-fluorophenyl)ethylene (*542*), and heating of *p*-dibromotetrafluorobenzene with copper dust at 200°C gives *p,p'*-dibromoperfluoroquaterphenyl (*422*).

**667**  $4 \; Br\!\!-\!\!\langle\text{ring}\rangle\!\!-\!\!Br \; \xrightarrow[290°, \, 80 \text{ hrs.}]{\text{Cu}} \; Br\!\!-\!\!\langle\text{rings}\rangle\!\!-\!\!Br \qquad [\textit{422}]$

## Dehydrohalogenation

Dehydrohalogenations from two adjacent carbon atoms lead to doubly- or triply-bonded compounds and are an important source of fluorinated olefins and acetylenes. Some dehydrohalogenations take place spontaneously, others require elevated temperatures, and the majority occur in a basic medium, which takes up the halide split-off. In addition to aqueous or alcoholic alkalies, organic bases such as pyridine, quinoline, triethylamine etc. are often used for this purpose.

Elimination of hydrogen halides from dihalogenofluoroparaffins gives fluorinated olefins and acetylenes (361).

$$CF_3—CH_2—CHCl_2 \xrightarrow[\text{reflux}]{\text{KOH}} CF_3—CH=CHCl \quad 52\%$$

**668**
$$10\%$$
$$+ \quad CF_3—C\equiv CH \quad [361]$$

$$CF_3—CH_2—CHBrI \xrightarrow[\text{reflux}]{\text{KOH}} CF_3—CH=CHBr \quad 66\%$$
$$11\%$$

From 2,2,3,3-tetrachloro-1,1-difluoropropane hydrogen chloride is liberated in such a manner that the hydrogen atom is detached from the carbon carrying two chlorine atoms, while the hydrogen bonded to the carbon with two fluorine atoms remains untouched (203).

**669**
$$CHCl_2—CCl_2—CHF_2 \xrightarrow{\text{NaOH, CH}_3\text{OH}} CCl_2=CCl—CHF_2 \quad [203]$$

Of theoretical interest and practical importance is the 1,1-elimination of hydrogen chloride from chlorodifluoromethane. It is carried out pyrolytically, best at 650–750°C and a slightly reduced pressure in a graphite, silver or platinum furnace. Under these conditions hydrogen chloride splits off exclusively (with only 0,5–3,4% of hydrogen fluoride depending on the material of the reaction tube). The intermediate divalent radical $—CF_2—$ dimerizes to tetrafluoroethylene, and to a lesser extent, polymerizes to compounds containing three, four and more carbon atoms (799).

**670**                                                                         [799]

$$CHClF_2 \xrightarrow[\text{0,5 atm.}]{650—800°} \begin{array}{c} \cdot[CF_2]\cdot \\ + \\ HCl \end{array} \rightarrow CF_2=CF_2 \quad \begin{array}{c} \text{conversion } 25-30\% \\ \text{yield } 90\% \end{array} + \frac{Cl(CF_2)_nCl}{n = 2 - 14}$$

If both hydrogen fluoride and hydrogen chloride can be split out, hydrogen chloride is eliminated preferentially (856).

**671** ⬡—CFCl—CHF_2 $\xrightarrow{\text{NaOH}}$ ⬡—CF=CF_2 17%        [856]

According to older literary data, hydrogen fluoride splits off spontaneously from some higher fluoroalkanes. More recent investigations in the field of fluoroparaffins find alkyl fluorides perfectly stable (*501, 821*). Hydrogen fluoride is however liberated on heating fluoroparaffins to high temperatures (*22*).

**672**    $CH_3—CF_2—CH_3 \xrightarrow{731°} CH_2{=}CF—CH_3$    conversion 42%    [*22*]
yield 90%

Alkaline elimination of hydrogen fluoride offers no difficulties. Its ease depends on the nature of the fluorinated compound. While fluorocyclohexane stands boiling with methanolic potassium hydroxide (*104*), 1H,2H-hexafluorocyclobutane is converted by aqueous potassium hydroxide at room temperature to 1H-pentafluoro-1-cyclobutene (*151*).

**673**    $\xrightarrow[15°, 5 \cdot 5 \text{ hrs.}]{\text{KOH aq.}}$ 64%    [*151*]

In some cases organic bases were used successfully for the elimination of hydrogen fluoride (*629*).

**674**    [*629*]

$\xrightarrow[\text{KOH, reflux}]{C_6H_5CH_2N(CH_3)_3OH}$ 29%

From 1,2-dihydrodecafluorocyclohexane (*263, 890, 961*) and 1,3-dihydrodecafluorocyclohexane (*264*), either one or two molecules of hydrogen fluoride can be split out.

**675**    $\xrightarrow[100°, 3 \text{ hrs.}]{\text{KOH aq.}}$ +    [*961*]
80%    16%

Elimination of hydrogen fluoride from 1,2,4,5-tetrahydrooctafluorocyclohexane affords pentafluorobenzene (*774, 967*) (Equation 230, p. 138), elimination of hydrogen fluoride from 1,2,4-trihydrononafluorocyclohexane gives hexafluorobenzene (*316, 1028*) (Equation 233, p. 139).

## Dehydration

The ease of elimination of water from fluorinated compounds is determined by the number and location of fluorine atoms in the molecule and by the distance of these atoms from hydrogen and hydroxyl. In polyfluoro-derivatives and perfluoro-derivatives, in which hydrogen and hydroxyl are close to $-CF_2-$ or $-CF_3$ groups, the elements of water are held with extraordinary obstinacy. This effect accounts for the extreme stability of the hydrates of fluorinated aldehydes and ketones and their difficult dehydration (479).

Also polyfluorinated alcohols withstand the action of strong dehydrating reagents even in drastic conditions. 1,1,1-Trifluoro-2-propanol was recovered in a 50% yield after being heated with concentrated sulphuric acid to 180–190°C, whereas the second half of the alcohol was transformed to its sulphate (1005). The same compound resisted heating with phosphorus pentoxide at 150–200°C (627).

676
$$CF_3-CH-CH_3 \quad \xrightarrow[\substack{180-190° \\ 1\cdot5 \text{ hrs}}]{H_2SO_4} \quad CF_3-CH-CH_3 \quad + \quad 50\% \quad \text{recovered} \qquad [1005]$$
$$\quad\quad\quad |\qquad\qquad\qquad\qquad\qquad\qquad | $$
$$\quad\quad\quad OH \qquad\qquad\qquad\qquad\qquad OSO_3H$$

Methylperfluoropropylcarbinol did not dehydrate even on heating with sulphuric acid and phosphorus pentoxide at 300°C (627).

A suitable method for the preparation of unsaturated compounds from polyfluoroalcohols is thermal decomposition of their acetates. In this way 1,1,1-trifluoro-2,4-pentanediol (454) and trifluoroacetone cyanohydrin (150) were smoothly converted to the corresponding olefinic compounds.

[454]

$$CF_3-CH-CH_2-CH-CH_3 \quad \xrightarrow[H_2SO_4]{(CH_3CO)_2O} \quad CF_3-CH-CH_2-CH-CH_3 \qquad 88\%$$
$$\quad\quad\;\; | \qquad\qquad | \qquad\qquad\qquad\qquad\qquad\qquad\quad | \qquad\qquad\quad |$$
$$\quad\quad\;\; OH \qquad\quad OH \qquad\qquad\qquad\qquad\qquad\; OCOCH_3 \;\; OCOCH_3$$

677

$$\xrightarrow[]{470°} \quad CF_3-CH=CH-CH=CH_2$$
$$75\%$$

## Decarboxylation

The loss of a carboxyl group from fluorinated acids may yield fluorinated paraffins, fluorinated olefins, fluorinated alkyl halides, or finally fluorinated paraffins containing twice as many carbon atoms as were in the original alkyl group. The results of the decarboxylation can be influenced, at least to a certain extent, by the choice of experimental conditions. Frequently, however, the reaction is determined by the nature of the starting material, especially by the cation of the acid.

Free perfluorocarboxylic acids lose hydrogen fluoride and carbon dioxide and afford a perfluoroolefin when heated to higher temperatures (876).

678
$$\langle\,\Phi\,\rangle-COOH \quad \xrightarrow[2 \text{ hrs.}]{550°} \quad \langle\,\overline{\overline{\Phi}}\,\rangle + CO_2 + HF \qquad [876]$$

Heating of free perfluorocarboxylic acids with bromine produces perfluoroalkyl bromides (611) (Equation 365, p. 182).

Electrolysis of fluorinated acids in the presence of sodium methoxide affords fluorinated paraffins with twice as many carbon atoms as the alkyl radical of the acid (*823*) (Equation 194, p. 127).

Silver salts of perfluorocarboxylic acids react with halogens with loss of carbon dioxide and formation of perfluoroalkyl halides (*364*) (Equations 366–370, p. 183–185). Treatment of silver perfluorocyclohexanecarboxylate with water leads mainly to perfluorocyclohexene besides smaller amounts of perfluorodicyclohexyl (*132*).

679   ⬡–COOAg   $\xrightarrow[20°,\,7\ \text{days}]{\text{H}_2\text{O}}$   ⬡   30%   +   ⬡–⬡   10%   [*132*]

Silver salts of perfluoroalkyl carboxylic acids decompose at higher temperatures to di(perfluoroalkyls), probably via the corresponding perfluoroacyl anhydrides (*380, 557, 609*).

680                                                                  [*557*]

$2\ CF_3-CF_2-CF_2-COOAg \xrightarrow{260-275°} CF_3-CF_2-CF_2-CF_2-CF_2-CF_3$   90—95%

Potassium salts of perfluorocarboxylic acids heated to 170–190°C in ethylene glycol afford 1-hydroperfluoroparaffins (*609*). Subjected to thermal decomposition at 165–270°C, they yield perfluoroolefins with terminal double bonds, or with double bond shifted inward (*130, 380, 386, 609*) (Equation 530, p. 232).

681   $\dfrac{\text{(CH}_2\text{OH)}_2}{170-190°}\Big\downarrow$   $CF_3-CF_2-CF_2-CF_2-COOK$   $\overline{\phantom{xx}165-200°\phantom{xx}}\Big\downarrow$   [*130,609*]

$CF_3-CF_2-CF_2-CHF_2$   84%                             $CF_3-CF_2-CF{=}CF_2$   7%

                                                          $CF_3-CF{=}CF-CF_3$   29%

The thermal decomposition of dry sodium salts of perfluorocarboxylic acids carried out at 200–450°C gives unambigously perfluoroolefins with terminal double bond (*130, 343, 364, 386, 609*).

The yields of this reaction can be increased by working at a reduced pressure (*364, 373*).

Heating of sodium trifluoroacetate with sodium hydroxide yields tetrafluoroethylene (*609*). The same product is obtained by thermal decomposition of dry sodium perfluoropropionate (*609*).

682   $CF_3-CF_2-COONa \xrightarrow{230-260°} CF_2{=}CF_2 \xleftarrow[270°]{\text{NaOH}} CF_3-COONa$   [*609*]

                                     90%        32%

*Procedure 34.*

*Decarboxylation of Sodium Salts of Perfluorocarboxylic Acids*

*Preparation of 1-Perfluorobutene from Sodium Perfluorovalerate (130)*

Sodium perfluorovalerate dried in a vacuum at 80–100°C (615 g, 2·15 moles) is heated in a bath in a one liter flask connected to a scrubber with 30% potassium hydroxide (to absorb the carbon dioxide evolved). The scrubber is further connected by means of a phosphorus-pentoxide tube to a dry-ice trap. The decomposition is carried out at 290–300°C and yields 386 g (1·93 moles) (90%) of a crude product whose redistillation gives 50% of pure 1-perfluorobutene, b.p. 0°.

Decomposition of sodium salts of perfluorodicarboxylic acids produces either unsaturated perfluorocarboxylic acids, or perfluorodiolefins (*373*).

$NaOCO-CF_2-CF_2-CF_2-CF_2-COONa$

$$\begin{array}{l} \mathbf{683} \left| \begin{array}{c} \overline{\phantom{160-450°}} \ \ \ \ \ \ \ \ \frac{160-450°}{10^{-2}\ mm} \ \ \ \ \ \ \ \downarrow\ \ 61\% \ \ \ \ \ \ [373] \\ \xrightarrow[10^{-2}\ mm]{450°} \ \ \ CF_2{=}CF-CF_2 \ \ \ CF_2-COONa \ + \ CF_2{=}CF-CF{=}CF_2 \\ \ \ \ \ \ \ \ \ \ \ \ \ \ \ \ \ \ \ \ \ 21\% \ \ \ \ \ \ \ \ \ \ \ \ \ \ \ \ \ \ \ \ \ \ \ \ \ 25-37\% \end{array} \right. \end{array}$$

The yields of the decomposition of some sodium and potassium salts of perfluorocarboxylic acids are listed in Table 49 (*609*).

TABLE 49

*Decomposition of Potassium Salts of Perfluorocarboxylic Acids in Ethylene Glycol at 170–190°, and of Dry Sodium Salts of Perfluorocarboxylic Acids at 230–260° (609)*

| Product of Decomposition of the Potassium Salt | Yield % | Perfluoro-carboxylic Acid | Product of Decomposition of the Sodium Salt | Yield % |
|---|---|---|---|---|
| $C_2F_5H$ | 98 | $C_2F_5COOH$ | $C_2F_4$ | 90 |
| $C_3F_7H$ | 97 | $C_3F_7COOH$ | $C_3F_6$ | 97 |
| $C_4F_9H$ | 84 | $C_4F_9COOH$ | $C_4F_8$ | 91 |
| $C_5F_{11}H$ | 80 | $C_5F_{11}COOH$ | $C_5F_{10}$ | 90 |
| $C_7F_{15}H$ | 60 | $C_7F_{15}COOH$ | $C_7F_{14}$ | 86 |

Salts other than those of sodium or potassium are unsuitable for the preparation of perfluoroolefins as shown in the results of decomposition of various perfluorobutyrates (*609*) (Table 50).

TABLE 50

*Decomposition Temperatures and Yields of Perfluoropropene in Heating Perfluorobutyrates*
*(609)*

| Cation | Decomposition Temp., °C | Yield of Perfluoropropene, % |
|---|---|---|
| Li | 240—250 | 20 |
| Na | 230—260 | 97 |
| K | 215—235 | 98 |
| Ca | 275—300 | <10 |
| Sr | 275—285 | 25 |
| Ba | 265—275 | 78 |
| Mg | 275—300 | < 5 |
| Pb | 300—305 | <10 |
| Al | 250 | < 5 |
| Ag | 300—320 | 45 |
| NH₄ | 180—200 | 0 |

## Decarbonylation

The elimination of carbon monoxide is similar to the elimination of carbon dioxide. Thus heating of trifluoroacetyl bromide (612), trifluoroacetyl iodide (359) and perfluorobutyryl bromide (612) to 500–650°C produces perfluoroalkylhalides as main reaction products.

**684** $\qquad CF_3-COBr \xrightarrow{650°} CF_3Br$ [612]

Heating of trifluoroacetylchloride and of some higher fluorinated acyl chlorides with potassium iodide at 200°C also results in the elimination of carbon monoxide and in the formation of fluorinated alkyl iodides (594).

**685** $\quad CF_3-COCl \xrightarrow[200°, 6\,hrs]{KI} CF_3I$ conversion 41%, yield 69% [594]

Both decarboxylation and decarbonylation takes place in thermal decomposition of fluoro- and difluoromaleic anhydride, which affords mono- and difluoroacetylene, respectively (706, 707) (Equations 221, 222, p. 135).

### PYROLYSIS

Polyfluoroparaffins and especially fluorocarbons and other perfluoroderivatives show remarkable heat resistance. They usually do not decompose at temperatures below 300°C; intentional decomposition, however, is carried out at temperatures of 500–800°C. Under such conditions all possible splits occur

and a complex mixture results which is difficult to separate. Perfluoroparaffins
(886), perfluorocycloparaffins (970) and perfluoroalkylamines (828) behave in
this way.

For preparation, pyrolyses, which yield only few products in acceptable
amounts, are of interest. Sometimes pyrolytic reactions of polyfluoro- and
perfluoro-derivatives, though carried out under drastic conditions, represent
the most feasible way to perfluoroolefins. Pyrolysis of perfluorocyclobutane
in a carbon tube leads to a mixture of perfluoropropene and perfluoroisobutylene (827).

$$
686 \quad
\begin{array}{c} CF_2\!-\!CF_2 \\ |\qquad| \\ CF_2\!-\!CF_2 \end{array}
\xrightarrow[15\ \text{min.}]{700-725^\circ}
\begin{array}{c} CF_3 \\ \diagdown \\ CF_3 \diagup \end{array}\!\!C\!\!=\!\!CF_2 \ + \ CF_3\!-\!CF\!\!=\!\!CF_2
\qquad [827]
$$

$$
70\% \qquad\qquad 20\%
$$

Under comparable conditions, perfluorocyclobutene is pyrolyzed on activated charcoal to perfluoro-1,3-butadiene (385). The dependence of this reaction on temperature is apparent from Table 51 (385). Conversion is very low.

$$
687 \quad
\begin{array}{c} CF_2\!-\!CF \\ |\qquad|| \\ CF_2\!-\!CF \end{array}
\xrightarrow[650^\circ]{\text{Act. C}}
\begin{array}{c} CF_2\!\!=\!\!CF \\ | \\ CF_2\!\!=\!\!CF \end{array}
\quad \text{conversion } 12\% \qquad [385]
$$

TABLE 51

*Dependence of the Equilibrium Between Perfluorocyclobutene and Perfluoro-1,3-butadiene
on Temperature (385)*

| Temp., °C | Content in the Equilibrium Mixture of: | |
|---|---|---|
|  | Perfluorocyclobutene, % | Perfluoro-1,3-butadiene, % |
| 550 |  | 3 |
| 580 | 80–89 | 8–10 |
| 600 | 82 | 10 |
| 640 | 78 |  |
| 650 |  | 12 |
| 690 | 84 |  |
| 700 |  | 6 |

Pyrolysis studies have often been applied to chlorotrifluoroethylene and
tetrafluoroethylene. Pyrolysis of chlorotrifluoroethylene at 560–590°C afforded
70–83% of a mixture containing both linear and cyclic dimer of chlorotrifluoroethylene and chlorofluoropropene (722) (Table 52).

TABLE 52

*Dependence of the Composition of the Product of Pyrolysis of Chlorotrifluoroethylene on Temperature (722)*

| Temperature °C | Total Conversion % | Composition of the Product, % | | |
|---|---|---|---|---|
| | | $CF_2 = CF - CFCl - CClF_2$ | $\begin{array}{c} CF_2 - CFCl \\ \vert \quad\quad \vert \\ CF_2 - CFCl \end{array}$ | Chloro-fluoro-propenes |
| 560 | 70 | 30 | 38 | 21 |
| 595 | 83 | 14 | 13 | 50 |

Pyrolysis of tetrafluoroethylene (722) shows a similar picture. Again linear and cyclic dimers are formed in addition to perfluoropropene (141) (Table 53).

TABLE 53

*Dependence of the Composition of the Product of Pyrolysis of Tetrafluoroethylene on Temperature (722)*

| Temp., °C | Composition of the Product, % | | |
|---|---|---|---|
| | $CF_2 = CF - CF_2 - CF_3$ | $\begin{array}{c} CF_2 - CF_2 \\ \vert \quad\quad \vert \\ CF_2 - CF_2 \end{array}$ | $CF_2 = CF - CF_3$ |
| 435 | 0 | 29 | 0 |
| 550 | 0 | 71 | 12 |
| 640 | 14 | 52 | 26 |
| 655 | 19 | 16 | 42 |
| 750 | 44 | 0 | 4 |

At temperatures higher than 450°C, polymeric tetrafluoroethylene, the most heat-resistant organic material, suffers pyrolysis. At 600–700°C, a mixture of monomeric tetrafluoroethylene, of a dimer, and of perfluoropropene formed by degradation is obtained (618). The composition of the product is almost independent of temperature in the range mentioned but it strongly depends on the pressure at which the pyrolysis is carried out. At atmospheric pressure, mainly dimers are formed, whereas at reduced pressures monomeric tetrafluoroethylene prevails. The reaction represents a suitable method for laboratory preparation of small amounts of tetrafluoroethylene (Table 54).

18*

Table 54

*Dependence of the Composition of the Product of Pyrolysis of Polymeric Tetrafluoro-ethylene at 600° on Pressure (618)*

| Pressure mm Hg | Composition of the Product, % | | |
|---|---|---|---|
| | $C_2F_4$ | $C_3F_6$ | $C_4F_8$ |
| 5 | 97 | — | — |
| 41 | 85·7 | 14·3 | <4 |
| 70 | 83 | 17 | <4 |
| 250 | 57·3 | 19 | 23·7 |
| 350 | 32·5 | 22·3 | 45·2 |
| 760 | 14·2 | 26 | 59·8 |

*Procedure 35.*

*Depolymerization of Polytetrafluoroethylene to Tetrafluoroethylene (618)*

The apparatus for thermal depolymerization of polytetrafluoroethylene consists of an electrically heated steel tube of 18 mm inner diameter and 1350 mm length fitted with a manometer and connected by means of two traps filled with glass wool (to stop entrained solid particles) to two traps immersed in liquid nitrogen. The last trap is attached to an aspirator which maintains the necessary vacuum.

Depolymerization of polytetrafluoroethylene is carried out by heating polytetra-fluoroethylene shavings in the steel tube at 600°C at 5 mm Hg. Under these conditions tetrafluoroethylene is formed almost exclusively and free from by-products. The depoly-merization is practically quantitative. The yield of tetrafluoroethylene collected in the traps approaches 97%.

Combined pyrolysis of polytetrafluoroethylene, first at 450°C, and thermal decomposition at 700°C of the gaseous product thus formed affords a mixture of 1-perfluorobutene, perfluoroisobutylene and perfluoropropene (751).

$$
\textbf{688} \quad [CF_2{-}CF_2]_n \xrightarrow{\ 450°\ } \text{a gas} \xrightarrow{\ 700°\ }
\begin{array}{ll}
CF_2{=}CF{-}CF_2{-}CF_3 & 9 \cdot 5\% \\
CF_2{=}C\big\langle{}^{CF_3}_{CF_3} & 53\% \\
CF_2{=}CF{-}CF_3 & 32\%
\end{array}
\quad [751]
$$

Other fluorinated polymers such as polychlorotrifluoroethylene and poly-$\alpha,\beta,\beta$-trifluorostyrene give the corresponding monomers and some oligomers on pyrolysis (685).

Remarkable results appear in the pyrolysis of compounds which are form-ed by the addition of nitrosoperfluoroalkanes to tetrafluoroethylene. Both the perfluoroalkyloxazetidine or the linear copolymer yield equivalent amounts of degradation products, perfluoroalkylmethyleneimine and carbonyl fluoride 43. 45).

**689** [45]

$$C_3F_7\text{—}N\text{—}CF_2 \xrightarrow{550\text{—}600°} \begin{matrix} 96\% & C_3F_7\text{—}N\text{=}CF_2 & 60\% \\ 102\% & O\text{=}CF_2 & 60\% \end{matrix} \xleftarrow[10^{-3}\text{mm}]{450\text{—}500°} \begin{matrix} | & | \\ C_3F_7\text{—}N & CF_2 \\ | & | \\ O\text{—}CF_2 \end{matrix}$$

The same degradation products result from the pyrolysis of di(perfluoro-alkyl)carbamyl fluorides (*1099*).

**690**

$$\begin{matrix} CF_3 \\ \phantom{x} \diagdown \\ \phantom{xx}N\text{—}COF \\ \phantom{x} \diagup \\ CF_3 \end{matrix} \xrightarrow{575°} \begin{matrix} CF_3 \\ \phantom{x} \diagdown \\ \phantom{xx}N \\ \phantom{x} \diagup \\ CF_2 \end{matrix} + COF_2 \qquad \begin{matrix} \text{conversion } 89\% \\ \\ \text{yield } 96\% \end{matrix} \qquad [1099]$$

Perfluoro-1,4-thiapyran tetrafluoride, when heated to 325–330°C, splits out sulphur tetrafluoride and closes a perfluorotetrahydrofuran ring (*221*).

**691**

$$\begin{matrix} & O & \\ & \diagup \phantom{x} \diagdown & \\ CF_2 & & CF_2 \\ | & & | \\ CF_2 & & CF_2 \\ & \diagdown \phantom{x} \diagup & \\ & S & \\ & F_4 & \end{matrix} \xrightarrow[9\text{—}12 \text{ atm, 10 hrs.}]{325\text{—}330°} \begin{matrix} & O & \\ & \diagup \phantom{x} \diagdown & \\ CF_2 & & CF_2 \\ | & & | \\ CF_2 & \text{—} & CF_2 \end{matrix} \quad 90\% + SF_4 \qquad [221]$$

Pyrolysis of α,α-dichlorotetrafluoroacetone produces in 22% conversion a mixture of chlorodifluoroacetyl chloride and tetrafluoroethylene which evidently results from an intermediate —CF₂— radical (*713*).

**692**

$$2 \text{ CF}_2\text{Cl—CO—CClF}_2 \xrightarrow[540°, 10 \text{ sec.}]{\text{Ni tube}} 2 \underset{77\%}{\text{CF}_2\text{Cl—COCl}} + \underset{47\%}{\text{CF}_2\text{=CF}_2} \qquad [713]$$

Pyrolytic reactions, in which fluorinated derivatives split out compounds which do not contain fluorine, were included in the section on elimination reactions (p. 263).

# VII

# FLUORINATED COMPOUNDS AS CHEMICAL REAGENTS

SOME of the fluorinated derivatives participate in the reactions of organic compounds only transiently and do not appear in the final products. They therefore act either as catalysts, or as true chemical intermediates. Other fluorinated derivatives are used for identification, and others are used as special solvents for synthetic or analytical reactions or for physical measurements. All these types of fluorinated compounds will be dealt with in the following chapter.

## TRIFLUOROACETIC ACID

Trifluoroacetic acid is useful in two respects: because of its excellent solvent power for polar compounds, and for its strong, almost mineral acidity.

Trifluoroacetic acid has been used in the diazotization of some aromatic amines, the diazonium trifluoroacetates thus formed being used in further reactions with other organic compounds, such as benzene (836, 838). This procedure does not seem to have any advantage over the more conventional methods.

By virtue of its acidity, trifluoroacetic acid is a good catalyst for polymerisation (1039) and esterification (750),

693    $\begin{matrix} CH_2OH \\ | \\ CH_2OH \end{matrix}$ + $(CH_3-CO)_2O$ $\xrightarrow[20° \rightarrow 100°]{0\cdot 1 \text{ mole } CF_3COOH}$ $\begin{matrix} CH_2-OCO-CH_3 \\ | \\ CH_2-OCO-CH_3 \end{matrix}$ 87%    [750]

an excellent condensing agent in the Friedel–Crafts synthesis (771),

694                                                     [771]

$CH_3O-\langle\rangle$ + $(CH_3-CO)_2O$ $\xrightarrow[60-70°]{CF_3COOH}$ $CH_3O-\langle\rangle-CO-CH_3$ 31%

and an efficient catalyst for the Beckmann rearrangement (513, 516).

695    $\langle\rangle-\underset{\underset{NOH}{\|}}{C}-CH_3$ $\xrightarrow[\text{reflux}]{CF_3COOH}$ $\langle\rangle-NH-CO-CH_3$ 91%    [513]

Procedure 36.

### The Beckmann Rearrangement in Trifluoroacetic Acid
### Preparation of Benzanilide from Benzophenone Oxime (516)*

Trifluoroacetic acid (5·4 g ,0·047 mole) is placed in 25 ml. flask and frozen by immersion in a dry-ice bath. Benzophenone oxime (1·35 g, 0·007 mole) is then added in one portion and the flask is shaken at room temperature until a clear solution is formed. (This way of dissolving the oxime is more convenient then the portionwise addition of the oxime to liquid trifluoroacetic acid). The flask is fitted with a reflux condenser and heated on the steam bath. After approximately 1 minute the mixture effervesces intensively and turns dark; after refluxing for half an hour, the trifluoroacetic acid is distilled off *in vacuo*: recovery 4·2 g (0·037 mole) (78%). The residue crystallizes as a light yellow mass, which after recrystallization from ethanol and working up of the mother liquors yields 1·2 g (0·006 mole) (88%) of benzanilide.

## TRIFLUOROACETIC ANHYDRIDE

Trifluoroacetic anhydride is prepared by distilling trifluoroacetic acid from an excess of phosphorus pentoxide (118). It is used as a medium for some nitrosations (838) and nitrations (882). Secondary formamides yield nitramines in this way. The function of the trifluoroacetic anhydride in these cases is not clear.

696
$$CH_3 \diagdown N—CHO \xrightarrow[0° to—30°]{100\% HNO_3, (CF_3CO)_2O} CH_3 \diagdown N—NO_2 \quad 89\% \qquad [882]$$

Trifluoroacetic anhydride acts as a catalyst in the Beckmann rearrangement of ketoximes (248).

697 [248]
$$CH_2 \diagdown CH—C—CH_3 \xrightarrow[2. KOH, (CH_2OH)_2, H_2O, reflux]{1. (CF_3CO)_2O, (CH_2OCH_3)_2, reflux 2 hrs.} CH_2 \diagdown CH—NH_2 \quad 77\%$$
with NOH below.

The most frequent application of trifluoroacetic anhydride is in the esterification of hydroxy compounds. Carboxylic acids are converted to mixed anhydrides with trifluoroacetic acid (120, 251).

698 $$(CF_3—CO)_2O + CH_3—COOH \xrightarrow[20°]{C_5H_5N} CF_3—COO—COCH_3 \quad 51\% \qquad [120]$$

Because the acidity of trifluoroacetic acid is much higher than that of the

* Carried out by the author.

common organic acids, these mixed anhydrides possess a pronounced tendency to split to a trifluoroacetate anion and an acyl cation (*117, 483*), which easily reacts with alcohols (*120, 870*). Alcohols, which resist esterification by normal procedure, such as nitroalcohols (*118*), polyfluoroalcohols (*1, 177*), and even phenols, may be esterified in this manner (*1*):

**699** [118]

$$NO_2-\langle\!=\!\rangle-CH_2OH + CH_3-COOH \xrightarrow[60°]{(CF_3CO)_2O} NO_2-\langle\!=\!\rangle-CH_2OCO-CH_3 \quad 83\%$$

**700** [1]

$$CF_3-CF_2-CF_2-CH_2OH + CH_2\!=\!CH-COOH \xrightarrow{(CF_3CO)_2O}$$

$$CF_3-CF_2-CF_2-CH_2OCO-CH\!=\!CH_2$$
$$85-90\%$$

Esterification of alcohols with sterically hindered acids also gives satisfactory results, when carried out in the presence of trifluoroacetic anhydride (*870*).

**701**
$$\underset{CH_3}{\overset{CH_3}{\langle\!=\!\rangle}}-COOH + CH_3OH \xrightarrow[20°,\ 2\ hrs.]{(CF_3CO)_2O} \underset{CH_3}{\overset{CH_3}{\langle\!=\!\rangle}}-COOCH_3 \quad 93\% \qquad [870]$$

The easy esterification by means of trifluoroacetic anhydride is used in carbohydrate chemistry for the preparation of trifluoroacetyl derivatives, in which the hydroxyl groups are temporarily protected (*1075*). The recovery of the original hydroxy derivative is unusually easy, since the trifluoroacetyl group is hydrolysed by mere contact with water or alcohol. This reaction is not accompanied by Walden inversion (*121*).

**702**

[121]

| | |
|---|---|
| OCH$_3$<br>\|<br>H—C———<br>\|<br>H—C—OH \|<br>\|<br>HO—CH   O<br>\|<br>H—C———O<br>\|<br>H—C<br>\|        \ CH—C$_6$H$_5$<br>CH$_2$———O | $\xrightarrow[\text{spont. reaction}]{(CF_3CO)_2O,\ CF_3COONa}$<br><br><br>$\xleftarrow[\substack{\text{or CH}_3\text{OH}\\20°,\ 18\ \text{hrs.}}]{H_2O}$ |

$$\begin{array}{l} OCH_3 \\ | \\ H-C \text{———} \\ | \\ H-C \text{———} OCOCF_3 \\ | \\ CF_3COO-CH\ O \\ | \\ H-C \text{———} O \\ | \qquad\qquad \backslash \\ H-C \qquad\quad CH-C_6H_5 \\ | \qquad\quad / \\ CH_2 \text{———} O \end{array}$$

The mixed anhydrides of organic acids and trifluoroacetic acid add to the multiple bonds of olefins and acetylenes (*117, 483*). The products easily split off trifluoroacetic acid and yield, as final products, unsaturated ketones (from olefins) (*483*),

**703**  [*483*]

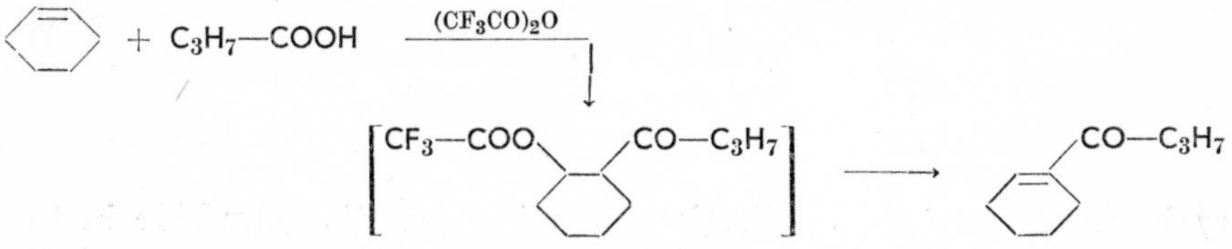

or β-diketones (from acetylenes) (483).

**704**  [*483*]

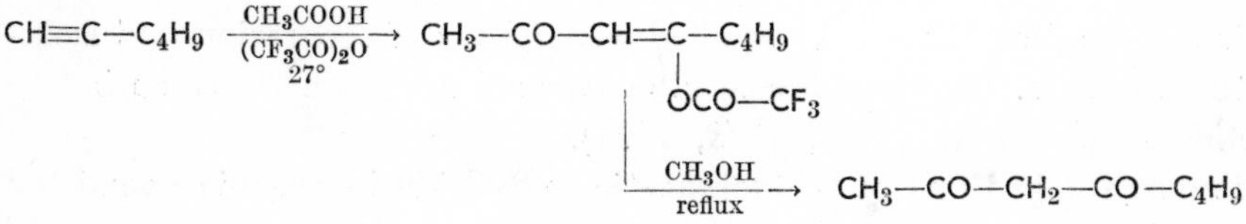

*Procedure 37.*

*Synthesis of 1,3-Diketones with Trifluoroacetic Anhydride as Condensing Agent
The Preparation of 2,4-Octanedione from 1-Hexyne and Acetic Acid (483)*

(*a*) *Preparation of Trifluoroacetic Anhydride (118).* Trifluoroacetic acid is mixed with phosphorus pentoxide (0·87 g per gram of the acid) and the mixture is distilled; a 74% yield of trifluoroacetic anhydride, b.p. 39°C, is obtained.

(*b*) *Condensation by Means of Trifluoroacetic Anhydride.* To a mixture of 5·7 ml. of acetic acid and 14·5 ml. of trifluoroacetic anhydride, 11·4 ml. of 1-hexyne is added. The mixture is allowed to stand for 18 hours at 27°C, treated with 25 ml. of methanol and refluxed for 1·5 hrs. in order to effect re-esterification of the diketone monoenol trifluoroacetate formed. Distillation of the mixture gives 2·41 g of 2,4-octanedione as the fraction boiling at 45–50°C/3·5–4 mm. From fraction boiling at 50–75°C some more of the diketone is obtained by redistillation. Total yield of the 2,4-octanedione is 2·84 g (20%).

In the reactions of carboxylic acids with sufficiently reactive aromatic systems such as benzene homologues, phenol ethers, furan and thiophene, trifluoroacetic anhydride has a similar function (*119*).

**705**  [*119*]

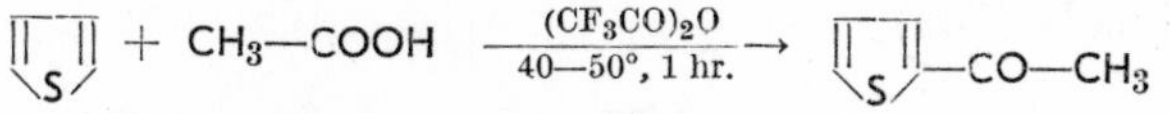

Where a five-membered or six-membered ring may be formed, trifluoroacetic anhydride readily promotes cyclization. γ-Phenylbutyric acid gives a quantitative yield of α-tetralone (*272*).

The reaction is not limited to aromatic systems. Even 5-hexenoic acid cyclizes to 2-cyclohexenone in a yield of 46% (*272*).

**706** [*272*]

quant.

Very important are the applications of trifluoroacetic anhydride in the field of peptide chemistry. Trifluoroacetylation of amino acids temporarily protects the amino groups, activates the carboxyl groups for syntheses, and increases the vapour pressure of the products to such an extent that sometimes distillation or sublimation can be applied as a means of separation and isolation of the products. The trifluoroacetyl groups are readily hydrolyzed before the peptide bonds break (*1072, 1073, 1074*).

Numerous applications of trifluoroacetic anhydride in organic chemistry are discussed in a review (*1032*).

## TRIFLUOROPERACETIC ACID

When 30% or preferably 90% hydrogen peroxide is dissolved in anhydrous trifluoroacetic acid (*249*) or better in trifluoroacetic anhydride (*246, 247, 250, 252, 253, 902*), trifluoroacetic acid is formed. The compound shows all the characteristic properties of organic peracids (*249, 902*).

Primary amines, nitroso derivatives and oximes are oxidized to nitrocompounds (*247, 253*),

**707** $\xrightarrow[\text{90\% H}_2\text{O}_2]{\text{(CF}_3\text{CO)}_2\text{O}}$ —NO$_2$  89% [*247*]

**708** =NOH $\xrightarrow{\text{CF}_3\text{CO}_3\text{H}}$ 60% [*253*]

olefins to *trans*-glycols (*254*)

**709** $\xrightarrow[\text{CF}_3\text{CO}_2\overset{\oplus}{\text{N}}(\text{C}_2\text{H}_5)_3]{\text{(CF}_3\text{CO)}_2\text{O, 90\% H}_2\text{O}_2}$ $\xrightarrow[\text{CH}_3\text{OH}]{\text{HCl}}$ 82% [*254*]

The intermediates in this oxidation, the epoxides, were isolated when the reaction was carried out in dichloromethane as the solvent (*252*):

710     $CH_2\!\!=\!\!CH\!-\!C_4H_9$   $\xrightarrow[\text{CH}_2\text{Cl}_2]{\text{(CF}_3\text{CO)}_2\text{O, 90\% H}_2\text{O}_2}}$   $\overset{\displaystyle CH_2\!-\!CH\!-\!C_4H_9}{\underset{O}{\diagdown\diagup}}$   91%     [*252*]

Unlike other peracids, trifluoroacetic acid oxidizes unsaturated tertiary amines preferentially at the double bond, so that the epoxide and not the aminoxide is formed, as shown in the reaction with acetyldehydrotropine (*282*).

711                                                                              [*282*]

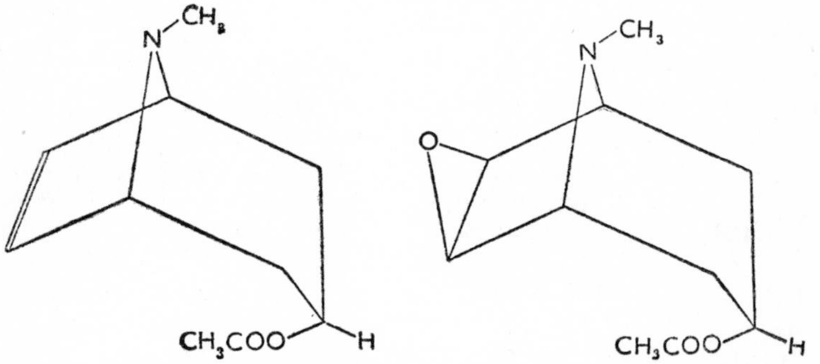

Another reaction typical for peracids—transformation of ketones to esters (*250, 415, 902*) — can also be effected by trifluoroperacetic acid. Aliphatic and cyclic ketones give good yields of esters and lactones, respectively. The oxidation of cyclohexanone gives, according to the reaction conditions, either ε-caprolactone, or with a large excess of trifluoroacetic anhydride, ε-trifluoro-acetoxycaproic acid (*902*).

712                                                                 [*250*]

$\overset{CH_3}{\underset{CH_3}{>}}CH\!-\!CH_2\!-\!CO\!-\!CH_3$   $\xrightarrow[\text{Na}_2\text{HPO}_4,\text{ CH}_2\text{Cl}_2,\text{ reflux}}{\text{(CF}_3\text{CO)}_2\text{O, 90\% H}_2\text{O}_2,}$   $\overset{CH_3}{\underset{CH_3}{>}}CH\!-\!CH_2\!-\!OCO\!-\!CH_3$   84%

713        $\xrightarrow[15°]{\text{CF}_3\text{CO}_3\text{H, CF}_3\text{COOH}}$   ⬡=O   $\xrightarrow[\;]{\text{CF}_3\text{CO}_3\text{H, (CF}_3\text{CO)}_2\text{O}}$               [*902*]

                                             $CF_3\!-\!COO(CH_2)_5COOH$

80·7−88·3%

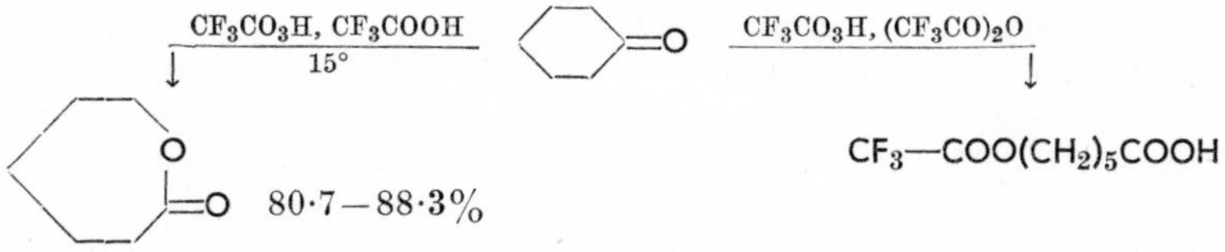

In the case of cycloheptanone and cyclooctanone, trifluoroperacetic acid affords good yields of the corresponding lactones (68 and 72%, respectively), while other oxidizing reagent give polyesters (521). Migration aptitudes of alkyls in the oxidation of alkyl phenyl ketones with trifluoroperacetic acid are apparent from Table 55 (415).

TABLE 55

*Migration Aptitudes of Alkyls in the Oxidation of Phenyl Alkyl Ketones by Trifluoro-peracetic Acid (415)*

| Starting Ketone | Yield | Composition of the Product, % | | Ratio Alkyl/Phenyl |
|---|---|---|---|---|
| $C_6H_5CO$-R | % | $C_6H_5OCOR$ | $C_6H_5COOR$ | |
| R = $CH_3$ | 90 | 90 | 0 | Very Small |
| $C_2H_5$ | 93 | 87 | 6 | 0·07 |
| $C_3H_7$ | 91 | 85 | 6 | 0·07 |
| $(CH_3)_2CH$ | 96 | 33 | 63 | 1·9 |
| $(CH_3)_3C$ | 90 (11% recovered) | 2 | 77 | 39 |
| $(CH_3)_3C.CH_2$ | 97 (4% recovered) | 84 | 9 | 0·1 |
| cyclo-$C_5H_9$ | 92 | 44 | 48 | 1·1 |
| cyclo-$C_6H_{11}$ | 100 | 25 | 75 | 3·0 |
| $C_6H_5CH_2$ | 90 | 39 | 51 | 1·3 |

The yields of the conversion of ketones to esters by trifluoroperacetic acid are so high that the reaction can be used for the semi-quantitative determination of aliphatic aldehydes and ketones (414).

*Procedure 38.*

   *Oxidation with Trifluoroperacetic Acid*

   *The Preparation of ε-Caprolactone from Cyclohexanone (902)*

   (a) *Preparation of Trifluoroperacetic Acid.* Hydrogen peroxide (85%) (17·45 ml., 0·6 mole) is added to 116·9 g (0·55 mole) of trifluoroacetic anhydride over a period of 1·5 hrs. while the temperature of the reaction mixture is maintained at 5–10°C.

   (b) *Oxidation of Cyclohexanone to ε-Caprolactone.* Cyclohexanone (94·5 g, 0·096 mole) is added in the course of 40 minutes to a 5% excess of a solution of trifluoroperacetic acid in trifluoroacetic acid, placed in a wide-necked flask fitted with a stirrer, a thermometer and a separating funnel, and immersed in an ice-salt bath to maintain the temperature at 10–15°C. The reaction mixture is diluted with 250 ml. of chloroform and poured, with stirring and cooling, into a 7% excess of a saturated solution of potassium carbonate. Sufficient water is added to dissolve the potassium trifluoroacetate which crystallizes, the organic layer is separated, the aqueous layer is extracted successively with 100, 50 and 50 ml. of chloroform, the combined organic layers are dried over magnesium sulphate, the solvent is evaporated, and the residue is distilled *in vacuo* to yield 97 g (0·085 mole) (88%) of ε-caprolactone and approximately 2–5% of acidic by-products.

## TRIFLUOROACETYLHYDROXYLAMINE

Trifluoroacetylhydroxamic acid and N,O-bis(trifluoroacetyl)hydroxylamine are prepared by refluxing hydroxylamine hydrochloride with two or three moles of trifluoroacetic anhydride, respectively. The treatment of aldehydes with any of these two reagents produces the corresponding nitriles in a one-step reaction in yields up to 87% (*848*).

714 ────────────────── NH$_2$OH·HCl ────────────────── [*848*]

CF$_3$—C(=O)NHOH    74%           CF$_3$—CO—NH—OCO—CF$_3$   80%

     R—CHO                 R—CHO

───────────────→ R—CN ←───────────────

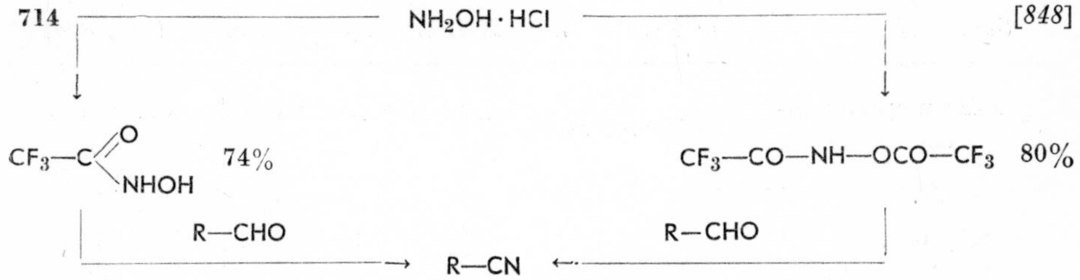

## TRIFLUOROACETYL HYPOHALITES

When silver trifluoroacetate is mixed with bromine or iodine in toluene, a spontaneous reaction sets in and bromotoluene and iodotoluene are formed in 73 and 84% yield, respectively (*490*). Similar reactions occur with benzene, benzoic acid, and even with the relatively unreactive nitrobenzene (*388*). In all these cases the halogenation itself is preceded by formation of trifluoroacetyl hypobromite or hypoiodite, respectively.

715         $\xrightarrow{I_2}$ CF$_3$—COOAg $\xrightarrow{Br_2}$            [*388*]

$\xrightarrow[150°]{C_6H_5COOH}$ CF$_3$—COOI         CF$_3$—COOBr $\xrightarrow[120°]{C_6H_5NO_2}$

COOH ... 84%             NO$_2$ ... 19%

(iodobenzoic acid)                 (bromonitrobenzene)

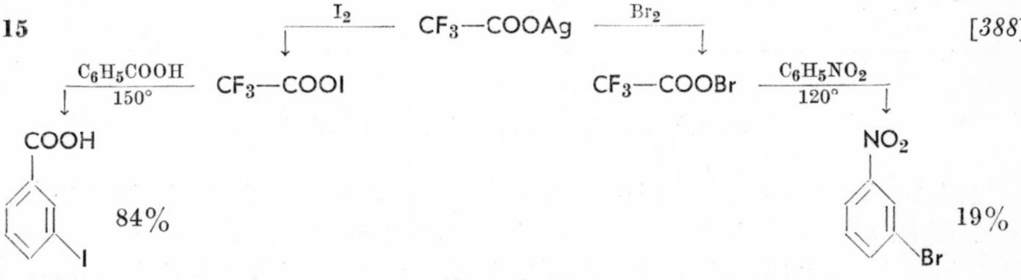

## BROMO FLUOROACYL AMIDES

Another group of halogenating reagents derived from fluorinated compounds are the N-bromoamides or imides of perfluorocarboxylic acids. They are prepared by treating the amides or the imides in trifluoroacetic acid with bromine in the presence of silver oxide (*489*).

716    CF$_2$—CO(NH)CF$_2$—CO $\xrightarrow[CF_3COOH]{Br_2, Ag_2O}$ CF$_2$—CO(NBr)CF$_2$—CO    yield 87%       [*489*]

                                                       purity 91%

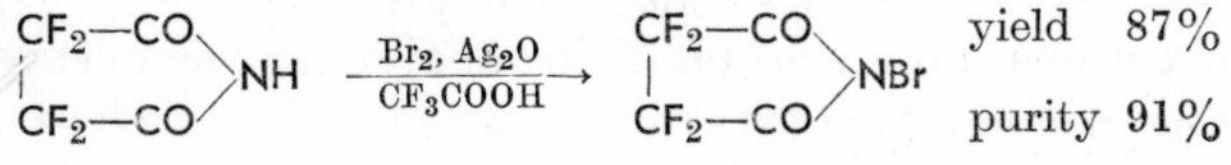

As a result of the strong electron-attracting effect of the perfluoroalkyl groups, the halogen is released from the nitrogen predominatly heterolytically as a cation, and only to a minor extent homolytically. This is evident from the course of bromination of toluene: ionic halogenation in the nucleus always predominates over radical side-chain halogenation (489, 804).

TABLE 56

*Course of Bromination of Toluene With N-Bromoperfluoro Amides (489, 804)*

| N-Bromoperfluoro-acylamide | Temperature, °C | $BrC_6H_4CH_3$ % | $C_6H_5CH_2Br$ % |
|---|---|---|---|
| $CF_3CONHBr$ | | 88 | 12 |
| $C_2F_5CONHBr$ | | 69·9 | 30·1 |
| $C_3F_7CONHBr$ | | 73·9 | 26·1 |
| $C_4F_9CONHBr$ | | 86·7 | 13·3 |
| $CF_2CO$ <br> $\quad\rangle NBr$ <br> $CF_2CO$ | 24 <br> 90 | 99·7 <br> 67 | 0·33 <br> 33 |
| 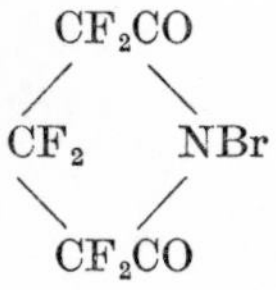 $CF_2CO$ <br> $CF_2\quad NBr$ <br> $CF_2CO$ | 50 | 98·8 | 1·2 |

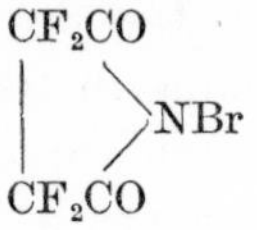

OTHER REAGENTS AND SOLVENTS

*2,4-Dinitrofluorobenzene* is much more reactive than its chlorinated analogue. With alcohols, it forms 2,4-dinitrophenyl ethers which can be used for identification of the alcohols (1078). Amines react with 2,4-dinitrofluorobenzene to give 2,4-dinitrophenylamino derivatives (167, 168).

*Trifluoromethylpicric acid*, prepared from *m*-hydroxybenzotrifluoride, has been proposed as a reagent for identifying amines and pyrylium compounds (1079).

*Perfluorocarboxylic acids* are suitable extraction reagents for the analytical separation of various cations (737).

*Furoyltrifluoroacetone* (683) *and thenoyltrifluoroacetone* (533) easily form chelates with cations and are used in analytical chemistry as valuable complex-forming reagents.

*Trichlorofluoromethane* is a suitable solvent for the preparation of alkyl hypochlorites from alcohols and hypochlorous acid at 0–10°C (*205*).

*Benzotrifluoride* is used as a solvent for cryoscopic measurements on fluorinated compounds (*342*), *o-chlorobenzotriflucride* (*341*), *2,5-dichlorobenzotrifluoride* (alone or mixed with diethyl phthalate) (*551*), *pentachloro-1,1,3-trifluoropropane* (*551*) and *1,1,2,2-tetrachlorotetrafluorocyclobutane* (*292*) find use as solvents for the fractional precipitation of chlorotrifluoroethylene polymers of various chain length.

# VIII

## PROPERTIES OF ORGANIC FLUORINE COMPOUNDS

IN the following paragraph physical, physico-chemical and biological properties will be discussed. Chemical properties have already been described in the sixth chapter.

To cut down the number of references, physical constants are cited from general handbooks and reference books (*196, 624, 833, 947, 948*).

### PHYSICAL PROPERTIES

Physical properties will be discussed in the sequence: melting points, boiling points, densities, refractive indices, dielectric properties, surface tensions, viscosities, and solubilities. While the data about the first four are quite complete, the last four have been measured in limited numbers only. For contrast homologous series of fluorinated derivatives will be compared with their non-fluorinated analogues where ever possible.

### Melting Points

Perfluoroparaffins melt higher than paraffins, except for the first member of the series. The most striking difference appears at the two and four carbon level. From the first member on melting points of perfluoroparaffins rise almost continuously, while paraffins show alternation (Fig. 19*, p. 289). In both series the minimum is at the three carbon level.

### Boiling points

Boiling points are the most commonly listed constants and comparison with non-fluorinated compounds is frequently possible.

Perfluoroparaffins with one to three carbon atoms have higher boiling points, perfluorobutane the same boiling point, and longer perfluoroparaffins lower boiling points than the corresponding paraffins (Fig. 20, p. 289).

---

\* In charts 19–28 and 31–34 line diagrams have been chosen to represent the dependences. Although not quite correct, they are more illustrative.

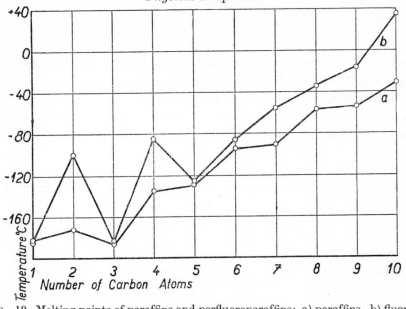

FIG. 19. Melting points of paraffins and perfluoroparaffins: a) paraffins, b) fluoro-carbons.

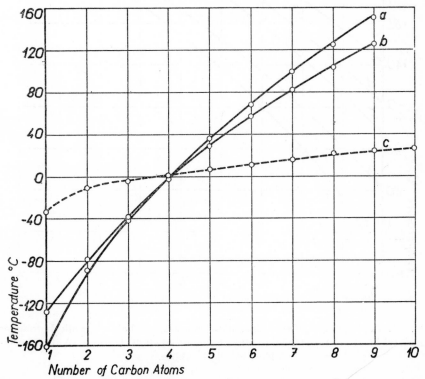

FIG. 20. Boiling points of paraffins and perfluoroparaffins: a) paraffins, b) per-fluoroparaffins, c) differences (b.p. of the paraffins — b.p. of the perfluoroparaffins).

Perfluoroolefins with terminal double bonds show a similar trend.

A system of parallels can be constructed, which estimates the boiling points of alkyl fluorides from that of the other halides (Fig. 21). For comparison the charts in Fig. 21 and 22 also show the boiling points of paraffins. The differences in individual series fade progressively with decrease of the relative importance of the halogen atoms (Fig. 22). This fact is particularly evident in the comparison of perfluoroparaffins with primary perfluoroalkyl halides (Fig. 23).

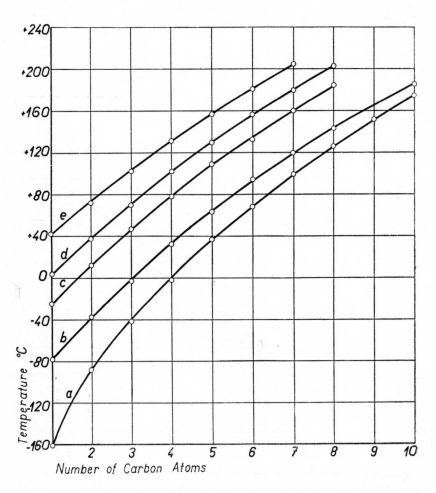

Fig. 21. Boiling points of paraffins, alkyl fluorides, alkyl chlorides alkyl bromides and alkyl iodides: a) paraffins, b) alkyl fluorides, c) alkyl chlorides, d) alkyl bromides, e) alkyl iodides.

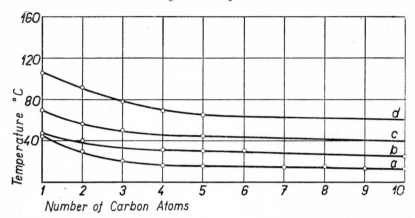

FIG. 22. Differences in the boiling points of alkyl fluorides from those of paraffins, alkyl chlorides, alkyl bromides and alkyl iodides; values which must be: a) added to b.p. of paraffins, b) subtracted from b.p. of alkyl chlorides, c) subtracted from b.p. of alkyl bromides, d) subtracted from b.p. of alkyl iodides, to obtain b.p. of alkyl fluorides.

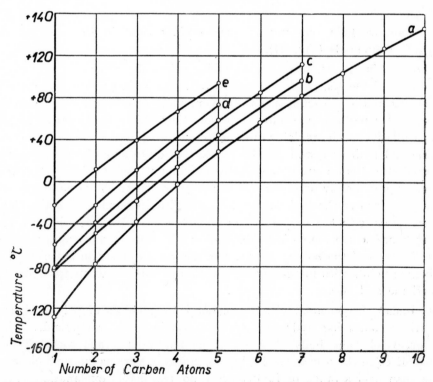

FIG. 23. Boiling points of perfluoroparaffins, perfluoroalkyl hydrides, perfluoro-alkyl chlorides, perfluoroalkyl bromides, and perfluoroalkyl iodides: a) perfluoro-paraffins, b) perfluoroalkyl hydrides, c) perfluoroalkyl chlorides, d) perfluoroalkyl bromides, e) perfluoroalkyl iodides.

19*

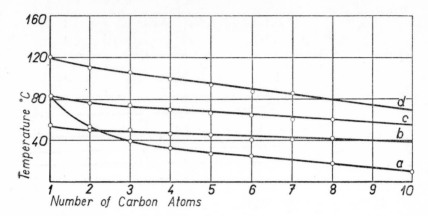

FIG. 24. Differences in the boiling points of perfluoroalkyl hydrides, perfluoroalkyl chlorides, perfluoroalkyl bromides and perfluoroalkyl iodides from those of perfluoroparaffins; values which must be added to b.p. of perfluoroparaffins to obtain a) b.p. of perfluoroalkyl hydrides, b) perfluoroalkyl chlorides, c) perfluoroalkyl bromides, d) perfluoroalkyl iodides.

The differences fade out much more rapidly than in the case of the $C_nH_{2n+1}X$ compounds (Fig. 24).

The chart in Fig. 23. also contains boiling points of perfluoroalkyl hydrides, compounds of the general formula $C_nF_{2n+1}H$. By analogy with the chart in Fig. 21, it could be expected that these compounds should boil lower than the corresponding perfluoroparaffins. However, they boil systematically higher. This peculiarity can be explained by the formation of hydrogen bonds between fluorine and hydrogen, and can be pursued in quite a few other instances. Fluorinated methanes and ethanes show maximum boiling points, when they contain approximately an equal number of hydrogen and fluorine atoms and when consequently the highest number of hydrogen bonds is possible. This fact contrasts with the course of the boiling points of chloro-, bromo- and iodomethanes where increasing substitution of halogens consistently results in the increase of the boiling points. Fig. 25 and 26 illustrate the change in the boiling points with increasing substitution of halogen atoms for hydrogen in methane and ethane, respectively.

A similar phenomenon, viz. the higher boiling points of polyfluorinated derivatives containing hydrogen, appears also in cyclic compounds. While perfluorocyclohexane boils at 50–52°C, monohydroundecafluorocyclohexane boils at 62°C. In the benzene series boiling points of various fluorinated derivatives span over a range of 16·5°C, with no reasonable explanation (Table 57, p. 294).

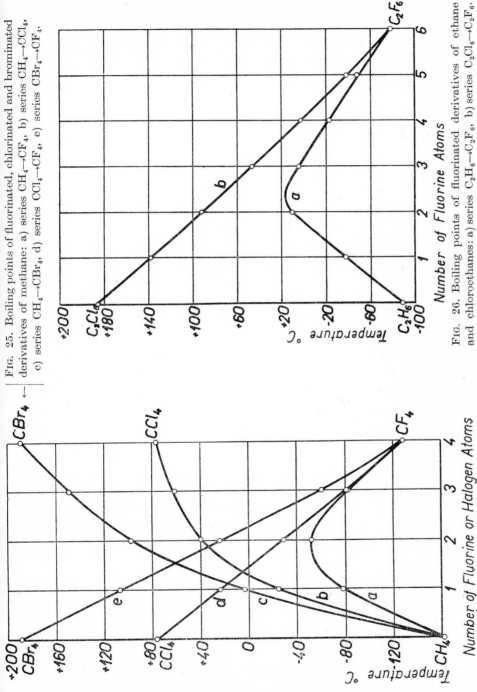

Fig. 25. Boiling points of fluorinated, chlorinated and brominated derivatives of methane: a) series $CH_4 \rightarrow CF_4$, b) series $CH_4 \rightarrow CCl_4$, c) series $CH_4 \rightarrow CBr_4$, d) series $CCl_4 \rightarrow CF_4$, e) series $CBr_4 \rightarrow CF_4$.

Fig. 26. Boiling points of fluorinated derivatives of ethane and chloroethanes: a) series $C_2H_6 \rightarrow C_2F_6$, b) series $C_2Cl_6 \rightarrow C_2F_6$.

TABLE 57

*Boiling Points of Benzene and Its Fluorinated Derivatives*

| Formula | B.P.°C | Formula | B.P.°C |
|---------|--------|---------|--------|
| | 80·1 | | 88 |
| | 84·8 | | 75 |
| | 91·5 | | 90 |
| | 82·5 | | 85 |
| | 88·5 | | 81·5 |

Evidently, in addition to the number of fluorine atoms, their mutual position plays an important role. This is true especially of some fluorinated ethane derivatives (Table 58).

TABLE 58

*Boiling Points of Fluorinated Ethane Derivatives. Effect of Isomerism*

| Formula | $CH_2FCH_2F$ | $CH_3CHF_2$ | $CCl_2FCCl_2F$ | $CCl_3CClF_2$ |
|---------|--------------|-------------|----------------|---------------|
| B.P. °C | $+10\cdot5$ | $-24\cdot7$ | $92\cdot8$ | $91$ |

| Formula | $CH_2FCHF_2$ | $CH_3CF_3$ | $CCl_2FCClF_2$ | $CCl_3CF_3$ |
|---------|--------------|------------|----------------|-------------|
| B.P. °C | $+5$ | $-46\cdot7$ | $47\cdot7$ | $45\cdot9$ |

Next to the fluorinated hydrocarbons, fluorinated acid derivatives supply most of the known physical data. Perfluorocarboxylic acids boil lower than the parent compounds by approximately 45°C (Fig. 27).

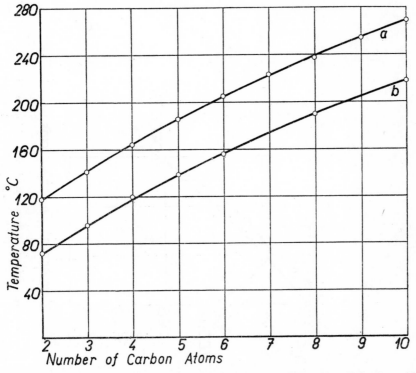

FIG. 27. Boiling points of carboxylic and perfluorocarboxylic aliphatic acids: a) carboxylic acids, b) perfluorocarboxylic acids.

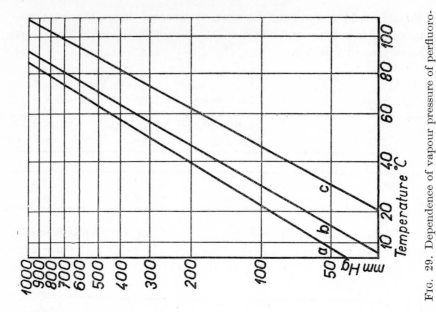

Fig. 29. Dependence of vapour pressure of perfluoro-paraffins on temperature: a) perfluoroheptane, b) perfluoro(methylcyclohexane), c) perfluoro (dimethyl-cyclohexane).

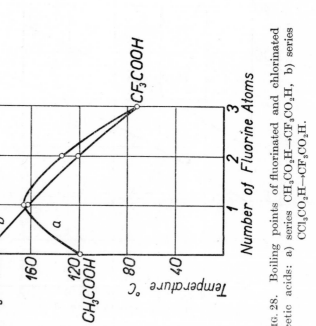

Fig. 28. Boiling points of fluorinated and chlorinated acetic acids: a) series $CH_3CO_2H \rightarrow CF_3CO_2H$, b) series $CCl_3CO_2H \rightarrow CF_3CO_2H$.

Fluorinated acetic acids containing both hydrogen and fluorine atoms in their methyl groups boil higher than acetic acid itself, whereas trifluoroacetic acid, where no fluorine-hydrogen bonds are possible, boils considerably lower. This is an essential difference from chlorine substitution, which causes a systematic rise in the boiling points (118°C–189°C–194°C–197·5°C for chlorinated acetic acids) (Fig. 28).

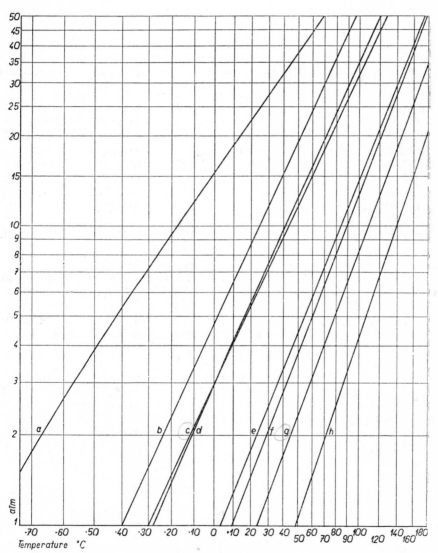

FIG. 30. Dependence of the vapour pressure of Freons on temperature: a) $CClF_3$, b) $CHClF_2$, c) $CCl_2F_2$, d) $C_2ClF_3$, e) $C_2Cl_2F_4$, f) $CHCl_2F$, g) $CCl_3F$, h) $C_2Cl_3F_3$.

Systematic measurements of vapour pressure of fluorinated derivatives are limited to commercially important products: freons and some perfluoro-paraffins and cycloparaffins. The dependences of the vapour pressure of some fluorinated compounds are given in Fig. 29 (p. 296) and 30 (p. 297).

A random selection of fluorinated compounds and their main physical constants is shown in Table 59. The monofluoro compounds usually have the highest, the perfluoro derivatives the lowest boiling points.

Table 59

*Physical Constants of Some Fluorinated Compounds*

| Formula | M.P.°C | B.P.°C | $d_4^t$ | $n_D^t$ |
|---|---|---|---|---|
| $CH_3CH_2OH$ | −114·6 | 78·4 | 0·7893/20 | 1·36242/18.35 |
| $CH_2FCH_2OH$ | − 43 | 103·5 | 1·040/20 | 1·3647/20 |
| $CHF_2CH_2OH$ | − 28·2 | 96 | 1·3084/17 | 1·3345/11.8 |
| $CF_3CH_2OH$ | − 43·5 | 74 | 1·3842/20 | 1·2907/22 |
| $CH_3COCH_3$ | − 95 | 56·5 | 0·792/20 | 1·35886/19.4 |
| $CH_2FCOCH_3$ | | 72 | 0·967/24 | 1·3693/21 |
| $CHF_2COCH_3$ | | 46·6 | 1·1644/20 | 1·3280/20 |
| $CF_3COCH_3$ | | 21·9 | 1·282/0 | |
| $CF_3COCF_3$ | −129 | −28 | | |
| $CH_3CH_2OCH_2CH_3$ | −116·3 | 34·6 | 0·7135/20 | 1·3497/24.8 |
| $CH_2FCH_2OCH_2CH_3$ | | 75 | | |
| $CHF_2CH_2OCH_2CH_3$ | | 66·5 | 1·039/15 | |
| $CF_3CH_2OCH_2CH_3$ | | 49·8 | 1·0910/20 | 1·3042/20 |
| $CHF_2CF_2OCH_2CH_3$ | | 57·5 | 1·1978/25 | 1·294/25 |
| $CH_3CF_2OCH_2CF_3$ | | 37·8 | | |
| $CHF_2CF_2OCH_2CF_3$ | | 56·7 | 1·4870/20 | 1·2728/20 |
| $CH_3SCH_3$ | − 83·2 | 37·8 | 0·8458/21 | |
| $CF_3SCH_2F$ | | 1 | | |
| $CF_3SCF_3$ | | −22·2 | | |
| $CH_3COOC_2H_5$ | − 83·6 | 77·15 | 0·901/20 | 1·37216/18.9 |
| $CH_2FCOOC_2H_5$ | | 120 | 1·093/20 | 1·3767/20 |
| $CHF_2COOC_2H_5$ | | 99·2 | 1·1800/17.5 | |
| $CF_3COOC_2H_5$ | | 62 | 1·1953/16.7 | 1·30783*/16.7 |
| $CH_3COCl$ | −112 | 51·5 | 1·1051/20 | 1·38976 |
| $CH_2FCOCl$ | | 71 | | 1·3835/27 |
| $CHF_2COCl$ | | 25 | | |
| $CF_3COCl$ | | −18·5 | | |
| $CH_3NHCH_3$ | − 96 | 7·4 | 0·6804/0 | 1·350/17 |
| $CF_3NHCF_3$ | | 6·3 | | |
| $CF_3NFCF_3$ | | −37 | | |
| $CH_3SO_3H$ | | 167/10 | 1·481 | |
| $CF_3SO_3H$ | | 162/760 | | |

*) For α-line.

## Density

The effort to find a relationship between the density and composition of fluorinated compounds is defeated by lack of sufficient data. Systematic measurements of density have been carried out only with perfluoroparaffins and perfluorocarboxylic acids (*546*). The dependences appear in Fig. 31 and 32 (p. 300) which show also nonfluorinated compounds (p. 299) for contrast.

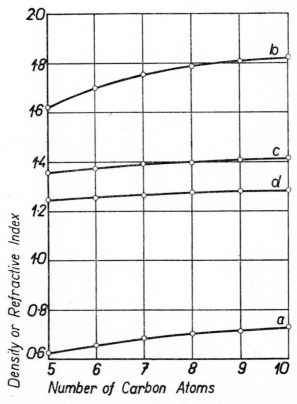

FIG. 31. Densities and refractive indices of paraffins and perfluoroparaffins: a) densities of paraffins, b) densities of perfluoroparaffins, c) refractive indices of paraffins, d) refractive indices of perfluoroparaffins.

The substitution of fluorine for hydrogen increases the density, the greatest change being produced by the first fluorine atom. This is apparent in Fig. 33.

The temperature coefficient of the density ranges from $-0 \cdot 0023$ to $-0 \cdot 0025$/per degree while paraffins have the much lower coefficient of approximately $-0 \cdot 0008$/per degree.

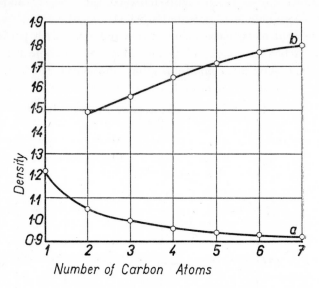

FIG. 32. Densities of carboxylic and perfluorocarboxylic acids: a) carboxylic acids, b) perfluorocarboxylic acids.

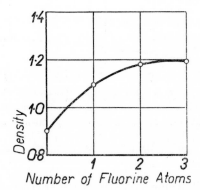

FIG. 33. Densities of fluorinated ethyl acetates.

## Refractive Index

As in the case of density, a precise relationship between the refractive indices and the fluorine content is difficult to establish for lack of necessary data. Fig. 31 (p.299) relates the index of the number of carbon atoms and shows the differences from the corresponding paraffins. The temperature coefficient is $-0 \cdot 0004$/per degree with perfluoroparaffins, approximately the same as that of paraffins ($-0 \cdot 0005$/per degree).

The refractive index of fluorinated derivatives is always lower than that of the non-fluorinated compounds. That of perfluoro-derivatives is exceedingly low: perfluoropentane with $n_D^{20}$ $1 \cdot 245$ has probably the lowest value ever measured. Since the refractive indices of fluorinated compounds very often lie considerably below those of water ($n_D^{20}$ $1 \cdot 33299$), many refractometers are unsuitable for their measurements.

The value of the refraction for the fluorine atom has not yet been determined with sufficient accuracy. It seems necessary to set various values for different types of compounds. Thus derivatives containing one or only few atoms of fluorine require a value other than that of the perfluorinated compounds. For calculations with the Eisenlohr values, the proposed atomic refractions for fluorine are $1 \cdot 082$ in saturated compounds (*981*), $0 \cdot 775$ in non-saturated compounds (*981*), $0 \cdot 997$ in aromatic compounds (*921*), $0 \cdot 95$ in monofluoro-derivatives (*330*), $0 \cdot 99$ in difluoro-derivatives (*327*), $1 \cdot 02$ in trifluoro-derivatives (*327*), $1 \cdot 08$ in tetrafluoro-derivatives (*327*), $1 \cdot 14$ in pentafluoro-derivatives (*327*), and $1 \cdot 23$ in perfluoroparaffins and their hydro-derivatives (*327*). For the aromatic fluorine derivatives, the following values have been derived for the C, D, F and G' lines (*921*): $0 \cdot 984$, $0 \cdot 997$, $1 \cdot 011$, and $1 \cdot 02$, respectively. For calculations with the Vogel values, the atomic refraction of fluorine has been computed as $0 \cdot 81$, $0 \cdot 81$, $0 \cdot 79$, and $0 \cdot 78$ for the above lines (*1055*). The last values, however, have been derived from a relatively small number of fluorinated compounds of similar types, so that the justification for their general use is still to be shown. Judging from the relatively large differences between the individual values, the situation in the determination of atomic refraction for fluorine is still unsatisfactory.

## Dielectric Constant

Dielectric constants have been measured only in a few cases. They largely depend on the ratio of hydrogen and fluorine atoms in the organic molecules. Pure fluorocarbons have a dielectric constant slightly lower than paraffins. On the other hand, polyfluoroparaffins containing a small number of hydrogen atoms show higher dielectric constants than the perfluoro-derivatives (*69*) (Table 60, p.302).

TABLE 60

*Dielectric Constant of Fluorinated Compounds (69)*

| Compounds | Dielectric Constant |
|:---:|:---:|
| $C_5F_9-C_2F_5$ | 1·80 |
| $C_6F_{11}-CF_3$ | 1·85 |
| $C_6F_{10}(CF_3)_2$ | 1·863 |
| $C_7F_{16}$ | 1·765 |
| $C_7HF_{15}$ | 2·47 |
| Isomer | 2·93 |
| $C_7H_2F_{14}$ | 3·18 |

## Surface Tension

A significant feature of fluorinated compounds is their low surface tension. This holds true especially for polyfluoro- and perfluoro-derivatives (Table 61). The value of the surface tension of perfluoroparaffins and cycloparaffins range from 10 to 20 dyne/cm (*947*) which is extraordinarily low (the surface tension of water at 20°C is 72,75 dyne/cm).

TABLE 61

*Surface Tension of Fluorocarbons at 20° (947)*

| Compound | Surface Tension dyne/cm |
|:---:|:---:|
| $C_5F_{12}$ | 10·0 |
| $C_6F_{14}$ | 12·0 |
| $C_7F_{16}$ | 12·06 |
| $C_8F_{18}$ | 13·6 |
| cyclo-$C_5F_{10}$ | 11·3 |
| cyclo-$C_6F_{12}$ | 12·4 |

The parachor values for fluorine range from 22·2 to 26·1, depending on the type of compounds and the manner of calculation (*274, 1055*) (Table 62).

TABLE 62

*Parachor of Fluorine in Fluorinated Compounds (274, 1055)*

| Calculation According to | Alkyl Fluorides and Fluoroacetates | Fluorocarbons | Pseudoperfluoroalcohols | Polyfluoroesters |
|:---|:---:|:---:|:---:|:---:|
| Sugden ........ | 25·7 | 24·4–24·8 | 24·9–25·9 | 24–25·8 |
| Mumford, Phillips | 25·5 | | | |
| Vogel ......... | 26·1 | | | 22·2–24·0 |

In addition to their own surface tension, fluorinated compounds can also considerably decrease the surface tension of other compounds: the addition of less than 1% of perfluorocarboxylic acids to water decreases its surface tension from 72 to 20 dyne/cm (*274, 425*).

## Viscosity

The absolute viscosity of perfluorocarbons is higher than that of the paraffins and lower than that of water (10 millipoise at 20°C). The differences in

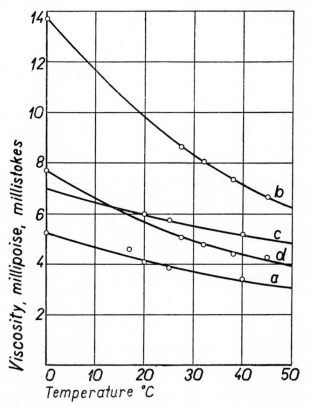

FIG. 34. Viscosity of heptane and perfluoroheptane: a) absolute viscosity of heptane, b) absolute viscosity of perfluoroheptane, c) kinematic viscosity of heptane, d) kinematic viscosity of perfluoroheptane.

viscosities between organic compounds and their perfluoroderivatives show best in a chart (Fig. 34). The viscosity index, i.e. the change of viscosity with temperature, is much higher with perfluoro compounds than with the paraffins.

Solubility

In the following paragraph, attention will be given to the solubility of fluorinated compounds in water, in inorganic solvents, in organic solvents, and to the solubility of some compounds in fluorinated derivatives.

The solubility of fluorinated derivatives in water has been systematically studied with Freons (17, 812). It is generally negligible. For example, only 0·026 g of dichlorodifluoromethane dissolves in 100 g of water at 27°C and 1 atm. The solubility of water in Freons is also very low (17) (Table 63).

TABLE 63

*Solubility of Water in Freons (17)*

| Compound | Solubility of Water in g per 100 g at | |
|---|---|---|
| | 0° | 30° |
| Dichlorodifluoromethane .................. | 0·0026 | 0·012 |
| Trichlorofluoromethane ................... | 0·0036 | 0·013 |

Higher polyfluoro- and perfluoroparaffins are only slightly soluble in water (327). On the other hand, polyfluoro- and perfluoroaldehydes, ketones and acids have very strong affinity for water. They easily form azeotropes and hydrates, which are dehydrated only with difficulty.

The solubility of fluorinated compounds in strong mineral acids, such as sulphuric acid or hydrofluoric acids, depends a great deal on their ability of forming oxonium, sulphonium or ammonium complexes. Fluorocarbons and perfluoroolefins, which do not have such possibility, are very slightly soluble in anhydrous hydrogen fluoride. While common ethers dissolve readily in concentrated sulphuric acid or anhydrous hydrogen fluoride, perfluorinated compounds such as perfluorotetrahydrofuran and perfluorotetrahydropyran are practically insoluble, because the basicity of their oxygen atoms which is responsible for the complex formation is decreased by the presence of fluorine atoms (476).

The solubility of fluorocarbons in organic solvents is usually much lower than that of the corresponding hydrocarbons. Very often two-phase systems are formed, which homogenize only at elevated temperatures. Relatively good solvents of fluorocarbons are acetone, chlorinated or fluorinated paraffins and aromatic hydrocarbons. The solubility of perfluorocyclohexane in various solvents is shown in Table 64 (327).

TABLE 64

*Solubility of Perfluorocyclohexane in Various Solvents (327)*

| Solvent | Solubility in % at °C | | |
|---|---|---|---|
| | 0 | 27 | 50 |
| Water.................... | | Insoluble | |
| Methanol ................ | | Insoluble | |
| Trichloroethylene .......... | | 0·2 | Miscible |
| Benzene ................. | | 2 | |
| Acetone ................. | | 8 | |
| Chloroform .............. | | 19 | Miscible |
| Carbon Tetrachloride ....... | 0·7 | Miscible | |
| Ethyl Ether ............. | | Miscible | |
| Benzotrifluoride ........... | Miscible | Miscible | |

Critical solution temperatures and composition of mixtures of fluorinated derivatives with non-fluorinated solvents are listed in Table 65 (*494*) and Table 66 (*600*).

TABLE 65

*Critical Solution Temperatures and Composition of Mixtures of Perfluoro(methylcyclo-hexane) with Organic Solvents (494)*

| Solvent | Critical Solution Temperature °C | Mole Fraction of Perfluoro-(methylcyclo-hexane) | Mole Fraction of the Second Component |
|---|---|---|---|
| Carbon Tetrachloride ........ | 26·8 | 0·3 | 0·7 |
| Chloroform ................ | 50·3 | 0·26 | 0·74 |
| Benzene ................... | 85·3 | 0·27 | 0·73 |
| Toluene.................... | 88·8 | 0·31 | 0·69 |
| Chlorobenzene.............. | 126·8 | 0·31 | 0·69 |

TABLE 66

*Critical Solution Temperatures and Compositions of Mixtures of Perfluoro Compounds with Non-fluorinated Solvents (600)*

| Solvent Pair | Consolution Temperature °K | Mole Fraction of the First Component |
|---|---|---|
| $C_7H_8-C_7F_{15}H$ | 309 | 0·75 |
| $C_7H_8-(C_4F_9)_3N$ | 415 | 0·82 |
| $C_7H_{16}-C_7F_{15}H$ | 305 | 0·61 |
| $C_7H_{16}-C_7F_{15}COOCH_3$ | 306 | 0·67 |
| $C_7H_{16}-(C_4F_9)_3N$ | 355 | 0·79 |
| $C_6H_{11}CH_3-C_7F_{15}H$ | 325 | 0·69 |
| $C_6H_{11}CH_3-C_7F_{15}COOCH_3$ | 318 | 0·75 |

The solubilities of fluorinated polymers have great importance for processing fluorinated plastics. Solubility and swelling of polychlorotrifluoroethylene in most organic solvents is negligible (p. 350). For the preparation of solutions of this compound special solvents are needed (p. 349).

### PHYSICO-CHEMICAL PROPERTIES

In the paragraph on physico-chemical properties, formation of molecular complexes, azeotropic mixtures, and the behaviour of fluorinated acids and bases will be discussed.

### Formation of Molecular Complexes and Azeotropes

Carbonyl compounds, with a large amount of fluorine atoms in the vicinity of the ketonic function, tend to withhold water and form crystalline hydrates (*523, 926, 1001*). The water is bound constitutionally, consequently its removal is very tedious and requires sometimes prolonged heating with phosphorus pentoxide (*926*).

The formation of hydrates was also observed with fluorinated acids. Thus perfluoropropionic acid forms a hydrate containing approximately 10% of water (*524*). Fluorinated acids and water form azeotropic mixtures having higher or lower boiling points than the acids alone. Boiling points and compositions of some azeotropes are listed in Table 67 (*524*).

TABLE 67

*Azeotropic Mixtures of Perfluorocarboxylic Acids with Water (524)*

| Acid | Content in the Mixture, % | B. P. of the Acid, °C | B. P. of the Mixture, °C |
|---|---|---|---|
| $CF_3CO_2H$ | 80 | 72·4 | 3–105 |
| $C_2F_5CO_2H$ | 91·8 | 96 | >96 |
| $C_3F_7CO_2H$ | 28 | 120 | 98 |

Surprisingly, some Freons form crystalline hydrates which readily decompose at elevated temperatures. Thus the hydrate of dichlorodifluoromethane is stable only below zero°C, and that of trichlorofluoromethane decomposes above 8°C (*17*).

## Fluorinated Compounds as Acids and Bases

Electronegative induction of fluorine atoms accumulated in one spot causes acidification of a neighbouring hydrogen atom. Thus hydrogen in bis(trifluoromethyl)amine is strongly positive (*383*). Also hydrogen in monohydroperfluoroparaffins $C_nF_{2n+1}H$ is reported to possess considerable acidity (*1044*), but this statement does not seem to be well substantiated.

Acidity of hydrogen in fluorinated diketones is responsible for their strong enolization (*800*). Enol content in some of such diketones is considerably higher than that of non-fluorinated diketones such as acetylacetone (88%) (*800*) (Table 68).

TABLE 68

*Enol Contents of Some Fluorinated Diketones* (*800*)

| Fluorinated Diketone | Enol Content, % |
|---|---|
| $CF_3COCH_2COCH_3$ | 92·5* |
| $CF_3CF_2COCH_2COCH_3$ | 120·3* |
| $CF_3CF_2CF_2COCH_2COCH_3$ | 118·5* |
| $CF_3CF_2CF_2CF_2COCH_2COCH_3$ | 120·0* |
| 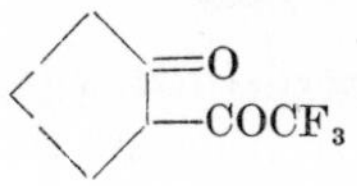 | 92·7* |
| 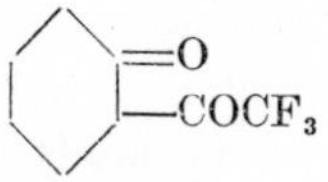 | 90·8* |
| 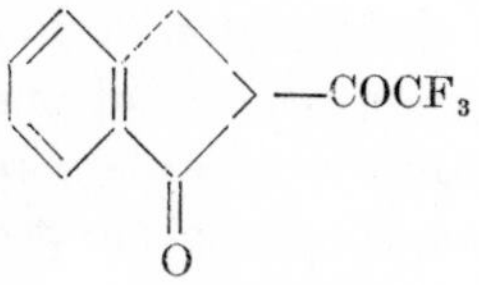 | 90·2 |
| 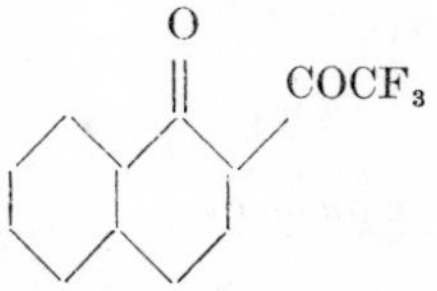 | 98·6 |

\* The maximum enol content is 200% as the compound contains two enolizable systems.

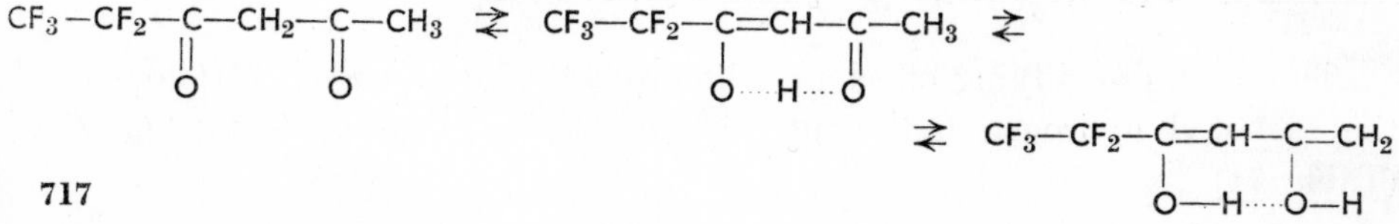

$$CF_3-CF_2-\underset{\underset{O}{\|}}{C}-CH_2-\underset{\underset{O}{\|}}{C}-CH_3 \rightleftarrows CF_3-CF_2-C=CH-\underset{\underset{O}{|}}{C}-CH_3 \rightleftarrows$$

$$\underset{O\cdots H\cdots O}{} $$

$$\rightleftarrows CF_3-CF_2-\underset{\underset{O-H\cdots O-H}{|}}{C}=CH-\underset{}{C}=CH_2$$

717

20*

Much attention has been focused on the study of the effect of the numqer of fluorine atoms upon the acidity of the hydroxyl hydrogen in fluorinated alcohols and acids. The acidity of alcohols increases with the number of per-fluoroalkyl groups bound to the carbinol carbon atom (467). Introduction of one more perfluoroalkyl group raises the dissociation constant approximately by one order (353, 448) (Table 69).

TABLE 69

*Dissociation Constants of Some Fluorinated Alcohols (353, 448, 467)*

| Fluorinated Alcohol | Dissociation Constant $K$ |
|---|---|
| $CF_3CH_2OH$ | $4 \ . \ 10^{-12}$ |
| $C_3F_7CH_2OH$ | $4 \cdot 3 \ . \ 10^{-12}$ |
| $C_3F_7CH(OH)C_3H_7$ | $4 \cdot 3 \ . \ 10^{-12}$ |
| $C_3F_7CH(OH)C_3F_7$ | $2 \cdot 2 \ . \ 10^{-11}$ |
| $CF_3C(OH)(CF_3)_2$ | $3 \ . \ 10^{-10}$ |
| $C_3F_7C(OH)(C_3F_7)_2$ | $1 \cdot 1 \ . \ 10^{-10}$ |

The acidity of tertiary alcohols carrying three perfluoroalkyl groups is comparable to that of phenols ($1 \cdot 3 \times 10^{-10}$). Substitution of fluorine for aromatic hydrogen atoms in phenol increases the dissociation constant considerably. Pentafluorophenol has $K = 4 \cdot 79 \times 10^{-6}$ (283), $2 \cdot 95 \times 10^{-6}$ (98).

Data accumulated in the field of fluorinated carboxylic acids allow us to draw conclusions as to the effect of the number and position of fluorine atoms upon the dissociation constant. In acetic acid, replacement of one hydrogen atom by fluorine increases the dissociation constant by two orders, replacement of the second and third atom of hydrogen always by one more order (444) (Table 70).

TABLE 70

*Effect of Replacement of Hydrogen by Fluorine upon the Acidity of Fluorinated Acids (444)*

| Acid | $CH_3CO_2H$ | $CH_2FCO_2H$ | $CHF_2CO_2H$ | $CF_3CO_2H$ |
|---|---|---|---|---|
| Dissociation Constant $K$ | $1 \cdot 8 \ . \ 10^{-5}$ | $2 \cdot 2 \ . \ 10^{-3}$ | $5 \cdot 7 \ . \ 10^{-2}$ | $5 \cdot 5 \ . \ 10^{-1}$ |

The strong acidifying effect of trifluoromethyl group upon the acidity of the carboxyl hydrogen fades out with increasing distance (444, 445, 664) (Table 71).

TABLE 71

*Effect of the Distance of a Perfluoroalkyl Group from the Carboxyl upon the Acidity of Fluorinated Acids (444, 445, 664)*

| Fluorinated Acid | Dissociation Constant $K$ |
|---|---|
| $CF_3CO_2H$ | $5 \cdot 5 \ . \ 10^{-1}$ |
| $CF_3CH_2CO_2H$ | $1 \ . \ 10^{-3}$ |
| $CF_3CH_2CH_2CO_2H$ | $7 \ . \ 10^{-5}$ |
| $CF_3CH_2CH_2CH_2CO_2H$ | $3 \cdot 2 \ . \ 10^{-5}$ |
| $CH_3CH_2CH_2CH_2CO_2H$ | $1 \cdot 56 \ . \ 10^{-5}$ |

$\alpha,\beta$-Unsaturated fluorinated acids are slightly stronger than the corresponding saturated acids (*664*) (Table 72).

TABLE 72

*Comparison of the Acidity of Saturated and Unsaturated Fluorinated Acids (664)*

| Saturated Fluoro-Acid | Dissociation Constant $K$ | Unsaturated Fluoro-acid | Dissociation Constant $K$ |
|---|---|---|---|
| $CF_3CH_2CH_2CO_2H$ | $7 \ . \ 10^{-5}$ | $CF_3CH=CHCO_2H$ | $33 \ . \ 10^{-5}$ |
| $C_3F_7CH_2CH_2CO_2H$ | $6 \cdot 6 \ . \ 10^{-5}$ | $C_3F_7CH=CHCO_2H$ | $59 \ . \ 10^{-5}$ |

The effect of the number and location of fluorine atoms on the acidity of $\alpha,\beta$-unsaturated acids appears in Table 73 (*446*).

TABLE 73

*Dissociation Constants of Halogenated Acrylic Acids (446)*

| Acrylic Acid | Dissociation Constant $K$ |
|---|---|
| $CH_2=CHCO_2H$ | $5 \cdot 56 \ . \ 10^{-5}$ |
| $CF_2=CHCO_2H$ | $68 \ . \ 10^{-5}$ |
| $CH_2=CFCO_2H$ | $280 \ . \ 10^{-5}$ |
| $CF_2=CFCO_2H$ | $1580 \ . \ 10^{-5}$ |
| $CCl_2=CClCO_2H$ | $6200 \ . \ 10^{-5}$ |

The fact that perchloroacrylic acid is four times as strong as perfluoroacrylic acid is somewhat surprising, since in other cases chlorine causes less acidity than fluorine.

Acidifying effect of fluorine appears also with sulphonic, phosphonic and arsonic acids. Trifluoromethanesulphonic acid is of the same or possibly greater acidity than sulphuric acid (*377*), and bis(trifluoromethyl)phosphonic acid almost as strong as perchloric acid (*241*). Relative acidities of some fluorinated acids and mineral acids are listed in Table 74 (*241*).

TABLE 74

*Relative Acidities of Mineral and Fluorinated Organic Acids (241)*

| Acid | Relative Acidity | Acid | Relative Acidity |
|---|---|---|---|
| $HNO_3$ | 1 | HCl | 9 |
| $CF_3CO_2H$ | 1 | $CF_3P(O)(OH)_2$ | 9 |
| $C_3F_7CO_2H$ | 1 | $H_2SO_4$ | 32 |
| $CF_3As(O)(OH)_2$ | 2·5 | HBr | 180 |
| $(CF_3)_2As(O)OH$ | 3·5 | $(CF_3)_2P(O)OH$ | 250 |
| | | $HClO_4$ | 360 |

The presence of fluorine atoms has a strong stabilizing effect upon malonic acid (*440*) and nitroacetic acid (*567*). The latter stands distillation under atmospheric pressure without decomposition, in strong contrast to its non-fluorinated analogue.

Accumulation of fluorine which enhances the dissociation of the carboxyl and hydroxyl hydrogen decreases the basicity of fluorinated amines. Here, too, the effect of the trifluoromethyl group fades off very rapidly with distance (*482*) (Table 75).

TABLE 75

*Dissociation Constants of Amines (482)*

| Amine | $K_B$ | Fluorinated Amine | $K_B$ |
|---|---|---|---|
| $CH_3CH_2NH_2$ | $4·5 \times 10^{-4}$ | $CF_3CH_2NH_2$ | $5 \times 10^{-9}$ |
| $CH_3CH_2CH_2NH_2$ | $4·5 \times 10^{-4}$ | $CF_3CH_2CH_2NH_2$ | $5 \times 10^{-6}$ |

The presence of fluorine in ethers considerably decreases the basicity of the ether oxygen. In perfluorotetrahydrofuran and perfluorotetrahydropyran the basicity of oxygen is so low that these compounds do not form oxonium complexes with strong mineral acids and do not dissolve in them (*476*).

BIOLOGICAL PROPERTIES

From the start much effort has been devoted to the study of the biological effects of fluorine in various types of organic compounds. In the majority of cases fluorinated analoga of biologically active compounds differ very slightly from the non-fluorinated compounds. Nevertheless a few exceptions are worth mentioning. An example is 1,1-bis(*p*-fluorophenyl)-2,2,2-trichloroethane or DFDT, which has higher knock-down power and lower toxicity toward warm-blooded animals than DDT. Generally speaking, single fluorine atoms contribute to biological activity, while accumulated fluorines tend to produce biological inertness.

Two groups of fluorinated compounds have aroused great interest: fluorophosphates and fluoroacetates. Both classes of compounds were subjected to systematic study during the second world war.

## Fluorophosphates

Fluorophosphates or phosphofluoridates, represented by *diisopropyl fluorophosphate* (*DFP*) (Equation 143, p. 110), *tetramethyldiamidofluorophosphate* (Equation 272, p. 156), *isopropyl methanefluorophosphonate* (*Sarin, Trilon 46*), and *methyl tert.-butylcarbinyl methanefluorophosphonate* (*Soman*)

$$
\textbf{718} \qquad
\underset{\underset{\displaystyle F}{|}}{\overset{\overset{\displaystyle O}{\|}}{CH_3-P}}-O-CH\!\!\begin{array}{l}{}^{\diagup CH_3}\\[-2pt]{}_{\diagdown CH_3}\end{array}
\qquad\qquad
\underset{\underset{\displaystyle F}{|}}{\overset{\overset{\displaystyle O}{\|}}{CH_3-P}}-O-CH\!\!\begin{array}{l}{}^{\diagup CH_3}\\ {}_{\diagdown C}\end{array}\!\!\begin{array}{l}{}^{\diagup CH_3}\\ {}^{\displaystyle |}_{\displaystyle CH_3}{}^{\diagdown CH_3}\end{array}
$$

are extremely toxic compounds, comparable to phosgene, cyanogen chloride, chloropicrin or hydrogen cyanide. They were produced in quantities during the second world war but fortunately were not used on either side. They inhibit cholinesterase, the enzyme which converts acetylcholine to choline in nerves, and thus impede the transmission of nerve impulses (*905, 908, 910, 911*). Outer symptoms of the action of this class of compounds are strong and long–lasting myosis, loss of visual accommodation, strong headaches and pain behind the eyes. Death occurs by asphyxia. A detailed description of the symptoms and pharmacological effects of this group of compounds has been published recently (*910*).

## Fluoroacetates

The second class is represented by the fluoroacetates. Surprisingly, their very high toxicity was discovered almost simultaneously in chemical laboratory and in nature. In South Africa, in the region of Pretoria, a shrub called "gifblaar" (*Dichapetalum cymosum*) was known to kill cattle. The toxic principle of this plant was isolated in the forties of this century and identified as *potassium fluoroacetate*. A kindred plant, "ratsbane" (*Dichapetalum toxicarium*) probably contains some higher fluorinated aliphatic acid. Not only salts but also esters of fluoroacetic acid are very toxic, especially *fluoroethyl fluoroacetate (143)*. Following the path of these compounds through the organism, it was found that fluoroacetic acid becomes involved in the Krebs cycle and produces *fluorocitric acid* by biosynthesis in some animal tissues. This compound blocks the cycle at the stage of tricarboxylic acisd by inhibiting the enzyme aconitase. This produces convulsions, disorder in respiration and heart trouble *(834, 910)*.

$$719 \qquad CH_2F—COOH \longrightarrow COOH—CH_2—\overset{\displaystyle OH}{\underset{\displaystyle COOH}{\overset{|}{\underset{|}{C}}}}—CHF—COOH \qquad [834]$$

A systematic study of this class of compounds revealed a connection to distance of the fluorine atom. It was observed that $\omega$-fluorocarboxylic acids having an even number of carbon atoms are very toxic whereas the toxicity of those having an odd number of carbons is almost hundred times lower *(909, 910)*. It was therefore concluded that only such acids that can be degraded by *beta*-oxidation to fluoroacetic acid, possess pronounced toxicity. (The acids with an odd number of carbon atoms do not yield fluoroacetic acid as a final oxidation product). This opinion has been supported by the fact that acids which have an even number of carbon atoms, but are incapable of undergoing *beta*-oxidation, show no toxicity. This is the case of ethyl $\gamma$-fluoro-$\beta,\beta$-dimethylbutyrate, where the *beta*-oxidation is stopped by the presence of a quanternary carbon atom *(910)*.

$$720 \qquad CH_2F—\overset{\displaystyle CH_3}{\underset{\displaystyle CH_3}{\overset{|}{\underset{|}{C}}}}—CH_2—COOC_2H_5 \qquad [910]$$

The regular alternation in toxicities of odd and even $\omega$-fluorocarboxylic acids is evident from Table 76 *(910)*.

Table 76

*Toxicities of $\omega$-Fluorocarboxylic Acids Tested on Mice by Injection in Propylene Glycol Solutions (910)*

| $\omega$-Fluorocarboxylate | $LD_{50}$mg/kg | $\omega$-Fluorocarboxylate | $LD_{50}$mg/kg |
|---|---|---|---|
| $FCH_2COOCH_3$ | 15 | $F(CH_2)_2COOC_2H_5$ | $>200$ |
| $F(CH_2)_3COOCH_3$ | Toxic | $F(CH_2)_4COOC_2H_5$ | $>160$ |
| $F(CH_2)_5COOC_2H_5$ | 4 | $F(CH_2)_{10}COOC_2H_5$ | $>100$ |
| $F(CH_2)_7COOC_2H_5$ | 9 | $FCH_2C(CH_3)_2CH_2COOC_2H_5$ | Non-toxic |
| $F(CH_2)_{11}COOC_2H_5$ | 20 | | |
| $FCH_2CH=CHCOOCH_3$ | Toxic | | |

## Other Fluorinated Derivatives

The same alternation in toxicities was noted in 1-fluoroparaffins *(821)* and other derivatives such as amino acids *(861)*.

Polyfluoro- and perfluoro-derivatives are generally biologically inert. Fluorinated methane and ethane halogen derivatives are among the least toxic gases and very high concentrations are considered safe still *(17, 708)*. These compounds can therefore be used without danger for aerosols in closed rooms.

On the other hand, a few perfluoro and polyfluoro-derivatives possess an extremely high toxicity: perfluoroisobutylene and unstated products appearing in the pyrolysis of Freons, difluorochloromethane *(799)*, or teflon *(350)*. Some derivatives of cyclopropane *(799)* are suspected to be toxic compounds.

721

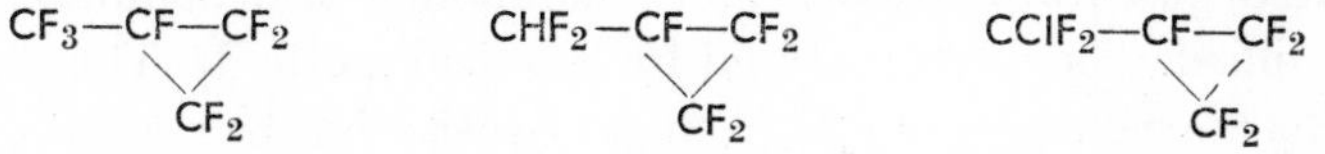

$$CF_3-CF-CF_2 \qquad CHF_2-CF-CF_2 \qquad CClF_2-CF-CF_2$$
$$\diagdown \diagup CF_2 \qquad\qquad \diagdown \diagup CF_2 \qquad\qquad \diagdown \diagup CF_2$$

*[799]*

# IX

## ANALYSIS OF ORGANIC FLUORIDES

ANALYSIS of organic fluorides is a vast and rather difficult problem. It involves gaseous, liquid and solid compounds of varied chemical behaviour.

Fluorinated gases are usually non-inflammable and combustion analysis cannot be applied. They are analysed by physical methods. The older method of *determining the molecular weight* from gas density has been replaced by *infra-red spectroscopy* (*210, 948*). Invaluable service has been rendered by the discovery of *gas chromatography* which allows an easy separation and identification of not only gaseous but also liquid fluorides. While analytical distillation required much time and relatively large amounts of material, today's gas, or better still *gas-liquid chromatography*, consumes only fractions of a gram and takes less than one hour to achieve accurate separation and identification (*262, 265, 317, 773, 871, 934, 968*).

To determine the fluorine content in organic compounds fluorine must be converted to the fluoride ion. Some organic compounds liberate fluoride ions relatively easily. When fluorine is bound to silicon, phosphorus or other metalloids, a short contact with water or alkalis hydrolyzes the fluorine as an anion. Hydrolytic removal of fluorine attached to carbon is possible only with some types of compounds. Acyl fluorides, sulphonyl fluorides and $\alpha$-fluoro-derivatives of carbonyl compounds or acids liberate fluorine on more or less intensive hydrolysis. Even the relatively tenacious trifluoromethyl group releases fluoride ions on drastic hydrolysis. Nevertheless this way of mineralizing organic fluorine is not general. Usually more energetic treatment is necessary to destroy the organic molecule: contact with alkaline metals or peroxides, especially at higher temperatures, or combustion in oxygen.

The determination of the fluoride ions is also much more intricate than the determination of other halogens. In fact, no entirely satisfactory method has been developed to date, as shown by the number of new analytical procedures offered every year. Details have been published in several reviews (*148, 625, 948*).

## QUALITATIVE TESTS FOR FLUORINE

*Gaseous fluorine* is estimated qualitatively by a primitive but dependable test, ignition of a wood splinter in contact with air and fluorine. Such test is carried out to ascertain whether a fluorine supply-line contains a sufficient amount of fluorine. If there are no sampling outlets, the line is disconnected and a 10–15 cm. long thin splinter of soft wood is prodded in the gas stream.

The *fluoride ion* is qualitatively detected either by precipitation of insoluble fluorides, or by bleaching of coloured lakes by formation of colourless complexes.

The first type of test is the detection of fluorides by means of *cerium nitrate (958)*:

Approximately one millilitre of a solution containing fluoride ions is acidified with acetic acid and added to an equal volume of the reagent prepared by dissolving 0·2 g of cerium-(III)-nitrate in 5 ml. of water. A milky gelatinous precipitate of cerium-(III)-fluoride is formed. Chlorine, bromine, iodine and cyanogen do not interfere (*958*).

A decoloration test is carried out by adding a fluoride to the acidic solution of red *ferric thiocyanate*, the colour of which disappears because of the formation of a stronger colourless complex of iron with the fluoride. This test is not specific, as it is given also by other complex forming anions such as the phosphate ions.

The reaction of fluorides with a *zirconium lake of alizarin* is very effective (*972*).

The reagent, alizarin lake, is prepared by dissolving 0·5 g of alizarin in 200 ml. of warm ethanol, adding a solution of 1·5 g of zirconium tetrachloride in 75 ml. of ethanol, allowing the lake to settle, pouring off the excess of ethanol, adjusting the volume of the suspension thus formed to 25 ml., and shaking 5 ml. of this suspension in 100 ml. of water until a colloidal solution is formed.

To a 2·5 ml. portion of the neutral or slightly acidic solution of fluoride is added an equal volume of concentrated hydrochloric acid and 0·5 ml. of the reagent and the mixture is stirred and allowed to stand not longer than 15 seconds. In the presence of 0·3 mg of fluoride the red colour turns yellow immediately, in the presence of 0·15 mg after 5 seconds, and in the presence of 0·03 mg after 15 seconds. Chlorates, bromates, and iodates interfere.

The decolouration reaction can be used for preparing papers for the detection of fluorides (*972*). Filter paper is soaked in a solution of 2 g of zirconium chloride in 100 ml. of ethanol, dried, immersed in a solution of 0·1 g of alizarin in 100 ml. of ethanol, dried again, and cut in strips. These are immersed in samples, to which equal volumes of concentrated hydrochloric acid have been added. In the presence of fluorides the paper becomes yellow in 5 to 10 seconds. An orange colouration cannot be considered as positive.

The most recent colour reaction for fluoride ions is the colour change of *cerium*[III] *alizarine complexon (1,2-dihydroxy-3-anthraquinonylmethylamine-N,, N'-diacetic acid)* from red to lilac blue *(58)*.

A direct test for fluorine in organic compounds is the etching of glass produced by hydrogen fluoride liberated by *heating the compound with concentrated sulphuric acid (948)*. A more dependable method seems to be mineralization of the organic compound by *fusion with metalic potassium (1030)*:

A 3–5 mg sample of the organic fluorine derivative is fused with a piece of potassium (or sodium) metal in a glass ampule. The ampule, while still hot, is immersed in 2 ml. of water, and the solution after the spontaneous decomposition of the excessive metal is filtered into 1 ml. of a solution of alizarin-zirconium lake, which turns yellow in the presence of fluorine.

The most elegant method of mineralization is heating the organic derivative with the strongly oxidizing product given by thermal decomposition of *silver permanganate*. In one sample it is possible to detect not only fluorine but also nitrogen, sulphur and other halogens. Special arrangements allow the carrying out of the test with a minimum amount of material and in a few minutes *(583, 584)*.

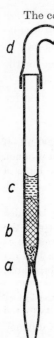

The combustion catalyst is prepared by heating a small amount of silver permanganate in a dry test-tube with a free flame until a spontaneous decomposition takes place and yields a brown-black voluminous residue. About 50 mg of the cooled mass is spread on a piece of glazed paper and mixed thoroughly with 0·1–1 mg of the compound to be tested. The mixture is then filled into a tube (Fig. 35) whose tapered part has been stuffed with shredded asbestos. Liquid samples are dropped from a capillary directly into the filled tube. The tube is then gently heated with a free flame until red glowing appears. Then 2–3 drops of distilled water is added and the liquid forced into a test tube containing 0·5 ml. of a solution of sodium alizarinsulphonate thorium lake. In the presence of fluorine, the pink colour changes to yellow.

The test solution is prepared by mixing 2 parts of 0·05% solution of sodium alizarin sulphonate with two parts of a buffer (containing 0·45 g of chloroacetic acid and 2 g of sodium hydroxide in 100 ml. of water) to which 4 drops of 0·05 N thorium nitrate solution has been added.

In the same sample, nitrogen can be detected by means of a 1% solution of diphenylamine in concentrated sulphuric acid, and sulphur with 5% solution of barium nitrate. By modifying the treatment of the sample, all the halogens can be qualitatively determined *(584)*.

FIG. 35. Tube for qualitative test for fluorine in organic compounds: a) asbestos plug, b) layer of combustion catalyst (decomposed AgMnO$_4$), c) layer of distilled water.

## QUANTITATIVE DETERMINATION OF THE FLUORIDE ION

For the quantitative determination of a fluoride a long series of gravimetric, volumetric and spectrophotometric methods has been and is still being devised. None is general or entirely satisfactory.

The most common gravimetric determination is based on the precipitation of *lead chlorofluoride* which forms an easily filterable precipitate. To a solution containing fluoride ions, an excess of lead acetate and as much hydrochloric acid or sodium chloride is added to obtain an equimolal ratio with the expected fluoride ion. If the fluoride content is unknown, a preliminary orientation determination is required. The lead chlorofluoride precipitate is washed with a saturated solution of lead chlorofluoride, dried and weighed. Instead of lead acetate, lead chloride or lead chloronitrate, obtained by boiling lead chloride and lead nitrate in water, may be used (*61*). Instead of weighing the insoluble lead chlorofluoride, its amount can be estimated *volumetrically*. The precipitate is dissolved in nitric acid and the chlorine ion is determined either according to Volhard (*719*), or according to Votoček (by titration with mercuric nitrate using sodium nitroprusside as a turbidity indicator) (*1052*). Lead chlorofluoride may also be dissolved in complexone III and the chloride determined argentometrically with Variamine Blue as indicator (*260*). Another possibility is the precipitation of lead chlorofluoride and determination of the excess of lead chloride by *complexometric titration* with ethylenediamine tetraacetic acid using Catechol Violet or Xylenol Orange as indicators (*1059*). This method, however, needs to be modified, as the precipitation of lead chlorofluoride under the conditions described does not seem to be quantitative. A new gravimetric method is based on precipitation of *lithium fluoride* by means of lithium chloride in aqueous-ethanolic solution (*155*). Fluorides also precipitate *calcium* and *cerium ions*. These precipitates are, however, unsuitable for gravimetric determination because of their gelatinous nature. Nevertheless the precipitation may be combined with complexometric determination of the excess of the precipitant (calcium chloride) by titration with ethylene diamine tetraacetic acid using Eriochrome Black T as indicator (*55*). When cerium nitrate is used for precipitation of cerium trifluoride, the excess of the precipitant is determined by titration with potassium permanganate at 80°C (*52*). A direct volumetric determination with cerium salts can be carried out in 50% aqueous methanol at pH 5–6 using cerium trichloride and Murexide as indicator (*140*). The titration can also be effected by cerium nitrate at 80°C using Amphomagenta or Methyl Red as indicators (*52*).

The most common volumetric determination of the fluoride ion is *titration with thorium nitrate* using sodium alizarin sulphonate as indicator. Thorium tetranitrate forms non-dissociated thorium tetrafluoride. The excess of thorium nitrate solution changes the yellow colour to a pink shade. Since the observation of the transition colour is extremely difficult, comparison is needed with a 50 ml. standard containing 0·04 ml. of 0·05% aqueous sodium alizarin sulphonate and 0·04–0·07 ml. of 0·001 M thorium nitrate (507). The titration is sensitive to acidity, it is best carried out at pH 3–4, in a buffer prepared from 1 part of sodium hydroxide and 2 parts of chloroacetic acid (507).

The described determination of fluorides is suitable for solutions containing no sulphates, phosphates or arsenates. If it is necessary to carry out the determination in their presence or in the presence of halide ions, the fluorine

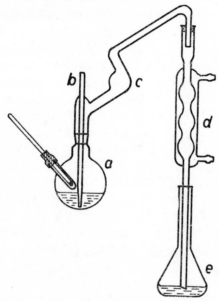

FIG. 36. Apparatus for distillation determination of fluorides:
a) distilling flask, b) steam inlet, c) distilling head, d) condenser,
e) volumetric flask.

ions must be separated by distillation. The *distillation method* is based on the formation of fluorosilicic acid from glass and hydrofluoric acid liberated from fluoride by sulphuric or perchloric acid, on the "distillation" of the fluorosilicic acid, and on its reconversion to fluoride by an excess of alkalies:

$$NaF + H_2SO_4 \longrightarrow NaHSO_4 + HF$$
722
$$4 HF + SiO_2 \longrightarrow SiF_4 + 2 H_2O$$
$$SiF_4 + 2 HF \longrightarrow H_2SiF_6$$
$$H_2SiF_6 + 6 NaHCO_3 \longrightarrow 6 NaF + SiO_2 + 4 H_2O + 6 CO_2$$

The distillation is carried out in a special apparatus (Fig. 36), whose distilling flask is heated to a temperature around 135°C, while steam is passed through the flask. Automatic regulation of the temperature at the desired value can be effected by surrounding the flask with a jacket filled with boiling *sym*-tetrachloroethane (*514*). The distillate is collected in an alkaline solution, and after acidification, is titrated with thorium nitrate to alizarin sulphonate (*507, 1088*).

For special purposes, especially for micro-scale determination, *spectrophotometric methods* based on decoloration by fluoride ions are very convenient (*57, 161, 162, 261, 591, 887*). Decolouration of the red complex of *trivalent iron with salicylic acid* (*887*) is considered specially suitable in connection with the Schöniger method of combustion of organic compounds (p. 321). A new colour reaction of fluoride ions is given by *thorium chloroanilate*. As little as 1 p.p.m. of fluoride can be determined by measuring absorption at 540 or 330 m$\mu$ (*13*).

## DETERMINATION OF FLUORINE BOUND TO CARBON

A quantitative conversion of organic fluorine to fluoride ion can be achieved by combustion of the organic compound or by its decomposition by alkaline metals, or by fusion with alkali peroxides.

*Combustion of organic fluorides* is carried out in standard quartz combustion tubes at a temperature of 900°C in a current of air or oxygen. Fluorine forms hydrogen fluoride and silicon tetrafluoride (quartz powder is sometimes spread on the bottom of the tube in order to minimize its etching). The tube is flushed with oxygen, nitrogen, hydrogen, and again nitrogen, the silicon tetrafluoride formed is hydrolysed in an alkaline solution of hydrogen peroxide to fluoride ions, and these are titrated with cerium-(III)-nitrate to Methyl Red and Bromocresol Green (*512*), or with thorium nitrate (*156*). If the organic compound also contained chlorine, its amount is determined in another aliquot by silver nitrate titration. Both methods are suitable for the analysis of halofluorocarbons.

A similar principle is the basis for the microdetermination of fluorine by combustion in a current of oxygen at 900°C in a quartz or Vycor (96% $SiO_2$) glass. A platinum gauze is used as a combustion contact, and silver gauze for absorption of halogens and sulphur. Combustion gases are absorbed in water and the fluoride ion is determined acidimetrically (*170*).

Very good results have been obtained by pyrolysis of organic fluoro-derivatives in a current of oxygen, completed by combustion of the decomposition products in an oxygen–hydrogen flame and absorption of the combustion gases. Originally, absorption was carried out with 1 N sodium hydroxide, the fluoride ions were transformed into lead chlorofluoride, and its amount was determined indirectly by the Volhard titration (*1083, 1084*) or titration with thorium nitrate (*1010*). The combustion gases can be absorbed in water instead of sodium hydroxide. This makes the final colour change more distinct (*582*).

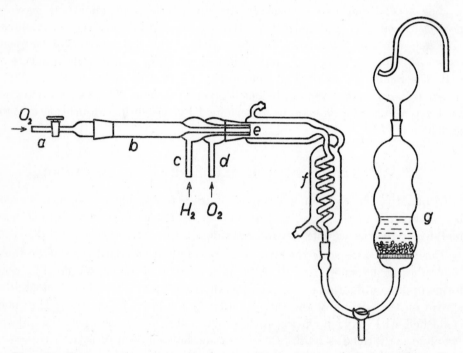

FIG. 37. Apparatus for the Wickbold determination of fluorine in organic compounds: a) oxygen inlet, b) combustion tube, c) hydrogen inlet, d) oxygen inlet, e) oxygen-hydrogen burner, f) water condenser, g) absorption vessel for hydrogen fluoride.

The combustion requires a special quartz apparatus (Fig. 37) which is vulnerable, and a steady-nerve worker; but the method is dependable, very rapid (15–20 minutes), and of sufficient accuracy for a semi-micro determination.

A 35–50 mg sample is pyrolyzed and burnt in the Wickbold apparatus (Fig. 37). The hydrogen fluoride formed is absorbed in the water formed by the hydrogen burning and condensed in the absorption vessel. The content of the vessel is transferred quantitatively to a titration flask so that the final volume after rinsing does not exceed 150 ml. Eight drops of 0·05% solution of sodium alizarin-sulphonate is added, the solution is neutralized with 0·5 N sodium hydroxide to the colour change to pink, 1 ml. of a buffer (prepared by dissolving 9·45 g of chloroacetic acid and 2 g of sodium hydroxide in 100 ml. of water) is added, and the yellow solution is titrated with 0·1 N thorium nitrate solution (containing 13·8 g of $Th(NO_3)_4$. $4 H_2O$ in 1 litre) to the first tint of pink colour (*583*).

The new (Schöniger) method, still in development, is based on the combustion of a fluorinated organic compound in a filter paper on a platinum gauze in oxygen in an Erlenmeyer flask (*887, 925*). Addition of sodium peroxide to the organic compound improves the combustion (*932*). The mineralized fluorine is determined volumetrically (*925*) or photometrically (*57, 887, 932*).

The solid compound is wrapped into a strip of filter paper (for liquids the method is less suitable), which is fastened to a platinum wire loop or to a platinum gauze mounted in a ground glass stopper of an Erlenmeyer flask. The flask is charged with a few millilitres of distilled water, filled with oxygen, the filter paper is ignited, and the stopper with the burning filter paper is quickly and tightly inserted into the neck of the flask. After the combustion is completed the content of the flask is stirred to speed up the gas absorption, 5 ml. of 0·01 N hydrochloric acid is added, the solution is boiled shortly to expel carbon dioxide, then neutralized with 0·01 N sodium hydroxide to the green shade of Bromothymol Blue, acid is added to change the colour to yellow (pH 5–6), the solution is diluted with an equal volume of methanol, and after adding 5 drops of 1% solution of murexide, the content of the flask is titrated with 0·01 N cerium chloride to the colour change from violet to orange (*925*).

For microscale operation, the Schöniger method is combined with spectrophotometric determination of fluoride by decolorizing the red complex of trivalent iron and salicylic acid (*887*). Spectrophotometric measurement of the colour change of cerium alizarin complexonate is suitable for determining as little as 10 $\gamma$ of fluoride (*57*).

Another kind of mineralization of organic fluorine is *decomposition by alkaline metals*. It is carried out either in liquid ammonia, or else by fusion in a bomb.

A sample containing about 0·1 g of fluorine is put into an ampoule which is placed in a Carius tube containing 5 ml. of ether and 0·5–1 g of sodium. The tube is cooled in a dry-ice bath and 10–15 ml. of ammonia is condensed into it. It is sealed and tumbled for 5 hours at room temperature. After cooling with dry ice the tube is opened, its contents poured into an Erlenmeyer flask containing 10 ml. of 95% ethanol, the tube is flushed several times with ethanol and distilled water, and the solution is diluted to 250 ml. An aliquot is analysed for fluorine by the lead chlorofluoride method combined with the Volhard determination of chlorine (*719*).

21

Fusion with alkali metals is carried out in a nickel micro-bomb fitted with a copper gasket (Fig. 38). The mineralization is then followed by the determination of the fluoride ion as lead chlorofluoride (*64*) or by thorium nitrate titration (*54, 900*). This last method has been worked out for micro-determination.

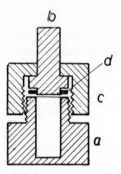

FIG. 38. Bomb for alkaline fusion of organic fluorides: a) body, fitted with thread, b) threaded lid (cap), c) union, d) gasket.

A sample containing 1–5 mg of fluorine is weighed into a little glass container or capillary tube, inserted into a Parr microbomb of 2·5 ml. capacity, a piece of sodium (30–50 mg) is added, the bomb sealed and heated with a Bunsen burner for 5–10 minutes. After cooling the bomb is opened, the lid washed with water into a 100 ml. beaker, the content of the bomb is carefully decomposed with a few drops of alcohol or water, more water is added, and the bomb flushed quantitatively into the beaker. Fluorine is determined by the distillation method and thorium nitrate titration (*626*).

Another modification is the fusion of an organic fluoride, held in a gelatin capsule with potassium in a nickel bomb for 2 hours at 500–550°C, followed by titration of the fluoride ion with thorium nitrate. If chlorine is present, another aliquot is analysed for chlorine by the Volhard method (*555*). Fusion with alkaline metals can be carried out even in special glass tubes (*796*).

Finally, mineralization can be achieved by *fusion of organic compounds with alkali peroxides*.

A sample containing about 0·06 g of fluorine is fused with 2·5 g of sodium peroxide and with (or without) 0·5 g of sugar in a nickel bomb. The content of the bomb is dissolved in 200 ml. of hot water, boiled to destroy the excess of peroxide, and fluorine is determined as lead chlorofluoride by the Volhard titration (*166*).

For most fluorine determination, standards of sufficient purity must be used: sodium fluoride for inorganic analysis, or fluoroacetamide for organic fluorine.

*Sodium fluoride* is obtained by adding sodium carbonate to analytical 40% hydrofluoric acid in a platinum dish, allowing the mixture to stand for 4 hours at room temperature, evaporating the content to dryness, repeating the evaporation after adding another portion of hydrofluoric acid, fusing the dry residue, grinding the cooled melt in an agate mortar, and drying the product at 110°C (*166*).

*Fluoroacetamide* is prepared by mixing 39 g of methyl fluoroacetate with 30 ml. of aqueous ammonia (d 0·89), allowing the mixture to stand overnight, filtering off the deposited crystals (16–21 g), and recrystallizing the crude product from chloroform, acetone or carbon tetrachloride. M.p. 108°C (*166*).

## DETERMINATION OF OTHER ELEMENTS IN ORGANIC FLUORINE COMPOUNDS

The presence of fluorine in organic compounds requires modification of methods used for the determination of other elements. Sometimes elementary analysis of a fluorinated compound can be carried out in the ordinary apparatus; nevertheless reliable results are obtained only if special precautions are taken in carrying out the determinations.

A simple *carbon–hydrogen determination* is based on the combustion of the compound in a quartz tube in a current of oxygen, finishing the oxidation on a platinum contact gauze, absorption of sulphur and halogen (including fluorine) in silver cotton, the silicon tetrafluoride formed on granulated sodium fluoride, and weighing of water and carbon dioxide by absorption in dehydrite and ascarite, respectively (*56*). For compounds containing nitrogen, a special filling must be used to stop the nitrogen oxides.

Another method of attractive simplicity is the combustion of the compound in a current of oxygen over a layer of the decomposition product of silver permanganate, and absorption of fluorine on a layer of minium (*506*).

The combustion tube of Supremax glass 35 cm long and 10 mm in diameter is filled as follows: A layer of asbestos, an 8 cm layer of the decomposition product of silver permanganate, a plug of asbestos, a 5 cm layer of minium on pumice, and a 2 cm plug of silver cotton. The combustion is carried out by means of traveling electric furnace at 550°C; the water is absorbed in anhydrone, and carbon dioxide in ascarite. With samples of 15–20 mg the combustion takes 10–15 minutes and the flushing of the apparatus with oxygen another 20 minutes. For the micro-scale determination, the tube is only 30 cm long, the layers of the silver permanganate and minium are cut down to 5 and 3·5 cm respectively, and the combustion takes only 5–7 minutes with samples of 3–4 mg.

Another modification of the classical carbon–hydrogen determination is absorption of the hydrogen fluoride liberated in the combustion by magnesium oxide placed around the sample in the combustion zone (*299, 300*).

The same principle, viz. the absorption of hydrogen fluoride by magnesium oxide, is adjusted for the *simultaneous determination of carbon, hydrogen and fluorine* from one sample. Magnesium oxide is placed in a quartz test-tube, ignited, and the test-tube is weighed. A weighed sample is placed in the test tube containing the magnesia, combustion and absorption are carried out as usual, and the fluorine content is computed from the weight increase of the magnesia test-tube (*300*). In another process, magnesia is packed in a perforated quartz cartridge. The combustion is carried out normally at 1000°C, carbon dioxide and water are absorbed in a conventional train, the magnesia cartridge is then treated with steam at 1000°C, and the hydrogen fluoride thus liberated determined by titration (*587*).

Another modification of simultaneous determination of carbon, hydrogen and fluorine consists in burning the fluorinated compounds in a quartz tube 800–850 mm long and 17–18 mm wide filled with a 50–60 mm layer of copper oxide and lead chromate; quartz powder 10–15 times the weight of the sample provides for the conversion of hydrogen fluoride to silicon tetrafluoride. The temperature in the sample zone is maintained at 450–500°C for one hour, at 500–550°C for half-an-hour, and the temperature in the filling zone is kept at 850–900°C. Water is absorbed in concentrated sulphuric acid, silicon tetrafluoride in a solution of potassium fluoride, and carbon dioxide in soda-lime asbestos (*775*).

In practice, the simultaneous determination of carbon, hydrogen and fluorine is hardly more convenient than two separate operations, the customary elementary analysis followed by a fluorine determination.

Simultaneous estimation of *carbon and fluorine* in volatile compounds is carried out by combustion in platinum tubes of a mixture prepared by evaporating the sample in a current of oxygen saturated with water vapour. The combustion is performed on a platinum gauze at 1250–1275°C, hydrogen fluoride is absorbed in water, and carbon dioxide in magnesium perchlorate, after drying by passage through concentrated sulphuric acid. The fluoride anion is determined by alkalimetric titration of hydrogen fluoride (*738*). Simultaneous determination of *carbon, fluorine and chlorine or bromine* in compounds containing no hydrogen, is carried out by combustion in a current of oxygen in a quartz tube at 1000°C over six platinum gauzes separated by layers of quartz sand. Fluorine is transformed into silicon tetrafluoride; the halogens are absorbed by silver at 295°C, silicon tetrafluoride in a mixture of aluminium oxide, sodium fluoride and drierite at 175°C, and carbon dioxide is caught in ascarite. The method is suitable for the semimicro-analysis of polyfluoro-halogen derivatives and fluorocarbons. Errors of the method may be rather high (*1036*).

If only *hydrogen* is to be determined in compounds containing fluorine or also chlorine, the sample is pyrolyzed at 1300°C in a stream of nitrogen in a platinum tube. Gaseous products of the pyrolysis are absorbed in water, the solution is filtered, and when only hydrogen fluoride is present, boiled for 5 minutes and titrated with sodium hydroxide to phenolphthalein. In the presence of chlorine, the filtrate is diluted to 500 ml., one aliquot is boiled for 5 minutes with 30% hydrogen peroxide and titrated to phenolphtalein with sodium hydroxide, while the other aliquot is treated with a titrated iodine solution, acidified with sulphuric acid, and the excess of iodine titrated with thiosulphate. The hydrogen content is determined from the difference of the two titrations (*718*).

For the determination of *nitrogen* in fluorinated compounds, the Dumas micromethod is recommended with the temperature of the stationary furnace at 800°C and that of the travelling furnace at 900°C (*900*), or else the Kirsten method with the furnace temperature of 1000–1050°C, time of combustion 25–35 minutes, permanent filling composed of nickel oxide, asbestos and hopcalite, and temporary filling of ground glass and nickel oxide in the ratio 1 : 2 (*59*).

*Halogens* in fluorinated compounds are determined simultaneously with fluorine (*555*, *1036*), or by the Carius method (*474*, *900*). A universal method suitable for the determination of all the halogens including fluorine, eventually in the presence of each other, is based on fusion of the sample with sodium in a nickel bomb, titration of one aliquot with thorium nitrate, and determining in other aliquots chlorine with mercuric oxycyanide, bromine after oxidation with hypochlorite and iodine after oxidation with bromine (*62*).

Determination of *sulphur* by the Carius method is surpassed by fusion with sodium in a bomb at 600°C, oxidation of the solution by hydrogen peroxide, precipitation of barium sulphate, dissolving of it in an excess of ethylene diamine tetraacetic acid, and back titration with magnesium chloride using Solochrome Black as indicator (*60*).

*Phosphorus* can be estimated as ammonium phosphomolybdate after the oxidation of the sample with fuming nitric acid at 300°C (*900*).

For the determination of *oxygen* in fluorine compounds, a known amount of oxygen isotope $^{18}O$ is added to the sample, and after exposure to a sufficiently high temperature for a time long enough to establish equilibrium, the ratio $^{18}O$ to $^{16}O$ is measured (*556*).


ANALYSIS OF FLUORINE AND HYDROGEN FLUORIDE

Crude elemental *fluorine* leaving electrolytic cells contains hydrogen fluoride, oxygen and nitrogen. The gaseous mixture is freed of hydrogen fluoride by passage over granulated sodium fluoride. The content of fluorine is estimated by passing the gas over anhydrous sodium chloride, the chlorine thus liberated is introduced into a 2 N solution of sodium hydroxide, and the hypochlorite formed is determined by iodometry. This method is suitable for gases containing more than 50% of fluorine. For mixtures below 10%, direct liberation of iodine from aqueous hydroiodic acid is

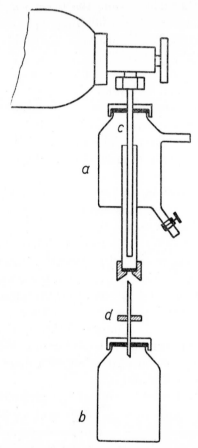

FIG. 39. Sampling of hydrogen fluoride for water content determination: a) sampling bottle, b) absorption and titration bottle, c) hydrogen fluoride inlet tube, d) connecting needle.

more convenient. Side reactions of fluorine with water-producing oxygen and oxygen fluoride are negligible and do not exceed 2% of the main reaction (*1049*).

A qualitative test for *fluorine in air* is based on the conversion of fluorescein to eosin by passing the sample gas through silica gel saturated with a solution of potassium bromide in potassium carbonate and fluorescein in potassium hydroxide (*831*).

Commercial *anhydrous hydrogen fluoride* contains, as a rule, small amounts of water, sulphur dioxide, sulphuric acid, fluorosulphonic acid, and fluorosilicic acid. Complete analysis is carried out in a polyethylene vessel in a sample diluted with five volumes of acetic acid. Sulphur dioxide is determined by iodine titration, hydrogen sulphide by precipitation as cadmium sulphide, decomposition of the sulphide and titration of the liberated hydrogen sulphide with iodine. Total sulphur content is determined from barium sulphate

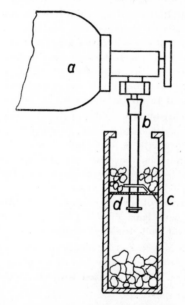

FIG. 40. Sampling of hydrogen fluoride for analysis: a) hydrogen fluoride cylinder, b) hydrogen fluoride inlet tube, c) weighing vessel, d) perforated bottom.

after oxidation with bromine, silicon dioxide is determined photometrically as silicomolybdate, and water is measured by the Karl Fischer method (*185*). Since the water content in anhydrous hydrogen fluoride is critical, the process for its determination will be described in detail (*797, 863*).

The most difficult problem in anhydrous hydrogen fluoride analysis is *sampling*. Therefore, special sampling vessels have been designed (*797, 938*) (Fig. 39, 40).

*Determination of Water in Anhydrous Hydrogen Fluoride* (*797*).

The device for the determination of water in anhydrous hydrogen fluoride (Fig. 39) consists of a sampling bottle (*a*), a titration bottle (*b*), an inlet tube (*c*), and a transfer hollow needle $(d)$. The sampling bottle is made of a standard 100 ml. polyethylene bottle with a screw cap. One side arm (for applying air pressure) is welded below the neck, another (for draining samples) at the bottom. A third polyethylene tube protruding through the bottom of the bottle and closed by a special rubber stopper serves as a measure for the hydrogen fluoride. The cap of the bottle is perforated in the centre and fitted with an inner gasket made of soft rubber, protected against the hydrogen fluoride vapours by a polyethylene or polytetrafluoroethylene sheet. The titration bottle is also a 100 ml. polyethylene bottle closed by a gasketed cap similar to that of the sampling bottle. The inlet tube for hydrogen fluoride is made of stainless steel, is inserted tightly into the hole in the cap of the sampling bottle, and is fastened to the hydrogen fluoride tank by means of a union. A transfer needle is a syringe needle fitted with a welded-on brass ring.

To take a sample of anhydrous hydrogen fluoride, the inlet tube is screwed to the tank, and the sampling bottle is so fastened that the inlet tube reaches almost to the bottom of the measure. A 0·5 atm pressure of air is applied to the upper side of the sampling bottle to prevent the hydrogen fluoride from boiling. With the lower draining arm pinched by a screw clamp, approximately 60–70 ml. of hydrogen fluoride is allowed to flow in to flush the bottle. It is drained, and the procedure repeated three times. Then hydrogen fluoride is introduced to fill the measure.

The titration bottle is charged with 50 ml. of the absorption solution, a 1:1 mixture of anhydrous pyridine and anhydrous methanol. It is weighed exactly, placed below the sampling bottle, and connected with it by stabbing one end of the needle through the lid of the titration bottle, and then the other end through the stopper of the measure in the sampling bottle. Thus hydrogen fluoride flows from the measure to the absorption solution of the titration bottle, This is then removed and weighed. The lid of the titration bottle is replaced by a rubber cap with a hole for the tapered end of a buret, and the solution is titrated with Karl Fischer reagent, till the brown colour (which is visible through the translucent polyethylene) is reached.

The Karl Fischer reagent is prepared by saturating 48 g of dry pyridine with sulphur dioxide to the weight increase of 14 g and by adding a solution of 15 g iodine in 150 ml. of dry methanol. The reagent is standardized towards methanol containing a known amount of water (*863*).

For a complete hydrogen fluoride analysis, the following procedure has been worked out (*15, 938*):

A sample of anhydrous hydrogen fluoride is drawn from the tank as illustrated in Fig. 40. The sampling vessel $(c)$ is placed on an accurate balance, its lower compartment is filled with 80 g of crushed ice (to compensate for the heat of dilution), the perforated bottom $(d)$ is covered with another 50 g of ice (for absorbing the hydrogen fluoride vapours), the metal tube $(b)$ is screwed to the tank valve $(a)$, and in the course of approximately 4 minutes, about 40 g of liquid hydrogen fluoride is allowed to flow into the vessel. The tube through which the sample flows must have its lower end sufficiently high, so that it does not reach the level of the liquid formed by mixing the sample with the ice. After reading the exact weight of the sample, the inlet tube is disconnected from the tank valve, allowed to sink into the sampling vessel, the vessel is closed by a paraffin-coated rubber stopper, and the content of the vessel is stirred. After opening the vessel the inlet tube is removed. (The hydrogen fluoride solution on its walls does not affect the analysis, as aliquots are taken for the individual determinations).

First, *sulphur dioxide* is determined. Into a 250 ml. paraffin coated beaker charged with 50 ml. of water and 10 ml. of 0·1 N solution of iodine, 50 g of the liquid from the sampling vessel is weighed with an accuracy of 0·5 g. The excess of iodine is titrated with 0·1 N sodium thiosulphate using starch as indicator (1 ml. of N sodium thiosulphate corresponds to 0·032 g of sulphur dioxide.

Another aliquot is analysed for *sulphuric and fluorosulphonic acid* (calculated together as sulphuric acid). A 50 g sample is evaporated in a platinum dish on the steam bath almost to dryness. 10 ml. of water is added, and the content evaporated again. This procedure is repeated (usually twice) as long as the hydrogen fluoride smell is noticeable. The residue is diluted with 25 ml. of water, and the solution is titrated to phenolphthalein with 0·1 N sodium hydroxide (1 ml. of N sodium hydroxide corresponds to 0·049 g of sulphuric acid).

To determine *fluorosilicic acid*, a 50 g sample is weighed into a platinum dish, 0·2 g of potassium chloride is added, the mixture is stirred with a platinum wire until the salt dissolves, and the solution is evaporated to dryness on the steam bath. The residue is dissolved in 25 ml. of water, 2 g of potassium chloride is added, and the solution is cooled in an ice bath below 10°C. After adding phenolphthalein, the solution is titrated with 0·5 N sodium hydroxide free of silicic acid and carbonates. The consumption of the alkali has no connection with the analysis; the titration serves only to cause exact neutralization of the acid potassium fluoride formed by the following equations:

$$H_2SiF_6 + 2\,KCl \longrightarrow K_2SiF_6 + 2\,HCl$$

**723**
$$2\,HF + KCl \longrightarrow KHF_2 + HCl$$

$$KHF_2 + KOH \longrightarrow 2\,KF + H_2O$$

$$K_2SiF_6 + 4\,KOH \longrightarrow 6\,KF + Si(OH)_4$$

The content of the dish is then heated to 60°C and while hot is titrated with 0·1 N hydroxide to the permanent rose colour. The consumption of the alkali at the elevated temperature corresponds to two thirds of the total amount necessary to neutralize the fluorosilicic acid (1 ml. of N sodium hydroxide corresponds to 0·024 of fluorosilicic).

*Total acidity* is determined by weighing 35–45 drops of the dilute sample in a platinum container, placing the container in a 250 ml. paraffin-coated beaker containing 100 ml. of water alkalized with 0·1 N sodium hydroxide to the light rose shade of phenolphthalein (1 ml.), and titrating the acid solution with 0·5 N sodium hydroxide to permanent rose colour. If the colour fades, the content of the beaker is heated on the platinum dish to 60°C, and the titration is continued to the permanent colour. Total acidity is calculated as hydrofluoric acid (1 ml. of N sodium hydroxide corresponds to 0·02 g of hydrogen fluoride).

The actual *content of hydrogen fluoride* in the sample is determined by subtracting from the total acidity:

$$0{\cdot}8330 \times \%\ H_2SiF_6,\quad 0{\cdot}4078 \times \%\ H_2SO_4,\quad \text{and}\quad 0{\cdot}6243 \times \%\ SO_2.$$

## ANALYSIS OF FREONS AND OTHER FLUORINE PRODUCTS

For evaluation of the quality of the Freons, the following factors are of importance: neutrality, freedom from water and chloride ions, non-absorbable gases, and boiling range.

The *acid content* in Freons is estimated by bubbling a sufficient amount of the Freons through a titrated sodium hydroxide solution and back-titrating.

The moisture, or *water content*, is determined by passing 300 g of the gas through towers filled with a mixture of asbestos and phosphorus pentoxide at a rate of 30 g per hour, and reading the increase in weight of the towers. If precise directions are followed, as little as 0·0035% of water can still be detected with an accuracy of 0·0001% (*829*).

Another method for the determination of water content is the Karl Fischer titration in 100 g samples of the Freons: The titration flask is charged with 25 ml. of methanol and a slight excess of Karl Fischer reagent, whose excess is back-titrated with a standard water solution. A known amount of Karl Fischer reagent is added, followed by a 100 g sample of the Freon, and the excess of the reagent is back-titrated (*17*).

For routine estimation of water content, an electrolytic method is more convenient. It is based on absorption of water contained in vaporized Freon by a phosphoric acid film, and electrolytic drying of this film to the standard dry condition (*17*).

Acceptable content of water in refrigeration grade Freon 12 is 10 p.p.m. by weight. Otherwise troubles due to hydrate formation are encountered in use.

The test for *chloride ions* is carried out by adding 3–4 drops of a saturated solution of silver nitrate in methanol to a mixture of 5 ml. of Freon and 5 ml. of methanol. No turbidity must be noticeable since even the slightest content of chlorides causes corrosion (*17*).

The *content of non-absorbable gases* which would cause inconvenience in refrigerator performance must be kept below 0·5% (by volume). It is determined as the amount of gas insoluble in tetrachloroethylene using a 100 ml. Freon sample (*17*).

The *boiling range* is the temperature difference after 5% and 85% of the sample have evaporated. It is determined by placing a 100 ml. Freon sample into a graduated distilling flask of conical shape (the Goetz flask), immersing the tip of the flask in a water bath kept at room temperature for dichlorodifluoromethane and at 75°C for trichlorofluoromethane, and reading the temperature (with the thermometer bulb in the liquid) after each successive 5 ml. have evaporated. The difference should not exceed 0·25°C.

After cooling the empty flask for 30 min. to 0°C for dichlorodifluoromethane, and to 50°C for trichlorofluoromethane, the volume of the high-boiling residue is read which should not exceed 0·01 volume% (*17*).

More accurate determination of the composition of a Freon or mixtures of Freons is possible by applying *gas-chromatography* (*535*) or *gas-liquid chromatography* (*17, 830, 847*).

The second method is particularly convenient for Freon analysis. Celite, impregnated with dibutyl phthalate (*847*) or dioctyl phthalate (*830*) or silicone oil (*847*), or Kieselguhr impregnated with dinonyl phthalate (*17*), have been used as stationary phases at temperatures of 20–100°C.

Of other practical compounds containing fluorine, the *complete analysis of fluorophosphates* should be mentioned. Determination of fluorine is carried out after hydrolysis of the fluorophosphates with water or alkalis (*906, 910*). In this way, distinction between the so called free fluorine (titrated with thorium nitrate immediately), hydrolysable fluorine (titrated after hydrolysis with water) and total fluorine (after treatment with sodium ethoxide) can be made.

# X

## PRACTICAL APPLICATIONS OF ORGANIC FLUORINE COMPOUNDS

THE first practical and industrial application of fluorinated organic compounds was the introduction of fluoro-derivatives of methane and ethane into refrigeration. The advantages of the new refrigerants, the "Freons", caused such a rapid increase in their manufacturing that it soon exceeded the demand which is limited in this particular field. In order to maintain the production, other uses for Freons had to be found. Nowadays approximately equal amounts are consumed in refrigeration and in aerosols. The important discovery of Teflon focused the attention of chemists on the field of fluorinated plastics which, together with refrigerants and propellants, constitute the most important industrial applications of fluorinated organic compounds.

The importance of the industrial applications is reflected in production which totals some 130,000 tons per year (*9*). The number of publications dealing with practical aspects of these compounds is very large, and consequently summing up of important data rather difficult. In the following chapter, only a very brief survey will be given since the topic is discussed in detail in special monographs (*17, 892, 919, 959*).

### THE FREONS

In 1930, Midgley and Henne noted that dichlorodifluoromethane, obtained by replacement of chlorine by fluorine in carbon tetrachloride, had convenient physical, chemical and biological properties for use as a refrigerant in compression refrigerators. This important discovery was followed by a systematic study of fluorinated halogen derivatives of methane and ethane. In a few publications and a long series of patents, preparation and properties of these compounds were thoroughly described. Commercially they are designated as *Freons* (Du Pont), *Genetrons* (General Chemicals), *Isotrons* (Pennsalt), *Ucons* (Union Carbide), *Arctons* (Imperial Chemical Industries), *Isceons* (Imperial Smelting Corporation), *Frigens* (Farbwerke Höchst), *Fluogens* (van Heyden),

*Alcofrens* (Montecatini), etc. Their chemical composition is now universally expressed by numerical symbols whose meaning has been discussed in chapter I (p. 23).

## Preparation of Freons

The original preparation of Freons is based on the Swarts reaction, the heating of chlorinated derivatives of methane and ethane with antimony trifluoride, activated by antimony pentachloride or fluorochloride (*708*). The antimony salts act as halogen carriers which replace the organic chlorine by fluorine. They must therefore be used in an amount at least equivalent to the desired degree of replacement. The antimony chloride formed by the reaction can be reconverted to antimony trifluoride by treatment with hydrogen fluoride.

This discontinuous way of preparation was later improved by using anhydrous hydrogen fluoride as a continuous source of fluorine and the antimony salts as mere "transfer catalysts" (*198, 200*). The chemical reaction is the same as that between the organic halogen derivative and antimony trifluoride, but the antimony fluoride is prepared *in situ* from antimony chloride and hydrogen fluoride, and the antimony chloride formed in the reaction is immediately reconverted to the fluoride by contact with hydrogen fluoride (Equation 69, p. 91). Also in this catalytic reaction at least a part of the antimony must be present in the pentavalent form, best as antimony pentachloride (*111, 133*).

The chemistry of the transformations of organic halogen derivatives to the Freons can be expressed by a somewhat idealized sequence of equations:

724

$$SbCl_5 + x\,HF \longrightarrow SbF_xCl_{(5-x)} + x\,HCl$$

$$CCl_4 + HF \xrightarrow{SbF_xCl_{(5-x)}} CCl_3F + HCl$$

$$CCl_3F + HF \xrightarrow{SbF_xCl_{(5-x)}} CCl_2F_2 + HCl$$

$$CCl_2F_2 + HF \xrightarrow{SbF_xCl_{(5-x)}} CClF_3 + HCl$$

The relations are far more complicated, and as a matter of fact, the exact knowledge of these reactions is still lacking. The halogen exchange for fluorine is stepwise. If the trichlorofluoromethane primarily formed is left in contact with the fluorinating medium, the second chlorine atom is replaced by fluorine and dichlorodifluoromethane is formed. The third chlorine atom is replaced only with difficulties and to a small extent. These circumstances are extremely favourable for the production of dichlorodifluoromethane, since by recycling of trichlorofluoromethane and, in a sufficiently long reaction

time, all of the carbon tetrachloride can be finally converted to dichlorodifluoromethane. All the products are easily separated by distillation, as their
boiling points differ from each other by approximately 50°C. The amount
of hydrogen fluoride used is approximately a 20% excess over the calculated
one. Commercial anhydrous hydrogen fluoride containing up to 2% of water
may be used without any detrimental effect on the reaction, but severe
corrosion of the steel apparatus occurs, unless the content of water is kept
lower than 0·3% (877). Antimony  pentachloride is used as the catalyst in
an amount of one tenth of the required one. If antimony trifluoride is used,
at least a part of it must be converted to the pentavalent stage by chlorine
addition. The catalyst is very robust and stands a number of batches without
an appreciable drop of activity. Its loss may be estimated as 1% per batch.
The catalyst may be renewed from time to time, or a small amount of new
material can be added to each following charge. Carbon disulphide, which
is sometimes present in commercial carbon tetrachloride, destroys the catalyst

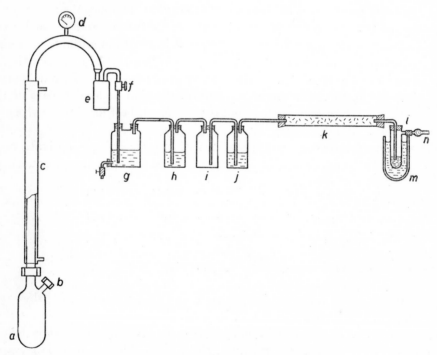

Fig. 41. Laboratory apparatus for the preparation of Freons: a) reactor, b) filling neck, c) water-cooled reflux condenser, d) manometer, e) safety trap, f) needle
valve, g) water scrubber, h) alkaline scrubber, i) trap, j) sulphuric acid drier,
k) potassium hydroxide drier, l) condensing trap, m) Dewar flask with dry ice,
n) calcium chloride tube.

by reducing the pentavalent salt to antimony trichloride. In such a case chlorine must be added in the amount required for converting all the present carbon disulphide to carbon tetrachloride (*539*).

## Procedure 39.

### Preparation of Dichlorodifluoromethane (539)*

A cooled reactor (*a*) in the apparatus shown in Fig. 41 is charged with 150 g (0·5 mole) of antimony pentachloride, 1020 g (6·6 moles) of carbon tetrachloride, and 300 g (15 moles) of anhydrous hydrogen fluoride. The filling neck (*b*) and needle valve (*f*) are closed, water is circulated through the reflux condenser (*c*), and the reactor is heated in an oil bath quickly to 100–120°C. During 1–2 hours the pressure rises to 30 atm. At this stage the needle valve (*f*) is opened and the gaseous products are discharged through the absorption train to the condensation vessel at such a rate that all the hydrogen chloride produced is absorbed entirely in the washing water of a scrubber (*g*). The pressure in the apparatus rises still for 1–2 more hours, reaches a maximum of 33–35 atm, drops continuously in the course of further 3–5 hours to 6 atm and stays at this value (vapour pressure of dichlorodifluoromethane at 15°C) for 3–4 hours. Then it drops rapidly to zero and the apparatus is emptied in the span of one hour.

The condensate is distilled in a low-temperature column. The distilling flask is immersed in cold water, and toward the end of the distillation the temperature of the bath is raised to 40–50°C. The head of the column is cooled with a dry-ice ethanol mixture to −80°C. After separating 0·5–1 ml. of a distillate at −34°C to −28°C, the main fraction distils at −28°C to −27°C. Approximately 750 g of dichlorodifluoromethane is thus collected (6·2 moles) (94%). After a short intermediate cut, about 2% of trichlorofluoromethane is obtained distilling between 0–24°C.

In addition to the classical pressure procedure (*133, 198*), conversion of carbon tetrachloride to the fluorinated derivatives can be effected at atmospheric or only slightly elevated pressure by feeding continuously carbon tetrachloride and an excess of gaseous anhydrous hydrogen fluoride in a proper ratio at a temperature of 50–60°C into a reactor containing the catalyst (*200, 877*).

The described process for the production of dichlorodifluoromethane is satisfactory for the laboratory preparation as well as for large scale production. However, some alternate procedures have been described and some of them developed. One of such is a catalytic vapour-phase process, working with carbon tetrachloride and anhydrous hydrogen fluoride at atmospheric pressure and at 300–350°C in the presence of activated charcoal alone (*199, 637*) or impregnated with ferric chloride (*637, 877*) (p. 90,91). Other catalysts are used nowadays, of which the composition is not disclosed. Conversion of methane to dichlorodifluoromethane by one-step reaction with chlorine and hydrogen fluoride presents the problem of disposing of large amounts of hydrogen chloride (*35*). In the new route to dichlorodifluoromethane from

---

* Carried out by L. Jarkovský.

phosgene and hydrogen fluoride (Equation 179, p. 123) (374), difficulties are encountered which are caused by strong corrosion. Another method for producing dichlorodifluoromethane is based on the addition of two moles of hydrogen fluoride to acetylene and high-temperature chlorinolysis of the 1,1-difluoroethane (35), (Equation 180, p. 123), or on the chlorinolysis of tetrachloro-1,1-difluoroethane (434) (Equation 375, p. 186).

The procedure described for the production of dichlorodifluoromethane from carbon tetrachloride is applicable to the preparation of trichlorofluoromethane (133, 198, 199, 200) and chlorotrifluoromethane (918), when an appropriate amount of hydrogen fluoride is used.

If the fluorinating reaction of hydrogen fluoride and antimony penta-chloride is applied to chloroform, dichlorofluoromethane, chlorodifluoromethane and fluoroform are obtained (198, 199). The most important is chlorodifluoromethane which is used increasingly in refrigeration and as a starting material for the production of tetrafluoroethylene (p. 343).

$$725 \qquad \begin{aligned} CHCl_3 &\xrightarrow[SbCl_5]{HF} CHCl_2F \\ CHCl_2F &\xrightarrow[SbCl_5]{HF} CHClF_2 \\ CHClF_2 &\xrightarrow[SbCl_5]{HF} CHF_3 \end{aligned}$$

The preparation of chlorodifluoromethane is somewhat intricate as compared with the preparation of dichlorodifluoromethane. Fluorination to the third stage, (i.e. to fluoroform) is much easier than in the case of carbon tetrachloride, and the formation of fluoroform, which is seldom utilized, represents a loss. The reaction is best carried out at approximately 50°C, and large amount of the catalyst and prolonged reaction time are to be avoided.

*Procedure 40.*

*Preparation of Chlorodifluoromethane (877)\**
For the fluorination under atmospheric pressure, a high vertical stainless steel reactor fitted with a mechanical stirrer, two inlet tubes leading to the bottom (one for chloroform, the other for hydrogen fluoride), an outlet tube for the products, and a water jacket is used. The reactor is filled with 20 kg of antimony pentachloride and heated to 60°C, and 10 moles of chloroform and 23 moles (15% excess) of anhydrous hydrogen fluoride per hour are fed simultaneously. The gaseous products are passed through a water scrubber (a tower filled with Raschig rings), dried with calcium chloride, and condensed in a dry-ice trap. The crude condensate is fractionated to give 75% of chlorodifluoromethane, 20% of dichlorofluoromethane and 5% of unreacted chloroform. The yield of the reaction is 90–95% based on the chloroform, or 66·5–70% based on the hydrogen fluoride consumed.

---

\* Carried out by V. Reinöhl.

In a similar fashion chlorofluoroethanes are obtained by treatment of hexachloroethane with hydrogen fluoride and the antimony catalyst (*200*). Some of these derivatives are important refrigerants; in addition, trichloro-trifluoroethane serves as a starting material for the production of chlorotri-fluoroethylene (p. 343).

The conversion of hexachloroethane to fluorinated derivatives is not as easy as the preparation of fluorinated methane derivatives. First of all, the starting material is a crystalline compound with a high melting point (187·2°C) and low solubility in the reaction medium. This difficulty can be obviated by using a mixture of liquid tetrachloroethylene and chlorine instead of hexachloroethane. Furthermore, the reaction is more complicated as there are much more possibilities of halogen exchange than in the methane series. Carrying out the reaction so as to obtain the desired product requires therefore much more care.

726

$$C_2Cl_6 \xrightarrow[\text{SbCl}_5]{\text{HF}} C_2Cl_5F \qquad \text{b.p. } 138°$$

$$C_2Cl_5F \xrightarrow[\text{SbCl}_5]{\text{HF}} C_2Cl_4F_2 \qquad \text{b.p. } 91°, 92·8°$$

$$C_2Cl_4F_2 \xrightarrow[\text{SbCl}_5]{\text{HF}} C_2Cl_3F_3 \qquad \text{b.p. } 47°, 47·7$$

$$C_2Cl_3F_3 \xrightarrow[\text{SbCl}_5]{\text{HF}} C_2Cl_2F_4 \qquad \text{b.p. } -2°, 3·8°$$

The main characteristic of the preparation of fluorinated ethane derivatives is higher reaction temperature (100–150°) and a larger proportion of antimony pentachloride, which acts at the same time as a solvent for hexachloroethane (*538*). If trichlorotrifluoroethane is the desired product, monofluoro and difluoro-derivatives must be recycled. At the same time, the trichlorotrifluoroethane formed must be kept out of the fluorinating medium, since otherwise its fluorination proceeds further. This is carried out by discharging the reaction products, after a certain time of heating of the reaction mixture, through the reflux condenser which is fed with water of 50–80° temperature. In this way trichlorotrifluoroethane passes out to absorption and condensation.

*Procedure 41.*

*Preparation of 1,1,2-Trichlorotrifluoroethane (538)\**
The transformation of hexachloroethane to trichlorotrifluoroethane is carried out in the apparatus shown in Fig. 41 in a similar fashion as in the case of dichlorodifluoromethane (Procedure 39, p. 335).

---

\* Carried out by L. Jarkovský.

A mixture of 720 g (2·4 moles) of antimony pentachloride, 1000 g (4·2 moles) of hexachloroethane, and 340 g (17 moles) of anhydrous hydrogen fluoride is heated at 150°C for 25 hours at a maximum pressure of 33 atm. Approximately 700 g of a crude product is collected mainly in the water scrubber, and only to a lesser extent in the dry-ice trap. Distillation affords 670 g (3·6 moles) (85%) of 1,1,2-trichlorotrifluoroethane boiling over a range of 45–48°C, and small amounts of tetrachlorodifluoroethane (2%) and dichlorotetrafluoroethane (4%).

In addition to the above–mentioned classical Freons, the following compounds of similar nature are produced: 1,1-difluoroethane (Genetron 100), obtained by the addition of hydrogen fluoride to acetylene, gives on chlorination 1-chloro-1,1-difluoroethane (Genetron 101) (*425*).

727          $CH{\equiv}CH \xrightarrow{\text{2 HF}} CH_3{-}CHF_2 \xrightarrow[h\nu]{\text{Cl}_2} CH_3{-}CClF_2$          [*425*]

Brominated fluoro-derivatives of methane, which proved to be excellent fire extinguishers, are manufactured either by fluorination of carbon tetrabromide, or by bromination of difluoromethane or trifluoromethane (*35*).

728          $CCl_4 \xrightarrow{\text{HBr/AlCl}_3} CBr_4 \xrightarrow{\text{HF}} CBr_2F_2 + CBrF_3$          [*35*]

729          $CH_2F_2 \xrightarrow{\text{Br}_2} CBr_2F_2$          [*35*]

730          $CHF_3 \xrightarrow{\text{Br}_2} CBrF_3$          [*35*]

Quite recently some attention has been concentrated on chloropentafluoroethane as a convenient refrigerant (*845*). Perfluorocyclobutane (Freon C 318) produced by pyrolysis of tetrafluoroethylene (p. 275) is a favourite propellant (*11*). 1-Bromo-1-chloro-2,2,2-trifluoroethane (Fluothane) is a modern inhalation anaesthetic (*976*). Its preparation is illustrated in Equation 191, (p. 125).

## Properties of Freons

Because of broad practical applications, physical and thermodynamic properties of the Freons have been thoroughly investigated (*17*). Special attention was devoted to the vapour pressure (*75*, *307*), specific and vapourization heats (*74*, *77*), vapour densities (*76*), viscosities (*78*) and some important thermodynamic functions, such as the Mollier diagram (*606*). Some of the selected values of the most frequently used Freons are listed in Table 77.

TABLE 77

*Physical and Thermodynamic Constants of Freons*

| Name | Formula | M.P.°C | B.P.°C | Heat of Evaporation at B.P. Kcal/ /mole | Specific Heat Kcal/ /mole | Critical Temp. °C | Critical Pressure atm. | Critical Density kg/l. |
|---|---|---|---|---|---|---|---|---|
| Freon 11 | $CCl_3F$ | −111·1 | 23·77 | 43·51 | 0·208 | 198·0 | 43·2 | 0·554 |
| Freon 12 | $CCl_2F_2$ | −155 | −29·8 | 39·9 | 0·204 | 111·5 | 40·95 | 0·555 |
| Freon 13 | $CClF_3$ | −180 | −81·5 | 35·65 | 0·203 | 28·8 | 39·4 | 0·581 |
| Freon 21 | $CHCl_2F$ | −135 | 8·92 | 57·85 | 0·246 | 178·5 | 51·0 | 0·522 |
| Freon 22 | $CHClF_2$ | −160 | −40·8 | 55·9 | 0·265 | 96·0 | 48·7 | 0·525 |
| Freon 23 | $CHF_3$ | −163 | −82·2 | 56 | 0·28 | 32·3 | 50·5 | |
| Freon 113 | $C_2Cl_3F_3$ | − 35 | 47·57 | 35·04 | 0·226 | 214·1 | 33·7 | 0·576 |
| Freon 114 | $C_2Cl_2F_4$ | − 94 | 3·55 | 32·8 | 0·232 | 145·7 | 32·1 | 0·582 |
| Freon 115 | $C_2ClF_5$ | −106 | −38 | | | | | |

For calculation of *vapour pressure* of some Freons, the following equations have been derived, and are valid for the decimal logarithm of absolute atmospheres and the Kelvin temperature scale:

$$\text{Freon } 12 : \log p = 31{\cdot}6315 - \frac{1816{\cdot}5}{T} - 10{\cdot}859 \log T + 0{\cdot}007175\,T \quad (307)$$

$$\text{Freon } 11 : \log p = 34{\cdot}8838 - \frac{2303{\cdot}95}{T} - 11{\cdot}7406 \log T + 0{\cdot}0064249\,T \quad [44]$$

$$\text{Freon } 21 : \log p = 38{\cdot}2974 - \frac{2367{\cdot}41}{T} - 13{\cdot}0295 \log T + 0{\cdot}0071731\,T \quad [44]$$

$$\text{Freon } 22 : \log p = 25{\cdot}1144 - \frac{1638{\cdot}32}{T} - 8{\cdot}1418 \log T + 0{\cdot}0051838\,T \quad [44]$$

$$\text{Freon } 113 : \log p = 29{\cdot}5335 - \frac{2406{\cdot}0}{T} - 9{\cdot}2635 \log T + 0{\cdot}0036970\,T \quad [44]$$

The graphic dependence of vapour pressure on temperature is plotted for some Freons in Fig. 30, p. 297.

The *solubility* of the Freons in water is very low (p. 304) except for chlorodifluoromethane and trifluoromethane. Hydrocarbons, halogenated hydrocarbons, monohydric alcohols, ketones, and some other solvents are miscible in all proportions with Freons. For Freon 12, the following solvents show unlimited miscibility (*17*): benzene, xylene, naphthalene, mineral oil, chloroform, carbon tetrachloride, methanol, ethanol, propanol, butanol, pentanol, diethyl ether, dioxane, cellosolve, acetone, acetic acid, ethyl acetate, dibutyl phthalate, pyridine, sulphur dioxide, and carbon disulphide. On the other hand, Freon 12 is insoluble in water, ethylene glycol, diethylene glycol, trimethylene glycol, carbitol, glycerol, ethylene chlorohydrin, thiodiglycol, triethanolamine, aniline, phenol, cresol, benzyl alcohol, cinnamyl alcohol, furfuryl alcohol, tetrahydrofurfuryl alcohol, cinnamaldehyde, benzophenone, fural, formamide, phenyl and nitromethane (*17*).

The Freons carrying hydrogen atoms dissolve in oxygen and nitrogen containing solvents by 50% more than they should according to Raoult's law (*1104*). Solubilities of Freon 12 and 21 in some ethers and esters are listed in Table 78 (*1103*).

Table 78

*Solubilities of Freon 12 and Freon 21 in Organic Solvents* (*1103*)

| Solvent | Solubility in g/ml. | |
|---|---|---|
| | Freon 12 at 2693 mm Hg | Freon 21 at 638 mm Hg |
| $C_2H_5OCH_2CH_2OCH_2CH_2OCOCH_3$ | 0·258 | 0·97 |
| $C_2H_5OCH_2CH_2OCH_2CH_2OC_2H_5$ | 0·38 | 1·05 |
| $CH_3O(CH_2CH_2O)_4CH_3$ | 0·258 | 1·04 |
| $C_2H_5O(CH_2CH_2O)_2OCOCH_2OCH_3$ | 0·162 | 0·89 |
| $ClCH_2CH_2CH_2OCH_2CH_2CH_2Cl$ | 0·22 | 0·49 |
| $(CH_3)_2CHCH_2CH(CH_3)OCOCH_3$ | 0·52 | 0·72 |
| $ClCH_2CH(CH_3)OCH(CH_3)CH_2Cl$ | 0·258 | 0·49 |
| $\alpha\text{-}C_{10}H_7F$ | 0·236 | 0·37 |

A significant property of Freons is their *chemical stability*. Dichlorodifluoromethane stood one month's heating at 175°C without change. Only contact with a free flame decomposes it to toxic products (*693, 709*). At room temperature Freons are stable toward concentrated sulphuric acid as well as toward concentrated alkaline hydroxides. Only those derived from chloroform react with alkalis (*877*). A peculiar behaviour of some Freons is the formation of hydrates (*17*) (p. 306).

Owing to the chemical inertness of dry Freons most of the common metals can be used for the construction of apparatus (*17, 606, 708*). Cast iron, steel, stainless steel, Monel metal, copper, brass, tin, zinc, lead and pure aluminum are perfectly resistant at ordinary temperatures. Magnesium and magnesium alloys, e.g. aluminum containing more than 2% of magnesium are to be avoided. Mild steel is slightly attacked at temperatures higher than 200°C (*17*).

From the biological point of view, the Freons show outstanding inertness. Practical tests revealed that the amount of Freon detrimental to animals considerably exceeds that of any other refrigerant. Dichlorodifluoromethane is tolerated without injury in concentrations of 20% by volume for a period of 2 hours, the safety limit for trichloromonofluoromethane is somewhat lower but far higher than that of methyl chloride (*17*). Toxicities and tolerable concentrations of some refrigerants are listed in Table 79 (*708*).

TABLE 79

*Toxicities of Common Refrigerants* (*708*)

| Refrigerant | Maximum Concentration Tolerable for Several Hours Without Danger to Life, % | Concentration Dangerous to Life After 30-60 Minute Exposure, % | Concentration Lethal to Animals in a Short Time % |
|---|---|---|---|
| Ammonia ........... | 0·01 | 0·25–0·45 | 0·5–1 |
| Methylchloride ...... | 0·05–0·1 | 2–4 | 15–30 |
| Carbon Dioxide ...... | 2–3 | 6–8 | 30 |
| Freon 12 ............ | 80 | 40 | Not reached |

## Applications of Freons

The Freons have been used primarily in *refrigeration*. They are particularly suitable for household refrigerators, because they are non-flammable, non-explosive and non-toxic, and therefore safe in the case of accidental leakage. Even in the presence of an open flame the Freons are not usually dangerous since the toxic products of their pyrolytic decomposition rarely reach harmful concentrations (*709*). Only prolonged or repeated exposition to such fumes may produce injury to the organism (*693*).

The relatively large number of Freons now available allows proper choice of a refrigerant suitable for any special purpose. For instance, household refrigerators which require cooling to temperatures around zero °C are best filled with Freon 12 or 22. Freezers cooled to temperatures below zero are filled with low-boiling Freons 13 and 23, and conversely some of the industrial air-conditioning units use higher boiling Freon 113, which does not necessarily require water-cooling. Unsuccessful attempts have been made to use Freons for absorption refrigeration with polyglycol ethers and esters. Thus compression refrigerators are the only type in use.

When markets for refrigerators approached saturation, other uses for the Freons had to be looked for to maintain production at high level. Today at least as much Freon is consumed as *aerosol* as goes into refrigeration. Large scale application of aerosol takes place for example in agronomy, where immense quantities of herbicides, insecticides and pesticides are to be dispersed. These compounds are dissolved in suitable Freons which are liquid under elevated pressures. When expanded to atmospheric pressure, the solvents evaporate leaving finely dispersed particles of the chemicals. An example is the insecticide aerosol produced by dissolving 2% of pyrethrum (containing 20% of pyrethrins) and 3% of DDT in a mixture of Freons 12 and 11 (*297*).

Another field for aerosols is cosmetics and perfumery. Large proportions of Freons are used as hair sprays and dressing propellants, for dispersing perfumes, shaving creams and deodorants. Recently Freons have been used for delivering food ingredients, whipped cream etc. In this connection especially perfluorocyclobutane, or Freon C318, is a favourite propellant because of its lack of smell and taste (*11, 411*).

The non-explosiveness and non-flammability of Freons accounts for their application as *fire extinguishers*. In this respect, the brominated products, dibromodifluoromethane (Freon12B2) and especially bromotrifluoromethane (Freon 13B1) are very efficient for extinguishing fires of any kind (*12, 411*). The latter compound is especially suitable for this purpose, since it is entirely non-toxic and even its decomposition products are devoid of toxicity (*411*).

Finally some percentage of Freons are used for the production of monomers in the field of plastics.

## FLUORINATED PLASTICS

In the total consumption of organic fluorides only 3–5% are fluorinated plastics. Such a small fraction, however, does not express the great importance of this group of compounds. Since the chance discovery by Plunkett of the singular properties of polytetrafluoroethylene (1938), a continuous search for similar compounds led to dozens of fluorinated plastics, of which a few are now produced in quantities. In addition to polytetrafluoroethylene (*Teflon, Fluon, Algaflon*), the most popular polymers are polychlorotrifluoroethylene (*Fluorothene, Kel-F, Polyfluoron, Genetron Plastic HL, Hostaflon, Teflex, Ekafluvin*), the copolymer of vinylidene fluoride with chlorotrifluoroethylene (*Kel-F Elastomer*), the copolymer of vinylidene fluoride with perfluoropropene (*Viton A*), poly(heptafluorobutyl acrylate (*1F4*), the polyester of 2,2,3,

3,4,4-hexafluoropentanediol with adipic acid, and 3,3,3-trifluoropropylmethyl-silicone (*Silastic LS 53*). All these polymers possess high thermal stability, and some have an extraordinary resistance to chemicals. In this respect, the first two compounds, and especially polytetrafluoroethylene, exceed in chemical stability any organic material known to date. The disadvantage is the tedious fabrication of the polymer, which requires special techniques.

In the following paragraphs, the preparation of the monomers, their polymerization and copolymerization, and the properties and uses of the polymers will be briefly discussed.

## Preparation of the Monomers

Various reactions leading to the important monomers have been mentioned in Chapter V on the preparation of organic fluorocompounds. Only the industrial methods will be described here.

*Tetrafluoroethylene* can be prepared by dehalogenation of 1,2-dichloro-tetrafluoroethane (*73, 621*), or by decarboxylation of sodium salts of trifluoroacetic (*609*) or perfluoropropionic acid (*609*), but it is produced commercially by the pyrolysis of chlorodifluoromethane according to the equation (*799*):

**731** $$2\ CHClF_2 \xrightarrow[0.5\ atm]{650°} CF_2{=}CF_2 + 2\ HCl \qquad [799]$$

The conditions and by-products of this reaction have been mentioned in the paragraph on elimination (p. 268).

*Chlorotrifluoroethylene* is obtained by dehalogenation of 1,1,2-trichloro-trifluoroethane with zinc in alcohol (*65, 621*).

**732** $$CClF_2{-}CCl_2F \xrightarrow[\text{gentle heating}]{Zn,\ C_2H_5OH} CF_2{=}CFCl\ \ 79{-}92\% \qquad [65]$$

The reaction is carried out in a vessel equipped with a gas-tight stirrer and a fractionating column the head of which is cooled to $-40°C$. The function of the column is to separate chlorotrifluoroethylene from the unreacted starting material and from the solvent used in the reaction. The dehalogenation is best carried out with zinc dust which is used in a slight excess. Methanol or ethanol are suitable solvents since they dissolve the starting material as well as the zinc chloride formed in the reaction. Water failed as a solvent. The reaction has an induction period which may be shortened by activating the zinc, either by treatment with 1% hydrochloric acid and decanting of the solution, or better by an alcoholic solution of zinc chloride. Heating is usually necessary, and the reaction temperature is approximately 40–50°C.

*Procedure 42.*

*Preparation of Chlorotrifluoroethylene from 1,1,2-Trichlorotrifluoroethane (65)\**

In a one-litre three-necked flask, fitted with a gas-tight stirrer and a low temperature distilling column whose head and receiver are cooled with a dry-ice alcohol mixture, 292 g (1·55 moles) of 1,1,2-trichlorotrifluoroethane is added to 144 g (2·2 moles) of zinc dust and 230 ml. of ethanol. The mixture is treated with a few ml. of a 10% alcoholic solution of zinc chloride, and heated to 30–50°C until a spontaneous reaction sets on. If necessary the mixture is heated to maintain reaction. Toward the end, a boiling steambath is applied for a short time. The reaction time is approximately 2·5 hours. During the reaction the reflux ratio of the distilling column is maintained at a value of 10:1, which allows sufficient separation of chlorotrifluoroethylene from the starting material and from the ethanol. The fraction boiling between −26°C and −24°C amounts to 160 g (1·37 moles) (88%) of pure chlorotrifluoroethylene.

Though the dehalogenation with zinc is quantitative and perfectly suitable for laboratory scale, other possibilities of dehalogenation are under investigation for industrial uses. Dehalogenation with iron does not work (*877*). Promising results have been given by experiments on catalytic removal of halogens by hydrogen (p. 265).

*Vinylidene fluoride* is produced either by dehydrohalogenation of 1-chloro-1,1-difluoroethane (Equation 200, p. 129) or by dehalogenation of 1,2-dichloro-1,1-difluoroethane (*644*), produced by the reaction of trichloroethylene with hydrogen fluoride (*349*) (Equation 199, p. 128).

*Perfluoropropylene* is available by decarboxylation of sodium perfluorobutyrate (*609*), but the commercial method is pyrolysis of tetrafluoroethylene (*618, 722*) (p. 274–276).

Preparation of fluorinated dienes, *perfluoro-1,3-butadiene* (*898*) (Equation 216, p. 133) and *1H,1H-tetrafluoro-1,3-butadiene* (*530*) (Equation 214, p. 133) has been mentioned in Chapter V.

The fluorinated monomers readily polymerize. In order to prevent undesired spontaneous polymerization, the content of oxygen in the monomers should be kept below 0·002% by careful redistillation (*226*). As the polymerization inhibitors, the main agents are olefinic compounds (*226*), amines such as dibutyl-and tributylamine (*347*), mercaptans such as butyl- or octylmercaptan (*138*), and the so called Terpene B, a mixture of dipentene and terpinolene, which in concentration of 0·5% prevents polymerization of tetrafluoroethylene at normal temperature for several months (*204*). Before starting the commercial polymerization, the stabilizer is removed by distilling of the monomer. The traces of Terpene B are removed by passing the distilled monomer through a column of silica gel (*345*).

Physical properties for some fluorinated monomers are listed in Table 80.

---

\* Carried out by L. Jarkovský and by the author.

Table 80

*Properties of Fluorinated Monomers (624, 878, 947, 948)*

| Monomer | M.P.°C | B.P.°C | $d_4^t$ | $n_D^t$ | Critical Temperature °C |
|---|---|---|---|---|---|
| Vinyl fluoride ....... | −160·5 | −72·2 | 0·675/26° | | |
| Vinylidenefluoride ... | | −83 | | | |
| Chlorotrifluoroethylene | −157·5 | −26·2 | 1·51/−40° | | |
| | −157·9 | −27·9 | 1·38/0° | | |
| Tetrafluoroethylene ... | −142·5 | −76·3 | 1·533/−80° | | 33·3 |
| | | | 1·1507 | | |
| Perfluoropropylene ... | −158·1 | −28·2 | 1·583/−40° | | |
| Perfluorobutadiene ... | − 61 | 0 | 1·602/−20° | 1·298/−20° | |
| | | +3 | | | |
| | | +6·6 | | | |
| 1,1-Dihydroperfluoro- butyl acrylate ..... | | 51·3/50 mm | 1·409/20° | 1·3317/20° | |
| 1,1-Dihydroperfluoro- hexyl acrylate .... | | 63·5/20 mm | 1·54/20° | 1·3296/20° | |

## Polymerization

Spontaneous polymerization of fluorinated olefins is undesirable, because it gives polymers of unreproducible properties. To obtain reproducible polymers, polymerization is carried out under rigorously controlled conditions, in block, in solution, or else in aqueous suspension. Different polymers result from the different kinds of polymerizations.

*Polymerization of chlorotrifluoroethylene* can give either low polymers (telomers), or else high polymers. The first type of polymerization is carried out in a chloroform solution, using large amounts of the polymerization catalyst (up to 5% of dibenzoyl peroxide, based on the weight of the monomer used). The resulting telomers are high boiling liquids, which find use as heat transfer media, often after a treatment with cobalt trifluoride, which replaces the chlorine atoms by fluorine (723, 733).

*Procedure 43.*

*Telomerization of Chlorotrifluoroethylene to High Boiling Oils (733)*

An evacuated autoclave cooled with dry ice is charged, by suction, with 2700 g of alcohol-free chloroform (distilled from phosphorus pentoxide), 300 g of redistilled chlorotrifluoroethylene, and 15 g of freshly prepared dibenzoyl peroxide. The autoclave is heated for 1·75 hours at 100°C in a bath. Thereafter the unreacted olefin is bled off through a dry-ice trap, and the content of the autoclave is distilled. After the evaporation of the chloroform low polymers of chlorotrifluoroethylene distil at 110–240°C under 0·3 mm.

High polymers of chlorofluoroethylene are white solids and are prepared by block or suspension polymerization. Prior to its polymerization, the monomer must be carefully purified by distillation in an oxygen-free atmosphere. The block polymerization is carried out by leaving the monomer in contact with 0·03% of bis(trichloroacetyl)peroxide at −16°C for seven days and gives a 30% conversion (723). Today chlorotrifluoroethylene is polymerized exclusively in an aqueous medium using a persulphate-bisulphite-silver nitrate system (345), or tert.-butyl perbenzoate-bisulphite-ferrocitrophosphate (234). The polymer is easily separated from the aqueous phase as it is hydrophobic.

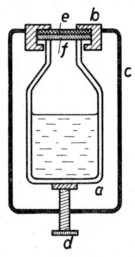

Fig. 42. Vessel for emulsion or suspension polymerization of fluoroolefins: a) thick-walled bottle, b) metal lid with a circular hole, c) metal frame, d) fixing screw, e) hard-rubber gasket f) soft-rubber gasket.

*Procedure 44.*

*Polymerization of Chlorotrifluoroethylene in Aqueous Suspension (234)*

Laboratory polymerization of chlorotrifluoroethylene is best carried out in sealed glass tubes or in thick-walled bottles with rubber stoppers, designed so that they allow injection of liquids into the bottles and drawing-off of samples by means of a syringe (Fig. 42) (877).

The polymerization vessel is charged with 54·5 parts of distilled water, 0·012 parts of sodium bisulphite, 0·0018 parts of hydrogen chloride, and 0·003 parts of the so-called soluble iron phosphate (ferrocitrophosphate, containing 12–15% of iron, 15% of phosphorus pentoxide and 45% of citric acid). The vessel is evacuated to a high vacuum while cooling in a liquid nitrogen bath, and flushed with nitrogen; 0·004 parts of *tert*-butyl perbenzoate is added in a thin-walled capillary, the vessel is again evacuated, and 7·5 parts of chlorotrifluoroethylene, free from stabilizer, is condensed into the prepared mixture. The polymerization vessel is sealed and heated to 25°C for 9 hours. After opening the vessel the polymer is separated, and an additional crop is obtained by treating the residual solution with sodium chloride. Total conversion is around 90%.

*Polymerization of tetrafluoroethylene* in a mixture of carbon tetrachloride and chloroform in the presence of large amounts of dibenzoyl peroxide affords low polymers of waxy nature and m.p. 200–210°C *(723)*.

High molecular polytetrafluoroethylene was first prepared by polymerization with zinc chloride or silver nitrate as catalysts *(846)*. This is no longer used. A quantitative polymerization is obtained at −16°C or 0°C by adding 0·5% of bis(trichloroacetyl)peroxide to a 10% solution of tetrafluoroethylene in trichloromethane *(723)*.

Suspension polymerization of tetrafluoroethylene in aqueous medium is carried out with the help of hydrogen peroxide *(348)*, *tert*.-butylperacetate *(597)* or some other organic peroxide compound. Usually three-component systems are used, comprising a peroxide, such as potassium persulphate, sodium bisulphite and traces of metals such as iron as activators. The exact formulations are usually not disclosed by the individual firms.

Other fluorinated olefins, *vinylidene fluoride* and *perfluoropropylene*, are rarely polymerized alone. Much more frequently copolymerization is carried out (see later).

Fluorinated dienes have been tentatively polymerized but so far with little success. *Perfluorobutadiene* gives, according to the conditions used, either viscous liquids, or rubbery material becoming plastic at 250–275°C *(724)*. Similar polymers are obtained by polymerizing *1H,1H-tetrafluoro-1,3-butadiene (530)*

*Polymerization of polyvinyl perfluorocarboxylates* affords transparent films *(875)*, that of *polyheptafluorobutyl acrylate (122, 177, 1029)* a rubber which remains elastic over the range of −50°C to +150°C and is resistant to many organic solvents *(1029)*.

**733** $CH_2{=}CH{-}OCO{-}C_nF_{2n+1}$ [*122*]

**734** $CH_2{=}CH{-}COO{-}CH_2{-}C_nF_{2n+1}$ [*875*]

Similar elastic materials are also obtained by polyesterification of *2,2,3,3,4,4-hexafluoro-1,5-pentanediol* with *adipic* acid *(930, 1029)*

**735** [*930*]

$-[\,O{-}CH_2{-}CF_2{-}CF_2{-}CF_2{-}CH_2{-}O{-}CO{-}CH_2{-}CH_2{-}CH_2{-}CH_2{-}CO{-}O\,]_n-$

and *polymerization of γ,γ,γ-trifluoropropyl methyl siloxane (1029)*.

736

$$-[\underset{\underset{CH_2-CH_2-CF_3}{|}}{\overset{\overset{CH_3}{|}}{Si}}-O]_n-$$

[*1029*]

Some fluorinated polymers show a strong tendency to crystallize in time, which is undesirable for a plastic material. This tendency is decreased by *copolymerization* or simultaneous polymerization of two or more components. In addition, some rather unexpected properties are encountered in the copoly-merization products. There is therefore no wonder that this modern means of preparing plastic materials is being subjected to a thorough investigation.

A great deal of various combinations have been tried and tested; e.g. copolymerization of chlorotrifluoroethylene with tetrafluoroethylene in block or solution (*728*) or in aqueous suspension (*225*), tetrafluoroethylene with per-fluoropropylene in block (*727*) or aqueous suspension (*907*), copolymerization of fluorinated halogenoolefins with non-fluorinated olefins and dienes (*855*), copolymerization of 1,1-dichloro-2,2-difluoroethylene with vinyl acetate, sty-rene, butadiene and isoprene (*644*), and even three component polymerization such as α,β,β-trifluorostyrene and styrene with chlorotrifluoroethylene (*620*). However, the most successful products, and so far the only ones produced commercially, are copolymerization products of *vinylidene fluoride with chloro-trifluoroethylene (Kel-F elastomer)*, and that of *vinylidene fluoride with per-fluoropropylene (Viton A)*.

Copolymerizations are carried out in aqueous medium. For example, 64·5 parts of chlorotrifluoroethylene and 35·5 parts of vinylidene fluoride are distilled into an auto-clave containing a frozen mixture of 140 parts of water, 20 parts of a 5% solution of potassium persulphate, 20 parts of a 2% solution of sodium pyrosulphite, and 20 parts of a 0·5% solution of ferrosulphate. After heating the autoclave for 48 hours at 25–35°C, a 96% yield of the copolymer is obtained (*1034*).

*Elastomer Kel-F 5500* contains in the final product 50% of vinylidene fluoride and 50% of chlorotrifluoroethylene units (*747*); *Kel-F 3700* has a 70 : 30 ratio (*747*). Similar copolymerization of vinylidene fluoride with per-fluoropropylene in aqueous suspension containing a persulphate, a bisulphite, and ammonium perfluorooctanoate as emulsifying agent affords a copolymer containing 30% of vinylidene fluoride and 70% perfluoropropylene units (*216*).

Both the homopolymers and copolymers are linear. They can be trans-formed to space polymers by *cross-linking (vulcanization)*. Copolymers with dienes which still contain double bonds are vulcanized by common vulcanizing agents. Polymers of chlorotrifluoroethylene can be cross-linked by intermolecular splitting-off of chlorine atoms by means of metal oxides (*346*), or organic bases (*346*). Good results have been obtained by using a mixture of zinc oxide, hexamethylene tetramine and methylene bis(*p*-phenyleneisocyanate) (*324*). The vinylidene fluoride-chlorotrifluoroethylene copolymer was successfully vulcanized by the addition of 5% of zinc oxide and 1·65% of 90% dibenzoyl peroxide (*505*); the vinylidene fluoride-perfluoropropylene copolymer was vulcanized by hexamethylene diammonium carbamate and dibenzoyl peroxide (*298, 971*). An example of a complete formulation for the Viton A vulcanization is as follows (*749*): 100 parts of vinylidene fluoride–perfluoropropylene copolymer, 15 parts of magnesium oxide, 20 parts of carbon black, 1 part of hexamethylene diamine carbamate, and 1·3 parts of disalicylidene-1,2-propylene diamine (accelerator).

More detailed data pertaining to polymerization (and copolymerization) of fluorinated olefins are contained in reviews (*569*).

## Properties of the Fluorinated Polymers

The description of the properties will be limited to the commercial products, i.e. polychlorotrifluoroethylene and polytetrafluoroethylene. The other fluorinated plastics will be mentioned only briefly.

*Polychlorotrifluoroethylene.* Low polymers of chlorotrifluoroethylene (*725*) prepared by solution polymerization have average molecular weights around 800, boiling points between 100°C and 200°C at 0·3 mm, density of 1·95 at 20°C and 1·85 at 100°C, refractive index of 1·40 at 20°C, and viscosity of 10–20 centistokes at 100°C. Their solubility in hydroxylated solvents is limited. At room temperature, absolute ethanol dissolves 33% of the polymer oil, 95% ethanol only 6%. Anhydrous hydrogen fluoride is practically insoluble in the liquid polymers (0·005 g in 100 g at 100°C and 400 mm Hg) (*725*).

Higher polymers of chlorotrifluoroethylene show some varying properties depending on the details of their preparation and further processing. For example, the final products of polychlorotrifluoroethylene which were allowed to cool slowly during their processing differ in some respects from those which were quenched suddenly (*726*).

*Molecular weights* of solid polymers of chlorotrifluoroethylene range from 20,000 to 200,000, as found both by osmometric measurements (*550*) and by measurements of viscosity of its solutions in 2,5-dichlorobenzotrifluoride or 1,1,3-trifluoropentachloropropane at 99°C (*551*). The density of the quenched polymer is 2·11, that of the slowly cooled polymer 2·13. Refractive index is 1·430, specific resistivity $5 \times 10^{17}$ ohm at room temperature (*726*).

*Polychlorotrifluoroethylene* swells and even dissolves in some solvents, especially in chlorinated olefins at higher temperatures. Table 81 shows some pertinent data about swelling.

TABLE 81

*Swelling of Polychlorotrifluoroethylene in Organic Solvents after 40 Hours at 45°C (726)*

| Solvent | Absorption of the Solvent by the Polymer (g per 100 g) | Solvent | Absorption of the Solvent by the Polymer (g per 100 g) |
|---|---|---|---|
| Hexane ............ | 0·05 | Trichloroethylene .... | 3·9 |
| Heptane ........... | 0 | Tetrachloroethylene .. | 1·75 |
| Methylcyclohexane .. | 0 | Perchlorobutadiene .. | 0 |
| Benzene ........... | 0·7 | Perfluoroheptane .... | 0 |
| Methylene Chloride... | 0·85 | Trichlorotrifluoro- | |
| Chloroform ........ | 1·35 | ethane .......... | 3·6 |
| Carbon Tetrachloride . | 1·0 | Ethyl Alcohol........ | 0 |
| | | Acetone ........... | 0·2 |

Good solvents of polytrifluorochloroethylene are: *o*-chlorobenzotrifluoride (*341*), 2,5-dichlorobenzotrifluoride (*551*), 1,1,3-trifluoropentachloropropane (*551*), and 1,1,2,2-tetrachlorotetrafluorocyclobutane (*292*). They are suitable for carrying out viscosometric measurements at higher temperatures.

The *chemical resistance* of chlorotrifluoroethylene is claimed to be very good. However, heating at higher temperatures in organic solvents, hydrocarbons, halogen derivatives, and especially amines promotes splitting-off of the chlorine, especially in the presence of metals or their salts (*314*) (Table 82).

TABLE 82

*Elimination of Chlorine from Polychlorotrifluoroethylene*
*Polychlorotrifluoroethylene (0·01 mole) Heated with 50 ml. of Isoamylnaphthalene at 227° (314)*

| Time of Heating in Minutes | 40 | 80 | 120 | 160 | 200 |
|---|---|---|---|---|---|
| Amount of Split-off Chlorine, % | 0·6 | 1·5 | 1·9 | 4·2 | 4·5 |

When heated, polymeric chlorotrifluoroethylene changes at 212–214°C from an opaque mass to a transparent gel (transition temperature), which, when cooled slowly at a rate of 1°C per minute shows signs of turbidity at 182–187°C. These two temperatures are characteristic for polychlorotrifluoroethylene and do not depend on the molecular weight (726). On the other hand, the molecular weight determines, at least to a certain extent, the mechanical behaviour of the polymer at higher temperature. The evaluation of the polymers is carried out either by determining the temperature at which a specially prepared molding strained by a constant load breaks (*No Strength Temperature — NST test*), or by determining the time required for breaking the molding at a certain temperature and under a standard load (*Zero Strength Time — ZST test*).

FIG. 43. Sample for NST or ZST tests.

A strip of the polymer having the following dimensions: $2 \times 1/8 \times 1/16$ inch is cut in the middle by two opposite notches so that the profile of the notched spot is $1/16 \times \times 3/64$ inch. The upper end of the strip is fastened to a clamp, the lower is loaded with 7·5 g. For the NST test, the strip is heated in a hollow copper block at a rate of 1·5°C per minute. The temperature is registered at which the strip breaks. For the ZST test a strip of $2 \times 3/16 \times 0·0625$ inch, having the notched profile $0·047 \times 0·0625$ inch is heated at 250°C, and time is measured which is necessary for breaking (549) (ASTM: D–1430–567). The latter test is gaining over the former because it is more convenient for routine measurements (can be carried out in a thermostat).

The NST values for a high-grade polychlorotrifluoroethylene range between 215–340°C, most generally between 300 and 325°C. A relationship between the NST values and molecular weight of the polymer is shown in Table 83 (550).

TABLE 83

*Relationship between the NST Value and Molecular Weight of Polychlorotrifluoroethylene* (550)

| NST Value | Molecular Weight |
|-----------|------------------|
| 235 | 56 000 |
| 273 | 76 000 |
| 317 | 100 000 |

*Tensile strength* and *elongation* are measured by increasing the load of a standard strip of the polymer till the strip breaks (726).

The *flow index* is determined by means of a plunger sinking in the molten polychlorotrifluoroethylene at 268°C.

*Polytetrafluoroethylene.* Polymeric tetrafluoroethylene is a white opaque mass with a waxy surface. Since it is absolutely insoluble, the molecular weight cannot be determined by standard methods. Using radioactive sulphur in compounds participating in the suspension polymerization of tetrafluoro-ethylene in aqueous medium, the *molecular weight of* various samples of the polymer was determined by the end groups method (142,000–534,000) (*94*). The density of polytetrafluoroethylene is 2·2–2·3, its refractive index 1·37–1·38 (*878*).

An outstanding property of polytetrafluoroethylene is its absolute *stability toward solvents* at any temperature up to the boiling point. Polytetrafluoro-ethylene is not attacked nor swelled by any of the listed solvents: water, alcohols, ethers, aldehydes, ketones, aliphatic hydrocarbons, aromatic hydro-carbons, halogenated hydrocarbons, phenols, esters, acids, amines, acyl chlo-rides, acid anhydrides, nitrocompounds or pyridines (*878*).

The *chemical resistance* of polytetrafluoroethylene is amazing. The polymer is attacked only by molten sodium at temperatures higher than 200°C, and by elementary fluorine at higher pressure or at atmospheric pressure and 150°C. Fluorine diluted by nitrogen has only a slight effect on the polymer (*878*). The following chemicals have been applied to polytetrafluoroethylene (tem-peratures in °C of action are in brackets) without producing an appreciable change: concentrated hydrochloric acid (50°, 100°). Aqua regia (50°), con-centrated nitric acid (25°, 85°), fuming nitric acid (60°), perchloric acid (25°), concentrated sulphuric acid (25°, 300°), fuming sulphuric acid (80°), concentrat-ed hydrofluoric acid (25°, 100°), anhydrous hydrogen fluoride (25°), organic acids (25°, 100°), liquid ammonia (25°), 50% sodium hydroxide (25°, 100°), 50% potassium hydroxide (25°, 100°), bromine (1 atm.) (25°, 100°), chlorine (1 atm). (25°, 100°), 5% potassium permanganate (25°, 100°), 30% hydrogen peroxide (25°), phosphorus pentachloride (100°), the complex of boron trifluoride with hydrogen fluoride (25°), sodium peroxide (100°), chlorosulphonic acid (25°), and ozone (25°) (*878*).

The *thermal resistance* of polytetrafluoroethylene is one of the greatest ever found in an organic material. The polymer stands heating up to 327°C without change which is the first order transition point. At this temperature the white opaque material changes to a transparent gel. Sintering of polymeric tetrafluoroethylene is carried out at 370–390°C without any detrimental effect. Prolonged heating to 250°C does not affect the mechanical properties at all; after one month's heating at 300°C a loss of 10–20% in the tensile strength was observed. At temperatures higher than 400°C, partial depolymerization takes place. The weight of the heated sample of material drops with rising temperature (*878*).

Another interesting property of polytetrafluoroethylene is an absolute *hydrophobia* and *oleophobia*. It does not become wet in contact with water, nor does it occlude water vapour. This is reflected in the electric properties of polytetrafluoroethylene. It has an extremely low loss factor and has therefore become indispensable in electrotechnique.

Comparison of polychlorotrifluoroethylene and polytetrafluoroethylene with polyethylene is made in Table 84.

*Fluorinated Copolymers.* The properties of fluorinated copolymers vary over a very wide range as they are determined not only by their qualitative but also quantitative composition. Published numerical data are very few (*42, 644, 743*). The main differences from pure homopolymers are the lower tendency to crystallize and the elastic properties of some of them. The copolymer of vinylidene fluoride with chlorotrifluoroethylene stands temperatures up to 200°C, the copolymer of vinylidene fluoride with perfluoropropylene up to 250°C, and eventually 300°C for short periods (*1029*).

TABLE 84

*Comparison of Polychlorotrifluoroethylene and Polytetrafluoroethylene with Polyethylene*
(*20, 425, 726, 878*)

| Property | Polyethylene | Polychloro-trifluoro-ethylene | Polytetra-fluoroethylene |
|---|---|---|---|
| Density | 0·92 | 2·11—2·13 | 2·2—2·3 |
| Refractive Index | 1·51 | 1·43 | 1·35 |
| Dielectric Constant ($10^3$ cycles) | 2·3 | 2·8 | 2·0—2·05 |
| Loss Factor | <0·0005 | 0·024 | 0·003 |
| Specific Resistivity (ohm/cm) | | $5 \cdot 10^{17}$ | |
| Transition Temperature (°C) | | 212—214 | 327 |
| Tensile Strength (kg/cm²) | 105—125 | 400 | 105—210 |
| Tensile Strength after Orientation (kg/cm²) | | 2100 | 1050 |
| Processability at Temperature (°C) | | 230—290° | 380 (sintering) |
| Moulding Temperature (°C) | 135—150 | 230—260° | |
| Moulding Pressure (atm) | 14 | 35—1050 | |

## Processing and Applications of Fluorinated Polymers

The outstanding thermal, chemical and electric properties of fluorinated polymers, especially of polytetrafluoroethylene, increase the demand for these compounds. Nevertheless, their applications are limited to technical material, since their price is still relatively high and cannot be compared with that of more common plastics.

Another reason preventing wider applications of fluorinated polymers is difficult processing, especially of polytetrafluoroethylene. While polychlorotrifluoroethylene can be worked up by common processes such as compression moulding, injection moulding, extrusion etc., at sufficiently high temperatures (230–280°C), polytetrafluoroethylene can be worked up only by a process comparable to powder metallurgy: the powdered polymer of a grain size of 30–50 mesh is compressed into the moulds by pressure round 140 atm, the moulding heated to 380°, and quenched (4).

A special process has been devised for the fabrication of polytetrafluoroethylene tapes. First a mash of polytetrafluoroethylene and a high-boiling hydrocarbon is made up from a polymer (500 g) prepared by suspension polymerization, methanol (1000 ml.) and decane, dodecane or hexadecane (110–125 g). The mixture is stirred for 1–2 minutes at 500 r.p.m., the precipitate is filtered off and allowed to dry (623). Polyisobutylene (2%) is sometimes added as a thickener (420).

The paste thus formed is extruded through a die as a tape. This is freed from the hydrocarbon by passing through a solvent, or by evaporation at temperatures of 300–325°C, shaped by passing through two rolls with a distance of 0·02–0·001 inch, sintered by passing through a furnace heated above 327°C, and quenched by immersing in a water bath (306, 623).

Gradually also coating and dipping techniques have been developed. Special aqueous dispersions of polychlorotrifluoroethylene (30–40% of solids) and of polytetrafluoroethylene (50% of solids) are applied to metal or ceramic surface and allowed to dry. Heating to the required temperature (327–370°C) gives resistant coating (622).

The composition and preparation of some dispersions of polychlorotrifluoroethylene and polytetrafluoroethylene is best illustrated by practical examples. An aqueous dispersion suitable for the fabrication of films is obtained by stirring 54·5 parts of finely ground polytetrafluoroethylene with 44·1 parts of water in the presence of 1·4 parts of sodium lauryl sulphate as an emulsifier (903). A non-aqueous dispersion is prepared by 85 hours' grinding of a mixture of 45 g of polychlorotrifluoroethylene, 71 g diisobutyl ketone, and 156 g of xylene in a pebble mill (1033). A dispersion is also obtainable by direct polymerization of tetrafluoroethylene under suitable conditions. Paraffin (m.p. 55–60°C) (4·86 parts) is melted in 44·6 parts of water, 0·44 parts of ammonium perfluorocaprylate and 0·1 part of persuccinic acid in 50 parts of water is added, and after the addition of a ferrous salt (10 p.p.m.), tetrafluoroethylene is fed to the autoclave at 85–90°C up to a pressure of 27 atm (596). The procedures taken from the patent literature must be judged with a certain amount of reserve.

Polychlorotrifluoroethylene and polytetrafluoroethylene cannot be modified by incorporating common softeners like many other polymers. The reason is negligible compatibility with common organic materials. A few possible softeners, which could be applied to the fluorinated polymers, are compounds produced by telomerization of chlorotrifluoroethylene and which are admixed in the amount of 4–10% based on the weight of the polymer (*814*).

Compression moulding, injection moulding and extrusion are used for the fabrication of sheets, gaskets, fittings and their parts, tubes, tapes, and containers; coating and dipping technique for the preparation of films and coatings inside or outside metal or ceramic vessels.

The properties of the product fabricated from polytetrafluoroethylene are determined by thermal treatment of the material after shaping. A moulding of polytetrafluoroethylene heated to the sintering temperature can be allowed to cool slowly, or can be shock-cooled (quenched) by immersing in a liquid bath. The second procedure gives products with better mechanical properties, especially higher tensile strength. This can be improved by cold stretching. Where internal stresses produced by quenching are detrimental, annealing is necessary. The article is slowly heated slightly above the temperature of the expected service but safely below the transition point, and kept at this temperature for some time.

Table 85 shows the effect of thermal and mechanical treatment on the quality of the polytetrafluoroethylene articles.

TABLE 85

*Tensile Strength of a Polytetrafluoroethylene Tape (623)*

| Treatment | Slow Cooling | Quenching | Not Oriented | Oriented by Cold Stretching |
|---|---|---|---|---|
| Tensile Strength kg/cm² (1b/sq. i.) | 235 (3370) | 320 (4540) | 400 (5760) | 1180 (16 800) |

## OTHER APPLICATIONS OF ORGANIC FLUORINE COMPOUNDS

In terms of volume of consumption, applications of fluorinated compounds other than refrigerants, propellants, fire extinguishers and plastics are insignificant.

*Fluorocarbons*, perfluoroheptane and perfluoromethyl- and dimethylcyclohexanes have been used in the Atomic Energy (*425*). Polychlorofluoroparaffins obtained by the telomerization of chlorotrifluoroethylene have been employed as liquids for heat transfer, as brake liquids and as greases (*425*). *Perfluoro ethers* and *perfluoroamines*, stable up to 400°C, are used as coolants in transformers and coaxial cables (*425*).

Some attention has been given to *fluorinated dyes*, because of their good light fastness (*213*). Some of them were used practically, e.g. Naphthol AS Red for German flags and Indanthrene Blue CLB for German Airforce uniforms (*425*).

737                                                                        [*425*]

Special *water-repellent and oil-resistant finish* is composed of a perfluoroacrylic ester (*Scotchgard FC-154*) and stearamidopyridinium chloride and shows a remarkable resistance to washing and dry-cleaning (*14*).

Quite a few compounds containing fluorine have found application in *pharmacy*. The use of 3-fluorotyrosine (*Pardinon*) (*425*) and 3-fluoro-4-hydroxyphenylacetic acid (*Capacin*) (*592*) for curing hyperthyreosis, *p,p'*-difluorodiphenyl (*Antitussin*) (*425*) for treating whooping cough, *p*-fluorophenetol (*Fluorrheumin*) (*425*) for curing rheumatism, and a mixture of *p,p'*-difluorodiphenyl with fluoropseudocumene (*Fluorepidermin*) for wound disinfection is outdated.

738   HO—⟨ ⟩—CH₂—CH—COOH          HO—⟨ ⟩—CH₂—COOH       [*425*]

739   F—⟨ ⟩—⟨ ⟩—F                 F—⟨ ⟩—OC₂H₅              [*425*]

Today some fluorinated compounds are used as *inhalation narcotics*. β,β,β-Trifluoroethyl vinyl ether (*941*) and especially 1-bromo-1-chloro-2,2,2-trifluoroethane (*Fluothane*) (*781*) are used for anesthesia, and bis(β,β,β-trifluoroethyl) ether (*Indoklon*) which develops cramps for curing schizzophrenia (*593*).

740   CF₃—CH₂—O—CH=CH₂        CF₃—CHBrCl        CF₃—CH₂—O—CH₂—CF₃

10-(γ-Dimethylaminopropyl)-2-trifluoromethylphenothiazine (*Vesprin*) is an efficient *ataracticum (tranquillizer)* (*1094*).

741                                                                        [*1094*]

The most recent discovery is the pharmacological activity of *fluorinated corticosteroids*, some of which are several hundred times more efficient in their anti-inflammatory and glucocorticoidal activity than the non-fluorinated compounds (*Triamcinolone*) (*351*). Usually fluorine is attached to the 6x- or 9x-positions (*517, 965*).

**742**

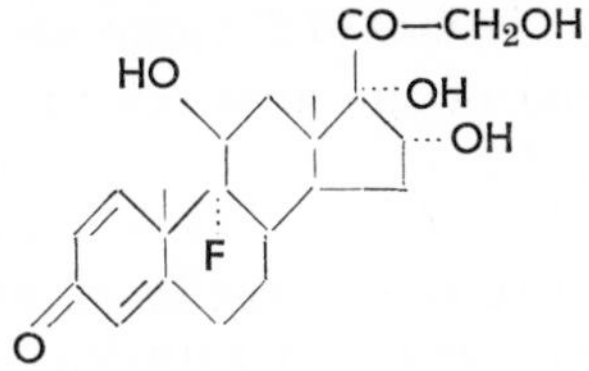

[*351*]

*Fluorinated derivatives of pyrimidine and purine* show retarding influences upon some kinds of cancer or leukemia. However the most successful compound in this respect, the 5-fluorouracil (*228*) has not yet gone further than clinical trials.

**743**

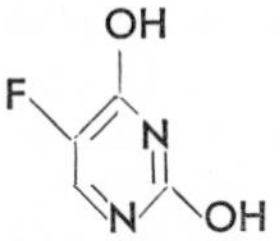

[*228*]

Sodium fluoroacetate (*rodenticide 1080*) is used as a very effective rodenticide for exterminating mice and rats (*165, 425*); the fluorinated analogue of DDT (*GIX, DFDT*), 1,1-bis(*p*-fluorophenyl)-2,2,2-trichloroethane, is an insecticide with higher knock-down power than DDT itself (*919*).

**744**

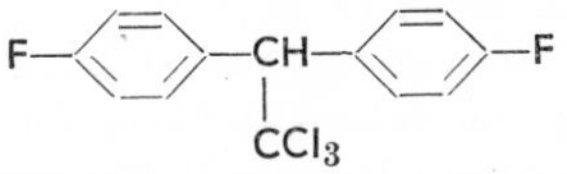

[*919*]

Fortunately, no application has been made of the *nerve gases*, isopropyl methanefluorophosphonate (*Sarin, Trilon 46*) and methyl-*tert.*-butylcarbinyl methanefluorophosphonate (*Soman*) (p. 311). Of this group of compounds, only *diisopropyl fluorophosphate* (*DFP*) (Equation 143, p. 110) has found application in medicine, especially in the treatment of glaucoma (*910*).

# LIST OF REFERENCES

A list of normalized abbreviations of the journals used is on p. 391. Numbers in square brackets at the end of the lines refer to the page of the book on which the reference is quoted.

1. AHLBRECHT A. H., CODDING D. W.: *J. Am. Chem. Soc.* **75**, 984 (1953); [141,280].
2. AIR REDUCTION Co.: *British Patent* 790824 (1958); *Chem. Abstracts* **53**, 1149b (1959); [141].
3. ALLISON J. A. C., CADY G. H.: *J. Am. Chem. Soc.* **81**, 1089 (1959); [84, 42]
4. ANON.: *Chem. Eng. News* **30**, 2688 (1952); [354].
5. ANON.: *Chem. Eng. News* **35**, No 13, 14 (1957); [34, 47].
6. ANON.: *Chem. Eng. News* **36**, No 13, 62 (1958); [60].
7. ANON.: *Chem. Eng. News* **36**, No 36, 32 (1958); [49].
8. ANON.: *Chem. Eng. News* **36**, No 38, 32 (1958); [44].
9. ANON.: *Chem. Eng. News* **36**, No 47, 19 (1958); [332].
10. ANON.: *Chem. Eng. News* **37**, No 22, 42 (1959); [48].
11. ANON.: *Chem. Eng. News* **37**, No 36, 124 (1959); [338, 342].
12. ANON.: *Chem. Eng. News* **37**, No 43, 90 (1959); [342].
13. ANON.: *Chem. Eng. News* **37**, No 45, 63 (1959); [319].
14. ANON.: *Chem. Eng. News* **37**, No 51, 40 (1959); [356].
15. ANON.: Harshaw Chemical Company, Technical Bulletin, 1949; [44,328].
16. ANON.: *Ind. Eng. Chem.* **39**, 241 (1947); [21].
17. ANON.: Isceon. Technical Bulletin, Imp. Smelting Corp., 1957; [304, 306, 313, 330, 331, 332, 338, 340, 341].
18. ANON.: *J. Chem. Soc.* **1953**, 5059; [21].
19. ANON.: Korosní Sborník. Průmyslové vydavatelství. Praha, 1952; [36,37].
20. ANON.: *Modern Plastics* **34**, Encyclopedia No (September) (1956); [353].
21. ANON.: Wiggin H. Co., Technical Bulletin V. Fluorine and Fluorine Compounds; [34, 35, 36].
22. AUSTIN P. R., COFFMAN D. D., HOEHN H. H., RAASCH M. S.: *J. Am. Chem. Soc.* **75**, 4834 (1953); [66, 269].
23. AVONDA F. P., GERVASI J. A., BIGELOW L. A.: *J. Am. Chem. Soc.* **78**, 2798 (1956); [71, 154].
24. AYNSLEY E. E., WATSON R. H.: *J. Chem. Soc.* **1955**, 576; [267].
25. BACON J. C., BRADLEY C. W., HOEBERG E. I., TARRANT P., CASSADAY J. T.: *J. Am. Chem. Soc.* **70**, 2653 (1948); [105, 111, 146, 152, 198].
26. BADGER G. M., STEPHENS J. F.: *J. Chem. Soc.* **1956**, 3637; [84].
27. BAER D. R.: *J. Org. Chem.* **23**, 1560 (1958); [195].
28. BALZ G., SCHIEMANN G.: *Ber.* **60**, 1186 (1927); [115, 137].
29. BANKS R. E., BARBOUR A. K., TIPPING A. E.: *Nature* **183**, 586 (1959); [266].
30. BANKS A. A., EMELÉUS H. J., HASZELDINE R. N., KERRIGAN V.: *J. Chem. Soc.* **1948**, 2188; [33, 104, 123].

31. BANUS J.: *J. Chem. Soc.* 1953, 3755; [192, 194].
32. BANUS J., EMELÉUS H. J., HASZELDINE R. N.: *J. Chem. Soc.* **1950**, 3041; [157, 166].
33. BANUS J., EMELÉUS H. J., HASZELDINE R. N.: *J. Chem. Soc.* **1951**, 60; [157, 166]
34. BARBEN I. K., SUSCHITZKY H.: *Chem. & Ind.* (London) **1957**, 1039; [117].
35. BARBOUR A. K.: Intern. Symposium Fluorine Chem., Birmingham, 1959; [122, 123, 129, 335, 336, 338].
36. BARBOUR A. K., BARLOW G. B., TATLOW J. C.: *J. Appl. Chem.* (London) **2**, 127 (1952;) [80, 81].
37. BARBOUR A. K., MACKENZIE H. D., STACEY M., TATLOW J. C.: *J. Appl. Chem. (London)* **4**, 347 (1954); [82, 134].
38. BARKLEY L. B., LEVINE R.: *J. Am. Chem. Soc.* **73**, 4625 (1951); [229].
39. BARKLEY L. B., LEVINE R.: *J. Am. Chem. Soc.* **75**, 2059 (1953); [229].
40. BARLOW G. B., TATLOW J. C.: *J. Chem. Soc.* **1952**, 4695; [166, 186].
41. BARNEY A. L., CAIRNS T. L.: *J. Am. Chem. Soc.* **72**, 3193 (1950); [205, 258].
42. BARNHART W. S.: *U.S. Patent* 2,753,328 (1956): *Chem. Abstr.* **50**, 15127i (1956); [353].
43. BARR D. A., HASZELDINE R. N.: *J. Chem. Soc.* **1955**, 1881; [154, 155, 192, 194, 256, 263, 277].
44. BARR D. A., HASZELDINE R. N.: *J. Chem. Soc.* **1955**, 2532; [68,].
45. BARR D. A., HASZELDINE R. N.: *J. Chem. Soc.* **1956**, 3416; [155, 192, 256, 277].
46. BARR D. A., HASZELDINE R. N.: *J. Chem. Soc.* **1956**, 3428; [154, 208, 235].
47. BARR D. A., HASZELDINE R. N.: *J. Chem. Soc.* **1957**, 30; [236].
48. BARR D. A., HASZELDINE R. N., WILLIS C. J.: *Proc. Chem. Soc.* **1959**, 230; [256].
49. BARR J. T., GIBSON J. D., LAFFERTY R. H. JR.: *J. Chem. Soc.* **74**, 4945 (1952); [186].
50. BARR J. T., RAPP K. E., PRUETT R. L., BAHNER C. T., GIBSON J. D., LAFFERTY R. H. JR.: *J. Am. Chem. Soc.* **72**, 4480 (1950); [249].
51. BARRICK P. L., CRAMER R. D.: *U.S. Patent* 2, 441, 128 (1948); *Chem. Abstr.* 42, 6847d (1948); [190].
52. BATCHELDER G., MELOCHE V. W.: *J. Am. Chem. Soc.* **53**, 2131 (1931); [317].
53. BEATY R. D., MUSGRAVE W. K. R.: *J. Chem. Soc.* **1952**, 875; [118, 140].
54. BELCHER R., CALDAS E. F., CLARK S. J., MACDONALD A.: *Mikrochim. Acta* 1953 283; [322].
55. BELCHER R., CLARK S. J.: *Anal. Chim. Acta* **8**, 222 (1953); [317].
56. BELCHER R., GOULDEN R.: *Mikrochemie* **36/37**, 679 (1951); [323].
57. BELCHER R., LEONARD M. A., WEST T. S.: *J. Chem. Soc.* **1959**, 3577; [319, 321].
58. BELCHER R., LEONARD M. A., WEST T. S.: *Talanta* **2**, 93 (1959); [316].
59. BELCHER R.. MACDONALD A. M. G.: *Mikrochim. Acta* **1956**, 1111; [325].
60. BELCHER R., MACDONALD A. M. G.: *Mikrochim. Acta* **1956**, 1187; [325].
61. BELCHER R., MACDONALD A. M. G.: *Mikrochim. Acta* **1957**, 510; [317].
62. BELCHER R., MACDONALD A. M. G.: NUTTEN A. J.: *Mikrochim. Acta* **1954**, 104; [325].
63. BELCHER R., SYKES A., TATLOW J. C.: *J. Chem. Soc.* **1957**, 2393; [182].
64. BELCHER R., TATLOW J. C.: *Analyst* **76**, 592 (1951); [322].
65. BELMORE E. A., EWALT W. M., WOJCIK B. H.: *Ind. Eng. Chem.* **39**, 338 (1947); [129, 264, 343, 344].
66. BENNINGTON F., SHOOP E. V., POIRIER R. H. *J. Org. Chem.* **18**, 1506 (1953); [153, 159].
67. BENKESER R. A., SEVERSON R. G.: *J. Am. Chem. Soc.* **71**, 3838 (1949); [241].
68. BENKESER R. A., SEVERSON R. G.: *J. Am. Chem. Soc.* **73**, 1353 (1951); [185].
69. BENNER R. G., BENNING A. F., DOWNING F. B., IRWIN C. F., JOHNSON K. C., LINCH

    A. L., PARMELEE H. M., WIRTH W. V.: *Ind. Eng. Chem.* **39**, 329 (1947); [80, 81' 134, 301, 302].

70. BENNETT F. W., BRANDT G. R. A., EMELÉUS H. J., HASZELDINE R. N.: *Nature* **166**, 225 (1950); [227].

71. BENNETT F.W., EMELÉUS H. J., HASZELDINE R. N.: *J.Chem. Soc.* **1953**,1565; [177].

72. BENNETT F. W., EMELÉUS H. J., HASZELDINE R. N.: *J. Chem. Soc.* **1954**, 3896: [209].

73. BENNING A. F., DOWNING F. B., PLUNKETT R. J.: *U.S. Patent* 2, 401, 897 (1946); *Chem. Abstr.* 40, 5066[8] (1946); [264, 343].

74. BENNING A. F., McHARNESS R. C.: *Ind. Eng. Chem.* **31**, 912 (1939); [338].

75. BENNING A. F., McHARNESS R. C.: *Ind. Eng. Chem.* **32**, 497 (1940); [338, 339].

76. BENNING A. F., McHARNESS R. C.: *Ind. Eng. Chem.* **32**, 698 (1940); [338].

77. BENNING A. F., McHARNESS R. C., MARKWOOD W. H. Jr., SMITH W. J.: *Ind. Eng. Chem.* **32**, 976(1940); [338].

78. BENNING A. F., MARKWOOD W. H.: *Refrig. Eng.* **37**, 243 (1939); [338].

79. BENNING A. F., PARK J. D.: *U.S. Patent* 2, 351, 390 (1944); *Chem. Abstr.* **38**, 5228[1] (1944); [169].

80. BENNING A. F., YOUNG E. G.: *U.S. Patent* 2, 615, 926 (1952); *Chem. Abstr.* **47**, 8770a (1953); [129, 159, 267].

81. BERGMANN E.: *J. Org. Chem.* 23, 476 (1958); [209].

82. BERGMANN E. D., BENTOV M.: *J. Org. Chem.* **19**, 1594 (1954); [117].

83. BERGMANN E. D., BERKOVIC S., IKAN R.: *J. Am. Chem. Soc.* 78, 6037 (1956); [117].

84. BERGMANN E. D., BLANK I.: *J. Chem. Soc.* **1953**, 3786; [105, 106, 107, 146].

85. BERGMANN E. D., IKAN R.: *Chem. & Ind.* (London) **1957**, 394; [68, 144].

86. BERGMANN E. D., SCHWARCZ J.: *J. Chem. Soc.* **1956**, 1524; [122, 228].

87. BERGMANN E. D., SHAHAK I.: *Chem. & Ind.* (London) **1958**, 157; [113].

88. BERGMANN E. D., SHAHAK I.: *J. Chem. Soc.* **1959**, 1418; [51].

89. BERGMANN E. D., SZINAI S.: *J. Chem. Soc.* **1956**, 1521; [240].

90. BERGMANN F., KALMUS A.: *J. Am. Chem. Soc.* **76**, 4137 (1954); [145, 161, 212, 213, 215].

91. BERGMANN F., KALMUS A., BREUER E.: *J. Am. Chem. Soc.* **79**, 4178 (1957); [190].

92. BERGMANN F., KALMUS A., VROMEN S.: *J. Am. Chem. Soc.* **77**, 2494 (1955); [228].

93. BERNSTEIN J., ROTH J. S., MILLER W. T. Jr.: *J. Am. Chem. Soc.* **70**, 2310 (1948); [136, 216].

94. BERRY K. L., PETERSON J. H.: *J. Am. Chem. Soc.* **73**, 5195 (1951); [352].

95. BEVAN C. W. L.: *J. Chem. Soc.* **1953**, 655; [241].

96. BIGELOW L. A., PEARSON J. H.: *J. Am. Chem. Soc.* **56**, 2773(1954); [28, 70].

97. BIRCHALL J. M., HASZELDINE R. N.: *J. Chem. Soc.* **1959**, 13; [123, 139, 141, 202, 207, 242, 266].

98. BIRCHALL J. M., HASZELDINE R. N.: *J. Chem. Soc.* **1959**, 3653: [308].

99. BISSELL E. R., FINGER M.: *J. Org. Chem.* **24**, 1256 (1959); [162].

100. BLANK I., BERGMANN E. D.: *Bull. Research Council Israel* **2**, No 1, 71 (1952): [105, 146].

101. BLANK I., MAGER J., BERGMANN E. D.: *J. Chem. Soc.* **1955**, 2190: [230, 244].

102. BLOMQUIST A. T., LALANCETTE E. A.: *Abstracts, Am. Chem. Soc.* 135th Meeting, April 1959; [203].

103. BLOOM B. M., BOGERT V. V., PINSON R. Jr.: *Chem. & Ind.* (London) **1959**, 1317; [83].

104. BOCKEMÜLLER W.: *Ann.* **506**, 20 (1933); [28, 48, 68, 72, 73, 200, 269].

105. BOCKEMÜLLER W.: *Ber.* **64**, 522 (1931); [69, 84].

106. BOCKEMÜLLER W., (KIBLER C. J.): Newer Methods of Preparative Organic Chemistry, Interscience, New York, 1948; [20].

107. BOOTH H. S., BIXBY E. M.: *Ind. Eng. Chem.* **24**, 637 (1932); [33, 96, 122, 201].

108. BOOTH H. S., BURCHFIELD P. E., BIXBY E. M., McKELVEY J. B.: *J. Am. Chem. Soc.* **55**, 2231 (1933); [99].

109. BOOTH H. S., CARNELL P. H.: *J. Am. Chem. Soc.* **68**, 2650 (1946); [95, 156].

110. BOOTH H. S., ELSEY H. M., BURCHFIELD P. E.: *J. Am. Chem. Soc.* **57**, 2066 (1935); [95, 136].

111. BOOTH H. S., MONG W. L., BURCHFIELD P. E.: *Ind. Eng. Chem.* **24**, 328 (1932); [125, 129, 333].

112. BOOTH H. S., SWINEHART C. F.: *J. Am. Chem. Soc.* **57**, 1333 (1935); [93].

113. BOOTH H. S., WILLSON K. S.: *J. Am. Chem. Soc.* **57**, 2273 (1935); *Inorg. Syntheses* **1**, 21 (1939); [54, 55].

114. BORDNER C. A.: *U.S. Patent* 2, 615, 925 (1952); *Chem. Abstr.* 47, 8770f (1953); [264].

115. BORNSTEIN J., BORDEN M. R.: *Chem. & Ind.* (London) **1958**, 441; [69, 157, 160, 234].

116. BOURNE E. J., HENRY S. H., TATLOW C. E. M., TATLOW J. C.: *J. Chem. Soc.* **1952**, 4014; [243].

117. BOURNE E. J., RANDLES J. E. B., STACEY M., TATLOW J. C., TEDDER J. M.: *J. Am. Chem. Soc.* **76**, 3206 (1954); [280, 281].

118. BOURNE E. J., STACEY M., TATLOW J. C., TEDDER J. M.: *J. Chem. Soc.* **1949**, 2976; [151, 279, 280, 281].

119. BOURNE E. J., STACEY M., TATLOW J. C., TEDDER J. M.: *J. Chem. Soc.* **1951**, 718; [281].

120. BOURNE E. J., STACEY M., TATLOW J. C., WORRALL R.: *J. Chem: Soc.* **1954**, 2006; [279, 280].

121. BOURNE E. J., TATLOW C. E. M., TATLOW J. C.: *J. Chem. Soc.* **1950**, 1367; [243, 280].

122. BOVEY F. A., ABERE J. F., ROTHMANN G. B., SANDBERG C. L.: *J. Polymer Sci.* **15**, 520 (1955); [347].

123. BOWERS A.: *Chem. Eng. News* **37**, No 37, 44 (1959); [70].

124. BOWERS A.: *J. Am. Chem. Soc.* **81**, 4107 (1959); [70].

125. BOWERS A., IBÁÑEZ L. C., RINGOLD H. J.: *J. Am. Chem. Soc.* **81**, 5991 (1959); [83, 118].

126. BRADSHER C. K., BOND J. B.: *J. Am. Chem. Soc.* **71**, 2659 (1949); [195].

127. BRAID M., ISERSON H., LAWLOR F. E.: *J. Am. Chem. Soc.* **76**, 4027 (1954); [143, 161].

128. BRANDT G. A. R., EMELÉUS H. J., HASZELDINE R. N.: *J. Chem. Soc.* **1952**, 2198; [205, 239].

129. BRAUNER B.: *J. Chem. Soc.* **41**, 68 (1882); **65**, 393 (1894); [57].

130. BRICE T. J., LAZERTE J. D., HALS L. J., PEARLSON W. H.: *J. Am. Chem. Soc.* **75**, 2698 (1953); [132, 271, 272].

131. BRICE T. J., PEARLSON W. H., SIMONS J. H.: *J. Am. Chem. Soc.* **71**, 2499 (1949); [186].

132. BRICE T. J., SIMONS J. H.: *J. Am. Chem. Soc.* **73**, 4016, 4017 (1951); [175, 185, 271].

133. BRITISH INTELLIGENCE OBJECTIVES SUBCOMMITTEE, *Report* 112; [91, 92, 333, 335, 336].

134. BROWN F., MUSGRAVE W. K. R.: *J. Chem. Soc.* **1952**, 5049; [143].

135. BROWN F., MUSGRAVE W. K. R.: *J. Chem. Soc.* **1953**, 2087; [235, 244].

136. BROWN H. C.: *J. Org. Chem.* **22**, 1256 (1957); [135, 257].
137. BROWN J. H., SUCKLING C. W., WHALLEY W. B.: *J. Chem. Soc.* **1949**, S 95; [89, 136].
138. BRUBAKER M. M.: *U. S. Patent* 2, 407, 396. (1946); *Chem. Abstr.* **41**, 2067b (1947); [344].
139. BRUCE W. T., HUBER R. DE V.: *J. Am. Chem. Soc.* **75**, 4668 (1953); [240].
140. BRUNISHOLZ G., MICHOD J.: *Helv. Chim. Acta* **37**, 598 (1954); [317].
141. BUCK F. A. M., LIVINGSTON R. L.: *J. Am. Chem. Soc.* **70**, 2817 (1948); [275].
142. BUCKLE F. J., HEAP R., SAUNDERS B. C.: *J. Chem. Soc.* **1949**, 913; [152, 243].
143. BUCKLE F. J., PATTISON F. L. M., SAUNDERS B. C.: *J. Chem. Soc.* **1949**, 1471; [312].
144. BURDON J., FARAZMAND I., STACEY M., TATLOW J. C.: *J. Chem. Soc.* **1957**, 2574; [84, 86].
145. BURDON J., GILMAN D. J., PATRICK C. R., STACEY M., TATLOW J. C.: *Nature* **186**, 231 (1960); [140, 266]
146. BURDON J., TATLOW J. C.: *J. Appl. Chem.* (London) **8**, 293 (1958); [151, 173, 174].
147. BURG A. B., MAHLER W., BILBO A. J., HABER C. P., HERRING D. L.: *J. Am. Chem. Soc.* **79**, 247 (1957); [156, 263].
148. BUSCH G. W., CARTER R. C., MCKENNA F. E.: Analytical Chemistry of the Manhattan Project (Rodden C. J.). McGraw-Hill, New York. 1950, p. 226 [314].
149. BUU HOI, HOAN NG., JACQUIGNON P.: *Rec. trav. chim.* **68**, 781 (1949); [210].
150. BUXTON M. W., STACEY M., TATLOW J. C.: *J. Chem Soc.* **1954**, 366; [158, 206, 228, 270].
151. BUXTON M. W., TATLOW J. C.: *J. Chem. Soc.* **1954**, 1177; [163, 269].
152. CADY G. H., GROSSE A. V., BARBER E. J., BURGER L. L., SHELDON Z. D.: *Ind. Eng. Chem.* **39**, 290 (1947); [28, 76, 77, 128].
153. CADY G. H., KELLOGG K. B.: *J. Am. Chem. Soc.* **75**. 2501 (1953); [152].
154. CADY G. H., KELLOGG K. B.: *U.S. Patent* 2,689, 254 (1954); *Chem. Abstr.* **49**, 11681i (1955); [71, 140].
155. CALEY E. R., KAHLE G. R.: *Anal. Chem.* **31**, 1880 (1959); [317].
156. CALFEE J. D., FUKUHARA N., YOUNG S. D., BIGELOW L. A.: *J. Am. Chem. Soc.* **62**, 267 (1940); [319].
157. CALLOWAY N. O.: *J. Am. Chem. Soc.* **59**, 1474 (1937); [211].
158. CALVIN M.: *U. S. Patent* 2, 493, 654 (1950); *Chem. Abstr.* **44**, 2559g (1950); [237].
159. CAMPBELL K. N., KNOBLOCH J. O., CAMPBELL B. K.: *J. Am. Chem. Soc.* **72**, 4380 (1950); [141, 145, 161, 162, 226].
160. CARBONI R. A., LINDSEY R. V., J., JR. *Am. Chem. Soc.* **80**, 5793 (1958); [246].
161. ČELECHOVSKÝ J.: *Acta Facultatis Pharmaceuticae Bohemoslovenicae* **1**, 59 (1958); *Chem. Abstr.* **53**. 8942f (1959); [319].
162. ČELECHOVSKÝ J., HOLER J.: *Chem. listy* **51**, 2129 (1957); [319].
163. CHABRIÉ C.: *Compt. rend.* **110**, 279 (1890); *Bull. soc. chim. France* (3), **7**, 18 (1892); [99].
164. CHANEY D. W.: *U.S. Patent* 2, 439, 505 (1948); *Chem. Abstr.* **42**, 7315f (1948); [150, 151].
165. CHAPMAN C., PHILLIPS M. A.: *J. Sci. Food Agr.* **6**, 231 (1955); [357].
166. CHAPMAN N. B., HEAP R., SAUNDERS B. C.: *Analyst* **73**, 434 (1948); [322, 323].
167. CHAPMAN N. B., PARKER R. E.: *J. Chem. Soc.* **1951**, 3301; [242, 286].
168. CHAPMAN N. B., PARKER R. E., SOANES P. W.: *Chem. & Ind.* **1951**, 148; [242, 286].
169. CHEEK P. H., WILEY R. H., ROE A.: *J. Am. Chem. Soc.* **71**. 1863 (1949); [118, 141].
170. CLARK H. S.: *Anal. Chem.* **23**, 659 (1951); [319].

171. CLARK J. W.: *U.S.Patent* 2, 704, 777(1955); *Chem. Abstr.* **50**, 15574h (1956); [129,264].
172. CLARK R. F., SIMONS J. H.: *J. Am. Chem. Soc.* **75**, 6305 (1953); [198].
173. CLARK R. F., SIMONS J. H.: *J. Am. Chem. Soc.* **77**, 6618 (1955); [238].
174. CLEAVER C. S., ENGLAND D. C.: *U. S. Patent* 2, 861, 990 (1958); *Chem. Abstr.* **53**, 14123d (1959); [249].
175. CLIFFORD A. F., EL-SHAMY H. K., EMELÉUS H. J., HASZELDINE R. N.: *J. Chem. Soc.* **1953**, 2372; [34, 72, 84, 86, 122, 152].
176. CODDING D. W.: *U.S. Patent* 2, 732, 370 (1956); *Chem. Abstr.* **50**, 8249c (1956); [199].
177. CODDING D. W., REID T. S., AHLBRECHT A. H., SMITH G. H., JR., HUSTED D. R.: *J. Polymer Sci.* **15**, 515 (1955); [141, 198, 280, 347].
178. COFFMAN D. D., BARRICK P. L., CRAMER R. D., RAASCH M. S.: *J. Am. Chem. Soc.* **71**, 490 (1949); [258].
179. COFFMAN D. D., CRAMER R., RIGBY G. W.: *J. Am. Chem. Soc.* **71**, 979 (1949); [250, 252].
180. COFFMAN D. D., RAASCH M. S., RIGBY G. W., BARRICK P. L., HANFORD W. E.: *J. Org. Chem.* **14**, 747 (1949); [152, 153, 196, 246, 250].
181. COHEN S., KALUSZYNER A., MECHOULAM R.: *J. Am. Chem. Soc.* **79**, 5979 (1957); [234].
182. COHEN S., LACHER J. R., PARK J. D.: *J. Am. Chem. Soc.* **81**, 3480 (1959); [203].
183. COHEN S. G., WOLOSINSKI H. T., SCHEUER P. J.: *J. Am. Chem. Soc.* **71**, 3439 (1949) [145, 190, 212]
184. COHEN S. G., WOLOSINSKI H. T., SCHEUER P. J.: *J. Am. Chem. Soc.* **72**, 3952 (1950); [137, 145].
185. COOK C. D., FINDLATER F. G.: *J. Soc. Chem. Ind.* (London) **66**, 169 (1947); [327].
186. COOK D. J., PIERCE O. R., McBEE E. T.: *J. Am. Chem. Soc.* **76**, 83 (1954); [215, 227].
187. COOPER W. D.: *J. Org. Chem.* **23**, 1382 (1958); [212].
188. COPPOLA P. P., HUGHES R. C.: *Anal. Chem.* **24**, 768 (1952); [28].
189. CRAWFORD G. H., SIMONS J. H.: *J. Am. Chem. Soc.* **75**, 5737 (1953); [183].
190. CUCULO J. A., BIGELOW L. A.: *J. Am. Chem. Soc.* **74**, 710 (1952); [71, 154].
191. DAHMLOS J.: *Angew. Chem.* **71**, 274 (1959); [111].
192. DALE J. W.: Intern. Symposium, Fluorine Chem., Birmingham, 1959; [71, 140].
193. DANNLEY R. L., TABORSKY R. G.: *J. Org. Chem.* **22**, 77 (1957); [157, 164].
194. DANNLEY R. L., TABORSKY R. G., LUKIN M.: *J. Org. Chem.* **21**, 1318 (1956); [157, 164].
195. DANNLEY R. L., YAMASHIRO D., TABORSKY R. G.: Abstracts, Am. Chem. Soc., 134th Meeting, 1958; [205].
196. D'ANS J., LAX E.: Taschenbuch für Chemiker und Physiker. Springer, Berlin. 1943; [288].
197. DARRALL R. A., SMITH F., STACEY M., TATLOW J. C.: *J. Chem. Soc.* **1951**, 2329; [228, 229].
198. DAUDT H. W., YOUKER M. A.: *U. S. Patent* 2, 005, 705, 2, 005,709 (1935); *Chem. Abstr.* **29**, 5123$^{4,8}$ (1935); [122, 333, 335, 336].
199. DAUDT H. W., YOUKER M. A.: *U. S. Patent* 2, 005 706, 2,005, 707 (1935); *Chem. Abstr.* **29**, 5123$^{8,9}$ (1935); [122, 335, 336].
200. DAUDT H. W., YOUKER M. A.: *U. S. Patent* 2, 005, 710 (1935); *Chem. Abstr.* **29** 5123$^4$ (1935); 2, 062, 743 (1936); *Chem. Abstr.* **31**, 700$^3$ (1937); [122, 123, 125, 126, 129, 333, 335, 336, 337].

201. DAVIES W., DICK J. H.: *J. Chem. Soc.* **1931**, 2104; **1932**, 483; [104].
202. DAVIS H. R., CHIANG S. H. K.: *U. S. Patent* 2,774, 798 (1956); *Chem. Abstr.* **51**, 12954h (1957); [264].
203. DAVIS H. W., WHALEY A. M.: *J. Am. Chem. Soc.* **72**, 4637 (1950); [268].
204. DEITRICH M. A., JOYCE R. M., JR.: *U. S. Patent* 2, 407, 405 (1946); *Chem. Abstr.* **41**, 2067c (1947); [344].
205. DEÑIVELLE L., FORT R., FAVRE J.: *Compt. rend.* **237**, 722 (1953); [287].
206. DENNIS L. M., VEEDER J. M., ROCHOW E. G.,: *J.Am. Chem. Soc.* **53**, 3263(1931); [45].
207. DÉSIRANT Y.: *Bull. classe sci. acad. roy. Belg.* (5), **15**, 966 (1929); [144].
208. DÉSIRANT Y.: *Bull. classe sci. acad. roy. Belg.* **41**, 759 (1955); [139, 266].
209. DÉSIRANT Y.: *Bull. soc. chim. Belges* **67**, 676 (1958); [139, 266].
210. DIAMOND W. J.: *Appl. Spectroscopy* **13**, 77 (1959); [314].
211. DICKEY B., TOWNE E. B.: *U. S. Patent* 2, 700, 686 (1955); *Chem. Abstr.* **50**, 5720i (1956); [189].
212. DICKEY J. B., TOWNE E. B., BLOOM M. S., TAYLOR G. J., HILL H. M., CORBITT R. A McCALL M. A., MOORE W. H.: *Ind. Eng. Chem.* **46**, 2213 (1954); [153, 239].
213. DICKEY J. B., TOWNE E. B., BLOOM M. S., TAYLOR G. J., HILL H. M., CORBITT R. A., McCALL M. A., MOORE W. H., HEDBERG D. G.: *Ind. Eng. Chem.* **45**, 1730 (1953); [356].
214. DIMROTH O., BOCKEMÜLLER W.: *Ber.* **64**, 516 (1931); 69, 73, 234.
215. DISHART K. T., LEVINE R.: *J. Am. Chem. Soc.* **78**, 2268 (1956); [141, 214].
216. DIXON S., REXFORD D. R., RUGG J. S.: *Ind. Eng. Chem.* **49**, 1687 (1957); [348].
217. DMITRIEV M. A., SOKOL'SKII G. A., KNUNYANTS I. L.: *Khim. nauka i prom.* **3**, 826 (1958); *Chem. Abstr.* **53**, 11211a (1959); *Doklady Akad. Nauk SSSR* **124**, 581 (1959); [196]
218. DOWNING R. C., BENNING A. F., DOWNING F. B., McHARNESS R. C., RICHARDS M. K., TOMKOWIT T. W.: *Ind. Eng. Chem.* **39**, 259 (1947); [34, 45, 46].
219. DRESDNER R.: *J. Am. Chem Soc.* **77**, 6633 (1955); [254].
220. DRESDNER R.: *J. Am. Chem. Soc.* **79**, 69 (1957); [254].
221. DRESDNER R. D., YOUNG J. A.: *J. Org. Chem.* **24**, 566 (1959); [277].
222. DRYSDALE J. J.: *U. S. Patent* 2, 861,095 (1958); *Chem. Abstr.* **53**, 9102a (1959); [234]
223. DRYSDALE J. J., GILBERT W. W., SINCLAIR H. K., SHARKEY W. H.: *J. Am. Chem. Soc.* **80**, 245 (1958); [234].
224. DUNKER M. F. W., STARKEY E. B.: *J. Am. Chem. Soc.* **61**, 3005 (1939); [226].
225. DUPONT DE NEMOURS E. I.: *British Patent* 606, 273 (1948); *Chem. Abstr.* **43**, 5637g (1949); [348].
226. DUPONT DE NEMOURS E. I.: *British Patent* 620, 296 (1949); *Chem. Abstr.* **43**, 6218h (1949); [344].
227. DUPONT DE NEMOURS E. I.: *British Patent* 732, 269 (1955); *Chem. Abstr.* **50**, 10122b (1956); [129].
228. DUSCHINSKY R., PLEVEN E.: *J. Am. Chem. Soc.* **79**, 4560 (1957); [357].
229. DYATKIN B. L., GERMAN L. S., KNUNYANTS I. L.: *Doklady Akad. Nauk SSSR* **114**, 320 (1957): *Chem. Abstr.* **52**, 251i (1958); [232].
230. EABORN C.: *J. Chem. Soc.* **1952**, 2846; [88, 156].
231. EDGELL W. F., PARTS L.: *J. Am. Chem. Soc.* **77**, 4899 (1955); [113, 122, 124, 126].
232. EDGELL W. F., PARTS L.: *J. Am. Chem. Soc.* **77**, 5515 (1955); [122, 157, 163].

233. EKSTRÖM B.: *Chem. Ber.* **92.** 749 (1959); [135, 257].
234. ELLIOTT J. R., MYERS R. L., ROEDEL G. F.: *Ind. Eng. Chem.* **45**, 1786 (1953); [346].
235. ELLIS J. F., MUSGRAVE W. K. R.: *J. Chem. Soc.* **1950**, 3608; [82].
236. ELLIS J. F., MUSGRAVE W. K. R.: *J. Chem. Soc.* **1953**, 1063; [82].
237. ELVING P. J., LEONE J. T.: *J. Am. Chem. Soc.* **79**, 1546 (1957); [157, 165].
238. EMELÉUS H. J., HASZELDINE R. N.: *J. Chem. Soc.* **1949**, 2948; [226].
239. EMELÉUS H. J., HASZELDINE R. N.: *J. Chem. Soc.* **1949**, 2953; [226].
240. EMELÉUS H. J., HASZELDINE R. N., PAUL R. C.: *J. Chem. Soc.* **1954**, 881; [156, 263].
241. EMELÉUS H. J., HASZELDINE R. N., PAUL R. C.: *J. Chem. Soc.* **1955**, 563; [310].
242. EMELÉUS H. J., HASZELDINE R. N., WALASCHEWSKI E. G.: *J. Chem. Soc.* **1953**, 1552; [161].
243. EMELÉUS H. J., HEAL H. G.: *J. Chem. Soc.* **1946**, 1126; [51].
244. EMELÉUS H. J., WILKINS C. J.: *J. Chem. Soc.* **1944**, 454; [111, 156, 203].
245. EMELÉUS H. J., WOOD J. F.: *J. Chem. Soc.* **1948**, 2183; [95, 146].
246. EMMONS W. D.: *J. Am. Chem. Soc.* **76**, 3468 (1954); [152, 282].
247. EMMONS W. D.: *J. Am. Chem. Soc.* **76**, 3470 (1954); [152, 169, 282].
248. EMMONS W. D.: *J. Am. Chem. Soc.* **79**, 6522 (1957); [279].
249. EMMONS W. D., FERRIS A. F.: *J. Am. Chem. Soc.* **75**, 4623 (1953); [152, 282].
250. EMMONS W. D., LUCAS G. B.: *J. Am. Chem. Soc.* **77**, 2287 (1955); [152, 282, 283].
251. EMMONS W. D., McCALLUM K. S., FERRIS A. F.: *J. Am. Chem. Soc.* **75**, 6047 (1953); [279].
252. EMMONS W. D., PAGANO A. S.: *J. Am. Chem. Soc.* **77**, 89 (1955); [152, 282, 283].
253. EMMONS W. D., PAGANO A. S.: *J. Am. Chem. Soc.* **77**, 4557 (1955); [152, 282].
254. EMMONS W. D., PAGANO A. S., FREEMAN J. P.: *J. Am. Chem. Soc.* **76**, 3472 (1954) [282].
255. ENGELBRECHT A., ATZWANGER H.: *Monatsh.* **83**, 1087 (1952) [60]
256. ENGLAND D. C.: *U.S. Patent* 2,852,554 (1958); *Chem. Abstr.* **53**, 2253i (1959); [196, 202]
257. ENGLAND D. C., LINDSEY R. V., JR., MELBY L. R.: *J. Am. Chem. Soc.* **80**, 6442 (1958); [255]
258. ENGLUND B.: *Org. Syntheses* **34**, 16 (1954); [141, 149, 204]
259. ENGLUND B.: *Org. Syntheses* **34**, 49 (1954); [32, 248].
260. ERDEY L., MÁZOR L., PÁPAY M.: *Mikrochim. Acta* **1958**, 482; [317].
261. ERLER K.: *Z. Anal. Chem.* **131**, 103 (1950); [319].
262. EVANS D. E. M., MASSINGHAM W. E., STACEY M., TATLOW J. C.: *Nature* **182**, 591 (1958); [314].
263. EVANS D. E. M., TATLOW J. C.: *J. Chem. Soc.* **1954**, 3779; [70, 134, 151, 269].
264. EVANS D. E. M., TATLOW J. C.: *J. Chem. Soc.* **1955**, 1184; [173, 269].
265. EVANS D. E. M., TATLOW J. C.: Vapour Phase Chromatography (Desty D. H., Harbourn C. L. A.). Butterworths, London, 1956; [314].
266. FAINBERG A. H., MILLER W. T. JR.: *J. Am. Chem. Soc.* **79**, 4170 (1957); [257, 267].
267. FARLOW M. W., MUETTERTIES E. L.: *U.S. Patent* 2,709,186 (1955); *Chem. Abstr.* **50**, 6499b (1956); [129].
268. FARLOW M. W., MUETTERTIES E. T.: *U.S. Patent* 2,732,410 (1956); *Chem. Abstr.* **50**, 15574c (1956); [129].
269. FARLOW M. W., MUETTERTIES E. T.: *U.S. Patent* 2,732,411 (1956); *Chem. Abstr.* **50**, 15574f (1956); [129].
270. FEAR E. J. P., THROWER J., VEITCH J.: *J. Chem. Soc.* **1956**, 3199; [212].

271. FERM R. L., VANDER WERF C. A.: *J. Am. Chem. Soc.* **72**, 4809 (1950); [114, 137].
272. FERRIER R. J., TEDDER J. M.: *J. Chem. Soc.* **1957**, 1435; [281, 282].
273. FILLER R., O'BRIEN J. F., FENNER J. V., HAUPTSCHEIN M.: *J. Am. Chem. Soc.* **75**, 966 (1953); [190, 198].
274. FILLER R., O'BRIEN J. F., FENNER J. V., MOSTELLER J. C., HAUPTSCHEIN M., STOKES C. S.: *Ind. Eng. Chem.* **46**, 544 (1954); [302, 303].
275. FINGER G. C.: Intern. Symposium, Fluorine Chem., Birmingham, 1959; [107].
276. FINGER G. C., KRUSE C. W.: *J. Am. Chem. Soc.* **78**, 6034 (1956); [106, 107, 155].
277. FINGER G. C., OESTERLING R. E.: *J. Am. Chem. Soc.* **78**, 2593 (1956); [116, 117, 140].
278. FINGER G. C., REED F. H., BURNESS D. M., FORT D. M., BLOUGH R. R.: *J. Am. Chem. Soc.* **73**, 145 (1951); [138, 140].
279. FINGER G. C., REED F. H., FINNERTY J. L.: *J. Am. Chem. Soc.* **73**, 153 (1951); [138, 140].
280. FINGER G. C., STARR L. D.: *J. Am. Chem. Soc.* **81**, 2674 (1959); [106].
281. FLORIN R. E., PUMMER W. J., WALL L. A.: *J. Research Natl. Bur. Standards* **62**, 119 (1959); [160].
282. FODOR G.: *Österr. chem. Ztg.* **57**, 286 (1956); [283].
283. FORBES E. J., RICHARDSON R. D., STACEY M., TATLOW J. C.: *J. Chem. Soc.* **1959**, 2019; [141, 242].
284. FORBES E. J., RICHARDSON R. D., TATLOW J. C.: *Chem. & Ind.* (London) **1958**, 630; [155, 195, 242].
285. FOSDICK L. S., CAMPAIGNE E. E.: *J. Am. Chem. Soc.* **63**, 974 (1941); [145].
286. FOWLER R. D., ANDERSON H. C., HAMILTON J. M., JR., BURFORD W. B., III. SPADETTI A., BITTERLICH S. B., LITANT I.: *Ind. Eng. Chem.* **39**, 343 (1947); [61, 82].
287. FOWLER R. D., BURFORD W. B., III, HAMILTON J. M., JR., SWEET R. G., WEBER C. E., KASPER J. S., LITANT I.: *Ind. Eng. Chem.* **39**, 292 (1947); [61, 80, 81, 128, 134, 142, 233].
288. FOWLER R. D., HAMILTON J. M., JR., KASPER J. S., WEBER C. E., BURFORD W.B., III, ANDERSON H. C.: *Ind. Eng. Chem.* **39**, 375 (1947); [207].
289. FREDENHAGEN K., CADENBACH G.: *Ber.* **67**, 928 (1934); [28, 74].
290. FREDERICK M. R.: *U. S. Patent* 2,724,004 (1955); *Chem. Abstr.* **50**, 4194b (1956); [125].
291. FRONING J. F., RICHARDS M. K., STRICKLIN T. W., TURNBULL S. G.: *Ind. Eng. Chem.* **39**, 275 (1947); [34, 46, 47, 48].
292. FUCHS O., STALLER A., SCHÄFF R.: *U. S. Patent* 2,863,844 (1958); *Chem. Abstr.* **53**, 10848a (1959); [287, 350].
293. FUKUHARA N., BIGELOW L. A.: *J. Am. Chem. Soc.* **60**, 427 (1938); [28, 88].
294. FUKUHARA N., BIGELOW L. A.: *J. Am. Chem. Soc.* **63**, 788 (1941); [74, 122, 144, 147].
295. FUKUHARA N., BIGELOW L. A.: *J. Am. Chem. Soc.* **63**, 2792 (1941); [74, 75].
296. FUKUI K., KITANO N.: *Japanese Patent* 7761('57); *Chem. Abstr.* **52**, 13773b (1958); [108].
297. FULTON R. A.: *Ind. Eng. Chem.* **40**, 699 (1948); [342].
298. GÁBRIS T.: *Kautschuk u. Gummi* **11**, WT 242 (1958); [349].
299. GEL'MAN N. E., KORSHUN M. O., CHUMACHENKO M. N., LARINA N. I.: *Doklady Akad. Nauk SSSR* **123**, 468 (1958); *Chem. Abstr.* **53**, 3985e (1959); [323].
300. GEL'MAN N. E., KORSHUN M. O., SHEVELEVA N. S.: *Zhur. anal. chim.* **18**, 526 (1957); *Chem. Abstr.* **52**, 1853d (1958); [323, 324].

301. GERMANO A., SÉCHAUD G.: *Helv. Chim. Acta.* **37**, 1343 (1954); [176].
302. GERVASI J. A., BROWN M., BIGELOW L. A.: *J. Am. Chem. Soc.* **78**, 1679 (1956); [74].
303. GETHING B., PATRICK C. R., STACEY M., TATLOW J. C.: *Nature* **183**, 590 (1959); [139, 266].
304. GEYER A. M., HASZELDINE R. N.: *J. Chem. Soc.* **1957**, 1038; [247].
305. GEYER A. M., HASZELDINE R. N., LEEDHAM K., MARKLOW R. J.: *J. Chem. Soc.* **1957**, 4472; [156, 247].
306. G. H. B.: *Ind. Eng. Chem.* **43**, No 10, 17A (1951); [354].
307. GILKEY W. K., GERARD F. W., BIXLER M. E.: *Ind. Eng. Chem.* **23**, 364 (1931); [338, 339].
308. GILMAN H., GORSICH R. D.: *J. Am. Chem. Soc.* **77**, 3919 (1955); [224].
309. GILMAN H., JONES R. G.: *J. Am. Chem. Soc.* **65**, 1458 (1943); [152, 156, 159, 182, 194, 196, 197, 243].
310. GILMAN H., JONES R. G.: *J. Am. Chem. Soc.* **65**, 2037 (1943); [188, 217, 265].
311. GILMAN H., JONES R. G.: *J. Am. Chem. Soc.* **70**, 1281 (1948); [140, 158, 159].
312. GILMAN H., SODDY T. S.: *J. Org. Chem.* **22**, 1715 (1957); [224].
313. GILMAN H., WOODS L. A.: *J. Am. Chem. Soc.* **66**, 1981 (1944); [223, 224].
314. GLADSTONE M. T.: *Ind. Eng. Chem.* **45**, 1555 (1953); [350].
315. GLEMSER O., SCHRÖDER H., HAESELER H.: *Z. anorg. u. allgem. Chem.* **282**, 80 (1955); [71, 155].
316. GODSELL J. A., STACEY M., TATLOW J. C.: *Nature* **178**, 199 (1956); [70, 139, 269].
317. GODSELL J. A., STACEY M., TATLOW J. C.: *Tetrahedron* **2**, 193 (1958); [82, 134, 314].
318. GOLDSTEIN H., GIDDEY A.: *Helv. Chim. Acta* **37**, 1121 (1954); [202].
319. GORIN G., PIERCE O. R., McBEE E. T.: *J. Am. Chem. Soc.* **75**, 5622 (1953); [199].
320. GRADISHAR F. J.: *U.S. Patent* 2, 875, 254 (1959); *Chem. Abstr.* **53**, 14000f (1959); [177].
321. GRAMSTAD T., HASZELDINE R. N.: *J. Chem. Soc.* **1956**, 173; [34, 84, 86, 109, 153, 202].
322. GRAMSTAD T., HASZELDINE R. N.: *J. Chem. Soc.* **1957**, 2640; [86, 153].
323. GRAMSTAD T., HASZELDINE R. N.: *J. Chem. Soc.* **1957**, 4069; [185].
324. GRIFFIS C. B., MONTERMOSO J. C.: *Rubber Age* (N.Y.) **77**, 559 (1955); [349].
325. GRILLOT G. F., AFTERGUT S., MARMOR S., CARROCK F.: *J. Org. Chem.* **23**, 366, 386 (1958); [229].
326. GRISLEY D. W. JR., GLUESENKAMP E. W., HEININGER S. A.: *J. Org. Chem.* **23**, 1802 (1958); [109].
327. GROSSE A. V., CADY G. H.: *Ind. Eng. Chem.* **39**, 367 (1947); [21, 301, 304, 305].
328. GROSSE A. V., LINN C. B.: *J. Am. Chem. Soc.* **64**, 2289 (1942); [67, 124, 126].
329. GROSSE A. V., LINN C. B.: *J. Org. Chem.* **3**, 26 (1938); [65].
330. GROSSE A. V., WACKHER R. C., LINN C. B.: *J. Phys. Chem.* **44**, 275 (1940); [301].
331. GRUBER W.: *Can. J. Chem.* **31**, 1020 (1953); [114].
332. GRYSZKIEWICZ-TROCHIMOWSKI E.: *Rec. trav. chim.* **66**, 427 (1947); [141, 213, 214].
333. GRYSZKIEWICZ-TROCHIMOWSKI E.: *Rec. trav. chim.* **66**, 430 (1947); [170].
334. GRYSZKIEWICZ-TROCHIMOWSKI E., GRYSZKIEWICZ-TROCHIMOWSKI A., LÉVY R. *Bull. soc. chim. France* **1953**, 462; [105, 146].
335. GRYSZKIEWICZ-TROCHIMOWSKI E., GRYSZKIEWICZ-TROCHIMOWSKI O.: *Bull. soc. chim. France* **1953**, 123; [108].
336. GRYSZKIEWICZ-TROCHIMOWSKI E., SPORZYŃSKI A., WNUK J.: *Rec. trav. chim.* **66**, 413 (1947); [33, 105, 107, 124, 126, 141, 146].
337. GRYSZKIEWICZ-TROCHIMOWSKI E., SPORZYŃSKI A., WNUK J.: *Rec. trav. chim.* **66**, 419 (1947); [110, 145, 146, 201, 212, 236].

338. GUNTHER F. A., BLINN R. C.: *J. Am. Chem. Soc.* **72**, 4282 (1950); [165].

339. GUNTHER F. A., BLINN R. C.: *J. Am. Chem. Soc.* **72**, 5770 (1950); [168, 170].

340. HADLEY E. H., BIGELOW L. A.: *J. Am. Chem. Soc.* **62**, 3302 (1940); [74, 75, 121].

341. HALL H. T., BRADY E. L., ZEMANY P. D.: *J. Am. Chem. Soc.* **73**, 5460 (1951); [287, 350].

342. HALS L. J., BRYCE H. G.: *Anal. Chem.* **23**, 1694 (1951); [287].

343. HALS L. J., REID T. S., SMITH G. H., JR.: *J. Am. Chem. Soc.* **73**. 4054 (1951); [271].

344. HAMADA M., NAGASAWA S.: *Botyu-Kyaku* **21**, 4 (1956); *Chem. Abstr.* **51**, 3518g [210].

345. HAMILTON J. M., JR.: *Ind. Eng. Chem.* **45**, 1347 (1953); [344, 346].

346. HAMLIN H. C.: *Chem. Abstr.* **50**, 10442h (1956); [349].

347. HANFORD W. E.: *U. S. Patent* 2, 407, 419 (1946); *Chem. Abstr.* **41**, 2067d (1947); [344].

348. HANFORD W. E., JOYCE R. M.: *J. Am. Chem. Soc.* **68**, 2082 (1946); [347].

349. HARMON J.: *U. S. Patent* 2, 399, 024 (1946); *Chem. Abstr.* **40**, 3765 (1946); [128, 344].

350. HARRIS D. K.: *Lancet* **261**, 1008 (1951); [313].

351. HART F. D., GOLDING J. R., BURLEY D.: *Lancet* **1958**, II, 495; [357].

352. HASZELDINE R. N.: *Ann. Repts.* **51**, 279 (1954); [20].

353. HASZELDINE R. N.: *J. Am. Chem. Soc.* **75**, 991 (1953); Footnote 4; [308].

354. HASZELDINE R. N.: *J. Chem. Soc.* **1949**, 2856; [104, 123, 252, 253].

355. HASZELDINE R. N.: *J. Chem. Soc.* **1950**, 1638; [153, 154].

356. HASZELDINE R. N.: *J. Chem. Soc.* **1950**, 1966; [153. 154].

357. HASZELDINE R. N.: *J. Chem. Soc.* **1950**, 2789; [173, 261, 262].

358. HASZELDINE R. N.: *J. Chem. Soc.* **1950**, 3037; [261, 262].

359. HASZELDINE R. N.: *J. Chem. Soc.* **1951**, 584; [184, 273].

360. HASZELDINE R. N.: *J. Chem. Soc.* **1951**, 588; [131, 136].

361. HASZELDINE R. N.: *J. Chem. Soc.* **1951**, 2495; [166, 178, 251, 268].

362. HASZELDINE R. N.: *J. Chem. Soc.* **1952**, 3423; [217, 218, 219, 220].

363. HASZELDINE R. N.: *J. Chem. Soc.* **1952**, 3490; [188, 261, 262].

364. HASZELDINE R. N.: *J. Chem. Soc.* **1952**, 4259; [123, 271].

365. HASZELDINE R. N.: *J. Chem. Soc.* **1952**, 4423; [133, 175, 176, 189, 267].

366. HASZELDINE R. N.: *J. Chem. Soc.* **1953**, 922; [164, 165, 206, 252, 261, 262].

367. HASZELDINE R. N.: *J. Chem. Soc.* **1953**, 1748; [145, 167, 217, 218, 219].

368. HASZELDINE R. N.: *J. Chem. Soc.* **1953**, 3371; [94, 127, 130].

369. HASZELDINE R. N.: *J. Chem. Soc.* **1953**, 3559; [252].

370. HASZELDINE R. N.: *J. Chem. Soc.* **1953**, 3565; [127, 172, 252].

371. HASZELDINE R. N.: *J. Chem. Soc.* **1953**, 3761; [252, 253, 254].

372. HASZELDINE R. N.: *J. Chem. Soc.* **1954**, 1273; [221].

373. HASZELDINE R. N.: *J. Chem. Soc.* **1954**, 4026; [133, 271, 272].

374. HASZELDINE R. N.: ISERSON H., : *J. Am. Chem. Soc.* **79**, 5801 (1957); [123, 336].

375. HASZELDINE R. N., JANDER J.: *J. Chem. Soc.* **1953**, 4172; [194].

376. HASZELDINE R. N., KIDD J. M.: *J. Chem. Soc.* **1953**, 3219; [71, 72, 122, 153].

377. HASZELDINE R. N., KIDD J. M.: *J. Chem. Soc.* **1954**, 4228; [169, 310].

378. HASZELDINE R. N., KIDD J. M.: *J. Chem. Soc.* **1955**, 2901; [165, 209].

379. HASZELDINE R. N., LEEDHAM K.: *J. Chem. Soc.* **1952**, 3483; [127, 251, 260, 261].

380. HASZELDINE R. N., LEEDHAM K.: *J. Chem. Soc.* **1953**, 1548; [104, 126, 150, 175, 261, 262, 271].

381. HASZELDINE R. N., LEEDHAM K.: *J. Chem. Soc.* **1954**, 1261; [136, 261, 262].

382. HASZELDINE R. N., LEEDHAM K.: STEELE B. R.: *J. Chem. Soc.* **1954**, 2040; [252].

383. HASZELDINE R. N., MATTINSON B. J. H.: *Chem. & Ind.* (London) **1956**, 81; [307].

384. HASZELDINE R. N., NYMAN F.: *J. Chem. Soc.* **1959**, 1084; [169].

385. HASZELDINE R. N., OSBORNE J. E.: *J. Chem. Soc.* **1955**, 3880; [133, 165, 274].

386. HASZELDINE R. N., OSBORNE J. E.: *J. Chem. Soc.* **1956**, 61; [163, 232, 252, 271].

387. HASZELDINE R. N., SHARPE A. G.: Fluorine and Its Compounds. Methuen Ltd., London. 1951; [20].

388. HASZELDINE R. N., SHARPE A. G.: *J. Chem. Soc.* **1952**, 993; [285].

389. HASZELDINE R. N., SMITH F.: *J. Chem. Soc.* **1950**, 2689; [77].

390. HASZELDINE R. N., SMITH F.: *J. Chem. Soc.* **1950**, 2787; [77, 134].

391. HASZELDINE R. N., SMITH F.: *J. Chem. Soc.* **1950**, 3617; [82, 134].

392. HASZELDINE R. N., SMITH F.: *J. Chem. Soc.* **1956**, 783; [154].

393. HASZELDINE R. N., STEELE B. R.: *Chem. & Ind.* (London) **1951**, 684; [250].

394. HASZELDINE R. N., STEELE B. R.: *J. Chem. Soc.* **1953**, 1592; [252, 254].

395. HASZELDINE R. N., STEELE B. R.: *J. Chem. Soc.* **1954**, 923; [189, 251].

396. HASZELDINE R. N., STEELE B. R.: *J. Chem. Soc.* **1954**, 3747; [187].

397. HASZELDINE R. N., STEELE B. R.: *J. Chem. Soc.* **1955**, 3005; [132, 150, 252, 254].

398. HASZELDINE R. N., STEELE B. R.: *J. Chem. Soc.* **1957**, 2193; [251, 252, 253].

399. HASZELDINE R. N., STEELE B. R.: *J. Chem. Soc.* **1957**, 2800; [251, 253].

400. HASZELDINE R. N., WALASCHEWSKI E. G.: *J. Chem. Soc.* **1953**, 3607; [145, 225].

401. HAUPTSCHEIN M., BIGELOW L. A.: *J. Am. Chem. Soc.* **72**, 3423 (1950); [68].

402. HAUPTSCHEIN M., BIGELOW L. A.: *J. Am. Chem. Soc.* **73**, 5591 (1951); [179].

403. HAUPTSCHEIN M., BRAUN R. A : *J. Am. Chem. Soc.* **77**, 4930 (1955); [145, 232, 262].

404. HAUPTSCHEIN M., GROSSE A. V:. *J. Am. Chem. Soc.* **73**, 5461 (1951); [156, 239].

405. HAUPTSCHEIN M., KINSMAN R. L., GROSSE A. V.: *J. Am. Chem. Soc.* **74**, 849 (1952); [184].

406. HAUPTSCHEIN M., NODIFF E. A., GROSSE A. V.: *J. Am. Chem. Soc.* **74**, 1347 (1952) [128].

407. HAUPTSCHEIN M., NODIFF E. A., SAGGIOMO A. J.: *J. Am. Chem. Soc.* **76**, 1051 (1954); [237].

408. HAUPTSCHEIN M., SAGGIOMO A. J., STOKES C. S.: *J. Am. Chem. Soc.* **77**, 2284 (1955); [207, 237].

409. HAUPTSCHEIN M., STOKES C. S., GROSSE A. V.: *J. Am. Chem. Soc.* **74**, 848 (1952); [128, 184].

410. HAUPTSCHEIN M., STOKES C. S., GROSSE A. V.: *J. Am. Chem. Soc.* **74**, 1974 (1952) [128, 184, 205].

411. HAVEN A. C., JR.: DuPont de Nemours E. I., Magazin, February-March 1959, 2; [342].

412. HAWORTH W. N., STACEY M.: *U. S. Patent* 2, 476,490 (1949); *Chem. Abstr.* **44**, 654i (1950); [147, 171].

413. HAWORTH W. N., STACEY M., APPLETON E.: *British Patent* 626, 449 (1949); *Chem. Abstr.* **44**, 3009e (1950); [147].

414. HAWTHORNE F.: *Anal. Chem.* **28**, 540 (1956); [284].

415. HAWTHORNE M. F., EMMONS W. D., McCALLUM K. S.: *J. Am. Chem. Soc.* **80**, 6393 (1958); [283, 284].

416. HEAP R., SAUNDERS B. C.: *J. Chem. Soc.* **1948**, 1313; [93].

417. HEATH E. G.: *Chem. & Ind.* (London) **1958**, 1111; [27].

418. HEBERLING J. W., JR.: *J. Org. Chem.* **23**, 615 (1958); [179].

24

419. HELIN A. F., SVEINBJORNSSON A., VANDERWERF C. A.: *J. Am. Chem. Soc.* **73**, 1189 (1951); [170].

420. HELLER G. W.: *U.S.Patent* 2, 752, 321(1956); *Chem. Abstr.* **50**, 15128a (1956); [354]

421. HELLMANN M., BILBO A. J.: *J. Am. Chem. Soc.* **75**, 4590 (1953); [180].

422. HELLMANN M., BILBO A. J., PUMMER W. J.: *J. Am. Chem. Soc.* **77**, 3650 (1955); [180, 181, 267].

423. HELLMANN M., PETERS E., PUMMER W. J., WALL L. A.: *J. Am. Chem. Soc.* **79**, 5654 (1957); [266].

424. HENBEST H. B., WRIGLEY T. I.: *J. Chem. Soc.* **1957**, 4765; [118].

425. HENDRICKS J. O.: *Ind. Eng. Chem.* **45**, 99 (1953); [303, 338, 353, 355, 356, 357].

426. HENNE A. L.: Intern. Symposium, Fluorine Chem., Birmingham, 1959; [169, 177, 178].

427. HENNE A. L.: *J. Am. Chem. Soc.* **59**, 1200 (1937); [96, 102, 122, 123, 177, 205].

428. HENNE A. L.: *J. Am. Chem. Soc.* **59**, 1400 (1937); [122].

429. HENNE A. L.: *J. Am. Chem. Soc.* **60**, 96 (1938); [45, 46].

430. HENNE A. L.: *J. Am. Chem. Soc.* **60**, 1569 (1938); [102, 103].

431. HENNE A. L.: *J. Am. Chem. Soc.* **60**, 2275 (1938); [217].

432. HENNE A. L.: Organic Chemistry (Gilman H.), p. 944—964; J. Wiley, New York, 1947 (2nd. Edition); [20].

433. HENNE A. L.: *Org. Reactions* **2**, 49—93 (1944); [20].

434. HENNE A. L.: Private Communication; [129, 169, 186, 246, 264, 336].

435. HENNE A. L.: Unpublished results (Sedlak J.: PhD Thesis, Ohio State Univ., 1960); [157, 164].

436. HENNE A. L., ALDERSON T., NEWMAN M. S.: *J. Am. Chem. Soc.* **67**, 918 (1945); [147, 149, 152, 172, 196, 197, 209].

437. HENNE A. L., ALM R. M., SMOOK M.: *J. Am. Chem. Soc.* **70**, 1968 (1948); [162].

438. HENNE A. L., ARNOLD R. C.: *J. Am. Chem. Soc.* **70**, 758 (1948); [66].

439. HENNE A. L., (BROWN R., KRAUS D., RENOLL M., GORDON J.): *J. Am. Chem. Soc.* **75**, 5750 (1953); [225, 267].

440. HENNE A. L., DEWITT E. G.: *J. Am. Chem. Soc.* **70**, 1548 (1948); [126, 151, 179, 310].

441. HENNE A. L., FINNEGAN W. G.: *J. Am. Chem. Soc.* **71**, 298 (1949); [135, 158, 232, 265].

442. HENNE A. L., FINNEGAN W. G.: *J. Am. Chem. Soc.* **72**, 3806 (1950); [183].

443. HENNE A. L., FLANAGAN J. V.: *J. Am. Chem. Soc.* **65**, 2362 (1943); [103].

444. HENNE A. L., FOX C. J.: *J. Am. Chem. Soc.* **73**, 2323 (1951); [308, 309].

445. HENNE A. L., FOX C. J.: *J. Am. Chem. Soc.* **75**, 5750 (1953); [308, 309].

446. HENNE A. L., FOX C. J.: *J. Am. Chem. Soc.* **76**, 479 (1954); [102, 150, 309].

447. HENNE A. L., FRANCIS W. C.: *J. Am. Chem. Soc.* **73**, 3518 (1951); [217, 218].

448. HENNE A. L., FRANCIS W. C.: *J. Am. Chem. Soc.* **75**, 991 (1953); [308].

449. HENNE A. L., FRANCIS W. C.: *J. Am. Chem. Soc.* **75**, 992 (1953); [217, 218, 219].

450. HENNE A. L., HAECKL F. W.: *J. Am. Chem. Soc.* **63**, 3476 (1941); [127, 175].

451. HENNE A. L., HINKAMP J. B.: *J. Am. Chem. Soc.* **67**, 1194 (1945); [179].

452. HENNE A. L., HINKAMP J. B.: *J. Am. Chem. Soc.* **67**, 1197 (1945); [65, 89, 126, 178].

453. HENNE A. L., HINKAMP J. B.: ZIMMERSCHIED W. J.: *J. Am. Chem. Soc.* **67**, 1906 (1945); [177, 178].

454. HENNE A. L., HINKAMP P. E.: *J. Am. Chem. Soc.* **76**, 5147 (1954); [140, 158, 270].

455. HENNE A. L., KAYE S.: *J. Am. Chem. Soc.* **72**, 3369 (1950); [158, 159, 187].

456. HENNE A. L., KRAUS D. W.: *J. Am. Chem. Soc.* **73**, 1791 (1951); [252].
457. HENNE A. L., KRAUS D. W.: *J. Am. Chem. Soc.* **76**, 1175 (1954); [252, 254].
458. HENNE A. L., LADD E. C.: *J. Am. Chem. Soc.* **58**, 402 (1936); [96, 179].
459. HENNE A. L., LATIF K. A.: *J. Am. Chem. Soc.* **76**, 610 (1954); [69].
460. HENNE A. L., MIDGLEY T., JR.: *J. Am. Chem. Soc.* **58**, 882 (1936); [201,203].
461. HENNE A. L., MIDGLEY T., JR.: *J. Am. Chem. Soc.* **58**, 884 (1936); [33, 54, 101, 102, 124, 125, 126].
462. HENNE A. L., NAGER M.: *J. Am. Chem. Soc.* **73**, 1042 (1951); [135, 175, 176, 265].
463. HENNE A. L., NAGER M.: *J. Am. Chem. Soc.* **73**, 5527 (1951); [251, 252].
464. HENNE A. L., NAGER M.: *J. Am. Chem. Soc.* **74**, 650 (1952); [188, 222, 260].
465. HENNE A. L., NEWBY T. H.: *J. Am. Chem. Soc.* **70**, 130 (1948); [130,131,175,232].
466. HENNE A. L., NEWMAN M. S., QUILL L. L., STANIFORTH R. A.: *J. Am. Chem. Soc.* **69**, 1819 (1947); [146, 229, 230].
467. HENNE A. L., PELLEY R. L.: *J. Am. Chem. Soc.* **74**, 1426 (1952); [147, 246, 308].
468. HENNE A. L., PELLEY R. L., ALM R. M.: *J. Am. Chem. Soc.* **72**, 3370 (1950); [143, 163].
469. HENNE A. L., PLUEDDEMAN E. P.: *J. Am. Chem. Soc.* **65**, 587 (1943); [67, 126].
470. HENNE A. L., PLUEDDEMAN E. P.: *J. Am. Chem. Soc.* **65**, 1271 (1943); [66, 125].
471. HENNE A. L., POSTELNEK W.: *J. Am. Chem. Soc.* **77**, 2334 (1955); [133, 267].
472. HENNE A. L., RENOLL M. W.: *J. Am. Chem. Soc.* **58**, 887 (1936); [102, 233].
473. HENNE A. L., RENOLL M. W.: *J. Am. Chem. Soc.* **58**, 889 (1936); [124].
474. HENNE A. L., RENOLL M. W.: *J. Am. Chem. Soc.* **59**, 2434 (1937); [325].
475. HENNE A. L., RENOLL M. W.: *J. Am. Chem. Soc.* **60**, 1060 (1938); [52, 101, 124].
476. HENNE A. L., RICHTER S. B.: *J. Am. Chem. Soc.* **74**, 5420 (1952); [141, 142, 304, 310].
477. HENNE A. L., RUH R. P.: *J. Am. Chem. Soc.* **69**, 279 (1947); [134, 257].
478. HENNE A. L., SCHMITZ J. V., FINNEGAN W. G.: *J. Am. Chem. Soc.* **72**, 4195 (1950); [261].
479. HENNE A. L., SHEPARD J. W., YOUNG E. J.: *J. Am. Chem. Soc.* **72**, 3577 (1950); [141, 213, 214, 262, 270].
480. HENNE A. L., SMOOK M. A.: *J. Am. Chem. Soc.* **72**, 4378 (1950); [141, 247, 263].
481. HENNE A. L., SMOOK M. A., PELLEY R. L.: *J. Am. Chem. Soc.* **72**, 4756 (1950); [238, 247].
482. HENNE A. L., STEWART J. J.: *J. Am. Chem. Soc.* **77**, 1901 (1955); [94, 127, 153, 235, 236, 243, 310].
483. HENNE A. L., TEDDER J. M.: *J. Am. Chem. Soc.* **1953**, 3628; [280, 281].
484. HENNE A. L., TROTT P.: *J. Am. Chem. Soc.* **69**, 1820 (1947); [97, 131, 147, 172].
485. HENNE A. L., WAALKES T. P.: *J. Am. Chem. Soc.* **67**, 1639 (1945); [69, 125].
486. HENNE A. L., WAALKES T. P.: *J. Am. Chem. Soc.* **68**, 496 (1946); [127, 187].
487. HENNE A. L., WHALEY A. M.: *J. Am. Chem. Soc.* **64**, 1157 (1942); [178].
488. HENNE A. L., WHALEY A. M., STEVENSON J. K.: *J. Am. Chem. Soc.* **63**, 3478 (1941); [95].
489. HENNE A. L., ZIMMER W. F., *J. Am. Chem. Soc.* **73**, 1103 (1951); [151, 152, 173, 182, 198, 285, 286].
490. HENNE A. L., ZIMMER W. F.: *J. Am. Chem. Soc.* **73**, 1362 (1951); [285].
491. HENNE A. L., ZIMMERSCHIED W. J.: *J. Am. Chem. Soc.* **67**, 1235 (1945); [151, 173].
492. HENNE A. L., ZIMMERSCHIED W. J.: *J. Am. Chem. Soc.* **69**, 281 (1947); [151, 173].

493. HENNION G. F., HINTON H. D., NIEWLAND J. A.: *J. Am. Chem. Soc.* **55**, 2858 (1933); [55].

494. HILDEBRAND J. H., COCHRAN D. R. F.: *J. Am. Chem. Soc.* **71**, 22 (1949); [305].

495. HILL H. M., TOWNE E. B., DICKEY J. B.: *J. Am. Chem. Soc.* **72**, 3289 (1950); [182].

496. HINE J., GHIRARDELLI R. G.: *J. Org. Chem.* **23**, 1550 (1958); [239].

497. HINE J., PORTER J. J.: *J. Am. Chem. Soc.* **79**, 5493 (1957); [238].

498. HINE J., TANABE K.: *J. Am. Chem. Soc.* **80**, 3002 (1958); [238].

499. HOFFMAN C. J., GUTOWSKY H. S.: *Inorg. Syntheses* **4**, 145 (1953); [57].

500. HOFFMANN F. W.: *J. Am. Chem. Soc.* **70**, 2596 (1948); [107, 140].

501. HOFFMANN F. W.: *J. Org. Chem.* **14**, 105 (1949); [50, 108, 124, 126, 200, 269].

502. HOFFMANN F. W.: *J. Org. Chem.* **15**, 425 (1950); [108, 124, 126, 240].

503. HOLLEMAN A. F., BEEKMAN J. W.: *Rec. trav. chim.* **23**, 253 (1904); [241].

504. HOLLEMAN M.: *Rec. trav. chim.* **24**, 26 (1905); [153, 504].

505. HONN F. J., SIMS W. M.: *U. S. Patent* 2,833,752 (1958); *Chem. Abstr.* **53**, 1804f (1959); [349].

506. HORÁČEK J., KÖRBL J.: *Chem. listy* **51**, 2132 (1957); *Collection Czechoslov. Chem. Communs.* **24**, 286 (1959) [323].

507. HOSKINS W. M., FERRIS C. A.: *Ind. Eng. Chem. Anal. Ed.* **8**, 6 (1936); [318, 319].

508. HOVE VAN T.: *Bull. acad. roy. Belg.* **1913**, 1074: [136].

509. HOWELL W. C., COTT W. J., PATTISON F. L. M.: *J. Org. Chem.* **22**, 255 (1957); [216].

510. HOWELL W. C., MILLINGTON J. E., PATTISON F. L. M.: *J. Am. Chem. Soc.* **78**, 3843 (1956); [153, 239].

511. HSI KWEI JIANG S.: *Hua Hsüeh Hsüeh Pao* **23**, 330 (1957); *Chem. Abstr.* **52**, 15493b (1958); [246].

512. HUBBARD D. M., HENNE A. L.: *J. Am. Chem. Soc.* **56**, 1078 (1934); [319].

513. HUBER M. L.: *U. S. Patent* 2, 721, 199 (1955); *Chem. Abstr.* **50**, 10762h (1956); [278].

514. HUCKABAY W. B., WELCH E. T., METLER A. V.: *Anal. Chem.* **19**, 154 (1947); [319].

515. HÜCKEL W.: *Nachr. Ges. Wiss. Gottingen, Math. phys. Kl.* **1946**, 36; [71, 154, 188].

516. HUDLICKÝ M.: *Chem. listy* **51**, 470(1957); *Collection Czechoslov. Chem. Communs.* **23**, 462 (1958); [278, 279].

517. HUDLICKÝ M.: *Chem. listy* **53**, 513 (1959); [357].

518. HUDLICKÝ M.: *Collection Czechoslov. Chem. Communs.* **25**, 1199 (1960); [95, 143, 170, 171].

519. HUDLICKÝ M.: *Collection Czechoslov. Chem. Communs.* **24**, 1414 (1961); [157, 165].

520. HUDLICKÝ M.: Unpublished observations; [52, 53, 159].

521. HUISGEN R., OTT H.: *Ang. Chem.* **70**, 312 (1958); [284].

522. HURKA V. R., *U.S. Patent* 2, 676, 983 (1954); *Chem, Abstr.* **49**, 5510g (1955); [169].

523. HUSTED D. R., AHLBRECHT A. H.: *J. Am. Chem. Soc.* **74**, 5422 (1952); [141, 158, 161, 162, 199, 208, 262, 306].

524. HUSTED D. R., AHLBRECHT A. H.: *J. Am. Chem. Soc.* **75**, 1605 (1953); [189, 235, 236, 306].

525. HUSTED D. R., KOHLHASE W. L.: *J. Am. Chem. Soc.* **76**, 5141 (1954); [189, 236].

526. ILLUMINATI G., MARINO G.: *J. Am. Chem. Soc.* **78**, 4975 (1956); [180].

527. INGOLD C. K., INGOLD E. H.: *J. Chem. Soc.* **1928**, 2249; [118, 136].

528. INMAN C. E., OESTERLING R. E., TYCZKOWSKI E. A.: *J. Am. Chem. Soc.* **80**, 5286 (1958); [60, 210].

529. INMAN C. E., OESTERLING R. E., TYCZKOWSKI E. A.: *J. Am. Chem. Soc.* **80**, 6533 (1958); [60, 73, 83].

530. Iserson H., Hauptschein M., Lawlor F. E.: *J. Am. Chem. Soc.* **81**, 2676 (1959) [133, 344, 347].

531. Jacobs T. L., Bauer R. S.: *J. Am. Chem. Soc.* **78**, 4815 (1956); [132]. [132].

532. Jacobs T. L., Bauer R. S.: *J. Am. Chem. Soc.* **81**, 606 (1959); [132].

533. James R. A., Bryan W. P.: *J. Am. Chem. Soc.* **76**, 1982 (1954); [286].

534. James W. R., Pearlson W. H., Simons J. H.: *J. Am. Chem. Soc.* **72**, 1761 [157, 160].

535. Janák J.: *Mikrochim. Acta* **1956**, 1038; [331].

536. Jander J., Haszeldine R. N.: *J. Chem. Soc.* **1954**, 912: [155, 192, 194, 263].

537. Jander J., Haszeldine R. N.: *J. Chem. Soc.* **1954**, 919; [155].

538. Jarkovský L., Hudlický M.: *Chem. listy* **51**, 978 (1957); *Collection Czechoslov. Chem. Communs.* **23**, 537 (1958); [35, 125, 129, 337].

539. Jarkovský L., Pešata V., Hudlický M.: *Chem. listy* **51**, 625 (1957); *Collection Czechoslov. Chem. Communs.* **22**, 1827 (1957); [122, 335].

540. Jarvie J. M. S., Fitzgerald W. E., Janz G. J.: *J. Am. Chem. Soc.* **78**, 978 (1956); [260].

541. Jenny E. F., Roberts J. D.: *Helv. Chim. Acta* **38**, 1248 (1955); [223].

542. Johnson L. V., Smith F., Stacey M., Tatlow J. C.: *J. Chem. Soc.* **1952**. 4710; [210].

543. Jones R. G.: *J. Am. Chem. Soc.* **70**, 143 (1948); [153, 158, 159, 213, 214, 230, 237].

544. Kaesz H. D., Stafford S. L., Stone F. G. A.: *J. Am. Chem. Soc.* **81**, 6336 (1959); [222].

545. Kaluszyner A.: *J. Org. Chem.* **24**, 995 (1959); [234].

546. Kauck E. A., Diesslin A. R.: *Ind. Eng. Chem.* **43**, 2332 (1951); [34, 84, 86, 147, 150, 299].

547. Kauck E. A., Simons J. H.: *U. S. Patent* 2, 616, 927 (1952); *Chem. Abstr.* **47**, 8771g (1953); [85, 153, 154].

548. Kauck E. A., Simons J. H.: *U. S. Patent* 2, 644, 823 (1953); *Chem. Abstr.* **48**, 6469i (1954); [86, 150].

549. Kaufman H. S., Kroncke C. O., Giannotta C. R.: *Modern Plastics* **32**, No 2, 146, 236 (1954); [351].

550. Kaufman H. S., Muthana M. S.: *J. Polymer Sci.* **6**, 251 (1951); [349, 351].

551. Kaufman H. S., Solomon E.: *Ind. Eng. Chem.* **45**, 1779 (1953); [287, 349, 350].

552. Kellogg K. B., Cady G. H.: *J. Am. Chem. Soc.* **70**, 3986 (1948); [71].

553. Kende A. S.: *Chem. & Ind.* (London) **1959**, 1346; [235].

554. Kharasch M. S., Weinhouse S., Jensen E. V.: *J. Am. Chem. Soc.* **77**, 3145 (1955); [120].

555. Kimball R. H., Tufts L. E.: *Anal. Chem.* **19**, 150 (1947); [322, 325].

556. Kirshenbaum A. D., Streng A. G., Grosse A. V.: *Anal. Chem.* **24**, 1361 (1952); [325].

557. Kirshenbaum A. D., Streng A. G., Hauptschein M.: *J. Am. Chem. Soc.* **75**, 3141 (1953); [271].

558. Kitano H., Fukui K.: *J. Chem. Soc. Japan* **58**, 352 (1955); [104, 106, 107, 136].

559. Kitano H., Fukui K., Nozu R., Osaka T.: *J. Chem. Soc. Japan* **58**, 224 (1955); [105, 146, 152].

560. Kitano H., Fukui K., Osaka T.: *J. Chem. Soc. Japan* **58**, 119 (1955); [109, 140].

561. Knížek M.: *Chem. listy* **54**, 383 (1960); [27].

562. Knunyants I. L.: *Reakcii i methody issledovania organicheskich soedinenii* **6**, 343—387 (1957); [87].
563. Knunyants I. L.: Intern. Symposium, Fluorine Chem., Birminhgam, 1959; [153, 155, 245, 249].
564. Knunyants I. L., Dyatkin B. L., German L. S.: *Doklady Akad. Nauk SSSR* **124**, 1065 (1959); *Chem. Abstr.* **53**, 14920e (1959); [152, 250].
565. Knunyants I. L., Dyatkin B. L., German L. S.: *Khim. Nauka i Prom.* **3**, 828 (1958); *Chem. Abstr.* **53**, 10207g (1959); [249, 250].
566. Knunyants I. L., Fokin A. V.: *Doklady Akad. Nauk SSSR* **111**, 1035 (1956); *Chem. Abstr.* **51**, 9472i (1957); [155, 191].
567. Knunyants I. L., Fokin A. V.: *Doklady Akad. Nauk SSSR* **112**, 67 (1957); *Chem. Abstr.* **51**, 11234e (1957); [155, 204, 310].
568. Knunyants I. L., Fokin A. V.: *Izvest. Akad. Nauk SSSR* **1955**, 705; *Chem. Abstr.* **50**, 7069 (1956); [246].
569. Knunyants I. L., Fokin A. V.: *Uspekhi Khimii* **20**, 410—429 (1951); [349].
570. Knunyants I. L., German L. S., Dyatkin B. L.: *Izvest. Akad. Nauk SSSR* **1956**, 1353; *Chem. Abstr.* **51**, 8037f (1957); [245, 247, 248, 250].
571. Knunyants I. L., Kildisheva O. V., Petrov I. P.: *Zhur. Obshchei Khim.* **19**, 95 (1949); *Chem. Abstr.* **43**, 6163b (1949); [118].
572. Knunyants I. L., Kisel J. M., Bykhovskaya E. G.: *Izvest. Akad. Nauk SSSR* **1956**, 377; *Chem. Abstr.* **50**, 15454c (1956); [68].
573. Knunyants I. L., Mysov E. I., Krasuskaya M. P.: *Izvest. Akad. Nauk SSSR* **1958**, 906; *Chem. Abstr.* **53**, 1102b (1959); [110, 157, 158, 159, 160].
574. Knunyants I. L., Pervova E. Y., Tyuleneva V. V.: *Doklady Akad. Nauk SSSR* **129**, 576 (1959); [250].
575. Knunyants I. L., Pervova E. Y., Tyuleneva V. V.: *Izvest. Akad. Nauk SSSR* **1956**, 843; *Chem. Abstr.* **51**, 1814g (1957); [176].
576. Knunyants I. L., Shchekotikhin A. I., Fokin A. V.: *Izvest. Akad. Nauk. SSSR* **1953**, 282; *Chem. Abstr.* **48**, 5787h (1954); [204, 205, 247].
577. Knunyants I. L., Sterlin R. N., Pinkina L. N., Dyatkin B. L.: *Izvest. Akad. Nauk SSSR* **1958**, 296; *Chem. Abstr.* **52**, 12754e (1958); [255].
578. Knunyants I. L., Sterlin R. N., Yatsenko R. D., Pinkina L. N.: *Izvest. Akad. Nauk SSSR* **1958**, 1345; *Chem. Abstr.* **53**, 6987g (1959); [222].
579. Kober E., Grundmann C.: *J. Am. Chem. Soc.* **81**, 3769 (1959); [94].
580. Kolditz L.: *Z. anorg. allgem. Chem.* **289**, 128 (1957); [58].
581. Kolesnikov G. S.: *Syntezi org. soedinenii* **2**, 110(1952); *Chem. Abstr.* **48**, 635b (1954); [168].
582. Körbl J.: Private Communication; [320].
583. Körbl J.: Unpublished results; [316, 320, 321].
584. Körbl J., Přibil R.: *Collection Czechoslov. Chem. Communs.* **21**, 955 (1956); [316].
585. Korshak V. V., Kolesnikov G. S.: *Synthesi org. soedinenii* **2**, 108 (1952); *Chem. Abstr.* **48**, 635f (1954); [213, 216].
586. Korshak V. V., Kolesnikov G. S.: *Synthesi org. soedinenii* **2**, 140 (1952); *Chem. Abstr.* **48**, 636i (1954); [143].
587. Korshun M. O., Gel'man N. E., Glazova K. I.: *Doklady Akad. Nauk SSSR* **111**, 1255 (1956); *Chem. Abstr.* **51**, 8581i (1957); [324].
588. Koshar R. J., Simmons T. C., Hoffmann F. W.: *J. Am. Chem. Soc.* **79**, 1741 (1957); [248].

589. KOSHAR R. J., TROTT P. W., LAZERTE J. D.: *J. Am. Chem. Soc.* **75**, 4595 (1953); [153, 196].

590. KOTON M. M., MOSKVINA E. P., FLORINSKII F. S.: *Zhur. Obschei Khim.* **21**, 1843 (1951); *Chem. Abstr.* **46**, 8027g (1952); [143].

591. KOUTNÝ O., SYNEK O.: *Spisy Vysoké školy veterinární* (Brno) **F 183**, 1 (1951); *Chem. Abstr.* **49**, 6028b (1955); [319].

592. KRAFT K.: *Ang. Chem.* **60A**, 248 (1948); [356].

593. KRANTZ J. C., JR., ESQUIBEL A., TRUITT E. B., JR., LING S. C., KURLAND A. A.: *J. Am. Med. Assoc.* **166**, 1555 (1958); [356].

594. KRESPAN C. G.: *J. Org. Chem.* **23**, 2016 (1958); [273].

595. KROGH L. C., REID T. S., BROWN H. A.: *J. Org. Chem.* **19**, 1124 (1954); [188, 194].

596. KROLL A. E.: *U. S. Patent* 2, 750, 350 (1956); *Chem. Abstr.* **50**, 13507i (1956); [354].

597. KROLL A. E., NELSON D. A.: *U. S. Patent* 2, 753, 329 (1956); *Chem. Abstr.* **50**, 15128c (1956); [347].

598. KUBICZEK G., NEUGEBAUER L.: *Monatsh.* **80**, 395 (1949); [153, 239].

599. KWASNIK W.: Field Information Agency, Technical **1114**, 1 (1947); [34, 35, 47].

✓ 600. KYLE B. G., REED T. M., III: *J. Am. Chem. Soc.* **80**, 6170 (1958); [305].

601. LACHER J. R., KIANPOUR A., PARK J. D.: *J. Phys. Chem.* **60**, 1454 (1956); [157, 160].

602. LANDAU R., ROSEN R.: *Ind. Eng. Chem.* **39**, 281 (1947); [24, 25, 34, 36, 48].

603. LANDAU R., ROSEN R.: Preparation, Properties and Technology of Fluorine and Organic Fluoro Compounds (Slesser C., Schram S. R.); McGraw-Hill, New York. 1951. P. 149; [25].

604. LANE E. S.: *J. Chem. Soc.* **1955**, 534; [239].

605. LATIF A.: *J. Ind. Chem. Soc.* **30**, 524 (1953); [97, 133].

606. LAWRENCE W. B.: *Refrig. Eng.* **24**, 286 (1932); [338, 340].

607. LAWTON E. A., LEVY A.: *J. Am. Chem. Soc.* **77**, 6083 (1955); [111].

608. LAZERTE J. D.: *U. S. Patent* 2, 704, 769 (1955); *Chem. Abstr.* **50**, 2654e (1956); [152, 245].

609. LAZERTE J. D., HALS L. J., REID T. S., SMITH G. H.: *J. Am. Chem. Soc.* **75**, 4525 (1953); [129, 131, 271, 272, 273, 343].

610. LAZERTE J. D., KOSHAR R. J.: *J. Am. Chem. Soc.* **77**, 910 (1955); [255].

611. LAZERTE J. D., PEARLSON W. H., KAUCK E. A.: *U. S. Patent* 2, 647, 933 (1953); *Chem. Abstr.* **48**, 7622f (1954); [182, 270].

612. LAZERTE J. D., PEARLSON W. H., KAUCK E. A.: *U. S. Patent* 2, 704, 776 (1955); *Chem. Abstr.* **50**, 2650a (1956); [190, 273].

613. LAZERTE J. D., RAUSCH D. A., KOSHAR R. J., PARK J. D., PEARLSON W. H., LACHER J. R.: *J. Am. Chem. Soc.* **78**, 5639 (1956); [208, 259].

614. LEECH H. R.: Intern. Symposium, Fluorine Chem., Birmingham, 1959; [34, 38, 41, 46, 47, 60].

615. LEEDHAM K., HASZELDINE R. N.: *J. Chem. Soc.* **1954**, 1634; [261, 262].

616. LEFAVE G. M.: *J. Am. Chem. Soc.* **71**, 4148 (1949); [207].

617. LEFFLER A. J.: *J. Org. Chem.* **24**, 1132 (1959); [134].

618. LEWIS E. E., NAYLOR M. A.: *J. Am. Chem. Soc.* **69**, 1968 (1947); [129, 131, 275, 276, 344].

619. LINHARD M., BETZ K.: *Ber.* **73**, 177 (1940); [68].

620. LIVINGSTON D. I., KAMATH P. M., CORLEY R. S.: *J. Polymer Sci.* **20**, 485 (1956); [348].

621. LOCKE E. G., BRODE W. R., HENNE A. L.: *J. Am. Chem. Soc.* **56**, 1726 (1934); [98, 125, 129, 264, 343].

622. Lontz J. F., Happoldt W. B., Jr.: *Ind. Eng. Chem.* **44**, 1800 (1952); [354].

623. Lontz J. F., Jaffe J. A., Robb L. E., Happoldt W. B., Jr,: *Ind. Eng. Chem.* **44**, 1805 (1952); [354, 355].

624. Lovelace A. M., Rausch D. A., Postelnek W.: Aliphatic Fluorine Compounds. Reinhold Publishing Corp., New York, 1958; [20, 288, 345].

625. Ma T. S.: *Anal Chem.* **30**, 1557(1958); *Microchem. J.* **2**, 91 (1958); [314].

626. Ma T. S., Gwirtsman J.: *Anal. Chem.* **29**, 140 (1957); [322].

627. McBee E. T.: Intern. Symposium, Fluorine Chem., Birmingham, 1959; [270].

628. McBee E. T., Bechtol L. D.: *Ind. Eng. Chem.* **39**, 380 (1947); [78].

629. McBee E. T., Bolt R. O.: *Ind. Eng. Chem.* **39**, 412 (1947); [141, 168, 238, 265, 269].

630. McBee E. T., Bolt R. O., Graham P. J., Tebbe R. F.: *J. Am. Chem. Soc.* **69**, 947 (1947); [91, 92].

631. McBee E. T., Burton T. M.: *J. Am. Chem. Soc.* **74**, 3902 (1952); [182].

632. McBee E. T., Campbell D. H., Kennedy R. J., Roberts C. W.: *J. Am. Chem. Soc.* **78**, 4597 (1956); [227].

633. McBee E. T., Campbell D. H., Roberts C. W.: *J. Am. Chem. Soc.* **77**, 3149 (1955); [190, 265].

634. McBee E. T., Filler R.: *J. Org. Chem.* **21**, 370 (1956); [198].

635. McBee E. T., Graham P. J.: *J. Am. Chem. Soc.* **72**, 4235 (1950); [241].

636. McBee E. T., Hass H. B., Bittenbender W. A., Weesner W. E., Toland W. G., Jr, Hausch W. R., Frost L. W.: *Ind. Eng. Chem.* **39**, 409 (1947); [124].

637. McBee E. T., Hass H. B., Frost L. W., Welch Z. D.: *Ind. Eng. Chem.* **39**, 404 (1947); [90, 91, 122, 335].

638. McBee E. T., Hass H. B., Hodnett E. M.: *Ind. Eng. Chem.* **39**, 389 (1947); [89, 139].

639. McBee E. T., Hass H. B., Tomas R. M., Toland W. G., Jr., Truchan A.: *J. Am. Chem. Soc.* **69**, 944 (1947); [119, 127].

640. McBee E. T., Hass H. B., Toland W. G., Jr., Truchan A.: *Ind. Eng. Chem.* **39**, 420 (1947); [125].

641. McBee E. T., Hass H. B., Weimer P. E., Rothrock G. M., Burt W. E., Robb R. M., Van Dyken A. R.: *Ind. Eng. Chem.* **39**, 298 (1947); [134, 136].

642. McBee E. T., Hathaway C. E., Roberts C. W.: *J. Am. Chem. Soc.* **78**, 3851 (1956); [228].

643. McBee E. T., Hathaway C. E., Roberts C. W.: *J. Am. Chem. Soc.* **78**, 4053 (1956); [241].

644. McBee E. T., Hill H. M., Bachman G. B.: *Ind. Eng. Chem.* **41**, 70 (1949); [128, 264, 344, 348, 353].

645. McBee E. T., Hodge E. B.: *Chem. Eng. News* **30**, 4513 (1952); [21].

646. McBee E. T., Hotten B. W., Evans L. R., Alberts A. A., Welch Z. D., Ligett W. B., Schreyer R. C., Krantz K. W.: *Ind. Eng. Chem.* **39**, 310 (1947); [34, 52, 61, 100, 111].

647. McBee E. T., Hsu C. G., Pierce O. R., Roberts C. W.: *J. Am. Chem. Soc.* **77**, 915 (1955); [135, 259, 260].

648. McBee E. T., Hsu C. G., Roberts C. W.: *J. Am. Chem. Soc.* **78**, 3389 (1956); [260].

649. McBee E. T., Kelley A. E., Rapkin E.: *J. Am. Chem. Soc.* **72**, 5071 (1950); [217].

650. McBee E. T., Leech R. E.: *Ind. Eng. Chem.* **39**, 393 (1947); [97, 136, 139].

651. McBee E. T., Ligett W. B., Lindgren V. V.: *U. S. Patent* 2, 586, 364 (1952); *Chem. Abstr.* **46**, 8675d (1952); [265].

652. McBee E. T., Lindgren V. V., Ligett W. B.: *Ind. Eng. Chem.* **39**, 378 (1947); [104, 138].

653. McBee E. T., Marzluff W. F., Pierce O. R.: *J. Am. Chem. Soc.* **74**, 444 (1952); [255].

654. McBee E. T., Meiners A. F., Roberts C. W.: *Proc. Indiana Acad. Sci.* **64**, 112 (1954); [218, 221].

655. McBee E. T., Pierce O. R., Bolt R. O.: *Ind. Eng. Chem.* **39**, 391 (1947); [140, 147, 148, 152, 243].

656. McBee E. T., Pierce O. R., Christman D. L.: *J. Am. Chem. Soc.* **77**, 1581 (1955); [226].

657. McBee E. T., Pierce O. R., Higgins J. F.: *J. Am. Chem. Soc.* **74**, 1736 (1952); [167, 213].

658. McBee E. T., Pierce O. R., Kilbourne H. W.: *J. Am. Chem. Soc.* **75**, 4091 (1953); [161, 205].

659. McBee E. T., Pierce O. R., Kilbourne H. W., Barone J. A.: *J. Am. Chem. Soc.* **75**, 4090 (1953); [157, 164, 232].

660. McBee E. T., Pierce O. R., Kilbourne H. W., Wilson E. R.: *J. Am. Chem. Soc.* **75**, 3152 (1953); [144, 230, 231].

661. McBee E. T., Pierce O. R., Marzluff W. F.: *J. Am. Chem. Soc.* **75**, 1609 (1953); [174].

662. McBee E. T., Pierce O. R., Meyer D. D.: *J. Am. Chem. Soc.* **77**, 83 (1955); [213, 223, 231].

663. McBee E. T., Pierce O. R., Meyer D. D.: *J. Am. Chem. Soc.* **77**, 917 (1955); [167, 213, 214].

664. McBee E. T., Pierce O. R., Smith D. D.: *J. Am. Chem. Soc.* **76**, 3722 (1954); [226, 308, 309].

665. McBee E. T., Rapkin E.: *J. Am. Chem. Soc.* **73**, 1366 (1951); [207].

666. McBee E. T., Robb R. M.: *U. S. Patent* 2, 487, 820 (1949); *Chem. Abstr.* **44**, 2563f (1950); [82].

667. McBee E. T., Robb R. M., Ligett W. B.: *U. S. Patent* 2, 493, 007 (1950); *Chem. Abstr.* **44**, 5375b (1950); [57].

668. McBee E. T., Roberts C. W., Curtis S. G.: *J. Am. Chem. Soc.* **77**, 6387 (1955); [168, 170, 225].

669. McBee E. T., Roberts C. W., Judd G. F., Chao T. S.: *J. Am. Chem. Soc.* **77**, 1292 (1955); [190, 217, 266].

670. McBee E. T., Roberts C. W., Judd G. F., Chao T. S.: *Proc. Indiana Acad. Sci.* **65**, 100 (1955); [178].

671. McBee E. T., Roberts C. W., Meiners A. F.: *J. Am. Chem. Soc.* **79**, 335 (1957); [221].

672. McBee E. T., Roberts C. W., Puerckhauer G. W. R.: *J. Am. Chem. Soc.* **79**, 2329 (1957); [156, 247].

673. McBee E. T., Roberts C. W., Wilson G., Jr.: *J. Am. Chem. Soc.* **79**, 2323 (1957); [228, 229].

674. McBee E. T., Sanford R. A.: *J. Am. Chem. Soc.* **72**, 5574 (1950); [224].

675. McBee E. T., Schreyer R. C., et al. (20): *Ind. Eng. Chem.* **39**, 305 (1947); [33, 98].

676. McBee E. T., Smith D. K., Ungnade H. E.: *J. Am. Chem. Soc.* **77**, 387 (1955); [258, 259].

677. McBee E. T., Truchan A.: *J. Am. Chem. Soc.* **70**, 2910 (1948); [178].

678. McBee E. T., Wiseman P. A,. Bachman G. B.: *Ind. Eng. Chem.* **39**, 415 (1947); [70, 98, 133, 134, 151, 153, 173].

679. McElvain S. M., Langston J. W.: *J. Am. Chem. Soc.* **66**, 1759 (1944); [65, 133].

680. McGrath T. F., Levine R.: *J. Am. Chem. Soc.* **77**, 3634 (1955); [223].

681. McGrath T. F., Levine R.: *J. Am. Chem. Soc.* **77**, 3656 (1955); [223].

682. McGrath T. F., Levine R.: *J. Am. Chem. Soc.* **77**, 4168 (1955); [223].

683. McIntyre R. T., Berg E. W., Campbell D. N.: *Anal. Chem.* **28**, 1316 (1956); [286].

684. McKay A. F., Vavasour G. R.: *Can. J. Chem.* **32**, 639 (1954); [153, 162].

685. Madorsky S. L., Straus S.: *J. Research Natl. Bur. Standards* **55**, 223 (1955); [276].

686. Maguire M. H., Shaw G.: *J. Chem. Soc.* **1957**, 2713; [237].

687. Malatesta P., D'Atri B.: *Ricerca sci.* **22**, 1589 (1952); [99, 152, 170, 171].

688. Mallory H. D.: *J. Am. Chem. Soc.* **74**, 839 (1952); [33, 97, 124].

689. Mamaev V. P.: *Zhur. Obshchei Khim.* **27**, 1290 (1957); *Chem. Abstr.* **52**, 2748 (1958); [228].

690. Mantell R. M., *U. S. Patent* 2, 697, 124(1954); *Chem. Abstr.* **50**, 2650g (1956); [121, 264, 265].

691. Marans N. S., Sommer L. H., Whitmore F. C.: *J. Am. Chem. Soc.* **73**, 5127 (1951); [112, 156].

692. Marks B. S., Schweiker G. C.: *J. Am. Chem. Soc.* **80**, 5789 (1958); [162].

693. Marti T.: *Chem. Abstr.* **43**, 2711d (1949); [340, 341].

694. Martin D. R.: *Inorg. Syntheses* **4**, 136 (1953); [51].

695. Martin D. R., Pizzolato P. J.: *J. Am. Chem. Soc.* **72**, 4584 (1950); [95, 156].

696. Martin E. L., Sharkey W. H.: *J. Am. Chem. Soc.* **81**, 5256 (1959); [133].

697. Mason C. T., Allain C. C.: *J. Am. Chem. Soc.* **78**, 1682 (1956); [103, 141].

698. Maxwell A. F., Fry J. S., Bigelow L. A.: *J. Am. Chem. Soc.* **80**, 548 (1958); [93, 94].

699. Mechoulam R., Cohen S., Kaluszyner A.: *J. Org. Chem.* **21**, 801(1956); [206].

700. Meier R. Böhler F.: *Chem. Ber.* **90**, 2344 (1957); [157, 160, 164].

701. Menefee A., Cady G. H.: *J. Am. Chem. Soc.* **76**, 2020 (1954); [182].

702. Meslans M.: *Ann. chim.* (7), **1**, 396 (1894); [205].

703. Meslans H.: *Compt. rend.* **111**, 882 (1890); [130].

704. Meslans M.: *Compt. rend.* **115**, 1080 (1892); [113].

705. Meulen van der P. A., Mater Van H. L.: *Inorg. Syntheses* **1**, 24 (1939); [56].

706. Middleton W. J.: *U. S. Patent* 2, 831, 835 (1958); *Chem. Abstr.* **52**, 14658f (1958); [135, 273].

707. Middleton W. J., Sharkey W. H.: *J. Am. Chem. Soc.* **81**, 803 (1959); [135, 273].

708. Midgley T., Jr., Henne A. L.: *Ind. Eng. Chem.* **22**, 542 (1930); [96, 122, 313, 340, 341].

709. Midgley T., Jr., Henne A. L.: *Ind. Eng. Chem.* **24**, 641 (1932); [340, 341].

710. Miller C. B.: *U. S. Patent* 2, 628, 989(1953); *Chem. Abstr.* **48**, 1406h (1954); [129],

711. Miller C. B., Bratton F. H,: *U. S. Patent* 2, 478, 932 (1949); *Chem. Abstr.* **44**. 2006h (1950); [233].

712. Miller C. B., Pearson J. H., Smith L. B.: *U. S. Patent* 2, 802, 887 (1957); *Chem. Abstr.* **52**, 1198d (1958); [159].

713. Miller C. B., Woolf C.: *U. S. Patent* 2, 741, 634 (1956): *Chem. Abstr.* **51**, 460h (1957); [277].

714. MILLER C. B., WOOLF C.: *U. S. Patent* 2, 827, 485; 2, 827, 486 (1958); *Chem. Abstr.* **52**, 13779d, f (1958); [209].

715. MILLER C. B., WOOLF C.: *U. S. Patent* 2, 853, 524 (1958); *Chem. Abstr.* **53**, 5133h (1959); *British Patent* 794, 075(1958); *Chem. Abstr.* **53**, 4137h (1959); [144, 233].

716. MILLER C. B., WOOLF C.: *U. S. Patent* 2, 870, 211 (1959); *Chem. Abstr.* **53**, 10038c (1959); [144].

717. MILLER H. C.: *Chem. Eng. News* **27**, 3854 (1949); [24, 25].

718. MILLER J. F., HUNT H., HASS H. B., MCBEE E. T.: *Anal. Chem.* **19**, 146 (1947); [325].

719. MILLER J. F., HUNT H., MCBEE E. T.: *Anal. Chem.* **19**, 148 (1947); [317, 321].

720. MILLER W. T., JR.: Intern. Symposium, Fluorine Chem., Birmingham, 1959; [66, 110].

721. MILLER W. T., JR.: Preparation, Properties and Technology of Fluorine and Organic Fluoro Compounds (Slesser C., Schram S. R.). McGraw-Hill, New York. 1951; P. 567; [257].

722. MILLER W. T., JR.: Ibid., p. 592—606; [134, 274, 275, 344].

723. MILLER W. T., JR.: Ibid., p. 606—623; [345, 346, 347].

724. MILLER W. T., JR.: Ibid., p. 624—630; [347].

725. MILLER W. T., JR.: Ibid., p. 630—635; [349].

726. MILLER W. T., JR.: Ibid., p. 636—661; [349, 350, 351, 353].

727. MILLER W. T.,: *U. S. Patent* 2, 598, 283 (1952); *Chem. Abstr.* **46**, 7824e (1952); [348].

728. MILLER W. T., *U. S. Patent* 2, 662, 072 (1953); *Chem. Abstr.* **48**, 4253d (1954); [348].

729. MILLER W. T.: *U. S. Patent* 2, 842, 603 (1958); *Chem. Abstr.* **53**, 6113c (1959); [110].

730. MILLER W. T., JR., BERGMAN E., FAINBERG A.: *J. Am. Chem. Soc.* **79**, 4159 (1957); [225].

731. MILLER W. T., JR., BERNSTEIN J.: *J. Am. Chem. Soc.* **70**, 3600 (1948); [201].

732. MILLER W. T., CALFEE J. D., BIGELOW L. A.: *J. Am. Chem. Soc.* **59**, 198 (1938); [88].

733. MILLER W. T., JR., DITTMAN A. L., EHRENFELD R. L., PROBER M.: *Ind. Eng. Chem.* **39**, 333 (1947); [345].

734. MILLER W. T., JR., FAGER E. W., GRISWALD P. H.: *J. Am. Chem. Soc.* **72**, 707 (1950) [233].

735. MILLER W. T., JR., FAINBERG A. H.: *J. Am. Chem. Soc.* **79**, 4164 (1957); [233].

736. MILLINGTON J. E., BROWN G. M., PATTISON F. L. M.: *J. Am. Chem. Soc.* **78**, 3846 (1956); [109, 153, 174, 196].

737. MILLS G. F., WETSEL H. B.: *J. Am. Chem. Soc.* **77**, 4690 (1955); [286].

738. MILNER O. I.: *Anal. Chem.* **22**, 315 (1950); [324].

739. MINNESOTA MINING AND MANUFACTURING COMPANY: *British Patent* 689, 425 (1953); *Chem. Abstr.* **48**, 3389a (1954); [154, 243].

740. MINNESOTA MINING AND MANUFACTURING COMPANY: *British Patent* 737, 164 (1955); *Chem. Abstr.* **50**, 13987d (1956); [208].

741. MIROSEVIC-SORGO P., SAUNDERS B. C.: *Tetrahedron* 5, 38 (1959); [201].

742. MISANI F., SPEERS L., LYON A. M.; *J. Am. Chem. Soc.* **78**, 2801 (1956); [256].

743. MOCHEL W. E., SALISBURY L. F., BARNEY A. L., COFFMAN D. D., MIGHON C. J.: *Ind. Eng. Chem.* **40**, 2285 (1948); [132, 353].

744. MOFFAT A., HUNT H.: *J. Am. Chem. Soc.* **79**, 54 (1957); [199].

745. MOISSAN H.: *Compt. rend.* **102**, 1543 (1886); **103**, 203 (1886); [44].

746. Moissan H.: *Compt. rend.* **140**, 407 (1905); [74].
747. Montermoso J. C., Griffis C. B., Wilson A., Oesterling J. F.: *Proc. Inst. Rubber Ind.* **5**, 97 (1958); [146, 348].
748. Montgomery R., Smith F.: *J. Chem. Soc.* **1952**, 258; [134].
749. Moran A. L., Kane R. P., Smith J. F.: *Ind. Eng. Chem.* **51**, 831 (1959); [97, 349].
750. Morgan P. W.: *Ind. Eng. Chem.* **43**, 2575 (1951); [278].
751. Morse A. T., Ayscough P. B., Leitch L. C.: *Can. J. Chem.* **33**, 453 (1955); [131, 145, 172, 276].
752. Muetterties E. L.: *U. S. Patent* 2, 709, 184 (1955); *Chem. Abstr.* **50** 6498e (1956); [119].
753. Muetterties E. L.: *U. S. Patent* 2, 729, 663 (1956); *Chem. Abstr.* **50**, 11362c (1956); [71, 72, 153].
754. Müller W., Walaschewsky E.: *German Patent* 947, 364 (1956); *Chem. Abstr.* **53**, 4299b (1959); [169].
755. Munter P. A., Aepli O. T., Kossatz R. A.: *Ind. Eng. Chem.* **39**, 427 (1947); [40].
756. Muray J.: *J. Chem. Soc.* **1959**, 1884; [103].
757. Murray R. L., Beanblossom W. S., Wojcik B. H.: *Ind. Eng. Chem.* **39**, 302 (1947); [134, 136].
758. Musgrave W. K. R.: *Quarterly Revs.* **8**, 331 (1954); [20].
759. Musgrave W. K. R., Smith F.: *J. Chem. Soc.* **1949**, 3021; [128, 134].
760. Musgrave W. K. R., Smith F.: *J. Chem. Soc.* **1949**, 3026; [76].
761. Nakanishi S., Morita K. I., Jensen E. V.: *J. Am. Chem. Soc.* **81**, 5259 (1959); [83].
762. Nakanishi S., Myers T. C., Jensen E. V.: *J. Am. Chem. Soc.* **77**, 3099 (1955); [56, 111, 120, 124, 126, 146].
763. Nakanishi S., Myers T. C., Jensen E. V.: *J. Am. Chem. Soc.* **77**, 5033 (1955); [120, 124, 126].
764. Nathan A. H., Magerlein B. J., Hogg J. A.: *J. Org. Chem.* **24**, 1517 (1959); [83].
765. Nerdel F.: *Naturwissenschaften* **39**, 209 (1952); [71, 154].
766. Nes W. R., Burger A.: *J. Am. Chem. Soc.* **72**, 5409 (1950); [231].
767. Nesmejanow A. N., Kahn E. J.: *Ber.* **67**, 370 (1934); [105, 151, 211].
768. Newcomer J. S., McBee E. T.: *J. Am. Chem. Soc.* **71**, 952 (1949); [69].
769. Newkirk A. E.: *J. Am. Chem. Soc.* **68**, 2467 (1946); [67, 124, 128].
770. Newkirk A. E.: *J. Am. Chem. Soc.* **68**, 2737 (1946); [156].
771. Newman M. S.: *J. Am. Chem. Soc.* **67**, 345 (1945); [278].
772. Newman M. S., Renoll M. W., Auerbach I.: *J. Am. Chem. Soc.* **70**, 1023 (1948); [164].
773. Nield E., Stephens R., Tatlow J. C.: *J. Chem. Soc.* **1959**, 159; [314].
774. Nield E., Stephens R., Tatlow J. C.: *J. Chem. Soc.* **1959**, 166; [138, 153, 176, 181, 196, 222, 269].
775. Nikolaev N. S.: *Bull. Acad. Sci. USSR, Classe sci. chim.* **1945**, 309; *Chem. Abstr.* **40**, 4316$^5$ (1946); *Chem. Age* (N.Y.) **54**, 309 (1946); [324].
776. Nischk G., Müller E.: *Ann.* **576**, 232 (1952); [114].
777. Nodzu R., Kitano H., Osaka T.: *J. Chem. Soc. Japan* **58**, 12 (1955); [105, 146].
778. Nodzu R., Osaka T., Kitano H., Fukui K.: *Nippon Kagaku Zasshi* **76**, 775 (1955); *Chem. Abstr.* **51**, 17793a (1957); [196].
779. Norton T. R.: *J. Am. Chem. Soc.* **72**, 3527 (1950); [97, 140, 147, 148, 197, 198].

780. NORTON T. R., MARTIN M. W.: *U. S. Patent* 2,723,982 (1955); *Chem. Abstr.* **50**, 4232h (1956); [146, 229, 230].

781. O'BRIEN H. D.: *Brit. Med. J.* **1956**, II, 969; [356].

782. OLÁH G. A.: Intern. Symposium, Fluorine Chem., Birmingham, 1959; [211].

783. OLÁH G., KUHN S.: *Chem. Ber.* **89**, 864 (1956); [143, 144, 244].

784. OLÁH G., KUHN S.: *Chem. Ber.* **89**, 866 (1956); [211].

785. OLÁH G., KUHN S.: *Chem. Ber.* **89**, 2211 (1956); [244].

786. OLÁH G., KUHN S., BEKE S.: *Chem. Ber.* **89**, 862 (1956); [109, 151].

787. OLÁH G., KUHN S., OLLAH J.: *J. Chem. Soc.* **1957**, 2174; [211].

788. OLÁH G., PAVLÁTH A.: *Acta Chim. Acad. Sci. Hung.* **3**, 191 (1953); [105, 146, 151].

789. OLÁH G., PAVLÁTH A.: *Acta Chim. Acad. Sci. Hung.* **3**, 199 (1953); [109, 140].

790. OLÁH G., PAVLÁTH A.: *Acta Chim. Acad. Sci. Hung.* **3**, 203 (1953); [162].

791. OLÁH G., PAVLÁTH A.: *Acta Chim. Acad. Sci. Hung.* **3**, 425 (1953); [211].

792. OLÁH G., PAVLÁTH A.: *Acta Chim. Acad. Sci. Hung.* **3**, 431 (1953); [110, 140, 143, 174].

793. OLÁH G., PAVLÁTH A.: *Acta Chim. Acad. Sci. Hung.* **4**, 119 (1954); [108].

794. OLÁH G., PAVLÁTH A., KUHN I., VARSÁNYI G.: *Acta Chim. Acad. Sci. Hung.* **7**, 431 (1955); [192].

795. OLÁH G., PAVLÁTH A., VARSÁNYI G.: *J. Chem. Soc.* **1957**, 1823; [179, 180].

796. OTTO K.: Private communication; [322].

797. OTTO K., UHLÍŘ M., PEŠA J.: *Chem. Prumysl* **9**, 587 (1959); [327].

798. PADBURY J. J., TARRANT P.: *U. S. Patent* 2,566,807 (1951); *Chem. Abstr.* **46**, 2561i (1952); [130].

799. PARK J. D., BENNING A. F., DOWNING F. B., LAUCIUS J. F., McHARNESS R. C.: *Ind. Eng. Chem.* **39**, 354 (1947); [129, 268, 313, 343].

800. PARK J. D., BROWN H. A., LACHER J. R.: *J. Am. Chem. Soc.* **75**, 4753 (1953); [307].

801. PARK J. D., HOPWOOD S. L., JR., LACHER J. R.: *J. Org. Chem.* **23**, 1169 (1958); [190]

802. PARK J. D., LARSEN E. R., HOLLER H. W. LACHER J. R.: *J. Org. Chem.* **23**, 1166 (1958); [235].

803. PARK J. D., LYCAN W. R., LACHER J. R.: *J. Am. Chem. Soc.* **73**, 711 (1951); [129].

804. PARK J. D., LYCAN W. R., LACHER J. R.: *J. Am. Chem. Soc.* **76**, 1388 (1954); [152, 182, 286].

805. PARK J. D., SEFFLE R. J., LACHER J. R.: *J. Am. Chem. Soc.* **78**, 59 (1956); [253, 256].

806. PARK J. D., SHARRAH M. L., LACHER J. R.: *J. Am. Chem. Soc.* **71**, 2337(1949); [249].

807. PARK J. D., SHARRAH M. L., LACHER J. R.: *J. Am. Chem. Soc.* **71**, 2339 (1949); [187].

808. PARK J. D., STRICKLIN B., LACHER J. R.: *J. Am. Chem. Soc.* **76**, 1387 (1954); [181].

809. PARK J. D., SWEENEY W. M., HOPWOOD S. L., JR., LACHER J. R.: *J. Am. Chem. Soc.* **78**, 1685 (1956); [248].

810. PARK J. D., VAIL D. K., LEA K. R., LACHER J. R.: *J. Am. Chem. Soc.* **70**, 1550 (1948); [248].

811. PARK J. D., WOLF D. R., SHAHAB M., LACHER J. R.: *J. Org. Chem.* **23**, 1474 (1958); [181].

812. PARMELEE H. M.: *Refrig. Eng.* **61**, 1341 (1953); [304].

813. PARSHALL G. W., ENGLAND D. C., LINDSEY R. V., JR.: *J. Am. Chem. Soc.* **81**, 4801 (1959); [246].

814. Passino H. J.: *U. S. Patent* 2,600,802 (1952); *Chem. Abstr.* **46**, 11774b (1952); [154, 355].

815. Pattison F. L. M., Cott W. J., Howell W. C., White R. W.: *J. Am. Chem. Soc.* **78**, 3484 (1956); [154, 192, 239, 240].

816. Pattison F. L. M., Howell W. C.: *J. Org. Chem.* **21**, 879 (1956); [150, 216].

817. Pattison F. L. M., Howell W. C., McNamara A. J., Schneider J. C., Walker J. F.: *J. Org. Chem.* **21**, 739 (1956); [107].

818. Pattison F. L. M., Howell W. C., White R. W.: *J. Am. Chem. Soc.* **78**, 3488 (1956); [153].

819. Pattison F. L. M., Hunt S. B. D., Stothers J. B.: *J. Org. Chem.* **21**, 883 (1956); [107, 150].

820. Pattison F. L. M., Millington J. E.: *Can. J. Chem.* **34**, 757 (1956); [113, 114].

821. Pattison F. L. M., Norman J. J.: *J. Am. Chem. Soc.* **79**, 2311 (1957); [269, 313].

822. Pattison F. L. M., Saunders B. C.: *J. Chem. Soc.* **1949**, 2745; [99, 150, 152, 159].

823. Pattison F. L. M., Stothers J. B., Woolford R. G.: *J. Am. Chem. Soc.* **78**, 2255 (1956); [127, 174, 271].

824. Patton R. H., Simons J. H.: *J. Am. Chem. Soc.* **77**, 2016, 2017 (1955); [244].

825. Paul R. C.: *J. Chem. Soc.* **1955**, 574; [209].

826. Pavláth A. E.: Intern. Symposium, Fluorine Chem., Birmingham, 1959; [83].

827. Pearlson W. H., Hals L. J.: *U. S. Patent* 2,617,836 (1952); *Chem. Abstr.* **47**, 8770e (1953); [131, 132, 145, 172, 274].

828. Pearlson W. H., Hals L. J.: *U. S. Patent* 2,643,267 (1953); *Chem. Abstr.* **48**, 6461c (1954); [274].

829. Pennington W. A.: *Anal. Chem.* **21**, 766 (1949); [330].

830. Percival W. C.: *Anal. Chem.* **29**, 20 (1957); [331].

831. Peregud E. A., Boikina B. S.: *Zhur Anal. Khim.* **12**, 513 (1957); *Chem. Abstr.* **52**, 1852e (1958); [327].

832. Perkins M. A., Irwin C. F.: *U. S. Patent* 2,410,358 (1946); *Chem. Abstr.* **41**, 574d (1947); [59].

833. Perry J. H.: Chemical Engineers' Handbook. McGraw-Hill, New York, 1950; [34, 36, 37, 288].

834. Peters R.: *Endeavour* **13**, 147 (1954); [312].

835. Petrov A. A., Tumanova A. V.: *Zhur. Obshchei Khim.* **26**, 2744 (1956); *Chem. Abstr.* **51**, 7325g (1957); [258].

836. Pettit M. R., Stacey M., Tatlow J. C.: *J. Chem. Soc.* **1953**, 3081; [278].

837. Pettit M. R., Tatlow J. C.: *J. Chem. Soc.* **1954**, 1071; [195].

838. Pettit M. R., Tatlow J. C.: *J. Chem. Soc.* **1954**, 1941; [278, 279].

839. Phillips M. A.: *Manufg. Chemist* **28**, 328 (1957); [146].

840. Pierce O. R., Kane T. G.: *J. Am. Chem. Soc.* **76**, 300 (1954); [161].

841. Pierce O. R., Levine M.: *J. Am. Chem. Soc.* **75**, 1254 (1953); [218, 219].

842. Pierce O. R., McBee E. T., Cline R. E.: *J. Am. Chem. Soc.* **75**, 5618 (1953); [187, 188, 217].

843. Pierce O. R., McBee E. T., Judd G. F.: *J. Am. Chem. Soc.* **76**, 474 (1954); [225].

844. Pierce O. R., Siegle J. C., McBee E. T.: *J. Am. Chem. Soc.* **75**, 6324 (1953); [167, 213].

845. Plank R.: *Kältetechnik* **8**, 127 (1956); [338].

846. Plunkett R. J.: *U. S. Patent* 2,230,654 (1941); *Chem. Abstr.* **35**, 3365[8] (1941); [347].

847. Pollard F. H., Hardy C. J.: *Anal. Chim. Acta* **16**, 135 (1957); [331].

848. POMEROY J. H., CRAIG C. A.: *J. Am. Chem. Soc.* **81**, 6340 (1959); [285].
849. PORTER R. S., CADY G. H.: *J. Am. Chem. Soc.* **79**, 5625 (1957); [255].
850. PRICE C. C., JACKSON W. G.: *J. Am. Chem. Soc.* **69**, 1065 (1947); [201].
851. PRIEST H. F.: *Inorg. Syntheses* **3**, 175 (1950); [61].
852. PRIEST H. F.: *Inorg. Syntheses* **3**, 176 (1950); [52].
853. PRIEST H. F.: *Inorg. Syntheses* **3**, 178 (1950); [119, 122].
854. PRITCHARD G. O., PRITCHARD H. O., TROTMAN-DICKENSON A. F.: *Chem. & Ind.* (London) **1955**, 564; [71, 155].
855. PROBER M.: *J. Am. Chem. Soc.* **72**, 1036 (1950); [348].
856. PROBER M.: *J. Am. Chem. Soc.* **75**, 968 (1953); [137, 145, 211, 212, 268].
857. PROBER M., MILLER W. T., JR.: *J. Am. Chem. Soc.* **71**, 598 (1949); [257].
858. PRUETT R. L., BARR J. T., RAPP K. E., BAHNER C. T., GIBSON J. D., LAFFERTY R. H. JR.: *J. Am. Chem. Soc.* **72**, 3646 (1950); [153, 249].
859. PUMMER W. J., WALL L. A.: *Science* **127**. 643 (1958); [242].
860. PUMMER W. J., WALL L. A., FLORIN R. E.: *Chem. Eng. News* **36**, No 48, 42, 44 (1958); [157, 160].
861. RAASCH M. S.: *J. Org. Chem.* **23**, 1567 (1958); [313].
862. RAASCH M. S., MIEGEL R. E., CASTLE J. E.: *J. Am. Chem. Soc.* **81**, 2678 (1959); [135].
863. RAINES M. M.: *Zavodskaya Lab.* **14**, 284 (1948); *Chem. Abstr.* **43**, 971g (1949); [327, 328].
864. RAPP K. E., BARR J. T., PRUETT R. L., BAHNER C. T., GIBSOBN J. D., LAFFERTY R. H., JR.: *J. Am. Chem. Soc.* **74**, 749 (1952); [173].
865. RAPP K. E., PRUETT R. L., BARR J. T., BAHNER C. T., GIBSON J. D., LAFFERTY R. H., JR.: *J. Am. Chem. Soc.* **72**, 3642 (1950); [249].
866. RAY F. E., ALBERTSON C. E.: *J. Am. Chem. Soc.* **70**, 1954 (1948); [54, 103].
867. RAY P. C., GOSWAMI H. C., RAY A. C.: *J. Ind. Chem. Soc.* **12**, 93(1935); [144, 145].
868. RAY P. C., SARKAR P. B., RAY A.: *Nature* **132**, 749 (1933); [144].
869. RAZUMOVSKII V. V., FRIDENBERG A. E.: *Zhur. Obshchei Khim.* **19**, 92(1949); *Chem. Abstr.* **43**, 6154c (1949); [113].
870. REED R., JR.: *J. Am. Chem. Soc.* **77**, 3403 (1955); [280].
871. REED T. M., WALTER J. F., CECIL R. R., DRESDNER R. D.: *Ind. Eng. Chem.* **51**, 271 (1959); [314].
872. REID J. C.: *J. Am. Chem. Soc.* **69**, 2069 (1947); [152, 196, 197].
873. REID J. C., CALVIN M.: *J. Am. Chem. Soc.* **72**, 2948 (1950); [229].
874. REID T. S.: *U.S. Patent* 2,706,733 (1955); *Chem. Abstr.* **50**, 2661f (1956); [243].
875. REID T. S., CODDING D. W., BOVEY F. A.: *J. Polymer Sci.* **18**, 417 (1955); [347].
876. REID T. S., SMITH G. H., PEARLSON W. H.: *U.S. Patent* 2,746,997 (1956); *Chem. Abstr.* **51**, 1260i (1957); [270].
877. REINÖHL V.: Private communication; [30, 33, 34, 58, 123, 126, 264, 334, 335, 336, 340, 346].
878. RENFREW M. M., LEWIS E. E.: *Ind. Eng. Chem.* **38**, 870 (1946); [48, 345, 352, 353].
879. RIVETT D. E. A.: *J. Chem. Soc.* **1953**, 3710; [230].
880. ROBERTS J. D., CURTIN D. Y.: *J. Am. Chem. Soc.* **68**, 1658 (1946); [224].
881. ROBINSON C. H., FINCKENOR L., OLIVETO E. P., GOULD D.: *J. Am. Chem. Soc.* **81**, 2191 (1959); [70].
882. ROBSON J. H.: *J. Am. Chem. Soc.* **77**, 107 (1955); [34, 279].
883. ROCHOW E. G., KUKIN I.: *J. Am. Chem. Soc.* **74**, 1615 (1952); [60].

884. ROE A.: *Org. Reactions* 5, 193 (1949); [56, 115, 137, 141, 145, 151, 153].

885. ROE A., HAWKINS G. F.: *J. Am. Chem. Soc.* 69, 2443 (1947); [117, 139, 140].

886. ROGERS G. C., CADY G. H.: *J. Am. Chem. Soc.* 73, 3523 (1951); [274].

887. ROGERS R. N., YASUDA S. K.: *Anal. Chem.* 31, 616 (1959); [319, 321].

888. ROSEN R.: Preparation, Properties and Technology of Fluorine and Organic Fluoro Compounds (Slesser C., Schram S. R.). McGraw-Hill, New York, 1951; P. 3; [80].

889. ROSS S. D., MARKARIAN M., SCHWARZ M.: *J. Am. Chem. Soc.* 75, 4967 (1953); [194].

890. ROYLANCE J., TATLOW J. C., WORTHINGTON R. E.: *J. Chem. Soc.* 1954, 4426; [163, 269].

891. RUCKER J. T., STORMON D. B.: *U. S. Patent* 2,760,997(1956); *Chem. Abstr.* 51, 3653i (1957); [264].

892. RUDNER M. A.: Fluorocarbons. Reinhold Publishing Corp., New York, 1958; [20, 332].

893. RUFF O.: Die Chemie des Fluors. Springer, Berlin, 1920; [27, 51, 52, 53, 57, 58, 59].

894. RUFF O., BRAIDA A.: *Z. anorg. allgem. Chem.* 220, 43 (1934); [61].

895. RUFF O., BRETSCHNEIDER O., LUCHSINGER W., MILTSCHITZKY G.: *Ber.* 69, 299 (1936); [101].

896. RUFF O., GIESE M.: *Z. anorg. allgem. Chem.* 219, 143 (1934); [52].

897. RUGGLI P., CASPAR E.: *Helv. Chim. Acta* 18, 1414 (1935); [55].

898. RUH R. P., DAVIS R. A., ALLSWEDE K. A.: *U. S. Patent* 2,777,004 (1957); *Chem. Abstr.* 51, 11370h (1957); [133, 344].

899. RUNNER M. E., BALOG G., KILPATRICK M.: *J. Am. Chem. Soc.* 78, 5183 (1956); [42].

900. RUSH C. A., CRUIKSHANK S. S., RHODES E. J. H.: *Mikrochim. Acta* 1956, 858; [322, 325].

901. RYSS I. G., POLYAKOVA E. M.: *Zhur. Obshchei Khim.* 19, 1596 (1949); *Chem. Abstr.* 44, 1235i (1950); [54].

902. SAGER W. F., DUCKWORTH A.: *J. Am. Chem. Soc.* 77, 188 (1955); [152, 282, 283, 284].

903. SANDERS P. F.: *U. S. Patent* 2,520,173 (1950); *Chem. Abstr.* 44, 11177i (1950); [354].

904. SARGENT J. W., CLIFFORD A. F., LEMMON W. R.: *Anal. Chem.* 25, 1727 (1953); [42].

905. SARTORI M. F.: *Chem. Revs.* 48, 225 (1951); [156, 311].

906. SASS S., BEITSCH N., MORGAN C. U.: *Anal. Chem.* 31, 1970 (1959); [331].

907. SAUER J. C.: *U. S. Patent* 2, 549, 935 (1951); *Chem. Abstr.* 45, 6874c (1951); [348].

908. SAUNDERS B. C.: *Endeavoue* 19, 36 (1960); [311].

909. SAUNDERS B. C.: *J. Chem. Soc.* 1949, 1279; [312].

910. SAUNDERS B. C.: Phosphorus and Fluorine. Some Aspects of the Chemistry and Toxic Action of Their Organic Compounds. Cambridge University Press, Cambridge, 1957; [20, 22, 146, 311, 312, 331, 357].

911. SAUNDERS B. C., STACEY G. J.: *J. Chem. Soc.* 1948, 695; [110, 156, 203, 311].

912. SAUNDERS B. C., STACEY G. J.: *J. Chem. Soc.* 1948, 1773; [33, 94, 105, 110, 146, 201, 230].

913. SAUNDERS B. C., STACEY G. J.: *J. Chem. Soc.* 1949, 916; [198].

914. SAUNDERS B. C., STACEY G. J., WILDING I. G. E.: *J. Chem. Soc.* 1949, 773; [109, 110, 111, 140, 143, 174].

915. SCHALLENBERG E. E., CALVIN M.: *J. Am. Chem. Soc.* 77, 2779 (1955); [243].

916. SCHERER H., HAHN H.: *German Patent* 924, 512 (1955); *Chem. Abstr.* 52, 7353d (1957); [106].

917. SCHERER O., KÜHN H.: *German Patent* 907, 173 (1954); *Chem. Abstr.* 52, 10141d (1958); [264].

918. SCHERER P.: Field Information Agency, Technical, 1114, 43; [20].

919. SCHIEMANN G.: Die organischen Fluorverbindungen in ihrer Bedeutung für die Technik. Steinkopff, Darmstadt, 1951; [20, 332, 357].
920. SCHIEMANN G.: *J. prakt. Chem.* **140**, 97 (1934); [138].
921. SCHIEMANN G.: *Naturwissenschaften* **19**, 706 (1931); [301].
922. SCHMIDT H., SCHMIDT H. D.: *J. prakt. Chem.* (4), **2**, 105 (1955); [85].
923. SCHMIDT H., SCHMIDT H. D.: *J. prakt. Chem.* (4), **2**, 250 (1955); [84].
924. SCHMIDT R.: *Glas-Email-Keramo-Technik* **7**, 77 (1936); [32].
925. SCHÖNIGER W.: *Mikrochim. Acta* **1956**, 869; [321].
926. SCHULTZ B. G., LARSEN E. M.: *J. Am. Chem. Soc.* **71**, 3250 (1949); [306].
927. SCHUMB W. C., ARONSON J. R.: *J. Am. Chem. Soc.* **81**, 806 (1959); [119, 122].
928. SCHUMB W. C., GAMBLE E. L.: *J. Am. Chem. Soc.* **54**, 583 (1932); [111, 156].
929. SCHUMB W. C., YOUNG R. C., RADIMER K. J.: *Ind. Eng. Chem.* **39**, 244 (1947); [46].
930. SCHWEIKER G. C., ROBITSCHEK P.: *J. Polymer Sci.* **24**, 33 (1957); [347].
931. SEEL F., JONAS H., RIEHL L., LANGER J.: *Ang. Chem.* **67**, 32 (1955); [109].
932. SENKOWSKI B. Z., WOLLISH E. G., SHAFER E. G. E.: *Anal. Chem.* **31**, 1574 (1959) [126, 321].
933. SENTEMENTES T. J., DE SESA M. A.: *Chemist Analyst* **44**, 54 (1955); [26, 42].
934. SERPINET J.: *Chim. anal.* **41**, 146 (1959); [314].
935. SEVERSON W. A., BRICE T. J.: *J. Am. Chem. Soc.* **80**, 2313 (1958); [170].
936. SHECHTER H., CONRAD F.: *J. Am. Chem. Soc.* **72**, 3371 (1950); [143, 169, 192, 208].
937. SHEPARD R. A., LOISELLE A. A.: *J. Org. Chem.* **23**, 2013 (1958); [99].
938. SHERRY W. B., SWINEHART C. F., DURPHY R. A., OGBURN S. C.: *Ind. Eng. Chem. Anal. Ed.* **16**, 483 (1944); [35, 327, 328].
939. SHOESMITH J. B., SOSSON C. E., SLATER R. H.: *J. Chem. Soc.* **1926**, 2761; [143].
940. SHOPPEE C. W., SUMMERS G. H. R.: *J. Chem. Soc.* **1957**, 4813; [113, 233].
941. SHUKYS J. G.: *U. S. Patent* 2,830,007 (1958); *Chem. Abstr.* **52**, 14651b (1958); [356].
942. SILVERSMITH E. F., KITAHARA Y., CASERIO M. C., ROBERTS J. D.: *J. Am. Chem. Soc.* **80**, 5840 (1958); [203].
943. SILVERSMITH E. F., ROBERTS J. D.: *J. Am. Chem. Soc.* **80**, 4083 (1958); [203].
944. SILVEY G. A., CADY G. H.: *J. Am. Chem. Soc.* **72**, 3624 (1950); [71, 72, 153].
945. SILVEY G. A., CADY G. H.: *J. Am. Chem. Soc.* **74**, 5792 (1952); [72, 153].
946. SIMONS J. H.: *Chem. Eng. News* **26**, 1317 (1948); [22].
947. SIMONS J. H. (Editor) Fluorine Chemistry. I. Academic Press, New York, 1950; [20, 42, 44, 48, 60, 288, 302, 345].
948. SIMONS J. H. (Editor): Fluorine Chemistry. II. Academic Press, New York, 1954; [20, 288, 314, 316, 345].
949. SIMONS J. H.: *Inorg. Syntheses* **1**, 134 (1939); [40].
950. SIMONS J. H.: *Inorg. Syntheses* **3**, 184 (1950); [60].
951. SIMONS J. H., BLOCH L. P.: *J. Am. Chem. Soc.* **61**, 2962 (1939); [119, 122].
952. SIMONS J. H., BLOCH W. T., CLARK R. F.: *J. Am. Chem. Soc.* **75**, 5621 (1953); [212].
953. SIMONS J. H., BOND R. L., McARTHUR R. E.: *J. Am. Chem. Soc.* **62**, 3477 (1940); [88, 122, 123].
954. SIMONS J. H., HERMAN D. F., PEARLSON W. H.: *J. Am. Chem. Soc.* **68**, 1672 (1946); 89].
955. SIMONS J. H., LEWIS C. J.: *J. Am. Chem. Soc.* **60**, 492 (1938); [89, 136].
956. SIMONS J. H., McARTHUR R. E.: *Ind. Eng. Chem.* **39**, 366 (1947); [89, 207].

957. SIMONS J. H., PEARLSON W. H., BRICE T. J., WILSON W. A., DRESDNER R. D.: J. Electrochem. Soc. **95**. 59 (1949); [150].

958. SIMONS J. H., RAMLER E. O.: J. Am. Chem. Soc. **65**, 389 (1943); [145, 181, 190, 207, 208, 212, 222, 229, 315].

959. SLESSER C., SCHRAM S. R.: Preparation, Properties and Technology of Fluorine and Organic Fluoro Compounds. McGraw-Hill, New York, 1951; [20, 332].

960. SMITH F., STACEY M., TATLOW J. C., DAWSON J. K., THOMAS B. R. J.: J. Appl. Chem. (London) **2**, 97, 127 (1952); [136].

961. SMITH R. P., TATLOW J. C.: J. Chem. Soc. **1957**, 2505; [70, 82, 134, 269].

962. SMITH W. C.: U. S. Patent 2,859,245 (1958); Chem. Abstr. **53**, 12236a (1959); [112, 127]

963. SMITH W. C.: U.S. Patent 2,862,029 (1958); Chem. Abstr. **53**, 9152g (1959); [112].

964. SMITH W. C., TULLOCK C. W., MUETTERTIES E. L., HASEK W. R., FAWCETT F. S., ENGELHARDT V. A., COFFMAN D. D.: J. Am. Chem. Soc. **81**, 3165 (1959); [60, 112, 127].

965. SPERO G. B., THOMPSON J. L., LINCOLN F. H., SCHNEIDER W. P., HOGG J. A.: J. Am. Chem. Soc. **79**, 1515 (1957); [357].

966. STACEY M., TATLOW 7. C., SHARPE A. G., Advances in Fluorine Chemistry Butterworths, London, 1900 [20]

967. STEPHENS R., TATLOW J. C.: Chem. & Ind. (London) **1957**', 821; [138, 176].

968. STEPHENS R., TATLOW J. C., WISEMAN E. A.: J. Chem. Soc. **1959**, 148; [314].

969. STERLIN R. N., SIDOROV V. A., KNUNYANTS I. L.: Izvest. Akad. Nauk SSR **1959**, 62; Chem. Abstr. **53**, 14916d (1959); [186].

970. STEUNENBERG R. K., CADY G. H.: J. Am. Chem. Soc. **74**, 4165 (1952); [274].

971. STIVERS D. A., HONN F. J., ROBB L. E.: Ind. Eng. Chem. **51**, 1465 (1959); [349].

972. STONE I.: J. Chem. Educ. **8**, 347 (1931); [315].

973. STONER G. G.: U. S. Patent 2,761,875 (1956); Chem. Abstr. **51**, 3657a (1957); [204].

974. STRUVE W. S., BENNING A. F., DOWNING F. B., LULEK R. N., WIRTH W. V.: Ind. Eng. Chem. **39**, 352 (1947); [52].

975. STUEWE A. H.: Chem. Eng. News **36**, No 51, 34 (1958); [39].

976. SUCKLING C. W.: Australian Patent 205, 979 (1955); [125, 338].

977. SUCKLING C. W., RAVENTOS J.: British Patent 767, 779 (1957); Chem. Abstr. **51**, 15547a (1957); U. S. Patent 2,849,502 (1958); Chem. Abstr. **52**, 18213d (1958); [125, 193].

978. SUSCHITZKY H.: J. Chem. Soc. **1953**, 3326; [155, 165, 195].

979. SUSCHITZKY H.: J. Chem. Soc. **1955**, 4026; [155].

980. SUTER C. M., LAWSON E. J., SMITH P. G.: J. Am. Chem. Soc. **61**, 161 (1939); [56].

981. SWARTS F.: Bull. acad. roy. Belg. (3), **34**, 293 (1897); Chem. Zentr. **1897**, II, 1042; [301].

982. SWARTS F.: Bull. acad. roy. Belg. (3), **34**, 307 (1897); Chem. Zentr. **1897**, II, 1098; [168].

983. SWARTS F.: Bull. acad. roy. Belg. (3), **35**, 375 (1898): Chem. Zentr. **1898**, II, 26; [199, 206].

984. SWARTS F.: Bull. acad. roy. Belg. (3), **35**, 849 (1898); Chem. Zentr. **1898**, II, 704; [94, 209].

985. SWARTS F.: Bull. acad. roy. Belg. **1900**, 414; Chem. Zentr. **1900**, II, 667; [164, 201, 204].

986. SWARTS F.: Bull. acad. roy. Belg. **1901**, 383; Chem. Zentr. **1901**, II, 804; [128, 264].

987. SWARTS F.: Bull. acad. roy. Belg. **1902**, 731; Chem. Zentr. **1903**, I, 437; [203].

988. SWARTS F.: *Bull. acad. roy. Belg.* **1903**, 597; *Chem. Zentr.* **1903**, II, 709; [149].

989. SWARTS F.: *Bull. acad. roy. Belg.* **1904**, 762, 955; *Chem. Zentr.* **1904**, II, 944, 1377; [153. 239].

990. SWARTS F.: *Bull. acad. roy. Belg.* **1906**, 42; *Chem. Zentr.* **1906**, I, 1237; [149, 204].

991. SWARTS F.: *Bull. acad. roy. Belg.* **1909**, 728; *Chem. Zentr.* **1909**, II, 1414; [94, 124].

992. SWARTS F.: *Bull. acad. roy. Belg.* **1910**, 113; *Chem. Zentr.* **1910**, I, 1868; [164, 165, 204].

993. SWARTS F.: *Bull. acad. roy. Belg.* **1911**, 563; *Chem. Zentr.* **1911**, II, 848; [99, 217].

994. SWARTS F.: *Bull. acad. roy. Belg.* **1913**, 241; *Chem. Zentr.* **1913**, II, 760; [114].

995. SWARTS F.: *Bull. acad. roy. Belg.* **1914**, 7; *Chem. Zentr.* **1914**, I, 1551; [109, 140, 141].

996. SWARTS F.: *Bull. acad. roy. Belg.* **1920**, 389; *Chem. Zentr.* **1921**, III, 32; [95, 136, 207].

997. SWARTS F.: *Bull. acad. roy. Belg.* **1920**, 399; *Chem. Zentr.* **1921**, III, 32; [157, 159].

998. SWARTS F.: *Bull. acad. roy. Belg.* (5), **7**, 438 (1921); *Chem. Zentr.* **1921**, III, 1457; [51, 98, 100, 101, 126].

999. SWARTS F.: *Bull. acad. roy. Belg.* (5), **8**, 331 (1923); *Chem. Zentr.* **1923**, I, 65; [170].

1000. SWARTS F.: *Bull. acad. roy. Belg.* (5), **8**, 343 (1923); *Chem. Zentr.* **1923**, I, 66; [147, 171].

1001. SWARTS F.: *Bull. acad. roy. Belg.* (5), **12**, 679 (1927); *Chem. Zentr.* **1927**, I, 1286; [230, 262, 306].

1002. SWARTS F.: *Bull. acad. roy. Belg.* (5), **22**, 105 (1936); *Chem. Zentr.* **1936**, I, 4899; [101, 133].

1003. SWARTS F.: *Bull. acad. roy. Belg.* (5), **22**, 122 (1936); *Chem. Zentr.* **1936**, I, 4900; [159].

1004. SWARTS F.: *Bull. acad. roy. Belg.* (5), **22**, 781 (1936); *Bull. soc. chim. Belg.* **13**, 10 (1937); *Chem. Zentr.* **1937**, I, 854, 4082; [101, 122].

1005. SWARTS F.: *Bull. soc. chim. Belg.* **38**, 99 (1929); *Chem. Zentr.* **1929**, II, 712; [140, 158, 270].

1006. SWARTS F.: *Bull. soc. chim. France* (4), **25**, 103 (1919); [101].

1007. SWARTS F.: *Bull. soc. chim. France* (4), **25**, 145 (1919); [265].

1008. SWARTS F.: *Bull. soc. chim. France* (4), **25**, 325 (1919); [228].

1009. SWARTS F.: *Compt. rend.* **197**, 1201 (1933); [158].

1010. SWEETSER P. B.: *Anal. Chem.* **28**, 1766 (1956); [320].

1011. SYKES A., TATLOW J. C., THOMAS C. R.: *Chem. & Ind.* (London) **1955**, 630; [213, 214].

1012. SYKES A., TATLOW J. C., THOMAS C. R.: *J. Chem. Soc.* **1956**, 835; [208, 214].

1013. SZMANT H. H., ANZENBERGER J. F., HARTLE R.: *J. Am. Chem. Soc.* **72**, 1419 (1950); [222].

1014. TALIPOV Š. T., ABDULLAEV D. A.: *Trudy Srednoaziat. Univ. Khim. Nauk* **5**, 65 (1953); *Chem. Abstr.* **49**, 6757e (1955); [51].

1015. TANNHAUSER P., PRATT R. J., JENSEN E. V.: *J. Am. Chem. Soc.* **78**, 2658 (1956); [99].

1016. TARRANT P.: Intern. Symposium, Fluorine Chem., Birmingham, 1959; [132].

1017. TARRANT P., ATTAWAY J., LOVELACE A. M.: *J. Am. Chem. Soc.* **76**, 2343 (1954); [89, 92].

1018. TARRANT P., BREY M. L., GRAY B. E.: *J. Am. Chem. Soc.* **80**, 1711 (1958); [252].

1019. TARRANT P., BROWN H. C.: *J. Am. Chem. Soc.* **73**, 1781 (1951); [248, 249].

1020. TARRANT P., BROWN H. C.: *J. Am. Chem. Soc.* **73**, 5831 (1951); [141, 204, 238, 248].

1021. TARRANT P., GILLMAN E. G.: *J. Am. Chem. Soc.* **76**, 5423 (1954); [252].

1022. TARRANT P., LILYQUIST M. R.: *J. Am. Chem. Soc.* **77**, 3640 (1955); [251].

1023. TARRANT P., LILYQUIST M. R., ATTAWAY J. A.: *J. Am. Chem. Soc.* **76**, 944 (1954); [94].

1024. TARRANT P., LOVELACE A. M.: *J. Am. Chem. Soc.* **76**, 3466 (1954); [251, 252].

1025. TARRANT P., LOVELACE A. M.: *J. Am. Chem. Soc.* **77**, 768 (1955); [164, 165, 251, 252, 254].

1026. TARRANT P., LOVELACE A. M., LILYQUIST M. R.: *J. Am. Chem. Soc.* **77**, 2783 (1955); [251].

1027. TARRANT P., WARNER D. A.: *J. Am. Chem. Soc.* **76**, 1624 (1954); [215].

1028. TATLOW J. C.: Intern. Symposium, Fluorine Chem., Birmingham, 1959; [138, 139, 160, 266, 269].

1029. TATLOW J. C.: *Rubber & Plastic Age* **39**, 33 (1958); [347, 348].

1030. TAYLOR N. F., KENT P. W.: *J. Chem. Soc.* **1958**, 872; [316].

1031. TEDDER J. M.: *Chem. & Ind.* (London) **1955**, 508; [73, 145].

1032. TEDDER J. M.: *Chem. Revs.* **55**, 787 (1955); [282].

1033. TEETERS W. O.: *U.S. Patent* 2,686,738 (1954); *Chem. Abstr.* **48**, 14294g (1954); [354].

1034. TEETERS W. O., PASSINO J., DITTMAN A. L.: *U.S. Patent* 2,770,606 (1956); *Chem. Abstr.* **51**, 7752a (1957); [348].

1035. TERRUZZI M.: *French Patent* 1,125,625 (1956); *U.S. Patent* 2,824,900 (1958); *Chem. Abstr.* **52**, 11886i (1958); [125].

1036. TESTON O'D. R., McKENNA F. E.: *Anal. Chem.* **19**, 193 (1947); [324, 325].

1037. THOMPSON J., EMELÉUS H. J.: *J. Chem. Soc.* **1949**, 3080; [153, 154, 203].

1038. THORNTON D. P., JR.: *Petroleum Processing* **6**, 488 (1951); [33].

1039. THROSSELL J. J., SOOD S. P., SZWARC M., STANNETT V.: *J. Am. Chem. Soc.* **78**, 1122 (1956); [278].

1040. TICHÝ V., TRUCHLÍK Š.: *Chem. Zvesti* **12**, 345 (1958); [110].

1041. TIERS G. V. D.: *J. Am. Chem. Soc.* **77**, 4837 (1955); [191].

1042. TIERS G. V. D.: *J. Am. Chem. Soc.* **77**, 6703 (1955); [191].

1043. TIERS G. V. D.: *J. Am. Chem. Soc.* **77**, 6704 (1955); [191].

1044. TISHCHENKO D.: *Zhur. Obshchei Khim.* **21**, 1625 (1951); *Chem. Abstr.* **46**, 4468c (1952) [307].

1045. TITOV A. I., BARYSHNIKOVA A. N.: *Zhur. Obshchei Khim.* **23**, 346 (1953); *Chem. Abstr.* **48**, 2623f (1954); [118].

1046. TRONOV B., KRÜGER E.: *Zhur. Russ. Fyz. Khim. Obshchestva* **58**, 1270 (1926); [444]. *Chem. Abstr.* **12**, 3887 (1927); [200, 201].

1047. TRUCE W. E., BIRUM G. H., McBEE E. T.,: *J. Am. Chem. Soc.* **74**, 3594 (1952); [28, 96, 152, 153, 171].

1048. TRUCE W. T., HOERGER F. D.: *J. Am. Chem. Soc.* **76**, 3230 (1954); [104, 202].

1049. TURNBULL S. G., BENNING A. F., FELDMANN G. W., LINCH A. L., McHARNESS R. P., RICHARDS M. K.: *Ind. Eng. Chem.* **39**, 286 (1947); [326].

1050. TYCZKOWSKI E. A., BIGELOW L. A.: *J. Am. Chem. Soc.* **75**, 3523 (1953); [28, 71, 72, 74, 122].

1051. TYCZKOWSKI A., BIGELOW L. A.: *J. Am. Chem. Soc.* **77**, 3007 (1955); [74, 125].

1052. ULLMANN J.: *Publs. fac. sci. univ. Masaryk*, No. **390**, 33 (1958); *Chem. Abstr.* **53**, 2933h (1959); [317].

1053. UMEDA T.: *Japanese Patent* 3874('52); *Chem. Abstr.* **47**, 10551a (1953); [94].

1054. VAN VLECK R. T.: *J. Am. Chem. Soc.* **71**, 3256 (1949); [94].

1055. VOGEL A. I.: *J. Chem. Soc.* **1948**, 644; [52, 301, 302].

1056. VOROZHTSOV N. N., JR., YAKOBSON G. G.: *Khim. Nauka i Prom.* **3**, 403 (1958); *Chem. Abstr.* **52**, 19988f (1958); [104].

1057. VOROZHTSOV N. N., JR., YAKOBSON G. G.: *Nauch. Doklady Vyshei Shkoly Khim. i Khim. Technol.* **1958**, No 1, 122; *Chem. Abstr.* **53**, 3110a (1959); [106].

1058. VOROZHTSOV N. N., JR., YAKOBSON G. G., KRIZHECHKOVSKAYA N. I.: *Khim. Nauka i Prom.* **3**, 404 (1958); *Chem. Abstr.* **52**, 19987h (1958); [185].

1059. VŘEŠŤÁL J., HAVÍŘ J., BRANŠTETR J., KOTRLÝ S.: *Chem. listy* **51**, 1677 (1957); [317].

1060. WÄCHTER R. *Ang. Chem.* **67**, 305 (1955); [147, 171].

1061. WALBORSKY H. M., BAUM M. E.: *J. Am. Chem. Soc.* **80**, 187 (1958); [241].

1062. WALBORSKY H. M., BAUM M. E.: *J. Org. Chem.* **21**, 538 (1956); [240].

1063. WALBORSKY H. M., BAUM M., LONCRINI D. F.: *J. Am. Chem. Soc.* **77**, 3637 (1955); [163].

1064. WALBORSKY H. M., SCHWARZ M.: *J. Am. Chem. Soc.* **75**, 3241 (1953); [153, 188, 245].

1065. WALLACH O.: *Ann.* **235**, 242, 255 (1880); [114].

1066. WALLACH O., HEUSLER F.: *Ann.* **243**, 219 (1888); [114].

1067. WALLENFELS K.: Intern. Symposium, Fluorine Chem., Birmingham, 1959; [106, 107]

1068. WALLENFELS K., DRABER W.: *Chem. Ber.* **90**, 2819 (1957); [104, 106].

1069. WARTENBERG H. **von**: *Z. anorg. allgem. Chem.* **244**, 337 (1940); [57, 59].

1070. WEBER C. E., BURFORD W. B., III. BITTERLICH S. B.: Preparation, Properties and Technology of Fluorine and Organic Fluoro Compounds (Slesser C., Schram S. R.). McGraw-Hill, New York, 1951; p. 184; [29].

1071. WEINLAND R. F., STILLE W.: *Ann.* **328**, 132 (1903); [112, 156].

1072. WEYGAND F., GEIGER R.: *Chem. Ber.* **89**, 647 (1956); [243, 282].

1073. WEYGAND F., GLÖCKLER U.: *Chem. Ber.* **89**, 653 (1956); [243, 282].

1074. WEYGAND F., LEISING E.: *Chem. Ber.* **87**, 248 (1954); [243, 282].

1075. WEYGAND F., RAUCH E.: *Chem. Ber.* **87**, 211 (1954); [280].

1076. WEYGAND F., RÖSCH A.: *Chem. Ber.* **92**, 2095 (1959); [243].

1077. WEYGAND F., STEGLICH W.: *Z. Naturforsch.* **14b**, 472 (1959); [243].

1078. WHALLEY W. B.: *J. Chem. Soc.* **1950**, 2241; [241, 286].

1079. WHALLEY W. B.: *J. Chem. Soc.* **1950**, 2792; [286].

1080. WHALLEY W. B.: *J. Chem. Soc.* **1951**, 665; [208, 212].

1081. WHALLEY W. B.: *J. Soc. Chem. Ind.* (London) **66**, 427 (1947); [59, 92, 122, 123, 125].

1082. WHALLEY W. B.: *J. Soc. Chem. Ind.* (London) **66**, 430 (1947); [92].

1083. WICKBOLD R.: *Ang. Chem.* **64**, 133 (1952); [320].

1084. WICKBOLD R.: *Ang. Chem.* **66**, 173 (1954); [32, 320].

1085. WIECHERT K.: *Chem. Technik* **5**, 80 (1952); [112].

1086. WIECHERT K., (JONES J. E.): Newer Methods of Preparative Organic Chemistry. Interscience Publishers, New York, 1948; P. 315-368; [44].

1087. WILKINS C. J., *J. Chem. Soc.* **1951**, 2726; [110, 156].

1088. WILLARD H. H., WINTER O. B.: *Ind. Eng. Chem. Anal. Ed.* **5**, 7 (1933); [319].

1089. WILSHIRE J. F. K., PATTISON F. L. M.: *J. Am. Chem. Soc.* **78**, 4996 (1956); [142, 200].

1090. WRIGHTSON J. M., DITTMAN A. L.: *U. S. Patent* 2,667,518 (1954); *Chem. Abstr.* **49**, 2478b (1955); [130].

1091. YAGUPOLSKII L. M., MARENETS M. S.: *Zhur. Obshchei Khim.* **24**, 887(1954); *Chem. Abstr.* **49**, 8172d (1955); [95].

1092. YAKUBOVICH A. Y., GOGOL V., BORZOVA I.: *Zhur. Priklad. Khim.* **32**, 451 (1959); *Chem. Abstr.* **53**, 13045i (1959); [130, 147, 172].

1093. YAKUBOVICH A. Y., SHPANSKII V. A., LEMKE A. L.: *Zhur. Obshchei Khim.* **24**, 2257(1954); *Chem. Abstr.* **50**, 206b (1956); [189].

1094. YALE H. L., SOWINSKI F., BERNSTEIN J.: *J. Am. Chem. Soc.* **79**, 4375 (1957); [356].

1095. YAROVENKO N. N., MOTORNYI S. P., KIRENSKAYA L. I.: *Zhur. Obshchei Khim.* **27**, 2796 (1957); *Chem. Abstr.* **52**, 8042a (1958): [145].

1096. YAROVENKO N. N., MOTORNYI S. P., KIRENSKAYA L. I., VASILEVA A. S.: *Zhur. Obshchei Khim.* **27**, 2243 (1957); [235].

1097. YAROVENKO N. N., RAKSHA M. A., SHEMANINA V. N., VASILEVA A. S.: *Zhur. Obshchei Khim.* **27**, 2246 (1957); *Chem. Abstr.* **52**, 6176a (1958); [149].

1098. YOUNG D. S., FUKUHARA N., BIGELOW L. A.: *J. Am. Chem. Soc.* **62**, 1171 (1940); [125].

1099. YOUNG J. A., SIMMONS T. C., HOFFMANN F. W.: *J. Am. Chem. Soc.* **78**, 5637 (1956); [154, 277].

1100. YOUNG J. A., TARRANT P.: *J. Am. Chem. Soc.* **71**, 2432 (1949); [204].

1101. YOUNG J. A., TARRANT P.: *J. Am. Chem. Soc.* **72**, 1860 (1950); [141, 204, 238, 247].

1102. ZAHN H., WÜRZ A.: *Ang. Chem.* **63**, 147 (1951); [155, 193].

1103. ZELLHOEFER G. F.: *Ind. Eng. Chem.* **29**, 548 (1937); [340].

1104. ZELLHOEFER G. F., COPLEY M. J., MARVEL C. S.: *J. Am. Chem. Soc.* **60**, 1337 (1938); [340].

1105. ZIMA G. E., DOESCHER R. N.: *Metal Progress* **59**, 660 (1951); [35].

# LIST OF JOURNALS

The list contains periodicals quoted in the book. The **boldface** printed letters represent the conventional abbreviations used in the list of references (p. 358)

**Acta** **Chim**ica **Acad**emiae **Scientiarum** **Hung**aricae
**Analytical** **Chemistry**
**Analytica** **Chimica** **Acta**
**Analyst**, The
**Angewandte** **Chem**ie
**Annalen** der **Chemie**, Justus Liebig's
**Annales** de **chimie** (**Paris**)
**Annual** **Reports** on the **Progress** of **Chemistry** (**Chemical** Society of **London**)
**Applied** **Spectroscopy**
**Berichte** der deutschen chemischen Gesellschaft
**British** **Medical** **Journal**
**British** **Patent**
**Bulletin** de l'**acad**émie **roy**ale de **Belgique**
**Bulletin** de la **classe** des sciences, **Acad**émie **roy**ale de **Belgique**
**Bulletin** de l'**acad**émie des **sciences** de l'**URSS**, **Classe** des sciences **chimiques**
**Bulletin** of the **Research** **Council** of **Israel**
**Bulletin** de la **société** **chimique** de **Belgique**
**Bulletin** de la **société** **chimique** de **France**
**Canadian** Journal of **Chemistry**
**Canadian** Journal of **Research**
**Chemical** **Abstracts**
**Chemical** **Age** (New York)
**Chemische** **Berichte**
**Chemical** **Engineering** **News**
**Chemist** **Analyst**
**Chemistry** & **Industry** (**London**)
**Chemické** **listy**
**Chemický** **průmysl**
**Chemical** **Reviews**
**Chemische** **Technik**, Die (**Berlin**)
**Chemisches** **Zentralblatt**
**Chemiker** **Zeitung**
**Chemické** **zvesti**
**Chimie** **analytique**
**Collection** of **Czechoslovak** Chemical **Communications**
**Comptes** **rendus** hebdomadaires des séances de l'académie des sciences
**Doklady** Akademii **Nauk** **SSSR**

Endeavour
French Patent
German Patent
Glas-, Email-, Keramo-Technik
Helvetica Chimica Acta
Industrial and Engineering Chemistry
Industrial and Engineering Chemistry, Analytical Edition
Inorganic Syntheses
Izvestiya Akademii Nauk SSSR
Journal of the American Chemical Society
Journal of the American Medical Association
Japanese Patent
Journal of Applied Chemistry (London)
Journal of Chemical Education
Journal of the Chemical Society (London)
Journal of the Chemical Society (Japan)
Journal of the Electrochemical Society
Journal of the Indian Chemical Society
Journal of Organic Chemistry
Journal of Physical Chemistry
Journal of Polymer Science
Journal für praktische Chemie
Journal of Research of the National Bureau of Standards
Journal of the Science of Food and Agriculture
Journal of the Society of Chemical Industry (London)
Kältetechnik
Kautschuk und Gummi
Khimicheskaya Nauka i promyshlenost
Lancet
Manufacturing Chemist
Metal Progress
Microchemical Journal
Mikrochemie
Mikrochimica Acta
Modern Plastics
Monatshefte für Chemie
Nachrichten der Gesellschaft der Wissenschaften zu Göttingen, Mathematisch-physi-
    kalische Klasse
Nature
Naturwissenschaften
Nippon Kagaku Zasshi
Österreichische Chemiker Zeitung
Organic Reactions (John Wiley, New York)
Organic Syntheses (John Wiley, New York)
Petroleum Processing
Proceedings of the Indiana Academy of Science
Proceedings of the Institute of Rubber Industry
Publications de la faculté des sciences de l'université Masaryk

Quarterly Reviews (London)
Reakcii i metody issledovanija organicheskich soedinenii
Recueil des travaux chimiques des Pays Bas
Refrigerating Engineering
Ricerca scientifica
Rubber Age (New York)
Rubber and Plastics Age, The
Science
Spisy vysoké školy veterinární, Brno
Talanta
Tetrahedron
Trudy Srednoaziatskogo Gosudarstvennogo Universitěta
United States Patent
Uspekhi Khimii
Zeitschrift für analytische Chemie, Fresenius
Zeitschrift für anorganische und allgemeine Chemie
Zeitschrift für Naturforschung
Zavodskaya Laboratoriya
Zhurnal Analyticheskoi Khimii
Zhurnal Obshchei Khimii
Zhurnal Prikladnoi Khimii
Zhurnal Russkogo Fiziko-Khimicheskogo Obshchestva

# LIST OF ILLUSTRATIONS

# LIST OF ABBREVIATIONS

| | | | |
|---|---|---|---|
| Act. | Activated | Dil. | Dilute, diluted |
| Alc. | Alcoholic | Equiv. | Equivalent |
| Alk. | Alkaline | $h\nu$ | Active illumination |
| Anhyd. | Anhydrous | (g) | Gaseous |
| Aq. | Aqueous | (l) | Liquid |
| Azeotr. | Azeotrope | Mol. | Molal, molecular |
| B.p. | Boiling point | Mol. wt. | Molecular weight |
| Catal. | Catalyst, catalytic, catalytic-ally | M.p. | Melting point |
| | | $n_D$ | Refractive index for D line |
| Concd. | Concentrated | Quant. | Quantitative, quantitatively |
| Concn. | Concentration | Sol. | Soluble |
| d | Density | Subl. | Sublimes, sublimation |
| Decomp. | Decomposition | | |

# AUTHOR INDEX

This index is a supplement to the alphabetical list of references on p. 358 and includes only those authors who are mentioned on the second or following place.

# FORMULA INDEX

Because of the intricacy of the nomenclature of organic compounds of fluorine the organic compounds mentioned in the book are listed according to their empirical formulas. The sequence of the elements in the formulas is that used by *Chemical Abstracts* (alphabetical). In addition to the conventional abbreviations the following symbols are used: Me methyl, Et ethyl, Pr propyl, isoPr isopropyl, Bu butyl, isoBu isobutyl, tert. Bu tert. butyl, Ph phenyl, Ac acetyl, and Bz benzoyl. The inorganic compounds are listed in the Subject Index.

## $C_1$

**C** *Carbon*
fluorination, 119

**$CAsF_3I_2$** *Trifluoromethylodiiodoarsine*
formation, 227, 263

**$CBrClF_2$** *Bromochlorodifluoromethane*
by Hunsdiecker's method, 123
free-radical addition, 251, 252, 254

**$CBrCl_2F$** *Bromodichlorofluoromethane*
by Hunsdiecker's method, 123

**$CBrCl_3$** *Bromotrichloromethane*
free-radical addition, 251, 252, 254

**$CBrF_3$** *Bromotrifluoromethane*
by bromination of $CHF_3$, 123, 177, 338
by bromolysis of $C_2F_6$, 186
by decarbonylation, 273
by degradation, 236
by reaction with $BrF_3$, 104, 123
by reaction with HF, 338
fire extinguisher, 342
nomenclature, 24

**$CBr_2ClF$** *Dibromochlorofluoromethane*
by Hunsdiecker's method, 123

**$CBr_2F_2$** *Dibromodifluoromethane*
addition to olefins, 132, 251, 252
by reaction with $Br_2$, 123, 338 with HF, 123, 338
fire extinguisher, 342
nomenclature, 24
reaction with PhOK, 238

## C₂

## C₃

**C₃HF₇**  *1H-heptafluoropropane*
   by decarboxylation, 272
   by decomposition of C₃F₇Li, 225
   by decomposition,
      of C₃F₇MgI, 220
   by degradation of C₃F₇CONH₂, 236
   by hydrolysis, 208, 209
   by reaction of C₃F₇ZnI, 225

**C₃H₂BrClF₄**  *1-Bromo-3-chloro-1,1,3,3-tetrafluoropropane*
   by free-radical addition, 254

**C₃H₂BrF₃**  *1-Bromo-3,3,3-trifluoropropene*
   by addition of HBr, 188
   by dehydrohalogenation, 268
   *2-Bromo-3,3,3-trifluoropropene*
   by dehydrohalogenation, 135

**C₃H₂BrF₃O**  *α-Bromo-α',α',α'-trifluoroacetone*
   by bromination, 182

**C₃H₂BrF₅**  *1,1-Dihydro-3-bromopentafluoropropane*
   by bromination, 178
   *1,3-Dihydro-1-bromopentafluoropropane*
   by bromination, 178

**C₃H₂Br₂F₄**  *1,3-Dibromo-1,1,3,3-tetrafluoropropane*
   dehydrohalogenation, 132

**C₃H₂Br₃F₃**  *2,2,3-Tribromo-1,1,1-trifluoropropane*
   by bromination, 135

**C₃H₂ClF₃**  *1-Chloro-3,3,3-trifluoropropene*
   by addition of HCl, 188
   by dehydrohalogenation, 268
   *3-Chloro-2,3,3-trifluoropropene*
   addition of HF, 187

**C₃H₂ClF₃O₂**  *β-Chloro-α,α,β-trifluoropropionic acid*
   formation, 255

**C₃H₂ClF₅**  *1-Chloro-1,1,2,2,3-pentafluoropropane*
   by chlorination, 178
   *1-Chloro-1,1,3,3,3-pentafluoropropane*
   by reaction with HgF₂
   *1-Chloro-1,2,2,3,3-pentafluoropropane*
   by chlorination, 178

**C₃H₂Cl₂F₂**  *3,3-Dichloro-2,3-difluoropropene*
   addition of HF, 187

**C₃H₂Cl₂F₃NO**  *α,β-Dichlorotrifluoropropionamide*
   formation, 151

**C₃H₂Cl₂F₄**  *1,2-Dichloro-2,3,3,3-tetrafluoropropane*
   by addition of Cl₂, 175
   *1,3-Dichloro-1,1,2,2-tetrafluoropropane*
   by reaction with HgF₂, 103

**C₃H₆ClNO** *Dimethylcarbamyl chloride*
electrochemical fluorination, 154

**C₃H₆Cl₂** *1,2-Dichloropropane*
chlorination, 127

**C₃H₆F₂** *1,3-Difluoropropane*
by reaction with KF, 108
*2,2-Difluoropropane*
by addition of HF, 66
by reaction with SF₄, 112
elimination, of HF 269

**C₃H₆F₃N** *3,3 3-Trifluoropropylamine*
by Curtius degradation, 235
by Hofmann degradation, 235, 236
dissociation constant, 310

**C₃H₆O** *Acetone*
Claisen condensation, 229
Grignard reaction, 221, 222
physical properties, 298
reaction with F₂, 74
with SF₄, 112
solubility of C₆F₁₂, 305
of polychlorotrifluoroethylene, 350
solvent for HF, 67
*Allyl alcohol*
addition to fluoroolefins, 248
*Propionaldehyde*
reaction with C₃F₇Li, 225

**C₃H₆O₂** *Methyl acetate*
hydrolysis, 199
*Propionic acid*
conversion to fluoride, 110

**C₃H₇BrMg** *Isopropylmagnesium bromide*
reactions, 167, 214
*Propylmagnesium bromide*
reactions, 214

**C₃H₇ClO₂** *Glycerol-α-chlorohydrin*
reaction with LiF, 110, 140

**C₃H₇ClO₂S** *1-Propanesulphonyl chloride*
electrochemical fluorination, 86

**C₃H₇F** *1-Fluoropropane (propyl fluoride)*
by reaction with KF, 113
*2-Fluoropropane (isopropyl fluoride)*
by addition of HF, 65
by cleavage of ethers, 249
by decomposition of fluoroformate, 120
by reaction with KF, 113
hydrogenation, 160

29*

**C₄H₆O₂**  *Crotonic acid*
    addition of HF, 68
    *Vinyl acetate*
        copolymerization, 348
        reaction with alcohols, 199

**C₄H₆O₃**  *Acetic anhydride*
    dehydrating agent, 270
    electrochemical, fluorination, 86
    esterification, 278
    Friedel–Crafts reaction, 145
    reaction with aldehydes, 199

**C₄H₆O₄**  *Methylmalonic acid*
    by hydrolysis, 206
    *Succinic acid*
        by oxidation, 170

**C₄H₇BrO₂**  *β-Bromoethyl acetate*
    reaction with AgF, 141
    *Ethyl bromoacetate*
        reaction with KF, 105
           with TlF, 110, 111
        Reformatsky synthesis, 226

**C₄H₇ClF₂**  *1-Chloro-1,1-difluorobutane*
    by reaction with HF, 65

**C₄H₇ClO**  *Butyryl chloride*
    reaction with F₂, 73

**C₄H₇ClO₂**  *Ethyl chloroacetate*
    reaction with KF, 105, 106

**C₄H₇Cl₂F**  *1,1-Dichloro-1-fluorobutane*
    by addition of HF, 65

**C₄H₇FO**  *Butyryl fluoride*
    by reaction with KHF₂, 110
    *γ-Fluorobutyraldehyde*
        by Neff's reaction, 142, 200
    *Isobutyryl fluoride*
        by reaction with KHF₂, 110
    *Methyl β-fluoroethyl ketone*
        by reaction with F₂, 73, 145
    nomenclature, 20

**C₄H₇FO₂**  *Ethyl fluoroacetate*
    alkylation, of 240
    by esterification, 198
    by reaction with KF, 105, **106**
        with TlF, 111
    Claisen condensation, 230, 231
    hydrolysis, 146, 147
    physical constants, 298

*Ag salt*
reaction with halogens, 184

**C₅H₂F₇I**   *1H,2H,1-iodoheptafluoropentene*
by addition of C₃F₇I, 262
oxidation, 173

**C₅H₂F₇NO₂**   *1H,2H-heptafluoro-1-nitropentene*
reaction with Grignard reagents, 215

**C₅H₂F₁₀**   *1,4-Dihydrodecafluoropentane*
nomenclature, 21

**C₅H₃ClF₃N**   *2-Chloro-2,3,3-trifluorocyclobutyl cyanide*
hydrolysis, 205

**C₅H₃ClF₄O**   *2,2,3,3-Tetrafluorocyclobutanecarboxylic acid, chloride*
by reaction with SOCl₂

**C₅H₃ClN₂O₂**   *2-Chloro-3-nitropyridine*
reaction with KF, 106
  *2-Chloro-5-nitropiridyne*
  reaction with KF, 106

**C₅H₃FN₂O₂**   *2-Fluoro-3-nitropyridine*
by reaction with KF, 106
  *2-Fluoro-5-nitropyridine*
  by reaction with KF, 106

**C₅H₃F₄N**   *2,2,3,3-Tetrafluorocyclobutylcyanide*
hydrolysis, 205

**C₅H₃F₅O**   *1-Methoxypentafluorocyclobutene*
by reaction with MeOH, 249

**C₅H₃F₇**   *3,3,4,4,5,5,5-Heptafluoropentene*
addition of HBr, 188

**C₅H₃F₇O**   *Methyl perfluoropropyl ketone*
hydrolysis, 208
preparation, 145, 225

**C₅H₃F₇O₂**   *Methyl perfluorobutyrate*
reaction with C₃F₇Li, 225
  with PhLi, 223

**C₅H₄BF₄N**   *α-Pyridinediazonium*
fluoroborate
decomposition, 139
stability, 117
  *β-Pyridinediazoniumfluoroborate*
  decomposition, 140

**C₅H₄BrF₇**   *1-Bromo-3,3,4,4,5,5,5-heptafluoropentane*
by addition of HBr, 188
reaction with Mg, 217

**C₅H₄BrF₇Mg**   *3,3,4,4,5,5,5-Heptafluoroamylmagnesium bromide*
formation, 217

**C₅H₄FN**   *α-Fluoropyridine*
preparation, 117, 139, 140
  *β-Fluoropyridine*
  preparation, 139, 140

## C₆

**C₇H₁₂FN**   *ω-Fluoroenanthonitrile*
    preparation, *240*

**C₇H₁₂FNS**   *1-Fluoro-6-thiocyanohexane*
    preparation, 239

**C₇H₁₂F₃N**   *1-Diethylamino-3,3,3-trifluoropropene*
    by addition of Et₂NH, 261

**C₇H₁₂O**   *Cycloheptanone*
    oxidation with CF₃CO₃H, 283

**C₇H₁₂O₂**   *Cyclohexanecarboxylic acid*
    by hydrogenation of fluorobenzoic acid, 159
   *Methyl cyclopentanecarboxylate*
    by Favorskii rearrangement, 235

**C₇H₁₂O₄**   *Diethyl malonate*
    Michael addition, *229*

**C₇H₁₃BrO₂**   *7-Bromoheptanoic acid*
    by Grignard synthesis, 216

**C₇H₁₃ClO₂**   *7-Chloroheptanoic acid*
    by Grignard synthesis, 216

**C₇H₁₃FO₂**   *Ethyl δ-fluorovalerate*
    physiological properties, 312
   *7-Fluoroheptanoic acid*
    by Grignard synthesis, 150, 216
    electrolysis, 127

**C₇H₁₃F₃O₂**   *3-Ethoxy-2-hydroxy-2-trifluoromethylbutane*
    by alcoholysis, 241

**C₇H₁₃IO₂**   *7-Iodoheptanoic acid*
    by Grignard synthesis, 216

**C₇H₁₄**   *Methylcyclohexane*
    solubility of polychlorotrifluoroethylene, 350
      of perfluoro compounds, 305

**C₇H₁₅Br**   *1-Bromoheptane*
    reaction with HgF, 100

**C₇H₁₅ClO₂S**   *1-Heptanesulphonylchloride*
    electrochemical fluorination, 86

**C₇H₁₅F**   *1-Fluoroheptane*
    by reaction with HgF, 100

**C₇H₁₆**   *Heptane*
    reaction with CoF₃, 81, 128, 233
      with F₂, 76, 128
    solubility of polychlorotrifluoroethylene, 350
      of perfluoro compounds, 305
    viscosity, 303

**C₇H₁₆FO₂P**   *Methyl tert.-butylcarbinyl methanefluorophosphonate (Soman)*
    applications, 357
    physiological properties, 311

## C₈

**C₈H₁₁FO**  *4-Fluoro-Δ3-tetrahydroacetophenone*
   by Diels–Alder reaction, 258
**C₈H₁₁FO₅**  *Diethyl fluorooxalacetate*
   by Claisen condensation, 230
   reduction, 165
**C₈H₁₁F₃O₃**  *1,2-Cyclohexanediol trifluoroacetate*
   by reaction with CF₃CO₃H, 282
**C₈H₁₁F₃O₄**  *ε-Hydroxycaproic acid  trifluoroacetate*
   formation, 283
**C₈H₁₁F₄NO**  *2H-tetrafluoro-3-butenoic acid diethylamide*
   formation, 250
**C₈H₁₂N₄**  *2-Cyano-2-azopropane (azobisisobutyronitrile)*
   polymerization initiator
**C₈H₁₂O₈Pb**  *Lead tetraacetate*
   oxidizing agent for HF, 69, 73
**C₈H₁₃BrO₂**  *Cyclohexyl bromoacetate*
   reaction with KF, 105
**C₈H₁₃ClO₂**  *Cyclohexyl chloroacetate*
   reaction with KF, 105
**C₈H₁₃FO₂**  *Cyclohexyl fluoroacetate*
   by reaction with KF, 105
**C₈H₁₃FO₃**  *Ethyl α-fluoro-β-hydroxy-α,β-dihydrosorbate*
   by Reformatsky reaction, 226
**C₈H₁₃FO₅**  *Ethyl fluoromalate*
   by reduction with NaBH₄, 165
**C₈H₁₃F₃O**  *Hexyl trifluoromethylketone*
   preparation, 226
**C₈H₁₄**  *1-Octyne*
   addition of HF, 67
**C₈H₁₄ClFO**  *ω-Fluorcapryloyl chloride*
   hydrogenation, 142
**C₈H₁₄O**  *Cyclooctanone*
   oxidation with CF₃CO₃H, 283
**C₈H₁₄O₂**  *2,4-Octanedione*
   preparation, 281
**C₈H₁₄O₄**  *Suberic acid*
   by Grignard synthesis, 216
**C₈H₁₄O₅**  *Ethyl malate*
   by reduction with Zn, AlHg, 165
**C₈H₁₅FO**  *8-Fluorooctanal*
   by hydrogenation, 142
**C₈H₁₅FO₂**  *Ethyl ε-fluorocaproate*
   physiological properties, 312
   *Ethyl γ-fluoro-β,β-dimethylbutyrate*
   physiological properties, 312
**C₈H₁₆**  *1,3-Dimethylcyclohexane*
   reaction with F₂, 77
   *Ethylcyclohexane*

# C₁₁

## $C_{15}$

## $C_{16}$

$$\mathbf{C_{25}-C_{34}}$$

# SUBJECT INDEX

The subject index contains names of reaction types, compound types, and inorganic compounds. Of organic compounds only those which act as reagents are listed. All the individual organic compounds are filed in the Formula Index (page 407).

## A

*Accelerators of vulcanization,* 349

*Acetylenes*
addition of acids in trifluoroacetic anhydride, 281
addition of hydrogen fluoride, 67

*Acetylenes, fluorinated,*
*see Fluoroacetylenes*

*Acid potassium fluoride*
physical properties, 62
preparation, 51
replacement of halogen by fluorine 109, 110

*Acid hydrolysis*
of fluoroderivatives, 203, 204, 206–208

*Acids*
by hydrolysis of trifluoromethyl derivatives

*Acids, fluorinated*
*see Fluoroacids*

*Acidity*
of fluoroacids, 307–310
of Freons, 330
of hydrogen in fluoroacids, 307, 310
in fluoroparaffins, 177

*Activation*
of antimony fluorides, 93

*Acyl anhydrides*
electrochemical fluorination, 86

*Acyl chlorides*
addition to fluoroolefins, 255
conversion to fluorides by potassium fluorosulphinate, 109
by sodium fluorosilicate, 111
electrochemical fluorination, 86

*Acyl halides*
conversion to fluorides, 87
by hydrogen fluoride, 89
by metal fluorides, 94, 105, 108, 109

*Acyl fluorides*
by reaction of acids with benzoyl chloride and acid potassium fluoride, 109, 110
of acid halides with sodium fluorosilicate, 111
Friedel–Crafts synthesis, 210, 211
hydrolysis, 202
liberation of fluorine, 314

*Acyl fluoroborates*
formation, 211

*Acylation*
by acyl fluorides, 210, 211
by fluoroacyl halides, 211–213
by fluoro compounds, 243, 244

*Addition*
of alcohols to fluoroolefins, 198, 208
of fluorine by electrochemical process 84
by treatment with
antimony pentafluoride, 70, 98
cobalt trifluoride, 69
elemental fluorine, 68
hydrogen fluoride and lead dioxide or tetraacetate, 69
of halogenes, reaction heats, 65, 66
of hydrogen fluoride
to acetylenes, 67
to olefins, 65, 66
of sodium bisulphite to fluoroolefins, 196

*Additions*
to double and triple bond, 244–263

RA
2-10-76